Birkhäuser

Progress in Mathematics

Volume 337

Progress in Mathematics is a series of books intended for professional mathematicians and scientists, encompassing all areas of pure mathematics. This distinguished series, which began in 1979, includes research level monographs, polished notes arising from seminars or lecture series, graduate level textbooks, and proceedings of focused and refereed conferences. It is designed as a vehicle for reporting ongoing research as well as expositions of particular subject areas.

More information about this series at https://link.springer.com/bookseries/4848

Jacob Greenstein • David Hernandez
Kailash C. Misra • Prasad Senesi
Editors

Interactions of Quantum Affine Algebras with Cluster Algebras, Current Algebras and Categorification

In honor of Vyjayanthi Chari on the occasion of her 60th birthday

Editors
Jacob Greenstein
Department of Mathematics
University of California, Riverside
Riverside, CA, USA

David Hernandez
IMJ-PRG
Université de Paris
Paris, France

Kailash C. Misra
Department of Mathematics
North Carolina State University
Raleigh, NC, USA

Prasad Senesi
Department of Mathematics
Catholic University of America
Washington D.C., WA, USA

ISSN 0743-1643 ISSN 2296-505X (electronic)
Progress in Mathematics
ISBN 978-3-030-63851-1 ISBN 978-3-030-63849-8 (eBook)
https://doi.org/10.1007/978-3-030-63849-8

Mathematics Subject Classification: 17Bxx, 17Cxx, 16Fxx, 16Txx, 16Wxx, 18Dxx

This book is published under the imprint Birkhäuser, www.birkhauser-science.com by the registered company Springer Nature Switzerland AG
The registered company address is: Gewerbestrasse 11, 6330 Cham, Switzerland

Preface

This volume is dedicated to Vyjayanthi Chari on the occasion of her 60th birthday. A summer school and a conference entitled *Interactions of quantum affine algebras with cluster algebras, current algebras and categorification* was held in her honour at the Catholic University of America, Washington DC, USA, in June 2018. The conference attracted over a hundred participants from Australia, Brazil, Canada, China, France, Germany, India, Israel, Italy, Japan, Russia, Singapore and the USA. Apart from two mini-courses and eighteen talks given by renowned researchers, several sessions of short talks were organized in order to provide young researchers with an opportunity to present their work.

Subsequently, distinguished experts in representation theory were invited to contribute survey and research articles, which comprise the present volume. The focus here is on various aspects of representation theory connected with affine Lie algebras and their quantum analogues, the area in which the impact of Vyjayanthi Chari's work is difficult to overestimate. The volume contains lecture notes from mini-courses, three survey articles and eight original research articles. All contributions underwent a rigorous refereeing process befitting any mathematical journal.

The detailed notes of lecture courses not only provide a welcome reminder of the content of these courses to the participants but also enable all those who did not attend the meeting to profit from insights of leading specialists into the latest developments in corresponding fields.

The first, *String Diagrams and Categorification*, presented by Alistair Savage (Ottawa), provides an exposition of the use of the diagrammatics and the definition of various strict monoidal categories by generators and relations. The simplicity, brevity and clarity of these notes should be appreciated by graduate students and more advanced researchers seeking a quick introduction to the field. It also takes as its focus the Heisenberg category, which is not so well-known.

The second, *Quantum affine algebras and cluster algebras*, presented by Bernard Leclerc (Caen) and written jointly by him and David Hernandez (Paris 7), is concerned with categorification of cluster algebras via monoidal categories of finite-dimensional representations of quantum affine algebras. The area progressed significantly since the original work of Hernandez and Leclerc (*Duke Mathematical*

Journal, 2010). The present notes provide an introduction to the subject, building up from seminal work of Chari and Pressley on classification of simple finite-dimensional modules over quantum affine algebras to the latest developments, providing connections with geometric representation theory as well as a discussion of some conjectures.

The second part (Surveys) of the volume opens with a survey of scientific contributions of Vyjayanthi Chari, written by Jacob Greenstein and David Hernandez.

In his survey *Steinberg Groups for Jordan Pairs: An Introduction with Open Problems*, Erhard Neher provides a concise introduction to the theory developed in his book (co-authored with Ottmar Loos). The paper contains a lot of examples and, while avoiding heavy proofs and cumbersome technical details, provides necessary definitions and ideas behind the proofs. It ends with a list of important open problems for Steinberg groups and related questions in the setting of Jordan pairs.

Maxim Gurevich in his survey *On the Hecke-Algebraic Approach for General Linear Groups over a p-Adic Field* presents some similarities and links between various objects of rather different nature corresponding to root data of type A. It connects the category of complex smooth representations of $GL_n(D)$, where D is a division algebra over a non-Archimedean local field, with representations of affine Hecke algebras. The Grothendieck ring of the latter is then connected to the positive part of the quantum version of $\mathfrak{sl}_N$ and, via Chari–Pressley version of the Schur–Weyl duality, with quantum affine algebras. The survey puts a particular emphasis on recent developments.

The original research papers contained in the last part of the volume cover a large variety of topics, which reflects the versatility and scope of the work of Vyjayanthi Chari and her influence on the field. It opens with *Categorical Representations and Classical p-Adic Groups* by Michela Varagnolo and Éric Vasserot, which is closely connected to some of the structures discussed in the preceding survey. The paper studies categorical action for p-adic groups of types B and C. The quiver Hecke algebra involved has been introduced in an earlier work of the authors (*Inventiones Mathematicae*, 2011) when the involution of the quiver does not have fixed points. In this paper, the authors gave some more general results on these algebras when the involution has fixed points. As in type A, these quiver Hecke algebras also categorify some Boson algebra. The authors constructed a categorical action of this Boson algebra on the p-adic groups of types B and C and compared it with the categorification given by quiver Hecke algebras.

The paper *Formulae of ı-Divided Powers in* $\mathbf{U}_q(\mathfrak{sl}_2)$, *II* by Weiqiang Wang and Collin Berman is concerned with coideal subalgebras of $\mathbf{U}_q(\mathfrak{sl}_2)$ and their $\imath$-canonical bases whose existence was established in an earlier work of the first author and Huanchen Bao (*Inventiones Mathematicae*, 2018). The coideal subalgebra $U^\imath$ in question depends on a parameter κ. The present work extends the results of the authors (*Journal of Pure and Applied Algebra*, 2018) for $\kappa = 0$ or 1 to the case when κ is an arbitrary polynomial in $q + q^{-1}$ and obtains explicit formulae for elements of the $\imath$-canonical basis in that case.

In *Longest Weyl Group Elements in Action*, Yiqiang Li and Yan Ling define, for each Cartan matrix, a projective system of Weyl-like groups indexed by nonnegative integers. In type A, these groups are related to reduced Burau representations and, in general, to complex reflection groups. This construction is inspired by graded quiver varieties and is expected to serve as a toll in the study of graded/cyclic quiver varieties in the spirit of recent work of the first author.

The next two papers are concerned with crystals, which proved to be an invaluable tool in representation theory since their appearance in the early nineties in works of Kashiwara and Lusztig.

A notoriously hard problem in theory of crystals is finding an explicit combinatorial description of Kashiwara's $B(\infty)$ crystal. Only partial results are available so far. Notably, Berenstein and Zelevinsky (*Inventiones Mathematicae*, 2001) described $B(\infty)$ for any simple finite-dimensional Lie algebra $\mathfrak{g}$, as a polyhedral subset of the set of integer points of an r-dimensional affine space, where r is the number of positive roots of $\mathfrak{g}$. The linear functions describing this polyhedral subset are given by "trails" in the fundamental modules of lowest weight of the Langlands dual of $\mathfrak{g}$, depend on a reduced decomposition $\mathbf{i}$ of the longest element of the Weyl group of $\mathfrak{g}$ and, in general, are very difficult to compute. In *Dual Kashiwara Functions for the $B(\infty)$ Crystal in the Bipartite Case*, Anthony Joseph used combinatorial tools introduced in his recent work to show that, in the special case when $\mathbf{i}$ is obtained from a power of a suitable Coxeter element, the set of trails is in natural bijection with the crystal of a suitable highest weight fundamental module (the existence of such a bijection was conjectured by Shmuel Zelikson).

The paper *Lusztig's t-Analogue of Weight Multiplicity via Crystals* by Cédric Lecouvey and Cristian Lenart is concerned with combinatorial descriptions of Kostant's generalized exponents (also known to coincide with Kazhdan–Lusztig polynomials in a special situation, as well as Lustig's t-analogues of weight multiplicities). These polynomials have been attracting attention of combinatorialists and representation theorists for quite a while, due to a large variety of their interpretations from different points of view. The present paper surveys earlier results obtained by authors, as well as new results and ideas. For instance, they introduce modified crystal operator and, for symplectic Lie algebras, obtain a combinatorial description of the generalized exponents based on the so-called distinguished vertices in crystals of type A_{2n-1} and, as a result, a combinatorial proof of the positivity of Lusztig t-analogues. They were also able to deduce known results in type A without using either combinatorics of semistandard tableaux or charge statistics.

More geometric aspects of representation theory are represented in this volume in *Conormal Varieties on the Cominuscule Grassmannian* by V. Lakshmibai and Rahul Singh. Let G be an almost simple, simply connected algebraic group over an algebraically closed field $\mathbf{k}$ either of characteristic zero or of a "good" positive characteristic. The paper is concerned with the study of certain special homogeneous spaces G/P, known as "cominuscule Grassmannians", where P is a maximal parabolic subgroup corresponding to omitting a cominuscule root. Specifically, the authors construct an open dense embedding of the cotangent bundle of G/P into a Schubert variety in a partial affine flag variety associated with the loop group $G\left(\mathbf{k}[t, t^{-1}]\right)$. This embedding is then used to study the geometry of conormal bundles of subvarieties of G/P.

Evaluation modules over affine Lie algebras and quantum affine algebras were featured prominently in Vyjayanthi Chari's work. For instance, in one of her early papers, she classified simple modules of that type over affine Lie algebras, while in her joint work with Andrew Pressley they described all simple modules over quantum affine $\mathfrak{sl}_2$ in terms of quantum evaluation map. However, evaluation homomorphism does not exist in quantum affine setting outside of type A. Morally, toroidal algebras can be obtained from quantum affine algebras by applying Drinfeld's affinization to quantum affine algebras, and Kei Miki showed (although without a complete proof) that the resulting algebra admits an evaluation homomorphism to the quantum loop algebra of $\mathfrak{gl}_n$. In *Evaluation Modules for Quantum Toroidal* $\mathfrak{gl}_n$ *Algebras*, Boris Feigin, Michio Jimbo and Evgeny Mukhin study homomorphism and, under some restriction on quantum parameters, modules over quantum toroidal algebras that can be built out of it. They also obtained the decomposition of the evaluation Wakimoto module with respect to a Gelfand–Zeitlin-type subalgebra of a completion of this quantum toroidal algebra.

The last paper in this volume, *Dynamical Quantum Determinants and Pfaffians* by Naihuan Jing and Jian Zhang, is concerned with the so-called dynamical quantum groups. The famous way of constructing quantum (semi)groups out of R-matrices, due to Faddeev, Reshetikhin and Takhtajan, was generalized by Etingof and Varchenko to yield the so-called dynamical quantum groups. Just like in the "ordinary" FRT setting, a dynamical quantum group is obtained from the corresponding quantum semigroup using localization with respect to the dynamical quantum determinant, which is a canonical group-like central element, and one can develop quantum dynamical linear algebra. In the present paper, the authors introduce dynamical quantum Pfaffians and study their properties ("usual" quantum Pfaffians were studied in authors' earlier work *Advances in Mathematics* 2014). Their main technique is to use quadratic algebras or quantum de Rham complexes.

Acknowledgments

The conference *Interactions of quantum affine algebras with cluster algebras, current algebras and categorification* was supported by the NSF grant DMS-1810211, the European Research Council under the European Union's Framework Programme H2020 with ERC Grant Agreement number 647353 QAffine, the Catholic University of America (Washington DC, USA) and the Department of Mathematics at the University of California Riverside (Riverside, CA, USA).

The editors thank all the authors who contributed to this volumes and all the referees for their meticulous work.

Riverside, CA, USA Jacob Greenstein
Paris, France David Hernandez
Raleigh, NC, USA Kailash C. Misra
Washington, DC, USA Prasad Senesi
August, 2020

With speakers of the conference “Interactions of quantum affine algebras with cluster algebras, current algebras and categorification” *First row left to right*: Monica Vazirani, Shrawan Kumar, Vyjayanthi Chari, Erhard Neher, David Hernandez and Cristian Lenart *Second row left to right*: Masaki Kashiwara, Georgia Benkart, Anthony Joseph, Eric Vasserot, Bernard Leclerc, Alistair Savage, Adriano Moura and Bogdan Ion *Third row left to right*: Yiqiang Li, Weiqiang Wang, Michael Lau and Aaron Lauda

With graduate students, postdocs and visiting students *First row left to right*: Lisa Schneider, Kayla Murray, Ronald Dolbin, Nathan Manning and Matthew Bennett *Second row left to right*: Vyjayanthi Chari, Justin Davis, Matthew Lee and Peri Shereen *Third row left to right*: Ghislain Fourier, Tanusree Khandai, Adriano Moura, Jeffrey Wand and Deniz Kus *Fourth row left to right*: Prasad Senesi, Ryan Moruzzi Jr., Rekha Biswal, Matheus Brito and R. Venkatesh

Publications of Vyjayanthi Chari

1. (with S. Ilangovan), *On the Harish-Chandra homomorphism for infinite-dimensional Lie algebras*, J. Algebra **90** (1984), no. 2, 476–490, https://doi.org/10.1016/0021-8693(84)90185-6.
2. *Annihilators of Verma modules for Kac-Moody Lie algebras*, Invent. Math. **81** (1985), no. 1, 47–58, https://doi.org/10.1007/BF01388771.
3. *Integrable representations of affine Lie algebras*, Invent. Math. **85** (1986), no. 2, 317–335, https://doi.org/10.1007/BF01389093.
4. (with A. Pressley), *New unitary representations of loop groups*, Math. Ann. **275** (1986), no. 1, 87–104, https://doi.org/10.1007/BF01458586.
5. (with A. Pressley), *Towards a classification of integrable representations of affine Lie algebras*, Topological and geometrical methods in field theory (Espoo, 1986), World Sci. Publ., Teaneck, NJ, 1986, pp. 17–28.
6. (with A. Pressley), *A new family of irreducible, integrable modules for affine Lie algebras*, Math. Ann. **277** (1987), no. 3, 543–562, https://doi.org/10.1007/BF01458331.
7. (with A. Pressley), *Unitary representations of the maps* $S^1 \to su(N, 1)$, Math. Proc. Cambridge Philos. Soc. **102** (1987), no. 2, 259–272, https://doi.org/10.1017/S0305004100067281.
8. (with A. Pressley), *Integrable representations of twisted affine Lie algebras*, J. Algebra **113** (1988), no. 2, 438–464, https://doi.org/10.1016/0021-8693(88)90171-8.
9. (with A. Pressley), *Unitary representations of the Virasoro algebra and a conjecture of Kac*, Compositio Math. **67** (1988), no. 3, 315–342.
10. (with A. Pressley), *Unitary representations of some infinite-dimensional Lie algebras*, Theta functions—Bowdoin 1987, Part 1 (Brunswick, ME, 1987), Proc. Sympos. Pure Math., vol. 49, Amer. Math. Soc., Providence, RI, 1989, pp. 239–257.
11. (with A. Pressley), *Integrable representations of Kac-Moody algebras: results and open problems*, Infinite-dimensional Lie algebras and groups (Luminy-Marseille, 1988), Adv. Ser. Math. Phys., vol. 7, World Sci. Publ., Teaneck, NJ, 1989, pp. 3–24.

12. (with A. Pressley), *An application of Lie superalgebras to affine Lie algebras*, J. Algebra **135** (1990), no. 1, 203–216, https://doi.org/10.1016/0021-8693(90)90158-K.
13. (with A. Pressley), *Yangians and* R*-matrices*, Enseign. Math. (2) **36** (1990), no. 3–4, 267–302.
14. (with A. Pressley), *Notes on quantum groups*, Nuclear Phys. B Proc. Suppl. **18A** (1990), 207–228, https://doi.org/10.1016/0920-5632(90)90650-J. Integrability and quantization (Jaca, 1989).
15. (with A. Pressley), *Fundamental representations of Yangians and singularities of* R*-matrices*, J. Reine Angew. Math. **417** (1991), 87–128.
16. (with A. Pressley), *Minimal cyclic representations of quantum groups at roots of unity*, C. R. Acad. Sci. Paris Sér. I Math. **313** (1991), no. 7, 429–434.
17. (with A. Pressley), *Introduction to quantum groups*, Proceedings of the Hyderabad Conference on Algebraic Groups (Hyderabad, 1989), Manoj Prakashan, Madras, 1991, pp. 81–122.
18. (with A. Pressley), *Quantum affine algebras*, Comm. Math. Phys. **142** (1991), no. 2, 261–283.
19. (with A. Pressley), *Fundamental representations of quantum groups at roots of* 1, Lett. Math. Phys. **26** (1992), no. 2, 133–146, https://doi.org/10.1007/BF00398810.
20. (with A. Pressley), *Representations of modular Lie algebras through quantum groups*, C. R. Acad. Sci. Paris Sér. I Math. **317** (1993), no. 8, 723–728.
21. (with A. Pressley), *Representations of quantum* so(8) *and related quantum algebras*, Comm. Math. Phys. **159** (1994), no. 1, 29–49.
22. (with A. Pressley), *Small representations of quantum affine algebras*, Lett. Math. Phys. **30** (1994), no. 2, 131–145, https://doi.org/10.1007/BF00939701.
23. (with A. Premet), *Indecomposable restricted representations of quantum* sl2, Publ. Res. Inst. Math. Sci. **30** (1994), no. 2, 335–352, https://doi.org/10.2977/prims/1195166137.
24. (with A. Pressley), *Minimal affinizations of representations of quantum groups: the nonsimply-laced case*, Lett. Math. Phys. **35** (1995), no. 2, 99–114, https://doi.org/10.1007/BF00750760.
25. (with A. Pressley), *Quantum affine algebras and their representations*, Representations of groups (Banff, AB, 1994), CMS Conf. Proc., vol. 16, Amer. Math. Soc., Providence, RI, 1995, pp. 59–78.
26. *Minimal affinizations of representations of quantum groups: the rank* 2 *case*, Publ. Res. Inst. Math. Sci. **31** (1995), no. 5, 873–911, https://doi.org/10.2977/prims/1195163722.
27. (with A. Pressley), *Minimal affinizations of representations of quantum groups: the irregular case*, Lett. Math. Phys. **36** (1996), no. 3, 247–266, https://doi.org/10.1007/BF00943278.
28. (with A. Pressley), *Minimal affinizations of representations of quantum groups: the simply laced case*, J. Algebra **184** (1996), no. 1, 1–30, https://doi.org/10.1006/jabr.1996.0247.

29. (with A. Pressley), *Quantum affine algebras and affine Hecke algebras*, Pacific J. Math. **174** (1996), no. 2, 295–326.
30. (with A. Pressley), *Yangians: their representations and characters*, Acta Appl. Math. **44** (1996), no. 1–2, 39–58, https://doi.org/10.1007/BF00116515. Representations of Lie groups, Lie algebras and their quantum analogues.
31. (with A. Pressley), *Yangians, integrable quantum systems and Dorey's rule*, Comm. Math. Phys. **181** (1996), no. 2, 265–302.
32. (with A. Pressley), *Quantum affine algebras at roots of unity*, Represent. Theory **1** (1997), 280–328, https://doi.org/10.1090/S1088-4165-97-00030-7.
33. (with A. Pressley), *Factorization of representations of quantum affine algebras*, Modular interfaces (Riverside, CA, 1995), AMS/IP Stud. Adv. Math., vol. 4, Amer. Math. Soc., Providence, RI, 1997, pp. 33–40.
34. (with A. Pressley), *Quantum affine algebras and integrable quantum systems*, Quantum fields and quantum space time (Cargèse, 1996), NATO Adv. Sci. Inst. Ser. B Phys., vol. 364, Plenum, New York, 1997, pp. 245–263.
35. (with A. Pressley), *Twisted quantum affine algebras*, Comm. Math. Phys. **196** (1998), no. 2, 461–476, https://doi.org/10.1007/s002200050431.
36. (with J. Beck and A. Pressley), *An algebraic characterization of the affine canonical basis*, Duke Math. J. **99** (1999), no. 3, 455–487, https://doi.org/10.1215/S0012-7094-99-09915-5.
37. (with N. Xi), *Monomial bases of quantized enveloping algebras*, Recent developments in quantum affine algebras and related topics (Raleigh, NC, 1998), Contemp. Math., vol. 248, Amer. Math. Soc., Providence, RI, 1999, pp. 69–81, https://doi.org/10.1090/conm/248/03818.
38. (with N. Jing), *Realization of level one representations of $\mathcal{U}_q(\hat{g})$ at a root of unity*, Duke Math. J. **108** (2001), no. 1, 183–197, https://doi.org/10.1215/S0012-7094-01-10816-8.
39. *On the fermionic formula and the Kirillov-Reshetikhin conjecture*, Internat. Math. Res. Notices **12** (2001), 629–654, https://doi.org/10.1155/S1073792801000332.
40. (with A. Pressley), *Weyl modules for classical and quantum affine algebras*, Represent. Theory **5** (2001), 191–223, https://doi.org/10.1090/S1088-4165-01-00115-7.
41. (with A. Pressley), *Integrable and Weyl modules for quantum affine* sl2, Quantum groups and Lie theory (Durham, 1999), London Math. Soc. Lecture Note Ser., vol. 290, Cambridge Univ. Press, Cambridge, 2001, pp. 48–62.
42. *Braid group actions and tensor products*, Int. Math. Res. Not. **7** (2002), 357–382, https://doi.org/10.1155/S107379280210612X.
43. (with M. Kleber), *Symmetric functions and representations of quantum affine algebras*, Recent developments in infinite-dimensional Lie algebras and conformal field theory (Charlottesville, VA, 2000), Contemp. Math., vol. 297, Amer. Math. Soc., Providence, RI, 2002, pp. 27–45, https://doi.org/10.1090/conm/297/05091.
44. (with J. Greenstein), *Quantum loop modules*, Represent. Theory **7** (2003), 56–80, https://doi.org/10.1090/S1088-4165-03-00168-7.

45. (with T. Le), *Representations of double affine Lie algebras*, A tribute to C. S. Seshadri (Chennai, 2002), Trends Math., Birkhäuser, Basel, 2003, pp. 199–219.
46. (with A. A. Moura), *Spectral characters of finite-dimensional representations of affine algebras*, J. Algebra **279** (2004), no. 2, 820–839, https://doi.org/10.1016/j.jalgebra.2004.01.015.
47. (with A. A. Moura), *Characters and blocks for finite-dimensional representations of quantum affine algebras*, Int. Math. Res. Not. **5** (2005), 257–298, https://doi.org/10.1155/IMRN.2005.257.
48. (with J. Greenstein), *Filtrations and completions of certain positive level modules of affine algebras*, Adv. Math. **194** (2005), no. 2, 296–331, https://doi.org/10.1016/j.aim.2004.06.008.
49. (with D. Jakelic' and A. A. Moura), *Branched crystals and the category* O, J. Algebra **294** (2005), no. 1, 51–72, https://doi.org/10.1016/j.jalgebra.2005.03.008.
50. (with J. Greenstein), *An application of free Lie algebras to polynomial current algebras and their representation theory*, Infinite-dimensional aspects of representation theory and applications, Contemp. Math., vol. 392, Amer. Math. Soc., Providence, RI, 2005, pp. 15–31, https://doi.org/10.1090/conm/392/07350.
51. (with A. Moura), *The restricted Kirillov-Reshetikhin modules for the current and twisted current algebras*, Comm. Math. Phys. **266** (2006), no. 2, 431–454, https://doi.org/10.1007/s00220-006-0032-2.
52. (with A. A. Moura), *Characters of fundamental representations of quantum affine algebras*, Acta Appl. Math. **90** (2006), no. 1–2, 43–63, https://doi.org/10.1007/s10440-006-9030-9.
53. (with S. Loktev), *Weyl, Demazure and fusion modules for the current algebra of* $\mathfrak{sl}_{r+1}$, Adv. Math. **207** (2006), no. 2, 928–960, https://doi.org/10.1016/j.aim.2006.01.012.
54. (with J. Greenstein), *Current algebras, highest weight categories and quivers*, Adv. Math. **216** (2007), no. 2, 811–840, https://doi.org/10.1016/j.aim.2007.06.006.
55. (with A. Moura), *Kirillov-Reshetikhin modules associated to* G_2, Lie algebras, vertex operator algebras and their applications, Contemp. Math., vol. 442, Amer. Math. Soc., Providence, RI, 2007, pp. 41–59, https://doi.org/10.1090/conm/442/08519.
56. (with J. Greenstein), *Graded level zero integrable representations of affine Lie algebras*, Trans. Amer. Math. Soc. **360** (2008), no. 6, 2923–2940, https://doi.org/10.1090/S0002-9947-07-04394-2.
57. (with G. Fourier and P. Senesi), *Weyl modules for the twisted loop algebras*, J. Algebra **319** (2008), no. 12, 5016–5038, https://doi.org/10.1016/j.jalgebra.2008.02.030.
58. (with J. Greenstein), *A family of Koszul algebras arising from finite-dimensional representations of simple Lie algebras*, Adv. Math. **220** (2009), no. 4, 1193–1221, https://doi.org/10.1016/j.aim.2008.11.007.
59. (with R. J. Dolbin and T. Ridenour), *Ideals in parabolic subalgebras of simple Lie algebras*, Symmetry in mathematics and physics, Contemp. Math., vol. 490,

Amer. Math. Soc., Providence, RI, 2009, pp. 47–60, https://doi.org/10.1090/conm/490/09586.

60. (with D. Hernandez), *Beyond Kirillov-Reshetikhin modules*, Quantum affine algebras, extended affine Lie algebras, and their applications, Contemp. Math., vol. 506, Amer. Math. Soc., Providence, RI, 2010, pp. 49–81, https://doi.org/10.1090/conm/506/09935.
61. (with G. Fourier and T. Khandai), *A categorical approach to Weyl modules*, Transform. Groups **15** (2010), no. 3, 517–549, https://doi.org/10.1007/s00031-010-9090-9.
62. (with J. Greenstein), *Minimal affinizations as projective objects*, J. Geom. Phys. **61** (2011), no. 3, 594–609, https://doi.org/10.1016/j.geomphys.2010.11.008.
63. *Representations of affine and toroidal Lie algebras*, Geometric representation theory and extended affine Lie algebras, Fields Inst. Commun., vol. 59, Amer. Math. Soc., Providence, RI, 2011, pp. 169–197.
64. (with M. Bennett, R. J. Dolbin, and N. Manning), *Square-bounded partitions and Catalan numbers*, J. Algebraic Combin. **34** (2011), no. 1, 1–18, https://doi.org/10.1007/s10801-010-0260-6.
65. (with M. Bennett, J. Greenstein, and N. Manning), *On homomorphisms between global Weyl modules*, Represent. Theory **15** (2011), 733–752, https://doi.org/10.1090/S1088-4165-2011-00407-6.
66. (with S. Loktev), *An application of global Weyl modules of* $\mathfrak{sl}_{n+1}[t]$ *to invariant theory*, J. Algebra **349** (2012), 317–328, https://doi.org/10.1016/j.jalgebra.2011.09.017.
67. (with A. Khare and T. Ridenour), *Faces of polytopes and Koszul algebras*, J. Pure Appl. Algebra **216** (2012), no. 7, 1611–1625, https://doi.org/10.1016/j.jpaa.2011.10.014.
68. (with M. Bennett and N. Manning), *BGG reciprocity for current algebras*, Adv. Math. **231** (2012), no. 1, 276–305, https://doi.org/10.1016/j.aim.2012.05.005.
69. (with M. Bennett), *Tilting modules for the current algebra of a simple Lie algebra*, Recent developments in Lie algebras, groups and representation theory, Proc. Sympos. Pure Math., vol. 86, Amer. Math. Soc., Providence, RI, 2012, pp. 75–97, https://doi.org/10.1090/pspum/086/1411.
70. (with A. Moura and C. Young), *Prime representations from a homological perspective*, Math. Z. **274** (2013), no. 1–2, 613–645, https://doi.org/10.1007/s00209-012-1088-7.
71. (with M. Bennett, A. Berenstein, A. Khoroshkin, and S. Loktev), *Macdonald polynomials and BGG reciprocity for current algebras*, Selecta Math. (N.S.) **20** (2014), no. 2, 585–607, https://doi.org/10.1007/s00029-013-0141-7.
72. (with A. Bianchi, G. Fourier, and A. Moura), *On multigraded generalizations of Kirillov-Reshetikhin modules*, Algebr. Represent. Theory **17** (2014), no. 2, 519–538, https://doi.org/10.1007/s10468-013-9408-0.
73. (with L. Schneider, P. Shereen, and J. Wand), *Modules with Demazure flags and character formulae*, SIGMA Symmetry Integrability Geom. Methods Appl. **10** (2014), Paper 032, 16, https://doi.org/10.3842/SIGMA.2014.032.

74. (with G. Fourier and D. Sagaki), *Posets, tensor products and Schur positivity*, Algebra Number Theory **8** (2014), no. 4, 933–961, https://doi.org/10.2140/ant.2014.8.933.
75. (with R. Venkatesh), *Demazure modules, fusion products and* Q-*systems*, Comm. Math. Phys. **333** (2015), no. 2, 799–830, https://doi.org/10.1007/s00220-014-2175-x.
76. (with B. Ion), *BGG reciprocity for current algebras*, Compos. Math. **151** (2015), no. 7, 1265–1287, https://doi.org/10.1112/S0010437X14007908.
77. (with B. Ion and D. Kus), *Weyl modules for the hyperspecial current algebra*, Int. Math. Res. Not. IMRN **15** (2015), 6470–6515, https://doi.org/10.1093/imrn/rnu135.
78. (with M. Bennett), *Character formulae and a realization of tilting modules for* $\mathfrak{sl}_2[t]$, J. Algebra **441** (2015), 216–242, https://doi.org/10.1016/j.jalgebra.2015.06.026.
79. (with R. Biswal, L. Schneider, and S. Viswanath), *Demazure flags, Chebyshev polynomials, partial and mock theta functions*, J. Combin. Theory Ser. A **140** (2016), 38–75, https://doi.org/10.1016/j.jcta.2015.12.003.
80. (with P. Shereen, R. Venkatesh, and J. Wand), *A Steinberg type decomposition theorem for higher level Demazure modules*, J. Algebra **455** (2016), 314–346, https://doi.org/10.1016/j.jalgebra.2016.02.008.
81. (with M. Brito and A. Moura), *Demazure modules of level two and prime representations of quantum affine* $\mathfrak{sl}_{n+1}$, J. Inst. Math. Jussieu **17** (2018), no. 1, 75–105, https://doi.org/10.1017/S1474748015000407.
82. (with R. Biswal and D. Kus), *Demazure flags,* q-*Fibonacci polynomials and hypergeometric series*, Res. Math. Sci. **5** (2018), no. 1, Paper No. 12, 34, https://doi.org/10.1007/s40687-018-0129-1.
83. (with D. Kus and M. Odell), *Borel–de Siebenthal pairs, global Weyl modules and Stanley-Reisner rings*, Math. Z. **290** (2018), no. 1-2, 649–681, https://doi.org/10.1007/s00209-017-2035-4.
84. (with M. Brito), *Tensor products and* q-*characters of HL-modules and monoidal categorifications*, J. Éc. polytech. Math. **6** (2019), 581–619, https://doi.org/10.5802/jep.101.

Preprints

85. (with M. Brito), *Resolutions and a Weyl Character formula for prime representations of quantum affine* $\mathfrak{sl}_{n+1}$, available at arXiv:1704.02520.
86. (with R. Biswal, P. Shereen, and J. Wand), *Macdonald Polynomials and level two Demazure modules for affine* $\mathfrak{sl}_{n+1}$, available at arXiv:1910.0548.
87. (with J. Davis and R. Moruzzi Jr.), *Generalized Demazure Modules and Prime Representations in Type* Dn, available at arXiv:1911.07155.

Book

88. (with A. Pressley), *A guide to quantum groups*, Cambridge University Press, Cambridge, 1994.

Edited Volumes

89. (with I. B. Penkov), *Modular interfaces*, AMS/IP Studies in Advanced Mathematics, vol. 4, American Mathematical Society, Providence, RI; International Press, Cambridge, MA, 1997. Proceedings of the conference held in honour of Richard E. Block at the University of California, Riverside, CA, February 18–20, 1995.
90. (with D. Babbitt and R. Fioresi), *Symmetry in mathematics and physics*, Contemporary Mathematics, vol. 490, American Mathematical Society, Providence, RI, 2009. Papers from the conference in honour of V. S. Varadarajan's birthday held at the University of California, Los Angeles, CA, January 18–20, 2008.
91. (with J. Greenstein, K. C. Misra, K. N. Raghavan, and S. Viswanath), *Recent developments in algebraic and combinatorial aspects of representation theory*, Contemporary Mathematics, vol. 602, American Mathematical Society, Providence, RI, 2013. Proceedings of the International Congress of Mathematicians Satellite Conference on Algebraic and Combinatorial Approaches to Representation Theory held in Bangalore, August 12–16, 2010, and its Follow-up Conference held at the University of California, Riverside, CA, May 18–20, 2012.

Students of Vyjayanthi Chari

Matthew Bennett
Angelo Bianchi (joint with Adriano Moura)
Donna Blanton
Samuel Chamberlin
Justin Davis
Ronald Dolbin
Tammy Fisher-Vasta
Thang Le
Matthew Lee
Suzanne Lindborg
Mathew Lunde
Nathan Manning
Ryan Moruzzi, Jr.
Kayla Murray
Matthew O'Dell
Timothy Ridenour
Lisa Schneider
Prasad Senesi
Peri Shereen
Jeffrey Wand

Contents

Part I
Courses

String Diagrams and Categorification

Alistair Savage

Dedicated to Vyjayanthi Chari on the occasion of her 60th birthday

Abstract These are lecture notes for a mini-course given at the conference *Interactions of Quantum Affine Algebras with Cluster Algebras, Current Algebras, and Categorification* in June 2018. The goal is to introduce the reader to string diagram techniques for monoidal categories, with an emphasis on their role in categorification.

1 Introduction

Categorification is rapidly becoming a fundamental concept in many areas of mathematics, including representation theory, topology, algebraic combinatorics, and mathematical physics. One of the principal ingredients in categorification is the notion of a monoidal category. The goal of these notes is to introduce the reader to these categories as they often appear in categorification. Our intention is to motivate the definitions as much as possible, to help the reader build an intuitive understanding of the underlying concepts.

We begin in Sect. 2 with the definition of a strict $\Bbbk$-linear monoidal category. Our treatment will center on the string diagram calculus for such categories. The importance of this formalism comes from both the geometric intuition it provides and the fact that string diagrams are the framework upon which the applications of categorification to other areas such as knot theory and topology are built.

In Sect. 3, we give a number of examples of strict $\Bbbk$-linear monoidal categories. Most mathematicians encounter monoidal categories as additional structure on

A. Savage (✉)
Department of Mathematics and Statistics, University of Ottawa, Ottawa, ON, Canada
e-mail: alistair.savage@uottawa.ca

J. Greenstein et al. (eds.), *Interactions of Quantum Affine Algebras with Cluster Algebras, Current Algebras and Categorification*, Progress in Mathematics 337,
https://doi.org/10.1007/978-3-030-63849-8_1

some concept they already study: sets, vector spaces, group representations, etc., all naturally form monoidal categories. However, we will define abstract monoidal categories via generators and relations. Even though this idea has been around for some time, it is still somewhat foreign to many mathematicians working outside of category theory. We will see how, using this approach, one can obtain extremely efficient descriptions of familiar objects such as symmetric groups, degenerate affine Hecke algebras, braid groups, and wreath product algebras.

The formalism of string diagrams is at its best when one has a *pivotal category*, and we turn to this concept in Sect. 4. We start by discussing dual objects in monoidal categories. Pivotal categories are categories in which all objects have duals, the duality data is compatible with the tensor product, and the right and left mates of morphisms are equal. In pivotal categories, isotopic string diagrams correspond to the same morphism, allowing for intuitive topological arguments, as well as deep connections to topology and knot theory.

In Sect. 5, we discuss the idea of categorification, beginning with what is perhaps the standard approach, involving the Grothendieck group/ring of an additive category. We then discuss the trace of a category, and how this gives rise to the second type of categorification. We also define the Chern character map, which relates the Grothendieck group to the trace, and the notion of idempotent completion, motivated by the concept of a projective module.

We conclude, in Sect. 6, with the example of Heisenberg categories. We define these categories using the ideas we have developed and explain their relationship to the Heisenberg algebra. Our discussion here is necessarily brief, aiming only to give the reader a taste of a current area of research. We point the interested reader to references for further reading.

2 Strict Monoidal Categories and String Diagrams

2.1 Definitions

Throughout this chapter, all categories are assumed to be locally small. In other words, we have a *set* of morphisms between any two objects.

A *strict monoidal category* is a category $\mathcal{C}$ equipped with

- a bifunctor (the *tensor product*) $\otimes : \mathcal{C} \times \mathcal{C} \to \mathcal{C}$ and
- a *unit object* $\mathbb{1}$

such that, for all objects X, Y, and Z of $\mathcal{C}$, we have

- $(X \otimes Y) \otimes Z = X \otimes (Y \otimes Z)$ and
- $\mathbb{1} \otimes X = X = X \otimes \mathbb{1}$,

and, for all morphisms f, g, and h of $\mathcal{C}$, we have

- $(f \otimes g) \otimes h = f \otimes (g \otimes h)$ and
- $1_{\mathbb{1}} \otimes f = f = f \otimes 1_{\mathbb{1}}$.

Here, and throughout the chapter, 1_X denotes the identity endomorphism of an object X.

Remark 2.1 Note that, in a (not necessarily strict) *monoidal category*, the equalities above are replaced by isomorphism, and one imposes certain coherence conditions. For example, suppose $\mathbb{k}$ is a field, and let $\mathrm{Vect}_{\mathbb{k}}$ be the category of finite-dimensional $\mathbb{k}$-vector spaces. In this category, one has isomorphisms $(U \otimes V) \otimes W \cong U \otimes (V \otimes W)$, but these isomorphisms are not equalities in general. Similarly, the unit object in this category is the one-dimensional vector space $\mathbb{k}$, and we have $\mathbb{k} \otimes V \cong V \cong V \otimes \mathbb{k}$ for any vector space V.

We will be building monoidal categories "from scratch" via generators and relations. Thus, we are free to require them to be strict. In general, Mac Lane's coherence theorem for monoidal categories asserts that every monoidal category is monoidally equivalent to a strict one. (For a proof of this fact, see [16, §VII.2] or [12, §XI.5].) So, in practice, we do not lose much by assuming that monoidal categories are strict. (See also [24].) □

Fix a commutative ground ring $\mathbb{k}$. A $\mathbb{k}$-*linear category* is a category $\mathcal{C}$ such that

- for any two objects X and Y of $\mathcal{C}$, the hom-set $\mathrm{Hom}_{\mathcal{C}}(X, Y)$ is a $\mathbb{k}$-module and
- composition of morphisms is bilinear:

$$f \circ (\alpha g + \beta h) = \alpha(f \circ g) + \beta(f \circ h),$$
$$(\alpha f + \beta g) \circ h = \alpha(f \circ h) + \beta(g \circ h),$$

  for all $\alpha, \beta \in \mathbb{k}$ and morphisms f, g, and h such that the above operations are defined.

The category of $\mathbb{k}$-modules is an example of a $\mathbb{k}$-linear category. For any two $\mathbb{k}$-modules M and N, the space $\mathrm{Hom}_{\mathbb{k}}(M, N)$ is again a $\mathbb{k}$-module under the usual pointwise operations. Composition is bilinear with respect to this $\mathbb{k}$-module structure.

A *strict* $\mathbb{k}$-*linear monoidal category* is a category that is both strict monoidal and $\mathbb{k}$-linear, and such that the tensor product of morphisms is $\mathbb{k}$-bilinear. Before discussing some examples, we mention the important *interchange law*. Suppose

$$X_1 \xrightarrow{f} X_2 \qquad \text{and} \qquad Y_1 \xrightarrow{g} Y_2$$

are morphisms in a strict $\mathbb{k}$-linear monoidal category $\mathcal{C}$. Then,

$$(1_{X_2} \otimes g) \circ (f \otimes 1_{Y_1}) = \otimes((1_{X_2}, g)) \circ \otimes((f, 1_{Y_1})) = \otimes((1_{X_2}, g) \circ (f, 1_{Y_1}))$$
$$= \otimes((f, g)) = f \otimes g,$$

where the second equality uses that the tensor product is a bifunctor. Similarly,

$$(f \otimes 1_{Y_2}) \circ (1_{X_1} \circ g) = f \otimes g.$$

Thus, the following diagram commutes:

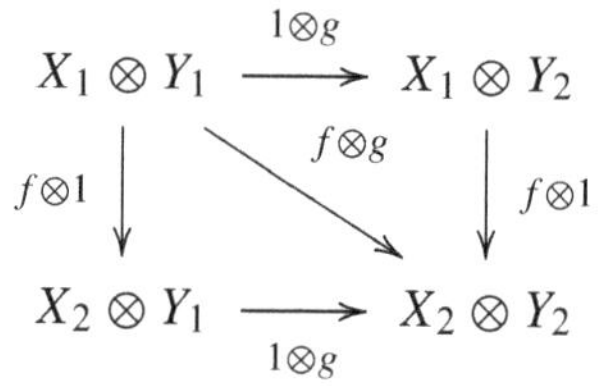

$$\begin{array}{ccc} X_1 \otimes Y_1 & \xrightarrow{1 \otimes g} & X_1 \otimes Y_2 \\ {\scriptstyle f \otimes 1} \downarrow & \searrow^{f \otimes g} & \downarrow {\scriptstyle f \otimes 1} \\ X_2 \otimes Y_1 & \xrightarrow[1 \otimes g]{} & X_2 \otimes Y_2 \end{array}$$

2.2 Examples

Let us consider a very simple strict monoidal category. Every monoidal category must have a unit object $\mathbb{1}$ by definition. But it is possible that this is the *only* object. The identity axiom for a strict monoidal category forces $\mathbb{1} \otimes \mathbb{1} = \mathbb{1}$. There is only one hom-set in this category, namely

$$\mathrm{End}(\mathbb{1}) := \mathrm{Hom}(\mathbb{1}, \mathbb{1}).$$

The associativity axiom for morphisms in a category implies that $\mathrm{End}(\mathbb{1})$ is a monoid under composition, with identity $1_{\mathbb{1}}$, the identity endomorphism of $\mathbb{1}$. The axioms of a strict monoidal category imply that $\mathrm{End}(\mathbb{1})$ is also a monoid under the tensor product. However, the interchange law forces these monoids to coincide and to be commutative! Indeed, for all $f, g \in \mathrm{End}(\mathbb{1})$, we have

$$f \circ g = (f \otimes 1_{\mathbb{1}}) \circ (1_{\mathbb{1}} \otimes g) = f \otimes g = (1_{\mathbb{1}} \otimes g) \circ (f \otimes 1_{\mathbb{1}}) = g \circ f. \tag{2.1}$$

Conversely, given any commutative monoid A, we have a strict monoidal category with one object $\mathbb{1}$, and $\mathrm{End}(\mathbb{1}) = A$. The composition and tensor product are both given by the multiplication in A.

Now, consider a strict $\Bbbk$-*linear* monoidal category with one object $\mathbb{1}$. Then, $\mathrm{End}(\mathbb{1})$ is an associative $\Bbbk$-algebra, and an argument exactly analogous to the one above shows that it is, in fact, commutative. Conversely, every commutative associative $\Bbbk$-algebra gives rise to a one-object strict $\Bbbk$-linear monoidal category.

Note that the above discussion actually shows that $\mathrm{End}(\mathbb{1})$ is a commutative monoid in *any* strict monoidal category and is a commutative $\Bbbk$-algebra in *any* strict

$\Bbbk$-linear monoidal category. The monoid/algebra $\mathrm{End}(\mathbb{1})$ is called the *center* of the category.

Example 2.2 (Center of $\mathrm{Vect}_\Bbbk$*)* Suppose $\Bbbk$ is a field and consider the category $\mathrm{Vect}_\Bbbk$ of finite-dimensional $\Bbbk$-vector spaces. This is not a *strict* monoidal category, but, as noted in Remark 2.1 (see, in particular, [24, Th. 4.3]), we can safely avoid this technicality. The unit object of $\mathrm{Vect}_\Bbbk$ is the one-dimensional vector space $\Bbbk$, and so the center of this category is $\mathrm{End}_\Bbbk(\Bbbk)$, which is canonically isomorphic, as a ring, to $\Bbbk$ via the isomorphism

$$\mathrm{End}_\Bbbk(\Bbbk) \xrightarrow{\cong} \Bbbk, \quad f \mapsto f(1). \tag{2.2}$$

2.3 String Diagrams

Strict monoidal categories are especially well suited to being depicted using the language of *string diagrams*. These diagrams, which are also sometimes called *Penrose diagrams*, have their origins in the work of Roger Penrose in physics [19]. Working with string diagrams helps build intuition. It also often makes certain arguments obvious, whereas the corresponding algebraic proof can be a bit opaque. We give here a brief overview of string diagrams, referring the reader to [27, Ch. 2] for a detailed treatment. Throughout this section, $\mathcal{C}$ will denote a strict $\Bbbk$-linear monoidal category.

We will denote a morphism $f: X \to Y$ by a strand with a coupon labeled f:

Note that we are adopting the convention that diagrams should be read from bottom to top. The *identity map* $1_X : X \to X$ is a string with no coupon:

We sometimes omit the object labels (e.g., X and Y above) when they are clear or unimportant. We will also sometimes distinguish identity maps of different objects by some sort of decoration of the string (orientation, dashed versus solid, etc.), rather than by adding object labels.

Composition is denoted by *vertical stacking* (recall that we read pictures bottom to top), and tensor product is *horizontal juxtaposition*:

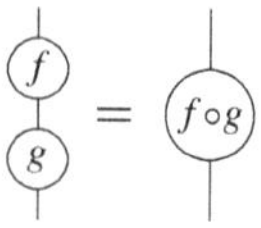

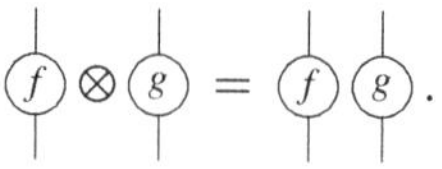

The *interchange law* then becomes

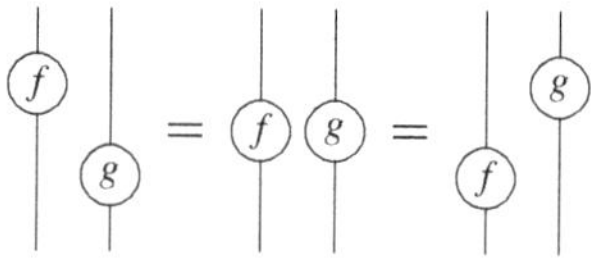

A general morphism $f : X_1 \otimes \cdots \otimes X_n \to Y_1 \otimes \cdots \otimes Y_m$ can be depicted as a coupon with n strands emanating from the bottom and m strands emanating from the top:

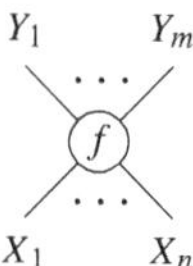

For the moment, let us denote the identity morphism $1_{\mathbb{1}}$ of the identity object $\mathbb{1}$ by a dashed line:

Then, the translation into diagrams of our argument from Sect. 2.2 that the center $\mathrm{End}(\mathbb{1})$ of the category is a commutative algebra becomes that, for all $f, g \in \mathrm{End}(\mathbb{1})$,

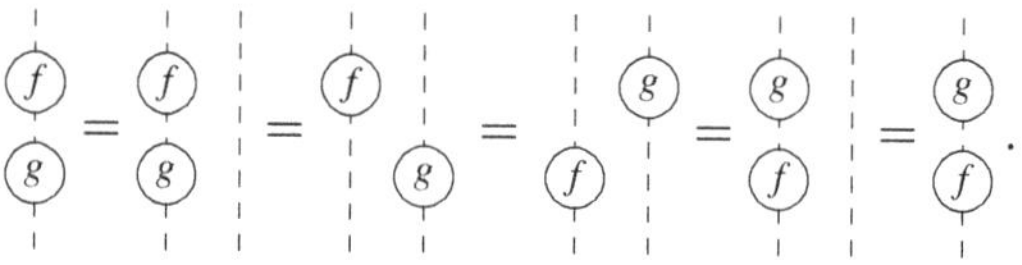

In fact, as we see above, the axioms of a strict ($\mathbb{k}$-linear) monoidal category make it natural to omit the identity morphism $1_{\mathbb{1}}$ of the identity object. So, we draw endomorphisms $f \in \mathrm{End}(\mathbb{1})$ of the identity as free-floating coupons:

By (2.1), the horizontal and vertical juxtapositions of such free-floating coupons coincide. So, we may slide these coupons around at will.

3 Monoidally Generated Algebras

3.1 *Presentations*

One should think of working with strict $\Bbbk$-linear monoidal categories as doing "two-dimensional" algebra with the morphisms. Besides the addition (which corresponds to formal addition of string diagrams), we have two flavors of "multiplication": horizontal (the tensor product) and vertical (the composition in the category).

Just as one can define associative algebras via generators and relations, one can also define strict $\Bbbk$-linear monoidal categories in this way. Recall that the free associative $\Bbbk$-algebra A on some set $\{a_i : i \in I\}$ of generators consists of formal finite $\Bbbk$-linear combinations of words in the generators. These words are of the form

$$a_{i_1}a_{i_2}\cdots a_{i_n}, \quad i_1, i_2, \ldots, i_n \in I.$$

Multiplication is given by concatenation of words, extended by linearity. The empty word corresponds to the multiplicative identity. If we wish to impose some set $R \subseteq A$ of relations, we then consider the algebra $A/\langle R\rangle$, where $\langle R\rangle$ is the two-sided ideal of A generated by the set R. What this means in practice is that we can make "local substitutions" in words using the relations. For example, if A is the algebra with generators a, b, c, and d and relations $ab = c$ and $d^2 = ba$, then $R = \{ab - c, d^2 - ba\}$, and we have

$$acabbcdda = ac(ab)bc(d^2)a = accbcbaa.$$

In a similar way, we can give presentations of strict $\Bbbk$-linear monoidal categories. (See [26, §I.4.2] for more details.) Now we should specify a set of generating objects, a set of generating morphisms, and some relations *on morphisms* (not on objects!). If $\{X_i : i \in I\}$ is our set of generating objects, then an arbitrary object in our category is a finite tensor product of these generating objects:

$$X_{i_1} \otimes X_{i_2} \otimes \cdots \otimes X_{i_n}, \quad i_1, i_2, \ldots, i_n \in I,\ n \in \mathbb{N}.$$

We think of $\mathbb{1}$ as being the "empty tensor product." If $\{f_j : j \in J\}$ is our set of generating morphisms, then we can take arbitrary tensor products and compositions (when domains and codomains match) of these generators, e.g.,

$$(f_{j_1} \otimes f_{j_2}) \circ \big((f_{j_3} \circ f_{j_4}) \otimes f_{j_5}\big), \quad j_1, j_2, j_3, j_4, j_5 \in J.$$

Working with string diagrams, our generating morphisms are diagrams, and we can vertically and horizontally compose them in any way that makes sense (i.e., making sure that domains and codomains match in vertical composition). Relations allow us to make "local changes" in our diagrams.

In the examples to be considered below, we will often have generating objects that we will denote by $\uparrow$ and $\downarrow$. We will always draw their identity morphisms as

$$\uparrow \quad \text{and} \quad \downarrow,$$

respectively.

3.2 The Symmetric Group

As a concrete example, define $\mathcal{S}$ to be the strict $\Bbbk$-linear monoidal category with

- one generating object $\uparrow$,
- one generating morphism,

$$: \uparrow \otimes \uparrow \to \uparrow \otimes \uparrow,$$

- two relations:

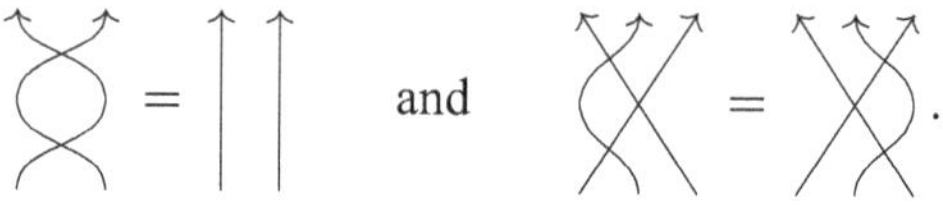

(3.1)

One could write these relations in a more traditional algebraic manner, if so desired. For example, if we let

$$s =$$  $$: \uparrow \otimes \uparrow \to \uparrow \otimes \uparrow,$$

then the two relations in (3.1) become

$$s^2 = 1_{\uparrow\otimes\uparrow} \qquad \text{and} \qquad (s \otimes 1_\uparrow) \circ (1_\uparrow \otimes s) \circ (s \otimes 1_\uparrow) = (1_\uparrow \otimes s) \circ (s \otimes 1_\uparrow) \circ (1_\uparrow \otimes s).$$

Now, in any $\Bbbk$-linear category (monoidal or not), we have an endomorphism algebra $\mathrm{End}(X)$ of any object X. The multiplication in this algebra is given by vertical composition. In $\mathcal{S}$, every object is of the form $\uparrow^{\otimes n}$ for some $n = 0, 1, 2, \ldots$. An example of an endomorphism of $\uparrow^{\otimes 4}$ is

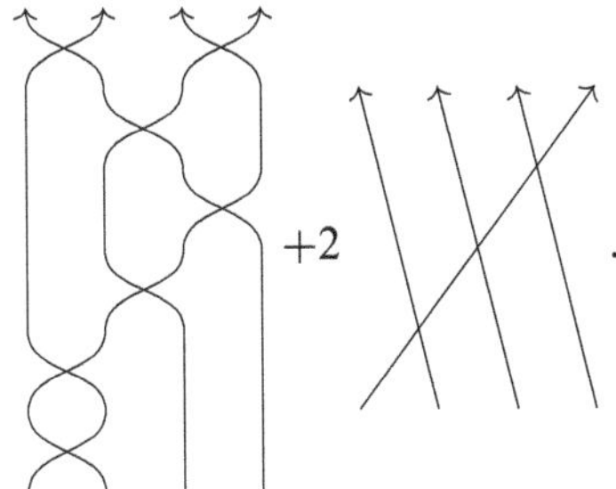

Using the relations, we see that this morphism is equal to

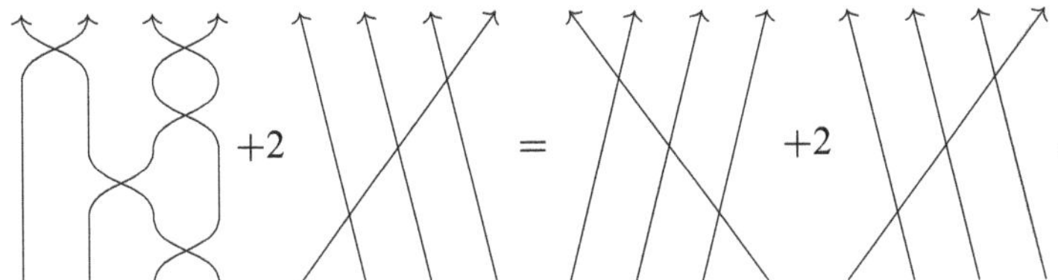

Fix a positive integer n, and recall that the group algebra $\Bbbk S_n$ of the symmetric group on n letters has a presentation with generators $s_1, s_2, \dots, s_{n-1}$ (the simple transpositions) and relations

$$s_i^2 = 1, \qquad 1 \le i \le n-1, \tag{3.2}$$

$$s_i s_{i+1} s_i = s_{i+1} s_i s_{i+1}, \qquad 1 \le i \le n-2, \tag{3.3}$$

$$s_i s_j = s_j s_i, \qquad 1 \le i, j \le n-1,\ |i-j| > 1. \tag{3.4}$$

Consider the map

$$\Bbbk S_n \to \mathrm{End}_{\mathcal{S}}(\uparrow^{\otimes n}),$$

where s_i is sent to the crossing of the ith and $(i+1)$th strands, labeled from right to left. In fact, this map is an isomorphism of algebras. So, the category $\mathcal{S}$ contains the group algebras of all of the symmetric groups!

Note that the presentation of $\mathcal{S}$ is much more efficient than the presentation of the algebras $\Bbbk S_n$. To define $\mathcal{S}$, we need only *one* generator and *two* relations, as opposed to the $n-1$ generators and relations (3.2)–(3.4) (whose number is of order n^2) in the algebraic presentation of $\Bbbk S_n$, for *each* n. This efficiency comes from the fact that we are generating the algebras *monoidally*, where we have both vertical and horizontal "multiplication." In particular, the "distant braid relation" (3.4) follows for free from the interchange law:

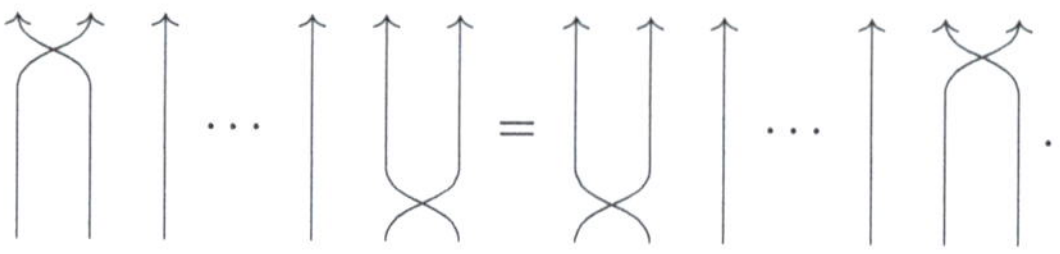

$$\cdots = \cdots .$$

3.3 Degenerate Affine Hecke Algebras

Let $\mathcal{AH}^{\rm deg}$ be the strict $\Bbbk$-linear monoidal category $\mathcal{S}$ defined in Sect. 3.2, but with an additional generating morphism, which we will call a *dot*,

$$: \uparrow \to \uparrow$$

and one additional relation:

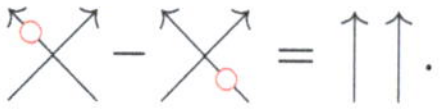

$$- = \uparrow\uparrow .$$

Now,

$$\operatorname{End}_{\mathcal{AH}^{\rm deg}}(\uparrow^{\otimes n})$$

is isomorphic to the *degenerate affine Hecke algebra* of type A_{n-1}. In the category $\mathcal{AH}^{\rm deg}$, the endomorphism algebra of $\uparrow$ is now infinite-dimensional, with basis given by

$$\Big\} m \text{ dots}, \quad m = 0, 1, 2, \dots .$$

3.4 The Braid Group

Consider another strict $\Bbbk$-linear monoidal category $\mathcal{B}$ with one generating object $\uparrow$ and one generating morphism

$$: \uparrow \otimes \uparrow \to \uparrow \otimes \uparrow . \tag{3.5}$$

We want to impose the relation that this morphism is invertible. Thus, we add another generating morphism that is inverse to (3.5). Let us denote this inverse by

$$: \uparrow \otimes \uparrow \to \uparrow \otimes \uparrow . \tag{3.6}$$

To say that (3.5) and (3.6) are inverse means that we impose the relations

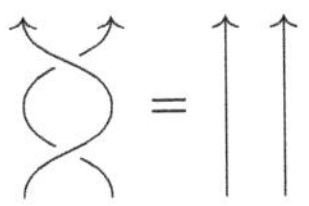 and 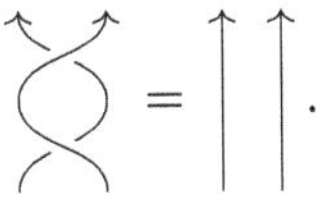 (3.7)

To complete the definition of $\mathcal{B}$, we impose one more relation:

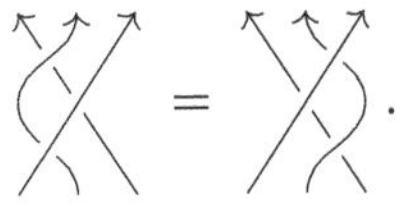 (3.8)

Then, $\mathrm{End}_{\mathcal{B}}(\uparrow^{\otimes n})$ is isomorphic to the group algebra of the braid group on n strands. Again, we see that generating these algebras monoidally is extremely efficient.

3.5 *Hecke Algebras*

Fix $z \in \Bbbk$. Let $\mathcal{H}(z)$ be the strict $\Bbbk$-linear monoidal category $\mathcal{B}$ defined in Sect. 3.4, but with one more relation:

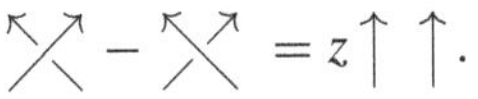 (3.9)

If $z = 0$, the relation (3.9) forces the generators (3.5) and (3.6) to be equal. The relations (3.7) and (3.8) then reduce to (3.1). Hence, $\mathcal{H}(0) = \mathcal{S}$. On the other hand, if $\Bbbk = \mathbb{C}(q)$ and $z = q - q^{-1}$, then $\mathrm{End}_{\mathcal{H}(z)}(\uparrow^{\otimes n})$ is isomorphic to the *Iwahori–Hecke algebra* of type A_{n-1}.

3.6 *Wreath Product Algebras*

Let A be an associative $\Bbbk$-algebra. Let us modify the category $\mathcal{S}$ from Sect. 3.2 by adding an endomorphism of $\uparrow$ for each element of A. More precisely, define the *wreath product category* $\mathcal{W}(A)$ to be the strict $\Bbbk$-linear monoidal category obtained from $\mathcal{S}$ by adding morphisms such that we have an algebra homomorphism

$A \to \mathrm{End}(\uparrow)$, $\quad a \mapsto$  .

In particular, this means that

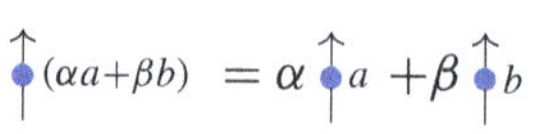

$(\alpha a+\beta b) = \alpha \, a + \beta \, b$ and $a, b = ab$ for all $\alpha, \beta \in \Bbbk$, $a, b \in A$.

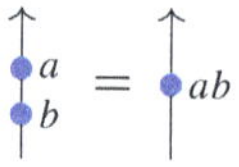

(3.10)

We call the closed circles appearing in the above diagrams as *tokens*. We then impose the additional relation

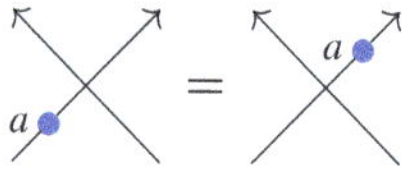

$$a \in A. \tag{3.11}$$

As an example of a diagrammatic proof, note that we can compose (3.11) on the top and bottom with a crossing to obtain

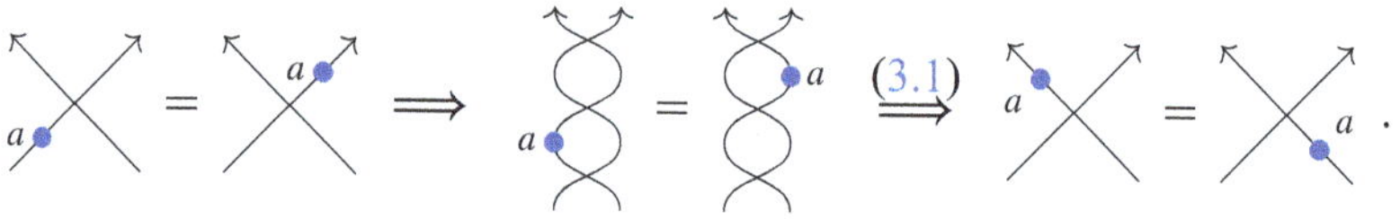

$$\Longrightarrow \quad \overset{(3.1)}{\Longrightarrow}$$

So, tokens also slide up left through crossings.

One can show that

$$\mathrm{End}_{\mathcal{W}(A)}(\uparrow^{\otimes n}) \cong A^{\otimes n} \rtimes S_n,$$

the nth *wreath product algebra* associated with A. As a $\Bbbk$-module,

$$A^{\otimes n} \rtimes S_n = A^{\otimes n} \otimes_{\Bbbk} \Bbbk S_n.$$

Multiplication is determined by

$$(a_1 \otimes \pi_1)(a_2 \otimes \pi_2) = a_1(\pi_1 \cdot a_2) \otimes \pi_1\pi_2, \quad a_1, a_2 \in A^{\otimes n}, \ \pi_1, \pi_2 \in S_n,$$

where $\pi_1 \cdot a_2$ denotes the natural action of $\pi_1 \in S_n$ on $a_2 \in A^{\otimes n}$ by permutation of the factors. Note that $\mathcal{W}(\Bbbk) = \mathcal{S}$, the symmetric group category.

3.7 *Affine Wreath Product Algebras*

The wreath product category $\mathcal{W}(A)$ is a generalization of the symmetric group category $\mathcal{S}$ that depends on a choice of associative $\Bbbk$-algebra A. We can generalize the degenerate affine Hecke category $\mathcal{AH}^{\mathrm{deg}}$ in a similar way as long as we have some additional structure on A. In particular, we suppose that we have a $\Bbbk$-linear *trace map*

$$\mathrm{tr}\colon A \to \Bbbk$$

and dual bases B and $\{b^\vee : b \in B\}$ of A such that

$$\mathrm{tr}(a^\vee b) = \delta_{a,b} \quad \text{for all } a, b \in B.$$

An algebra with such a trace map is called a *Frobenius algebra*. We will assume for simplicity here that the trace map is symmetric:

$$\mathrm{tr}(ab) = \mathrm{tr}(ba) \quad \text{for all } a, b \in A.$$

It is an easy linear algebra exercise to verify that the element

$$\sum_{b \in B} b \otimes b^\vee \in A \otimes A$$

is independent of the choice of basis B. Also, for all $x \in A$, we have

$$\begin{aligned}\sum_{b\in B} bx \otimes b^\vee &= \sum_{a,b\in B} \mathrm{tr}(a^\vee bx) a \otimes b^\vee = \sum_{a,b\in B} a \otimes \mathrm{tr}(a^\vee bx) b^\vee \\ &= \sum_{a,b\in B} a \otimes \mathrm{tr}(xa^\vee b) b^\vee = \sum_{a\in B} a \otimes xa^\vee.\end{aligned} \tag{3.12}$$

We define the *affine wreath product category* $\mathcal{AW}(A)$ to be the strict $\Bbbk$-linear monoidal category obtained from $\mathcal{W}(A)$ by adding a generating morphism

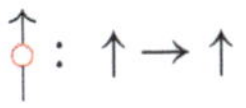

$$: \uparrow \to \uparrow$$

and the additional relations

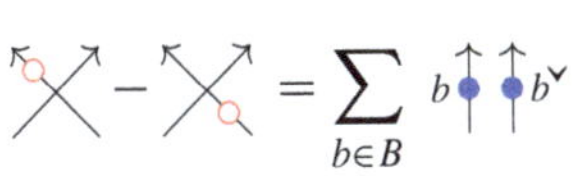

$$- = \sum_{b\in B} {}_{b}\ {}_{b^\vee} \qquad \text{and} \qquad {}_{a} = {}_{a}, \qquad \text{for all } a \in A. \tag{3.13}$$

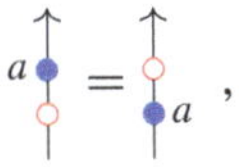

To motivate this definition, examine what happens if we add a token labeled $a \in A$ to the bottom of the left strand of the diagrams involved in the first relation in (3.13). For the diagrams on the left side, the token slides up to the top of the right strand:

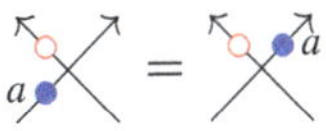

$$_a = {}^a \qquad \text{and} \qquad _a = {}^a.$$

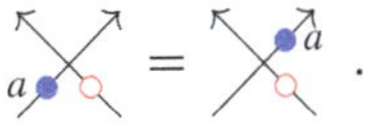

Since diagrams are linear in the token labels (see (3.10)), (3.12) tells us that the exact same thing happens with the term on the right side of the first relation in (3.13):

$$\sum_{b\in B} \overset{\uparrow}{\underset{a}{\overset{b}{\bullet}}}\ \overset{\uparrow}{\bullet}\,b^\vee = \sum_{b\in B} ba\,\overset{\uparrow}{\bullet}\ \overset{\uparrow}{\bullet}\,b^\vee \overset{(3.12)}{=} \sum_{b\in B} b\,\overset{\uparrow}{\bullet}\ \overset{\uparrow}{\bullet}\,ab^\vee = \sum_{b\in B} b\,\overset{\uparrow}{\bullet}\ \overset{\uparrow}{\underset{b^\vee}{\overset{a}{\bullet}}}\ .$$

Loosely speaking, tokens can "teleport" across the sum appearing in the left relation in (3.13).

The name *affine wreath product category* comes from the fact that

$$\mathrm{End}_{\mathcal{AW}(A)}(\uparrow^{\otimes n})$$

is isomorphic to an *affine wreath product algebra*. See [22] for a detailed discussion of these algebras. Note that $\mathcal{AW}(\Bbbk)$ is the degenerate affine Hecke category $\mathcal{AH}^{\mathrm{deg}}$. (Here, we take the trace map $\mathrm{tr}\colon \Bbbk \to \Bbbk$ to be the identity.)

3.8 Quantum Affine Wreath Product Algebras

One can also define affine versions of the Hecke category $\mathcal{H}(z)$ and generalizations of these categories depending on a Frobenius algebra. We refer the reader to [6, 8, 20] for further details.

4 Pivotal Categories

We give here a brief overview of pivotal categories. Further details can be found in [27, §1.7, §2.1].

4.1 Duality

Suppose a strict monoidal category has two objects, $\uparrow$ and $\downarrow$. Recalling our convention that we do not draw the identity morphism of the unit object $\mathbb{1}$, a morphism $\mathbb{1} \to \downarrow \otimes \uparrow$ would have string diagram

$$\smile\!\!\!\nearrow : \mathbb{1} \to \downarrow \otimes \uparrow,$$

where we may decorate the cup with some symbol if we have more than one such morphism. The fact that the bottom of the diagram is empty space indicates that the domain of this morphism is the unit object $\mathbb{1}$. Similarly, we can have

$$\frown\!\!\!\searrow : \uparrow \otimes \downarrow \to \mathbb{1}.$$

We say that $\downarrow$ is *right dual* to $\uparrow$ (and $\uparrow$ is *left dual* to $\downarrow$) if we have morphisms

$$: \mathbb{1} \to \downarrow \otimes \uparrow \qquad \text{and} \qquad : \uparrow \otimes \downarrow \to \mathbb{1}$$

such that

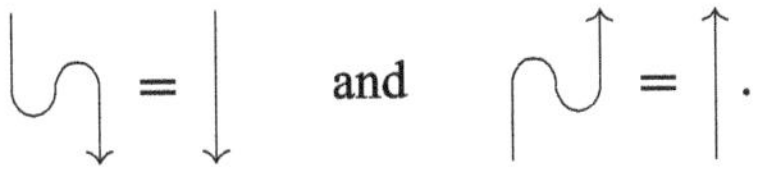

$$= \qquad \text{and} \qquad = . \tag{4.1}$$

(The above relations are analogous to the unit–counit formulation of adjunction of functors.) A monoidal category in which every object has both left and right duals is called a *rigid*, or *autonomous*, category.

If $\uparrow$ and $\downarrow$ are both left and right dual to each other, then, in addition to the above, we also have

$$: \mathbb{1} \to \uparrow \otimes \downarrow \qquad \text{and} \qquad : \downarrow \otimes \uparrow \to \mathbb{1}$$

such that

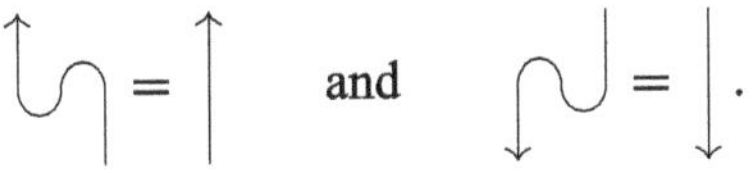

$$= \qquad \text{and} \qquad = . \tag{4.2}$$

To give a concrete example of duality in a monoidal category, consider the category $\mathrm{Vect}_{\mathbb{k}}$ of finite-dimensional $\mathbb{k}$-vector spaces, where $\mathbb{k}$ is a field (see Example 2.2). In this category, the unit object is $\mathbb{k}$. We claim that, if V is any finite-dimensional $\mathbb{k}$-vector space, the dual vector space V^* is both right and left dual to V in the sense defined above. Indeed, fix a basis B of V, and let $\{\delta_v : v \in B\}$ denote the dual basis of V^*. Viewing V as $\uparrow$ and V^* as $\downarrow$, we define

$$: \mathbb{k} \to V^* \otimes V, \quad \alpha \mapsto \alpha \sum_{v \in B} \delta_v \otimes v,$$

$$: V \otimes V^* \to \mathbb{k}, \quad \sum_{i=1}^{n} v_i \otimes f_i \mapsto \sum_{i=1}^{n} f_i(v_i),$$

$$: \mathbb{k} \to V \otimes V^*, \quad \alpha \mapsto \alpha \sum_{v \in B} v \otimes \delta_v,$$

$$: V^* \otimes V \to \mathbb{k}, \quad \sum_{i=1}^{n} f_i \otimes v_i \mapsto \sum_{i=1}^{n} f_i(v_i).$$

Let us check the right-hand relation in (4.1). The left-hand side is the composition

$$V \cong V \otimes \mathbb{k} \xrightarrow{1_V \otimes} V \otimes V^* \otimes V \xrightarrow{\otimes 1_V} \mathbb{k} \otimes V \cong V,$$

$$w \mapsto w \otimes 1 \mapsto \sum_{v \in B} w \otimes \delta_v \otimes v \mapsto \sum_{v \in B} \delta_v(w) \otimes v \mapsto \sum_{v \in V} \delta_v(w) v = w.$$

Thus, this composition is precisely the identity map 1_V, and so the right-hand relation in (4.1) is satisfied. The verification of the left-hand equality in (4.1) and both equalities in (4.2) are analogous and are left as an exercise for the reader.

If $\uparrow$ and $\downarrow$ are both left and right dual to each other, then we may form closed diagrams of the form

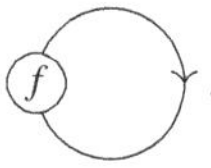
$$, \quad f \in \mathrm{End}(\uparrow). \tag{4.3}$$

Such closed diagrams live in the center $\mathrm{End}(\mathbb{1})$ of the category. Let us consider such a diagram in the category $\mathrm{Vect}_{\Bbbk}$, where we know from Example 2.2 that the center of the category is isomorphic to $\Bbbk$. If $f \in \mathrm{End}_{\Bbbk}(V)$, then the diagram (4.3) is the composition

$$\Bbbk \xrightarrow{\cup} V \otimes V^* \xrightarrow{f \otimes 1_{V^*}} V \otimes V^* \xrightarrow{\cap} \Bbbk,$$

$$\alpha \mapsto \alpha \sum_{v \in B} v \otimes \delta_v \mapsto \alpha \sum_{v \in B} f(v) \otimes \delta_v \mapsto \alpha \sum_{v \in B} \delta_v(f(v)) = \alpha \mathrm{tr}(f), \tag{4.4}$$

where $\mathrm{tr}(f)$ is the usual trace of the linear map f. Therefore, under the isomorphism (2.2), the diagram (4.3) corresponds to $\mathrm{tr}(f)$.

4.2 Mates

Suppose that an object X in a strict monoidal category has a right dual X^*. Since we will now need to consider multiple objects with duals, we will denote the identity endomorphisms of X and X^* by upward and downward strands labeled X:

$$1_X = \underset{X}{\uparrow} \qquad \text{and} \qquad 1_{X^*} = \underset{X}{\downarrow}\,.$$

As explained in Sect. 4.1, the fact that X^* is right dual to X means that we have morphisms

$$: \mathbb{1} \to X^* \otimes X \qquad \text{and}$$
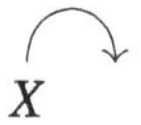
$$: X \otimes X^* \to \mathbb{1} \tag{4.5}$$

such that

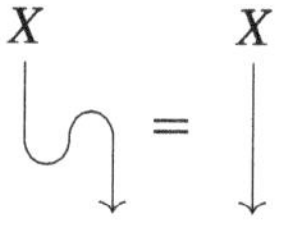
and
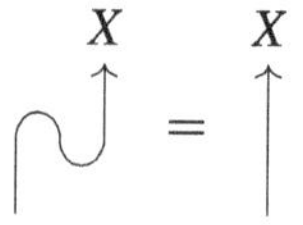
(4.6)

Here, we again label the strands with X to distinguish between the cups and caps for different objects.

Suppose X and Y have right duals X^* and Y^*, respectively. Then, every

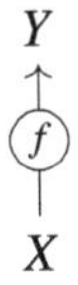

$$f \in \operatorname{End}(X, Y) \qquad \text{has } \textit{right mate} \qquad \in \operatorname{End}(Y^*, X^*).$$

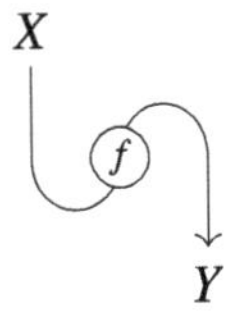

Now, suppose $\mathcal{C}$ is a strict monoidal category in which every object has a right dual. Consider the map $R \colon \mathcal{C} \to \mathcal{C}$ that sends every object to its right dual and every morphism to its right mate. How does R behave with respect to vertical composition? Omitting object labels, we have

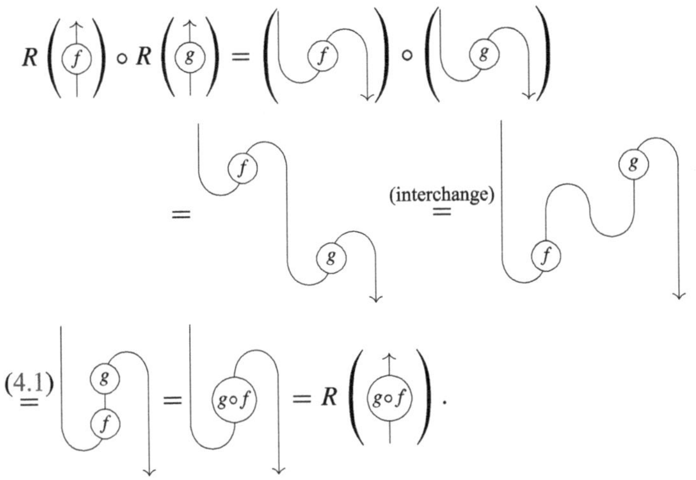

$$R(f) \circ R(g) = (\,f\,) \circ (\,g\,) = \overset{\text{(interchange)}}{=} \overset{(4.1)}{=} \; = R(g \circ f).$$

It follows that R is a *contravariant functor*. In particular, for every object X, the functor R induces a monoid anti-automorphism $\operatorname{End}(X) \to \operatorname{End}(X^*)$. (If $\mathcal{C}$ is strict $\Bbbk$-linear monoidal, then this is an algebra anti-automorphism.)

In a manner analogous to the above, if X^* and Y^* are *left* dual to X and Y, respectively, then every

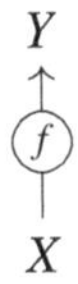

$$f \in \operatorname{End}(X, Y) \qquad \text{has } \textit{left mate} \qquad \in \operatorname{End}(Y^*, X^*).$$

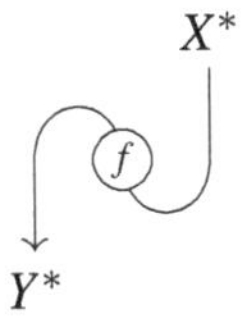

This gives another contravariant endofunctor of $\mathcal{C}$.

4.3 Pivotal Categories

Let $\mathcal{C}$ be a strict monoidal category. Suppose that all objects have right duals, and $(X^*)^* = X$ for every object X. It then follows that X^* is also *left* dual to X for every object X. In particular, in the language of Sect. 4.2, we define a left cup and cap labeled X to be a right cup and cap, respectively, labeled X^*:

X := X* and := .
X X*

If we also add the requirement that the duality data (i.e., the cups and caps) be compatible with the tensor product and that right mates always equal left mates, we get the following definition.

Definition 4.1 (Strict Pivotal Category) A strict monoidal category $\mathcal{C}$ is a *strict pivotal category* if every object X has a right dual X^* with (fixed) morphisms (4.5) satisfying (4.6) and the following three additional conditions:

(a) For all objects X and Y in $\mathcal{C}$,

$$(X^*)^* = X, \quad (X \otimes Y)^* = Y^* \otimes X^*, \quad \mathbb{1}^* = \mathbb{1}.$$

(b) For all objects X and Y in $\mathcal{C}$, we have

X ⊗ Y = X Y and = .
X ⊗ Y X Y

(c) For every morphism $f : X \to Y$ in $\mathcal{C}$, its right and left mates are equal:

X X
f = f .
Y Y

(4.7)

It is important to note that a strict pivotal structure is extra data on the category $\mathcal{C}$, namely the morphisms in (4.5). □

If $\mathcal{C}$ is strict pivotal, and

Y
f ∈ Hom(X, Y),
X

then we typically *define* the corresponding coupon on a downward strand to be the right (equivalently, left) mate:

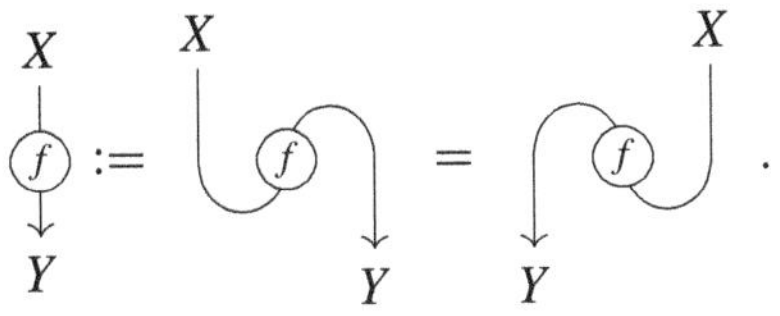

X X X

f := f = f .

Y Y Y

Suppose that a strict monoidal category $\mathcal{C}$ is defined in terms of generators and relations and that each generating object X has a right dual generating object X^*, with $(X^*)^* = X$. Then, in order to show that $\mathcal{C}$ is pivotal, it suffices to show that the right and left mates of each generating morphism are equal. The axioms of a strict pivotal category then uniquely determine the duality data for arbitrary objects, which are tensor products of the generating objects.

In a strict pivotal category, isotopic string diagrams represent the same morphism! (See [27, §2.4] for a detailed discussion.) This allows us to use geometric intuition and topological arguments in the study of such categories. In some places in the literature, the strict pivotal nature of a category is implicit in the definition. More precisely, categories where morphisms consist of planar diagrams *up to isotopy* are strict pivotal by definition. This is the case, for example, for the Heisenberg categories defined in [9, 13, 21] and the categorified quantum group of [14].

One of the simplest examples of a strict pivotal category is the *Temperley–Lieb category* $\mathcal{TL}(\delta)$, $\delta \in \Bbbk$. This is a strict $\Bbbk$-linear monoidal category on one generating object X. We make this object self-dual by adding generating morphisms

$$: \mathbb{1} \to X \otimes X, \qquad : X \otimes X \to \mathbb{1}$$

and relations

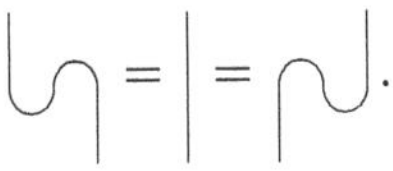

$$= \; = \; .$$

We also impose the relation

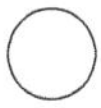

$$= \delta.$$

The endomorphism algebra $\mathrm{End}_{\mathcal{TL}(\delta)}(X^{\otimes n})$ is the *Temperley–Lieb algebra* $TL_n(\delta)$.

5 Categorification

5.1 Additive Categories

A $\Bbbk$-linear category is said to be *additive* if it admits all finitary biproducts (including the empty biproduct, which is a *zero object*). For example, if A is an associative $\Bbbk$-algebra, then the category of left modules over A is additive, with biproduct given by the direct sum $\oplus$ of modules.

Given a $\Bbbk$-linear category $\mathcal{C}$, we can enlarge it to an additive category by taking its *additive envelope* $\mathrm{Add}(\mathcal{C})$. The objects of $\mathrm{Add}(\mathcal{C})$ are formal finite direct sums

$$\bigoplus_{i=1}^{n} X_i$$

of objects X_i in $\mathcal{C}$. Morphisms

$$f \colon \bigoplus_{i=1}^{n} X_i \to \bigoplus_{j=1}^{m} Y_j$$

are $m \times n$ matrices, where the (j, i)-entry is a morphism

$$f_{i,j} \colon X_i \to Y_j.$$

Composition is given by the usual rules of matrix multiplication.

Example 5.1 Let A be an associative $\Bbbk$-algebra, and let $\mathcal{C}$ be the $\Bbbk$-linear category of free rank-one left A-modules. Then, $\mathrm{Add}(\mathcal{C})$ is equivalent to the category of free left A-modules. □

5.2 The Grothendieck Ring

Suppose $\mathcal{C}$ is an additive $\Bbbk$-linear category. Let $\mathrm{Iso}_{\mathbb{Z}}(C)$ denote the free abelian group generated by isomorphism classes of objects of $\mathcal{C}$, and let $[X]_{\cong}$ denote the isomorphism class of an object X. Let J denote the subgroup generated by the elements

$$[X \oplus Y]_{\cong} - [X]_{\cong} - [Y]_{\cong}, \qquad X, Y \text{ objects of } \mathcal{C}.$$

The *split Grothendieck group* of $\mathcal{C}$ is

$$K_0(\mathcal{C}) := \mathrm{Iso}_{\mathbb{Z}}(\mathcal{C})/J.$$

In general, the split Grothendieck group is simply an abelian group. However, if $\mathcal{C}$ is an additive $\Bbbk$-linear *monoidal* category, then $K_0(\mathcal{C})$ is a ring with multiplication given by

$$[X]_{\cong} \cdot [Y]_{\cong} = [X \otimes Y]_{\cong}$$

for objects X and Y (with the multiplication extended to all of $K_0(\mathcal{C})$ by linearity).

The process of passing to the split Grothendieck ring is a form of *decategorification*. The process of *categorification* is a one-sided inverse to this procedure. Namely, to categorify a ring R is to find monoidal category $\mathcal{C}$ such that $K_0(\mathcal{C}) \cong R$ as rings.

Example 5.2 (Categorification of the Ring of Integers) Suppose $\Bbbk$ is a field, and let $\mathrm{Vect}_{\Bbbk}$ be the category of finite-dimensional $\Bbbk$-vector spaces. This is an additive $\Bbbk$-linear category under the usual direct sum of vector spaces. Up to isomorphism, every vector space is determined uniquely by its dimension. Thus,

$$\mathrm{Iso}_{\mathbb{Z}}(\mathbb{C}) = \mathrm{Span}_{\mathbb{Z}}\{[\Bbbk^n]_{\cong} : n \in \mathbb{N}\}.$$

Now, for an n-dimensional vector space V, we have

$$V \cong \Bbbk^{\oplus n},$$

and so $[V]_{\cong} = n[\Bbbk]_{\cong}$ in $K_0(\mathrm{Vect}_{\Bbbk})$. It follows that we have an isomorphism

$$K_0(\mathrm{Vect}_{\Bbbk}) \xrightarrow{\cong} \mathbb{Z}, \quad \sum_{i=1}^{n} a_i [V_i]_{\cong} \mapsto \sum_{i=1}^{n} a_i \dim V_i. \tag{5.1}$$

Since, for finite-dimensional vector spaces U and V, we have $\dim(U \otimes V) = (\dim U)(\dim V)$, the isomorphism (5.1) is one of the rings. In other words, $\mathrm{Vect}_{\Bbbk}$ is a categorification of the ring of integers. □

5.3 The Trace

There is another common method of decategorification, which we now explain. (See also [27, §2.6].) Suppose $\mathcal{C}$ is a $\Bbbk$-linear category. The *trace*, or *zeroth Hochschild homology*, of $\mathcal{C}$ is the $\Bbbk$-module

$$\mathrm{Tr}(\mathcal{C}) := \left(\bigoplus_X \mathrm{End}_{\mathcal{C}}(X) \right) / \, \mathrm{Span}_{\Bbbk}\{f \circ g - g \circ f\},$$

where the sum is over all objects X of $\mathcal{C}$, and f and g run through all pairs of morphisms $f : X \to Y$ and $g : Y \to X$ in $\mathcal{C}$. We let $[f] \in Tr(\mathcal{C})$ denote the class of an endomorphism $f \in \mathrm{End}_{\mathcal{C}}(X)$.

If the category $\mathcal{C}$ is strict pivotal, we can think of the trace as consisting of diagrams on an annulus. In particular, if

is an endomorphism in $\mathcal{C}$, then we picture $[f]$ as

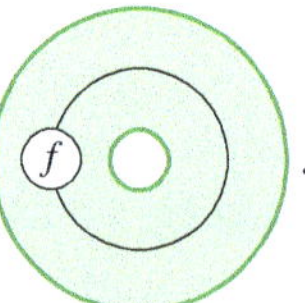

The fact that $[f \circ g] = [g \circ f]$ in $\mathrm{Tr}(\mathcal{C})$ then corresponds to the fact that we can slide diagrams around the annulus:

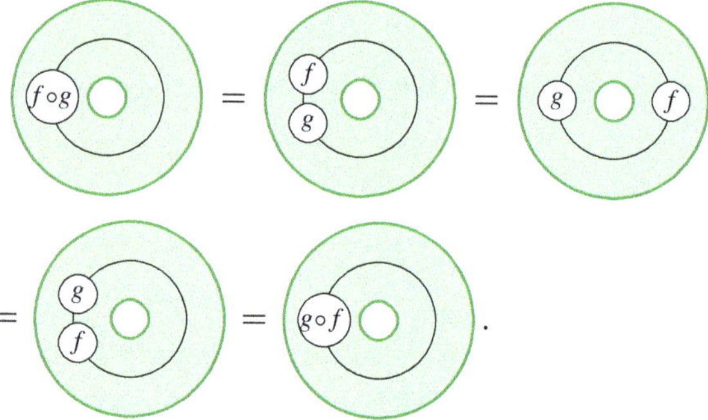

If $\mathcal{C}$ is a $\Bbbk$-linear monoidal category, then $\mathrm{Tr}(\mathcal{C})$ is a ring, with the multiplication given by

$$[f] \cdot [g] = [f \otimes g].$$

If $\mathcal{C}$ is strict pivotal and we view elements of the trace as diagrams on the annulus, then this multiplication corresponds to nesting of annuli:

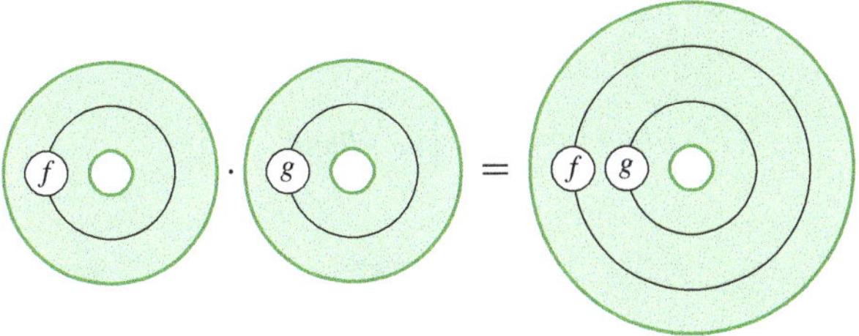

We see from the above that the trace gives another method of *decategorification*. We thus have another corresponding notion of *categorification*. To categorify a $\Bbbk$-algebra, R can mean to find a $\Bbbk$-linear monoidal category $\mathcal{C}$ such that $\operatorname{Tr}(\mathcal{C}) \cong R$ as $\Bbbk$-algebras.

To justify the use of the term *trace*, consider the category $\mathrm{Vect}_{\Bbbk}$ of finite-dimensional vector spaces over a field $\Bbbk$. Let V be a $\Bbbk$-vector space of finite dimension n. Then, we have an isomorphism

$$g \colon V \xrightarrow{\cong} \Bbbk^n.$$

For $1 \leq a \leq n$, define the inclusion and projection maps

$$i_a \colon \Bbbk \to \Bbbk^n, \quad \alpha \mapsto (\underbrace{0, \dots, 0}_{a-1}, \alpha, \underbrace{0, \dots, 0}_{n-a}),$$

$$p_a \colon \Bbbk^n \to \Bbbk, \quad (\alpha_1, \dots, \alpha_n) \mapsto \alpha_a.$$

Note that

$$p_b \circ i_a = \delta_{a,b} 1_{\Bbbk} \qquad \text{and} \qquad \sum_{a=1}^{n} i_a \circ p_a = 1_{\Bbbk^n}.$$

Now suppose $f \colon V \to V$ is a linear map. For $1 \leq a, b \leq n$, define

$$f_{a,b} = p_a \circ g \circ f \circ g^{-1} \circ i_b \colon \Bbbk \to \Bbbk.$$

Then, we have

$$f = g^{-1} \circ g \circ f \circ g^{-1} \circ g = \sum_{a,b=1}^{n} g^{-1} i_a f_{a,b} p_b g.$$

Hence, in $\mathrm{Tr}(\mathrm{Vect}_{\Bbbk})$, we have

$$[f] = \sum_{a,b=1}^{n} \left[g^{-1} i_a f_{a,b} p_b g\right] = \sum_{a,b=1}^{n} \left[f_{a,b} p_b g g^{-1} i_a\right]$$
$$= \sum_{a,b=1}^{n} \left[f_{a,b} p_b i_a\right] = \sum_{a=1}^{n} \left[f_{a,a}\right]. \tag{5.2}$$

So, the class of $[f]$ is equal to the sum of classes of endomorphisms of $\Bbbk$ given by its diagonal components in some basis. By (2.2), it follows that we have an isomorphism of rings

$$\mathrm{Tr}(\mathrm{Vect}_{\Bbbk}) \cong \Bbbk, \quad [f] \mapsto \mathrm{tr}(f). \tag{5.3}$$

In particular, $\mathrm{Vect}_{\Bbbk}$ is a trace categorification of the field $\Bbbk$.

5.4 *Action of the Trace on the Center*

Suppose $\mathcal{C}$ is a strict pivotal $\Bbbk$-linear monoidal category. We have seen that the trace $\mathrm{Tr}(\mathcal{C})$ can be thought of diagrams on the annulus, while the center $\mathrm{End}_{\mathcal{C}}(\mathbb{1})$ can be thought of as closed diagrams. There is then a natural action of the trace on the center given by placing a closed diagram inside the inner boundary of the annulus and then viewing the resulting diagram as a closed diagram. In particular, if $z \in \mathrm{End}_{\mathcal{C}}(\mathbb{1})$ is a closed diagram and $f \in \mathrm{End}_{\mathcal{C}}(X)$, then the action of $[f] \in \mathrm{Tr}(\mathcal{C})$ on z is

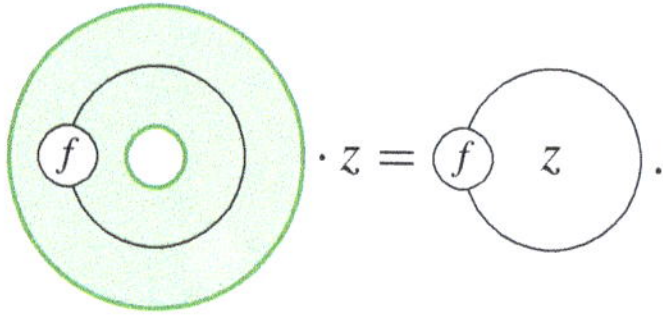

For example, (4.3) is the action of $[f]$ on the identity $1_{\Bbbk}$ (the empty diagram) of the center $\mathrm{End}_{\Bbbk}(\Bbbk)$ of $\mathrm{Vect}_{\Bbbk}$. This explains the connection between (4.4) and (5.3).

5.5 *The Chern Character*

There is a nice relationship between the split Grothendieck group of a category and the trace of that category, as we now explain. Suppose $\mathcal{C}$ is an additive $\Bbbk$-linear category.

Lemma 5.3 ([2, Lem. 3.1]) *If $f : X \to X$ and $g : Y \to Y$ are morphisms in $\mathcal{C}$, then*

$$[f \oplus g] = [f] + [g]$$

in $\mathrm{Tr}(\mathcal{C})$. □

Proof We have $f \oplus g = (f \oplus 0) + (0 \oplus g) : X \oplus Y \to X \oplus Y$. Thus,

$$[f \oplus g] = [f \oplus 0] + [0 \oplus g].$$

Let

$$i : X \to X \oplus Y \quad \text{and} \quad p : X \oplus Y \to X$$

denote the obvious inclusion and projection. Then,

$$[f \oplus 0] = [ifp] = [pif] = [f].$$

Similarly, $[0 \oplus g] = [g]$. This completes the proof. □

If X and Y are objects of $\mathcal{C}$, then we have

$$1_{X \oplus Y} = 1_X \oplus 1_Y.$$

Thus, by Lemma 5.3, we have

$$[1_{X \oplus Y}] = [1_X \oplus 1_Y] = [1_X] + [1_Y].$$

It follows that we have a well-defined map of abelian groups

$$h_{\mathcal{C}} : K_0(\mathcal{C}) \to \mathrm{Tr}(\mathcal{C}), \quad h_{\mathcal{C}}([X]_{\cong}) = [1_X]. \tag{5.4}$$

The map (5.4) is called the *Chern character map*. If $\mathcal{C}$ is an additive strict $\Bbbk$-linear monoidal category, then $h_{\mathcal{C}}$ is a homomorphism of rings.

In general, the Chern character map may not be injective, and it may not be surjective. See, for example, [2, Examples 8–10]. However, there are some situations when it is an isomorphism. A $\Bbbk$-linear category $\mathcal{C}$ is called *semisimple* if

- it has finite direct sums;
- idempotents split (i.e. $\mathcal{C}$ has subobjects); and
- there exist objects X_i, $i \in I$, such that $\mathrm{Hom}_{\mathcal{C}}(X_i, X_j) = \delta_{i,j}\Bbbk$ (such objects are called *simple*) and such that for any two objects V and W in $\mathcal{C}$, the natural composition map

$$\bigoplus_{i \in I} \mathrm{Hom}_{\mathcal{C}}(V, X_i) \otimes \mathrm{Hom}_{\mathcal{C}}(X_i, W) \to \mathrm{Hom}_{\mathcal{C}}(V, W)$$

is an isomorphism.

If $\mathcal{C}$ is an abelian category that is semisimple in the above sense and has a zero object, then $\mathcal{C}$ is semisimple in the usual sense (i.e., all short exact sequences split). If $\Bbbk$ is algebraically closed, then the two notions are equivalent for abelian categories. See [18, p. 89] for details.

Proposition 5.4 *If $\mathcal{C}$ is a semisimple additive $\Bbbk$-linear category, then the map*

$$h_{\mathcal{C}} \otimes 1 \colon K_0(\mathcal{C}) \otimes_{\mathbb{Z}} \Bbbk \to \mathrm{Tr}(\mathcal{C})$$

is an isomorphism. □

Proof Choose representatives X_i, $i \in I$, of the isomorphism classes of the simple objects in $\mathcal{C}$. As explained in [18, p. 89], every object in $\mathcal{C}$ is a finite direct sum of the X_i. Thus, $K_0(\mathcal{C}) \cong \bigoplus_{i \in I} \mathbb{Z}[X_i]_{\cong}$. Suppose Y is an object of $\mathcal{C}$. By assumption, there exist $\beta_{i,j} \in \mathrm{Hom}_{\mathcal{C}}(Y, X_i)$ and $\alpha_{i,j} \in \mathrm{Hom}_{\mathcal{C}}(X_i, Y)$ such that $1_Y = \sum_{i \in I} \sum_j \alpha_{i,j} \beta_{i,j}$. Thus,

$$[1_Y] = \sum_{i \in I} \sum_j [\alpha_{i,j} \beta_{i,j}] = \sum_{i \in I} \sum_j [\beta_{i,j} \alpha_{i,j}] \in \sum_{i \in I} [\mathrm{End}_{\mathcal{C}}(X_i)].$$

Since $\mathrm{Hom}_{\mathcal{C}}(X_i, X_j) = \delta_{i,j} \Bbbk$, it follows that $\mathrm{Tr}(\mathcal{C}) = \bigoplus_{i \in I} \Bbbk[1_{X_i}]$. It follows that $h_{\mathcal{C}} \otimes 1$ is an isomorphism. □

There are other conditions on a category that imply that the Chern character map is injective; see, for example, [3, Prop. 2.5].

5.6 *Idempotent Completions*

Suppose R is a ring. Throughout this subsection, we will assume that all R-modules are finitely generated left modules. The category of free R-modules is quite easy to work with, since all such modules are isomorphic to R^n for some $n \in \mathbb{N}$. However, we often want to work with the larger category consisting of *projective* R-modules. Fortunately, there is a natural relationship between these two categories.

Recall that an R-module is projective if and only if it is a direct summand of a free module. In other words, an R-module M is projective if and only if there exists another R-module N and $n \in \mathbb{N}$ such that

$$M \oplus N \cong R^n \text{ as } R\text{-modules.} \tag{5.5}$$

Now, given the isomorphism (5.5), let $p\colon R^n \twoheadrightarrow M$ denote the projection onto M, and let $i\colon M \hookrightarrow R^n$ denote the inclusion of M into R. Then,

$$i \circ p \in \mathrm{End}_R(R^n)$$

is an idempotent endomorphism of R^n that corresponds to projection onto a submodule of R^n isomorphic to M.

Conversely, if $e \in \mathrm{End}_R(R^n)$ is an idempotent (i.e., $e^2 = e$), then we have

$$R^n = eR^n \oplus (1-e)R^n.$$

So, the image eR^n of e is a projective R-module.

We see from the above that projective R-modules are precisely the images of idempotent morphisms of free R-modules. So, if we start with the category of free R-modules, we can enlarge this to the category of projective R-modules by adding objects corresponding to the images of idempotents. This motivates the following definition.

Definition 5.5 (Idempotent Completion) The *idempotent completion* (or the *Karoubi envelope*) of a category $\mathcal{C}$ is the category $\mathrm{Kar}(\mathcal{C})$ whose objects are pairs of the form (X, e), where X is an object of $\mathcal{C}$ and $e\colon X \to X$ is an idempotent in $\mathcal{C}$, and whose morphisms are triples

$$(e, f, e')\colon (A, e) \to (A', e'),$$

where $f\colon A \to A'$ is a morphism of $\mathcal{C}$ such that $f = e' \circ f \circ e$. □

One should think of the idempotent completion as a way of formally adding in images of idempotents, where (A, e) is thought of as the image of the idempotent e. The original category $\mathcal{C}$ embeds fully and faithfully into $\mathrm{Kar}(\mathcal{C})$ by mapping an object A of $\mathcal{C}$ to $(A, 1_A)$. The idempotent completion of the category of free R-modules is equivalent to the category of projective R-modules.

6 Heisenberg Categories

We conclude with some examples of categories, defined using the concepts introduced above, that are the focus of current research.

6.1 Categorification of Symmetric Functions

Recall the strict $\Bbbk$-linear monoidal category $\mathcal{S}$ from Sect. 3.2. We assume in this subsection that $\Bbbk$ is a field of characteristic zero. The objects of $\mathcal{S}$ are precisely $\uparrow^{\otimes n}$,

$n = 0, 1, 2, \ldots$, where $\uparrow^{\otimes 0} := \mathbb{1}$. Note that all of our generating morphisms are endomorphisms, that is, their domain and codomain are equal. Hence,

$$\mathrm{Hom}_{\mathcal{S}}(\uparrow^{\otimes n}, \uparrow^{\otimes m}) = 0 \quad \text{for } m \neq n.$$

In particular, $\uparrow^{\otimes n}$ is not isomorphic to $\uparrow^{\otimes m}$ for $m \neq n$. It follows that

$$K_0(\mathrm{Add}(\mathcal{S})) \cong \mathbb{Z}[x], \quad [\uparrow]_{\cong} \mapsto x,$$

is an isomorphism of rings.

Now, let us consider the idempotent completion $\mathrm{Kar}(\mathcal{S})$. To describe all the objects of $\mathrm{Kar}(\mathcal{S})$, we need to know all the idempotents of

$$\mathrm{End}_{\mathcal{S}}(\uparrow^{\otimes n}) \cong \Bbbk S_n \quad \text{for } n = 0, 1, 2, 3, \ldots.$$

Fortunately, the idempotents in the algebra $\Bbbk S_n$ are well known. For each partition λ of n, we have the corresponding *Young idempotent*

$$e_\lambda \in \Bbbk S_n.$$

For example, the Young idempotents for the partitions (n) and (1^n) are the complete symmetrizer and antisymmetrizer:

$$e_{(n)} = \frac{1}{n!} \sum_{\pi \in S_n} \pi, \qquad e_{(1^n)} = \frac{1}{n!} \sum_{\pi \in S_n} (-1)^{\ell(\pi)} \pi,$$

where $\ell(n)$ is the length of the permutation $\pi \in S_n$.

It follows that the indecomposable objects in $\mathrm{Kar}(\mathcal{S})$ are, up to isomorphism,

$$(\uparrow^{\otimes n}, e_\lambda), \quad n = 0, 1, 2, \ldots, \ \lambda \vdash n.$$

One can show that

$$K_0(\mathrm{Kar}(\mathrm{Add}(\mathcal{S}))) \cong \mathrm{Sym},$$

the ring of symmetric functions. The isomorphism is given explicitly by

$$\left[(\uparrow^{\otimes n}, e_\lambda)\right]_{\cong} \mapsto s_\lambda,$$

where s_λ is the *Schur function* corresponding to the partition $\lambda \vdash n$.

6.2 Base Category

Recall the strict $\Bbbk$-linear monoidal category $\mathcal{AH}^{\deg}$ from Sect. 3.3. This category has one generating object $\uparrow$, generating morphisms

$$: \uparrow \otimes \uparrow \to \uparrow \otimes \uparrow \quad \text{and} \quad : \uparrow \to \uparrow,$$

and relations

$$= \quad , \qquad = \quad , \qquad \text{and} \qquad - \quad = \quad .$$

Let us add another generating object $\downarrow$ that is right dual to $\uparrow$. As noted in Sect. 4.1, this means that we have morphisms

$$: \mathbb{1} \to \downarrow \otimes \uparrow \quad \text{and} \quad : \uparrow \otimes \downarrow \to \mathbb{1}$$

such that

$$= \quad \text{and} \quad = \quad .$$

Let us define a right crossing by

$$:= \tag{6.1}$$

We will now define various categories by imposing one additional relation involving this right crossing.

6.3 Affine Oriented Brauer Category

Suppose that, in addition to the above generating objects, morphisms, and relations, we impose the additional relation that the right crossing (6.1) is invertible. This means that it has a two-sided inverse, which we will denote

$$. \tag{6.2}$$

The assertion that (6.1) and (6.2) are two-sided inverses is precisely the statement that

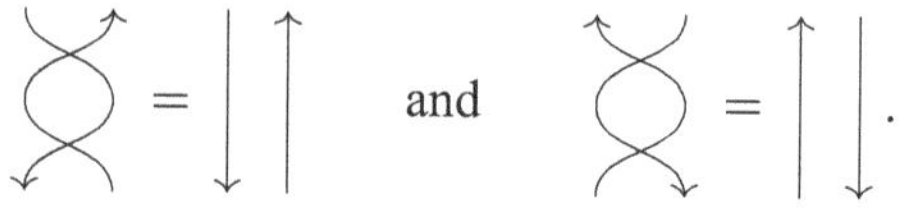

and .

Up to reflecting diagrams in a vertical axis, the resulting category $\mathcal{AOB}$ is the *affine oriented Brauer category* defined in [1]. One can show that it is strict pivotal (see [5, Th. 1.3]). The left cups and caps are defined by

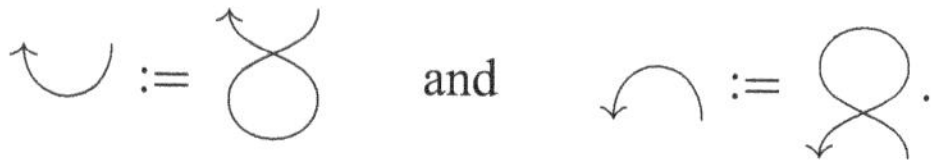

$:=$ and $:=$.

Omitting the dot generator yields the *oriented Brauer category* $\mathcal{OB}$, which is the free symmetric monoidal category on a pair of dual objects. It is obtained from the strict $\Bbbk$-linear monoidal category $\mathcal{S}$ of Sect. 3.2 by adding a right dual object and inverting the right crossing as in Sect. 6.2.

The categories $\mathcal{OB}$ and $\mathcal{AOB}$ encode much of the representation theory of $\mathfrak{gl}_n(\Bbbk)$. Let $\mathfrak{gl}_n(\Bbbk)$-mod denote the monoidal category of $\mathfrak{gl}_n(\Bbbk)$-modules, and let End($\mathfrak{gl}_n(\Bbbk)$-mod) be the monoidal category of endofunctors of $\mathfrak{gl}_n(\Bbbk)$-mod. Objects are functors $\mathfrak{gl}_n(\Bbbk)\text{-mod} \to \mathfrak{gl}_n(\Bbbk)\text{-mod}$, and morphisms are natural transformations. For functors F, G, F', and G' and natural transformations $\eta\colon F \to F'$ and $\xi\colon G \to G'$, we define the tensor product by

$$F \otimes G := F \circ G, \qquad \eta \otimes \xi := \eta\xi\colon F \circ G \to F' \circ G'.$$

Let V be the natural n-dimensional representation of $\mathfrak{gl}_n(\Bbbk)$, with dual representation V^*. We have a monoidal functor

$$\mathcal{OB} \to \mathfrak{gl}_n(\Bbbk)\text{-mod} \tag{6.3}$$

defined as follows. On objects,

$$\uparrow \mapsto V, \qquad \downarrow \mapsto V^*,$$

and, on morphisms,

$$\mapsto \Big(\Bbbk \to V^* \otimes V, \quad a \mapsto a \sum_{v \in B} \delta_v \otimes v\Big),$$

$$\mapsto \Big(V \otimes V^* \to \Bbbk, \quad v \otimes f \mapsto f(v)\Big),$$

$$\mapsto \Big(V \otimes V \to V \otimes V, \quad u \otimes v \mapsto v \otimes u\Big),$$

where B is a basis of V, and $\{\delta_v : v \in B\}$ is the dual basis.

Now, we have a natural monoidal functor

$$\mathfrak{gl}_n(\Bbbk)\text{-mod} \to \mathrm{End}(\mathfrak{gl}_n(\Bbbk)\text{-mod}), \quad M \mapsto M \otimes -, \quad f \mapsto f \otimes 1.$$

Composition with (6.3) yields a monoidal functor.

$$\mathcal{OB} \to \mathrm{End}(\mathfrak{gl}_n(\Bbbk)\text{-mod}).$$

This can be extended to a monoidal functor

$$\mathcal{AOB} \to \mathrm{End}(\mathfrak{gl}_n(\Bbbk)\text{-mod})$$

by defining

$$\overset{\uparrow}{\circ} \mapsto \left(V \otimes - \to V \otimes -, \quad v \otimes w \mapsto \sum_{i,j=1}^{n} e_{i,j} v \otimes e_{j,i} w \right),$$

where $e_{i,j}$ is the matrix with a 1 in the (i, j) position and a 0 in all other positions. This functor sends the center of $\mathcal{AOB}$ to End(Id), which can be identified with the center of $U(\mathfrak{gl}_n(\Bbbk))$.

6.4 *Heisenberg Categories*

Fix $k \in \mathbb{Z}_{<0}$. Let us return to the category of Sect. 6.2, but now impose the relation that the following matrix is an isomorphism in the additive envelope:

$$\left[\times \ \cup \ \overset{\circ}{\cup} \cdots \overset{\circ}{\cup}^{-k-1} \right] : (\uparrow \otimes \downarrow) \oplus \mathbb{1}^{\oplus(-k)} \to \downarrow \oplus \uparrow . \tag{6.4}$$

We denote the resulting category by $\mathcal{H}eis_k$.

The inversion relation imposed above means that there is some $k \times 1$ matrix of morphisms in $\mathcal{H}eis_k$ that is a two-sided inverse to (6.4). A thorough analysis of this category involves introducing notation for the entries of this inverse matrix and then deducing a simplified presentation of the category using the relations that arise from our two matrices being two-sided inverses to each other. This is done in [5], where it is shown that this category is strict pivotal and isomorphic to the *Heisenberg category* introduced in [17]. In the case $k = -1$, this category was originally defined by Khovanov in [13]. In [13, 17], the category was defined in terms of planar diagrams up to isotopy, and so it was strict pivotal by definition. The Heisenberg category encodes much of the representation theory of the symmetric group (when $k = -1$) and other degenerate cyclotomic Hecke algebras. The affine oriented Brauer category can be thought of the $k = 0$ version of the Heisenberg category.

Let us now explain why this category is called the *Heisenberg* category. The infinite-dimensional Heisenberg Lie algebra $\mathfrak{h}$ is the Lie algebra with generators

$$p_n^{\pm},\ n = \mathbb{Z}_{>0}, \quad \text{and} \quad c,$$

and relations

$$[p_n^+, p_m^+] = [p_n^-, p_m^-] = [c, p_n^{\pm}] = 0, \quad [p_n^+, p_m^-] = \delta_{n,m} nc, \qquad n, m \in \mathbb{Z}_{>0}.$$

Since the element c is central, it acts as a constant on any irreducible representation. If we fix a *central charge* $k \in \mathbb{Z}$, we can consider the associative algebra $U(\mathfrak{h})/\langle c-k \rangle$, where $U(\mathfrak{h})$ is the universal enveloping algebra of $\mathfrak{h}$. Representations of $U(\mathfrak{h})/\langle c-k \rangle$ are equivalent to representations of $\mathfrak{h}$ on which the central element c acts as multiplication by k. In particular, in $U(\mathfrak{h})/\langle c-k \rangle$, we have

$$p_1^+ p_1^- - k = p_1^- p_1^+. \tag{6.5}$$

Now, the inversion relation (6.4) implies that, in the Grothendieck group of the additive envelope of $\mathcal{H}eis_k$, we have

$$[\uparrow]_{\cong}[\downarrow]_{\cong} + (-k)[\mathbb{1}]_{\cong} = [\downarrow]_{\cong}[\uparrow]_{\cong}. \tag{6.6}$$

We see that (6.5) and (6.6) are the same relation after replacing $p_1^+ \leftrightarrow [\uparrow]_{\cong}$, $p_1^- \leftrightarrow [\downarrow]_{\cong}$, $1 \leftrightarrow [\mathbb{1}]_{\cong}$. In fact, if $\Bbbk$ is a field of characteristic zero, then we have an isomorphism of algebras

$$U(\mathfrak{h})/\langle c-k \rangle \cong K_0(\mathrm{Kar}(\mathrm{Add}(\mathcal{H}eis_k))). \tag{6.7}$$

Injectivity was proved in [13, Th. 1] in the case $k = -1$ and in [17, Th. 4.4] in the general case $k < 0$. It was conjectured in [13, Conj. 1] and [17, Conj. 4.5] that (6.7) is an isomorphism. This conjecture was recently proved in [7, Th. 1.1]. Earlier, an analog of the conjecture was proved when one enlarges the Heisenberg category by adding in additional generating morphisms corresponding to elements of a graded Frobenius algebra with nontrivial grading. See [9, Th. 1], [21, Th. 10.5], and [23, Th. 1.5]. The trace of the Heisenberg category has also been related to W-algebras in [11].

One can also define a *quantum Heisenberg category* by replacing the symmetric group relations by the Hecke algebra relations (3.7)–(3.9). See [8, 15]. The resulting category also categorifies the Heisenberg algebra. Its trace has been related to elliptic Hall algebras in [10]. The quantum analog of the oriented Brauer category is the *HOMFLY-PT skein category* introduced by Turaev in [25, §5.2], where it was called the *Hecke category*. See also [4, 8].

Acknowledgments This work was supported by Discovery Grant RGPIN-2017-03854 from the Natural Sciences and Engineering Research Council of Canada.

References

1. J. Brundan, J. Comes, D. Nash, and A. Reynolds. A basis theorem for the affine oriented Brauer category and its cyclotomic quotients. *Quantum Topol.*, 8(1):75–112, 2017. arXiv:1404.6574. doi:10.4171/QT/87.
2. A. Beliakova, Z. Guliyev, K. Habiro, and A. D. Lauda. Trace as an alternative decategorification functor. *Acta Math. Vietnam.*, 39(4):425–480, 2014. arXiv:1409.1198.
3. Anna Beliakova, Kazuo Habiro, Aaron D. Lauda, and Ben Webster. Current algebras and categorified quantum groups. *J. Lond. Math. Soc. (2)*, 95(1):248–276, 2017. arXiv:1412.1417. doi:10.1112/jlms.12001.
4. J. Brundan. Representations of the oriented skein category. arXiv:1712.08953, 2017.
5. J. Brundan. On the definition of Heisenberg category. *Algebraic Combinatorics*, 1(4):523–544, 2018. arXiv:1709.06589. doi:10.5802/alco.26.
6. J. Brundan, A. Savage, and B. Webster. Quantum Frobenius Heisenberg categorification. *J. Pure Appl. Algebra*, 226(1):106792, 2022. http://arxiv.org/abs/2009.06690
7. J. Brundan, A. Savage, and B. Webster. The degenerate Heisenberg category and its Grothendieck ring. arXiv:1812.03255, 2018.
8. J. Brundan, A. Savage, and B. Webster. On the definition of quantum Heisenberg category. *Algebra Number Theory*, 14(2):275–321, 2020. arXiv:1812.04779, doi:10.2140/ant.2020.14.275.
9. S. Cautis and A. Licata. Heisenberg categorification and Hilbert schemes. *Duke Math. J.*, 161(13):2469–2547, 2012. arXiv:1009.5147. doi:10.1215/00127094-1812726.
10. S. Cautis, A. D. Lauda, A. M. Licata, P. Samuelson, and J. Sussan. The elliptic Hall algebra and the deformed Khovanov Heisenberg category. *Selecta Math. (N.S.)*, 24(5):4041–4103, 2018. arXiv:1609.03506,. doi:10.1007/s00029-018-0429-8.
11. S. Cautis, A. D. Lauda, A. Licata, and J. Sussan. W-algebras from Heisenberg categories. *J. Inst. Math. Jussieu*, 17(5):981–1017, 2018. arXiv:1501.00589. doi:10.1017/S1474748016000189.
12. C. Kassel. *Quantum groups*, volume 155 of *Graduate Texts in Mathematics*. Springer-Verlag, New York, 1995. doi:10.1007/978-1-4612-0783-2.
13. M. Khovanov. Heisenberg algebra and a graphical calculus. *Fund. Math.*, 225(1):169–210, 2014. arXiv:1009.3295. doi:10.4064/fm225-1-8.
14. M. Khovanov and A. D. Lauda. A categorification of quantum sl(n). *Quantum Topol.*, 1(1):1–92, 2010. arXiv:0807.3250. doi:10.4171/QT/1.
15. A. Licata and A. Savage. Hecke algebras, finite general linear groups, and Heisenberg categorification. *Quantum Topol.*, 4(2):125–185, 2013. arXiv:1101.0420. doi:10.4171/QT/37.
16. S. MacLane. *Categories for the working mathematician*. Springer-Verlag, New York-Berlin, second edition, 1998. Graduate Texts in Mathematics, Vol. 5.
17. M. Mackaay and A. Savage. Degenerate cyclotomic Hecke algebras and higher level Heisenberg categorification. *J. Algebra*, 505:150–193, 2018. arXiv:1705.03066. doi:10.1016/j.jalgebra.2018.03.004.
18. M. Müger. From subfactors to categories and topology. I. Frobenius algebras in and Morita equivalence of tensor categories. *J. Pure Appl. Algebra*, 180(1-2):81–157, 2003. doi:10.1016/S0022-4049(02)00247-5.
19. R. Penrose. Applications of negative dimensional tensors. In *Combinatorial Mathematics and its Applications (Proc. Conf., Oxford, 1969)*, pages 221–244. Academic Press, London, 1971.
20. D. Rosso and A. Savage. Quantum affine wreath algebras. *Doc. Math.*, 25:425–456, 2020. http://arxiv.org/abs/1902.00143, https://doi.org/10.25537/dm.2020v25.425-456
21. D. Rosso and A. Savage. A general approach to Heisenberg categorification via wreath product algebras. *Math. Z.*, 286(1-2):603–655, 2017. arXiv:1507.06298. doi:10.1007/s00209-016-1776-9.

22. A. Savage. Affine wreath product algebras. *Int. Math. Res. Not. IMRN*, (10):2977–3041, 2020. http://arxiv.org/abs/1709.02998, https://doi.org/10.1093/imrn/rny092
23. A. Savage. Frobenius Heisenberg categorification. *Algebr. Comb.*, 2(5):937–967, 2019. arXiv:1802.01626, doi:10.5802/alco.73.
24. P. Schauenburg. Turning monoidal categories into strict ones. *New York J. Math.*, 7:257–265, 2001. http://nyjm.albany.edu:8000/j/2001/7_257.html.
25. V. G. Turaev. Operator invariants of tangles, and R-matrices. *Izv. Akad. Nauk SSSR Ser. Mat.*, 53(5):1073–1107, 1135, 1989. doi:10.1070/IM1990v035n02ABEH000711.
26. V. G. Turaev. *Quantum invariants of knots and 3-manifolds*, volume 18 of *De Gruyter Studies in Mathematics*. De Gruyter, Berlin, 2016. Third edition. doi:10.1515/9783110435221.
27. V. Turaev and A. Virelizier. *Monoidal categories and topological field theory*, volume 322 of *Progress in Mathematics*. Birkhäuser/Springer, Cham, 2017. doi:10.1007/978-3-319-49834-8.

Quantum Affine Algebras and Cluster Algebras

David Hernandez and Bernard Leclerc

To Vyjayanthi Chari on her birthday

Abstract This article is an extended version of the minicourse given by the second author at the summer school of the conference *Interactions of quantum affine algebras with cluster algebras, current algebras and categorification*, held in June 2018 in Washington. The aim of the minicourse, consisting of three lectures, was to present a number of results and conjectures on certain monoidal categories of finite-dimensional representations of quantum affine algebras, obtained by exploiting the fact that their Grothendieck rings have the natural structure of a cluster algebra.

1 A Forerunner: Chari and Pressley's Paper on $U_q(\widehat{\mathfrak{sl}}_2)$

In [3], Chari and Pressley launched a systematic study of tensor categories of finite-dimensional representations of quantum affine algebras by investigating in detail the case of $U_q(\widehat{\mathfrak{sl}}_2)$. They gave a classification of simple objects, as well as a concrete description of them as tensor products of evaluation modules. They also gave a necessary and sufficient condition for such tensor products to be irreducible, and they described the composition factors of a reducible tensor product of two evaluation representations.

D. Hernandez
Université de Paris, Univ Paris Diderot, CNRS Institut de Mathématiques de Jussieu-Paris Rive Gauche UMR 7586, Paris, France

Institut Universitaire de France, Paris, France
e-mail: david.hernandez@imj-prg.fr

B. Leclerc (✉)
Normandie Univ, UNICAEN, CNRS, LMNO, Caen, France
e-mail: bernard.leclerc@unicaen.fr

J. Greenstein et al. (eds.), *Interactions of Quantum Affine Algebras with Cluster Algebras, Current Algebras and Categorification*, Progress in Mathematics 337,
https://doi.org/10.1007/978-3-030-63849-8_2

In retrospect, these results may be seen as providing a cluster algebra structure on the Grothendieck ring of this category, predating by 10 years the invention of cluster algebras by Fomin and Zelevinsky [14]. We will therefore start our lectures by reviewing these results.

1.1 The Hopf Algebra $U_q(\widehat{\mathfrak{sl}_2})$

Throughout the paper, we fix $q \in \mathbb{C}^*$ not a root of unity. The algebra $U_q(\widehat{\mathfrak{sl}_2})$ is generated over $\mathbb{C}$ by

$$E_0,\ F_0,\ K_0,\ K_0^{-1},\ E_1,\ F_1,\ K_1,\ K_1^{-1},$$

subject to the following relations:

$$K_i K_i^{-1} = 1, \tag{1.1}$$

$$K_i K_j = K_j K_i, \tag{1.2}$$

$$K_i E_i K_i^{-1} = q^2 E_i, \tag{1.3}$$

$$K_i E_j K_i^{-1} = q^{-2} E_j, \tag{1.4}$$

$$K_i F_i K_i^{-1} = q^{-2} F_i, \tag{1.5}$$

$$K_i F_j K_i^{-1} = q^2 F_j, \tag{1.6}$$

$$E_i F_i - F_i E_i = \frac{K_i - K_i^{-1}}{q - q^{-1}}, \tag{1.7}$$

$$E_i F_j - F_j E_i = 0, \tag{1.8}$$

$$E_i^3 E_j - (q^2+1+q^{-2}) E_i^2 E_j E_i + (q^2+1+q^{-2}) E_i E_j E_i^2 - E_j E_i^3 = 0, \tag{1.9}$$

$$F_i^3 F_j - (q^2+1+q^{-2}) F_i^2 F_j F_i + (q^2+1+q^{-2}) F_i F_j F_i^2 - F_j F_i^3 = 0, \tag{1.10}$$

where $i \neq j$ are indices in $\{0, 1\}$. Moreover $U_q(\widehat{\mathfrak{sl}_2})$ is a Hopf algebra, with comultiplication Δ given by

$$\Delta(E_i) = E_i \otimes K_i + 1 \otimes E_i,$$

$$\Delta(F_i) = F_i \otimes 1 + K_i^{-1} \otimes F_i,$$

$$\Delta(K_i) = K_i \otimes K_i.$$

It follows that a tensor product of finite-dimensional $U_q(\widehat{\mathfrak{sl}_2})$-modules is again a $U_q(\widehat{\mathfrak{sl}_2})$-module.

1.2 Simple Finite-Dimensional $U_q(\widehat{\mathfrak{sl}}_2)$-Modules

Let E, F, K, K^{-1} denote the generators of $U_q(\mathfrak{sl}_2)$. (They are subject to the same relations as (1.1), (1.3), (1.5) and (1.7).)

For every $a \in \mathbb{C}^*$, we have a surjective algebra homomorphism $\mathrm{ev}_a : U_q(\widehat{\mathfrak{sl}}_2) \to U_q(\mathfrak{sl}_2)$ such that

$$\mathrm{ev}_a(E_1) = E, \quad \mathrm{ev}_a(F_1) = F, \quad \mathrm{ev}_a(E_0) = q^{-1}aF, \quad \mathrm{ev}_a(F_0) = qa^{-1}E.$$

Hence, every simple finite-dimensional $U_q(\mathfrak{sl}_2)$-module M becomes a finite-dimensional $U_q(\widehat{\mathfrak{sl}}_2)$-module $M(a)$ by pull-back through ev_a.

It is well-known that the simple finite-dimensional $U_q(\mathfrak{sl}_2)$-modules[1] are classified by their dimension: for every $n \in \mathbb{Z}_{\geq 0}$ there is a unique (up to isomorphism) simple module V_n with dimension $n+1$. Therefore, pulling back by the evaluation morphisms ev_a, we get for all $n \in \mathbb{Z}_{\geq 0}$ a one-parameter family of simple $U_q(\widehat{\mathfrak{sl}}_2)$-modules $V_n(a)$ $(a \in \mathbb{C}^*)$ with dimension $n+1$. The representations $V_0(a)$ are all equal to the trivial representation. Otherwise, for $n \geq 1$, the simple modules $V_n(a)$ and $V_n(b)$ are non-isomorphic if $a \neq b$. The modules $V_n(a)$ are called *evaluation modules*.

Theorem 1.1 ([3]) *Every non-trivial simple finite-dimensional $U_q(\widehat{\mathfrak{sl}}_2)$-module M is isomorphic to a tensor product of evaluation modules, that is,*

$$M \simeq V_{n_1}(a_1) \otimes \cdots \otimes V_{n_k}(a_k)$$

for some $k \in \mathbb{Z}_{>0}$, $n_1, \ldots, n_k \in \mathbb{Z}_{>0}$, and $a_1, \ldots, a_k \in \mathbb{C}^*$.

Note that tensor products of evaluation modules are *not* always irreducible. The next task is therefore to find some necessary and sufficient condition of irreducibility. In order to formulate this condition we introduce the notion of a *string*. This is a subset of $\mathbb{C}^*$ of the form:

$$\Sigma(n, a) := \{aq^{-n+1}, aq^{-n+3}, \ldots, aq^{n-1}\}, \qquad (n \in \mathbb{Z}_{\geq 0}, a \in \mathbb{C}^*).$$

(In fact, $\Sigma(n, a)$ is nothing else than the set of roots of the Drinfeld polynomial of $V_n(a)$.) We say that two strings Σ_1 and Σ_2 are *in general position* if and only if

(i) $\Sigma_1 \cup \Sigma_2$ is not a string, or
(ii) $\Sigma_1 \subseteq \Sigma_2$ or $\Sigma_2 \subseteq \Sigma_1$.

[1] In these lectures, we will only consider type I representations of quantum enveloping algebras. All representations can be obtained from the type I representations by twisting with some signs, see e.g. [4, §10.1].

Theorem 1.2 ([3]) *The tensor product $V_{n_1}(a_1) \otimes \cdots \otimes V_{n_k}(a_k)$ is irreducible if and only if for every $(i, j) \in \{1, \ldots, k\}^2$ the strings $\Sigma(n_i, a_i)$ and $\Sigma(n_j, a_j)$ are in general position.*

Two strings which are not in general position are called *in special position*. What can we say about the tensor product $V_{n_1}(a_1) \otimes V_{n_2}(a_2)$ when the strings $\Sigma_1 := \Sigma(n_1, a_1)$ and $\Sigma_2 := \Sigma(n_2, a_2)$ are in special position? It turns out that in this case the tensor product always has two non-isomorphic composition factors. These two irreducible modules are, by Theorems 1.1 and 1.2, parametrized by two collections of strings in general position. Here is how to obtain them from Σ_1 and Σ_2.

Because of Theorem 1.2 (i), $\Sigma_3 := \Sigma_1 \cup \Sigma_2$ is a string. Clearly, $\Sigma_4 := \Sigma_1 \cap \Sigma_2$ is also a string, contained in Σ_3. Removing from Σ_3 the points of Σ_4 together with its two nearest neighbours, we are left with the union of two strings Σ_5 and Σ_6. It is easy to see that the two pairs of strings (Σ_3, Σ_4) and (Σ_5, Σ_6) are in general position. For instance, if

$$\Sigma_1 = \{1, q^2, q^4, q^6, q^8\}, \qquad \Sigma_2 = \{q^6, q^8, q^{10}, q^{12}, q^{14}, q^{16}\},$$

then

$$\Sigma_3 = \{1, q^2, q^4, q^6, q^8, q^{10}, q^{12}, q^{14}, q^{16}\}, \qquad \Sigma_4 = \{q^6, q^8\}$$

and

$$\Sigma_5 = \{1, q^2\}, \qquad \Sigma_6 = \{q^{12}, q^{14}, q^{16}\}.$$

We can then state:

Proposition 1.3 ([3, Proposition 4.9]) *Let Σ_1 and Σ_2 be two strings in special position. With the above notation, in the Grothendieck ring the following relation holds:*

$$[V(\Sigma_1) \otimes V(\Sigma_2)] = [V(\Sigma_3) \otimes V(\Sigma_4)] + [V(\Sigma_5) \otimes V(\Sigma_6)]. \tag{1.11}$$

Here, $V(\Sigma_i)$ denotes the evaluation module whose associated string is Σ_i.

1.3 Relation with Cluster Algebras

A reader familiar with the definition of a cluster algebra will recognize in (1.11) an *exchange relation*. Let us make this more precise. First note that if two strings are in special position, all their points belong to the same class in $\mathbb{C}^*/q^{2\mathbb{Z}}$, that is, they all are of the form aq^k for some fixed $a \in \mathbb{C}^*$ and some $k \in 2\mathbb{Z}$. This motivates the following definition:

Definition 1.4 ([22]) Let $a \in \mathbb{C}^*$ and $\ell \in \mathbb{Z}_{>0}$. Let $\mathcal{C}_{a,\ell}$ be the full subcategory of the category of finite-dimensional $U_q(\widehat{\mathfrak{sl}_2})$-modules whose objects V satisfy:

Every composition factor of V is of the form $V_{n_1}(a_1) \otimes \cdots \otimes V_{n_k}(a_k)$ where all strings $\Sigma(n_i, a_i)$ are contained in $S := \{a, aq^{-2}, \dots, aq^{-2\ell}\}$.

The category $\mathcal{C}_{a,\ell}$ depends only on ℓ up to isomorphism. We can therefore restrict ourselves to the case $a = 1$, and write $\mathcal{C}_{1,\ell} = \mathcal{C}_\ell$. Then Theorems 1.1, 1.2 and Proposition 1.3 yield the following reformulation:

Theorem 1.5 ([22]) *The category $\mathcal{C}_\ell$ is a monoidal category, and its Grothendieck ring $K_0(\mathcal{C}_\ell)$ has the structure of a cluster algebra of finite type A_ℓ in the Fomin–Zelevinsky classification. More precisely, the cluster variables of $K_0(\mathcal{C}_\ell)$ are the classes of the evaluation modules contained in $\mathcal{C}_\ell$, the class $[V_{\ell+1}(q^{-\ell})]$ being the only frozen variable. The cluster monomials are equal to the classes of the simple modules in $\mathcal{C}_\ell$. Two cluster variables are* compatible *(i.e. belong to the same cluster) if and only if the corresponding strings are in general position. Otherwise they form an exchange pair with exchange relation given by (1.11).*

1.4 How Can We Generalize?

In an attempt to extend these results from $U_q(\widehat{\mathfrak{sl}_2})$ to other quantum affine algebras, Chari and Pressley introduced in [7] the notion of a *prime* module: this is a simple finite-dimensional module that cannot be factored as a tensor product of modules of smaller dimension. It follows from Theorem 1.1 that the prime $U_q(\widehat{\mathfrak{sl}_2})$-modules are precisely the evaluation modules. This is no longer true for $U_q(\widehat{\mathfrak{sl}_3})$, and Chari and Pressley have constructed an infinite class of prime $U_q(\widehat{\mathfrak{sl}_3})$-modules which are not evaluation modules, see [7]. In view of this, the following problems naturally arise.

Let $\mathfrak{g}$ be a simple Lie algebra over $\mathbb{C}$, and let $U_q(\widehat{\mathfrak{g}})$ denote the corresponding untwisted quantum affine algebra.

(P1) What are the prime $U_q(\widehat{\mathfrak{g}})$-modules?
(P2) Which tensor products of prime $U_q(\widehat{\mathfrak{g}})$-modules are simple?

It is known that Kirillov–Reshetikhin modules (see below Definition 2.6) are prime. The *minimal affinization* modules introduced by Chari and Pressley [5] as replacements for evaluation modules which do not exist outside type A, are also prime. But this is not a complete list as we shall see below. Problem (P2) is also completely open. In [22] we proposed to use cluster algebras to shed new light on these questions.

2 Reminder on Finite-Dimensional $U_q(\widehat{\mathfrak{g}})$-Modules

2.1 Cartan Matrix

Let $C = (c_{ij})_{i,j \in I}$ be the Cartan matrix of $\mathfrak{g}$. There is a diagonal matrix $D = \mathrm{diag}(d_i \mid i \in I)$ with entries in $\mathbb{Z}_{>0}$ such that the product

$$B = DC = (b_{ij})_{i,j \in I}$$

is symmetric. We normalize D so that $\min\{d_i \mid i \in I\} = 1$, and we put $t := \max\{d_i \mid i \in I\}$. Thus

$$t = \begin{cases} 1 \text{ if } C \text{ is of type } A_n,\, D_n,\, E_6,\, E_7 \text{ or } E_8, \\ 2 \text{ if } C \text{ is of type } B_n,\, C_n \text{ or } F_4, \\ 3 \text{ if } C \text{ is of type } G_2. \end{cases}$$

Example 2.1 The Lie algebra $\mathfrak{g} = \mathfrak{so}_7$, of type B_3 in the Cartan–Killing classification, has Cartan matrix

$$C = \begin{pmatrix} 2 & -1 & 0 \\ -1 & 2 & -1 \\ 0 & -2 & 2 \end{pmatrix}$$

We have $D = \mathrm{diag}(2, 2, 1)$ and the symmetric matrix B is given by

$$B = \begin{pmatrix} 4 & -2 & 0 \\ -2 & 4 & -2 \\ 0 & -2 & 2 \end{pmatrix}$$

We denote by α_i $(i \in I)$ the simple roots of $\mathfrak{g}$, and by ϖ_i $(i \in I)$ the fundamental weights. They are related by

$$\alpha_i = \sum_{j \in I} c_{ji} \varpi_j. \tag{2.1}$$

2.2 Classification

By Cartan–Killing theory, the simple finite-dimensional $\mathfrak{g}$-modules are in one-to-one correspondence with their highest weight, an element of the positive cone of integral dominant weights:

$$P_+ := \bigoplus_{i \in I} \mathbb{N}\varpi_i.$$

We denote by $L(\lambda)$ the simple $\mathfrak{g}$-module with highest weight $\lambda \in P_+$.

Chari and Pressley have obtained a similar classification of simple finite-dimensional $U_q(\widehat{\mathfrak{g}})$-modules. To formulate it, we introduce the cone

$$\widehat{P}_+ := \bigoplus_{i \in I,\, a \in \mathbb{C}^*} \mathbb{N}(\varpi_i, a)$$

of *dominant loop-weights*.

Theorem 2.2 ([4]) *Up to isomorphism, the simple finite-dimensional $U_q(\widehat{\mathfrak{g}})$-modules are in one-to-one correspondence with their highest loop-weight, an element of $\widehat{P}_+$.* □

We denote by $L(\widehat{\lambda})$ the simple $U_q(\widehat{\mathfrak{g}})$-module with highest loop-weight $\widehat{\lambda} \in \widehat{P}_+$.

Example 2.3 Let $\mathfrak{g} = \mathfrak{so}_8$, of type D_4. Thus $I = \{1, 2, 3, 4\}$, where we denote by 3 the trivalent node of the Dynkin diagram. Then the $\mathfrak{g}$-module $L(\varpi_3) = \bigwedge^2 \mathbb{C}^8$ is of dimension 28.

For $a \in \mathbb{C}^*$, the $U_q(\widehat{\mathfrak{g}})$-module $L(\varpi_3, a)$ has dimension 29. This is a *minimal affinization* of $L(\varpi_3)$ in the sense of [5].

2.3 q-Characters

Finite-dimensional $\mathfrak{g}$-modules M are characterized by their character

$$\chi(M) := \sum_{\mu \in P} \dim(M_\mu) e^\mu,$$

where $P := \oplus_{i \in I} \mathbb{Z}\varpi_i$ is the weight lattice, $M := \oplus_{\mu \in P} M_\mu$ is the weight space decomposition of M, and e^μ is a formal exponential. So $\chi(M)$ is a Laurent polynomial in the variables $y_i := e^{\varpi_i}$, $(i \in I)$.

Similarly, finite-dimensional $U_q(\widehat{\mathfrak{g}})$-modules $\widehat{M}$ have a loop-weight space decomposition

$$\widehat{M} := \oplus_{\widehat{\mu} \in \widehat{P}} \widehat{M}_{\widehat{\mu}},$$

where $\widehat{P} := \oplus_{i \in I,\, a \in \mathbb{C}^*} \mathbb{Z}(\varpi_i, a)$. Frenkel–Reshetikhin introduced the *q-character*

$$\chi_q(\widehat{M}) := \sum_{\widehat{\mu} \in \widehat{P}} \dim(\widehat{M}_{\widehat{\mu}}) e^{\widehat{\mu}},$$

a Laurent polynomial in the variables $Y_{i,a} := e^{(\varpi_i, a)}$, $(i \in I,\ a \in \mathbb{C}^*)$.

Theorem 2.4 ([13]) *Let $\widehat{M}$ and $\widehat{N}$ be two finite-dimensional $U_q(\widehat{\mathfrak{g}})$-modules. The following are equivalent:*

(i) $\chi_q(\widehat{M}) = \chi_q(\widehat{N})$*;*
(ii) $[\widehat{M}] = [\widehat{N}]$ *in the Grothendieck ring* $K_0(\text{-mod}(U_q(\widehat{\mathfrak{g}})))$*;*
(iii) $\widehat{M}$ *and* $\widehat{N}$ *have the same composition factors with the same multiplicities.*

In particular, q-characters characterize simple *$U_q(\widehat{\mathfrak{g}})$-modules up to isomorphism.*

Note also that, because of Theorem 2.4, χ_q descends to an injective ring homomorphism from $K_0(\text{-mod}(U_q(\widehat{\mathfrak{g}})))$ to the ring of Laurent polynomials $\mathbb{Z}[Y_{i,a}^{\pm 1} \mid i \in I,\ a \in \mathbb{C}^*]$.

Example 2.5 Let $\mathfrak{g} = \mathfrak{sl}_2$, of type A_1. Then, as is well-known, we have

$$\begin{aligned} \chi(L(\varpi_1)) = \chi(V_1) &= y_1 + y_1^{-1} \\ \chi(L(2\varpi_1)) = \chi(V_2) &= y_1^2 + 1 + y_1^{-2} \\ \cdots \quad & \cdots \end{aligned}$$

On the other hand, for any $a \in \mathbb{C}^*$,

$$\begin{aligned} \chi_q(L(\varpi_1, a)) &= \chi_q(V_1(a)) = Y_{1,a} + Y_{1,aq^2}^{-1} \\ \chi_q(L(2(\varpi_1, a))) &= Y_{1,a}^2 + 2Y_{1,a}Y_{1,aq^2}^{-1} + Y_{1,aq^2}^{-2} \\ \chi_q(L((\varpi_1, a) + (\varpi_1, aq^2)) &= \chi_q(V_2(aq)) \\ &= Y_{1,a}Y_{1,aq^2} + Y_{1,a}Y_{1,aq^4}^{-1} + Y_{1,aq^2}^{-1}Y_{1,aq^4}^{-1}. \end{aligned}$$

This shows that $L((\varpi_1, a) + (\varpi_1, aq^2))$ is a minimal affinization of $L(2\varpi_1)$, but $L(2(\varpi_1, a))$ is not.

Definition 2.6 For $i \in I$, $k \in \mathbb{N}$, $a \in \mathbb{C}^*$, set

$$\widehat{\lambda}_{k,a}^{(i)} := \sum_{j=0}^{k-1} (\varpi_i, aq^{2d_i j}) \in \widehat{P}_+.$$

The simple $U_q(\widehat{\mathfrak{g}})$-module $L\left(\widehat{\lambda}_{k,a}^{(i)}\right)$ is called a *Kirillov–Reshetikhin module*. We often write for short $W_{k,a}^{(i)} = L\left(\widehat{\lambda}_{k,a}^{(i)}\right)$. The modules $W_{1,a}^{(i)} = L((\varpi_i, a))$ are the *fundamental* $U_q(\widehat{\mathfrak{g}})$-modules.

2.4 T-Systems

With the quantum affine algebra $U_q(\widehat{\mathfrak{g}})$ is associated a system of difference equations called a T-system [33]. Its unknowns are denoted by

$$T_{k,r}^{(i)}, \qquad (i \in I,\ k \in \mathbb{N},\ r \in \mathbb{Z}).$$

We fix the initial boundary condition

$$T_{0,r}^{(i)} = 1, \qquad (i \in I,\ r \in \mathbb{Z}). \tag{2.2}$$

If $\mathfrak{g}$ is of type A_n, D_n, E_n, the T-system equations are

$$T_{k,r+1}^{(i)} T_{k,r-1}^{(i)} = T_{k-1,r+1}^{(i)} T_{k+1,r-1}^{(i)} + \prod_{j:\ c_{ij}=-1} T_{k,r}^{(j)}, \qquad (i \in I,\ k \geq 1,\ r \in \mathbb{Z}). \tag{2.3}$$

If $\mathfrak{g}$ is not of simply-laced type, the T-system equations are more complicated. They can be written in the form

$$T_{k,r+d_i}^{(i)} T_{k,r-d_i}^{(i)} = T_{k-1,r+d_i}^{(i)} T_{k+1,r-d_i}^{(i)} + S_{k,r}^{(i)}, \qquad (i \in I,\ k \geq 1,\ r \in \mathbb{Z}), \tag{2.4}$$

where $S_{k,r}^{(i)}$ is defined as follows. If $d_i \geq 2$, then

$$S_{k,r}^{(i)} = \prod_{j:\ c_{ji}=-1} T_{k,r}^{(j)} \prod_{j:\ c_{ji}\leq -2} T_{d_i k,\ r-d_i+1}^{(j)}. \tag{2.5}$$

If $d_i = 1$ and $t = 2$, then

$$S_{k,r}^{(i)} = \begin{cases} \displaystyle\prod_{j:\ c_{ij}=-1} T_{k,r}^{(j)} \prod_{j:\ c_{ij}=-2} T_{l,r}^{(j)} T_{l,r+2}^{(j)}, & \text{if } k = 2l, \\ \displaystyle\prod_{j:\ c_{ij}=-1} T_{k,r}^{(j)} \prod_{j:\ c_{ij}=-2} T_{l+1,r}^{(j)} T_{l,r+2}^{(j)} & \text{if } k = 2l+1. \end{cases} \tag{2.6}$$

Finally, if $d_i = 1$ and $t = 3$, that is, if $\mathfrak{g}$ is of type G_2, denoting by j the other vertex we have $d_j = 3$ and

$$S_{k,r}^{(i)} = \begin{cases} T_{l,r}^{(j)} T_{l,r+2}^{(j)} T_{l,r+4}^{(j)} & \text{if } k = 3l, \\ T_{l+1,r}^{(j)} T_{l,r+2}^{(j)} T_{l,r+4}^{(j)} & \text{if } k = 3l+1, \\ T_{l+1,r}^{(j)} T_{l+1,r+2}^{(j)} T_{l,r+4}^{(j)} & \text{if } k = 3l+2. \end{cases} \tag{2.7}$$

Example 2.7 Let $\mathfrak{g}$ be of type B_2. The Cartan matrix is

$$C = \begin{pmatrix} 2 & -1 \\ -2 & 2 \end{pmatrix}$$

and we have $d_1 = 2$ and $d_2 = 1$. The T-system reads

$$\begin{aligned}
T^{(1)}_{k,r+2}T^{(1)}_{k,r-2} &= T^{(1)}_{k-1,r+2}T^{(1)}_{k+1,r-2} + T^{(2)}_{2k,r-1}, && (k \geq 1,\ r \in \mathbb{Z}),\\
T^{(2)}_{2l,r+1}T^{(2)}_{2l,r-1} &= T^{(2)}_{2l-1,r+1}T^{(2)}_{2l+1,r-1} + T^{(1)}_{l,r}T^{(1)}_{l,r+2}, && (l \geq 1,\ r \in \mathbb{Z}),\\
T^{(2)}_{2l+1,r+1}T^{(2)}_{2l+1,r-1} &= T^{(2)}_{2l,r+1}T^{(2)}_{2l+2,r-1} + T^{(1)}_{l+1,r}T^{(1)}_{l,r+2}, && (l \geq 0,\ r \in \mathbb{Z}).
\end{aligned}$$

It was conjectured in [33], and proved in [39] (for $\mathfrak{g}$ of type A, D, E) and [21] (general case), that the q-characters of the Kirillov–Reshetikhin modules of $U_q(\widehat{\mathfrak{g}})$ satisfy the corresponding T-system. More precisely, we have

Theorem 2.8 ([21, 38]) *For $i \in I,\ k \in \mathbb{N},\ r \in \mathbb{Z}$,*

$$T^{(i)}_{k,r} = \chi_q\left(W^{(i)}_{k,q^r}\right),$$

is a solution of the T-system in the ring $\mathbb{Z}\left[Y^{\pm 1}_{i,q^r} \mid (i,r) \in I \times \mathbb{Z}\right]$. Equivalently, by Theorem 2.4,

$$T^{(i)}_{k,r} = \left[W^{(i)}_{k,q^r}\right],$$

is a solution of the T-system in the Grothendieck ring K_0(-mod$(U_q(\widehat{\mathfrak{g}}))$).

Remark 3.9

(i) For $\mathfrak{g} = \mathfrak{sl}_2$, Theorem 2.8 is a particular case of Proposition 1.3.
(ii) Theorem 2.8 allows to calculate inductively q-characters of Kirillov–Reshetikhin modules in terms of q-characters of fundamental modules. Note, however, that it is not straightforward to compute the q-characters of the fundamental modules in type E_8 or F_4, say. An algorithm has been obtained by Frenkel and Mukhin [12]. Another one, based on cluster mutation, is described in [23].
(iii) The T-system in the Grothendieck ring comes from a non-split short exact sequence

$$0 \to \mathcal{S}^{(i)}_{k,r} \to W^{(i)}_{k,q^{r-d_i}} \otimes W^{(i)}_{k,q^{r+d_i}} \to W^{(i)}_{k-1,q^{r+d_i}} \otimes W^{(i)}_{k+1,q^{r+d_i}} \to 0,$$

where the module $\mathcal{S}^{(i)}_{k,r}$ is defined as the tensor product of Kirillov–Reshetikhin modules associated with the factors of $S^{(i)}_{k,r}$ above (it does not depend on the order of the tensor product up to isomorphism). The representation $W^{(i)}_{k,q^{r-d_i}} \otimes$

$W^{(i)}_{k,q^{r+d_i}}$ is of length 2. By a general result of Chari on tensor products of Kirillov–Reshetikhin modules [2], it is cyclic generated by the tensor product of the highest weight vectors and so it is indecomposable.

3 Quivers, Subcategories, and Cluster Algebras

Following [25], we attach an infinite quiver to $U_q(\widehat{\mathfrak{g}})$, and we define some subcategories of the category of finite-dimensional $U_q(\widehat{\mathfrak{g}})$-modules. We then introduce cluster algebras corresponding to finite segments of this infinite quiver.

3.1 *Quivers*

Put $\widetilde{V} = I \times \mathbb{Z}$. We introduce a quiver $\widetilde{\Gamma}$ with vertex set $\widetilde{V}$. Recall the symmetric matrix $B = (b_{ij})_{i,j\in I}$ of Sect. 2.1. The arrows of $\widetilde{\Gamma}$ are given by

$$((i,r) \to (j,s)) \quad \Longleftrightarrow \quad (b_{ij} \neq 0 \quad \text{and} \quad s = r + b_{ij}).$$

It is easy to check that the oriented graph $\widetilde{\Gamma}$ has two isomorphic connected components. We pick one of them and call it Γ. The vertex set of Γ is denoted by V. Examples in type A_3 and B_2 are shown in Fig. 1.

3.2 *Subcategories*

First, using the vertex set V, we introduce

$$\widehat{P}_{+,\mathbb{Z}} := \bigoplus_{(i,r)\in V} \mathbb{N}(\varpi_i, q^{r+d_i}),$$

a discrete subset of the positive cone $\widehat{P}_+$ of loop-weights.

Definition 3.1 Let $\mathcal{C}_\mathbb{Z}$ be the full subcategory of the category of finite-dimensional $U_q(\widehat{\mathfrak{g}})$-modules whose objects M satisfy:

Every composition factor of M is of the form $L(\widehat{\lambda})$ with $\widehat{\lambda} \in \widehat{P}_{+,\mathbb{Z}}$.

By [23], $\mathcal{C}_\mathbb{Z}$ is a monoidal subcategory, i.e. it is stable under tensor products. Moreover, it is known that every simple finite-dimensional $U_q(\widehat{\mathfrak{g}})$-module can be written as a tensor product of simple objects of $\mathcal{C}_\mathbb{Z}$ with spectral shifts. Therefore it is enough to study the simple objects of $\mathcal{C}_\mathbb{Z}$.

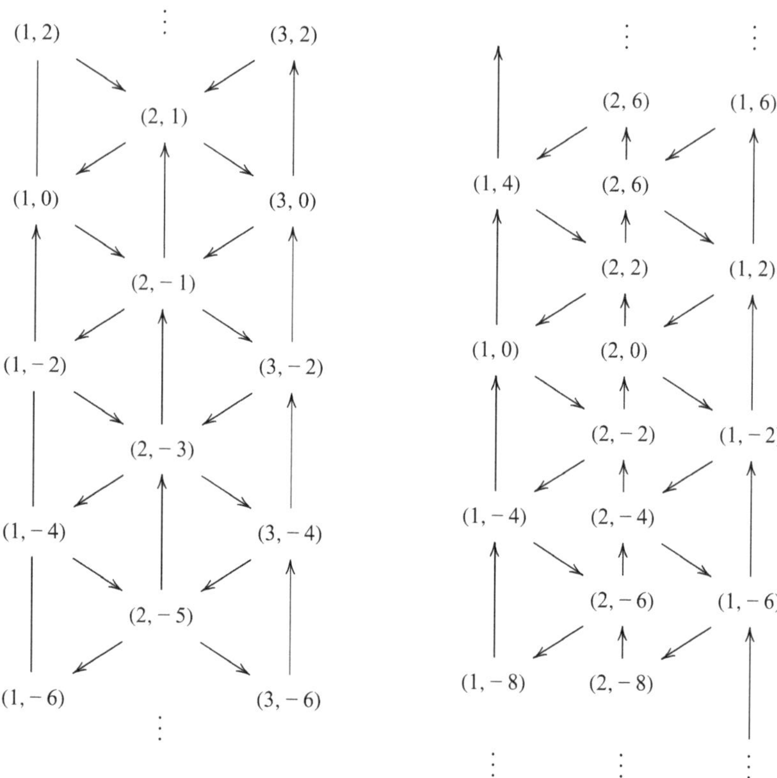

Fig. 1 The quivers Γ in type A_3 and B_2

In order to relate $\mathcal{C}_{\mathbb{Z}}$ with cluster algebras of finite rank, we need to "truncate" it, as we did in Sect. 1.3 for $\mathfrak{g} = \mathfrak{sl}_2$. Fix $\ell \in \mathbb{N}$, and put

$$\widehat{P}_{+,\ell} := \bigoplus_{(i,r)\in V,\ -2\ell-1\leq r+d_i\leq 0} \mathbb{N}(\varpi_i, q^{r+d_i}).$$

Definition 3.2 Let $\mathcal{C}_\ell$ be the full subcategory of the category of finite-dimensional $U_q(\widehat{\mathfrak{g}})$-modules whose objects M satisfy:

Every composition factor of M is of the form $L(\widehat{\lambda})$ with $\widehat{\lambda} \in \widehat{P}_{+,\ell}$.

Again by [23], $\mathcal{C}_\ell$ is a monoidal category and its Grothendieck ring is a polynomial ring in finitely many variables, namely, the classes of the fundamental modules contained in $\mathcal{C}_\ell$:

$$K_0(\mathcal{C}_\ell) = \mathbb{Z}\Big[[L\left((\varpi_i, q^{r+d_i})\right)] \mid (i,r) \in V,\ -2\ell - 1 \leq r + d_i \leq 0\Big].$$

3.3 Cluster Algebras

We refer the reader to [16] and [19] for an introduction to cluster algebras, and for any undefined terminology.

Let Γ_ℓ denote the full subquiver of Γ with vertex set

$$V_\ell := \{(i,r) \in V \mid -2\ell - 1 \le r + d_i \le 0\}.$$

Let $\mathbf{z}_\ell := \{z_{(i,r)} \mid (i,r) \in V_\ell\}$ be a set of commuting indeterminates indexed by V_ℓ. The pair $(\mathbf{z}_\ell, \Gamma_\ell)$ can be regarded as a *seed*, in the sense of [16].

Definition 3.3 Let $\mathcal{A}_\ell \subset \mathbb{Q}(\mathbf{z}_\ell)$ be the cluster algebra with initial seed $(\mathbf{z}_\ell, \Gamma_\ell)$, where we consider the variables $z_{i,r}$ with $r - d_i < -2\ell - 1$ as *frozen variables.* □

4 Main Conjecture

4.1 Statements and Examples

The category $\mathcal{C}_\ell$ and the cluster algebra $\mathcal{A}_\ell$ are related as follows:

Theorem 4.1 ([23]) *For $(i,r) \in V_\ell$ put $m_{i,r} := \max\{k \mid r + (2k+1)d_i \le 0\} + 1$. The assignment $z_{i,r} \mapsto \left[W^{(i)}_{m_{i,r},q^{r+d_i}}\right]$ extends to a ring isomorphism $\iota_\ell : \mathcal{A}_\ell \to K_0(\mathcal{C}_\ell)$.*

We can now formulate our main conjecture from [22, 23].

Conjecture 4.2

(i) The isomorphism ι_ℓ maps the subset of cluster monomials of $\mathcal{A}_\ell$ into the subset of classes of simple objects of $\mathcal{C}_\ell$.
(ii) The isomorphism ι_ℓ maps the subset of cluster variables of $\mathcal{A}_\ell$ into the subset of classes of *prime* simple objects of $\mathcal{C}_\ell$.

In the situation of Theorem 4.1 and Conjecture 4.2, we say that $\mathcal{C}_\ell$ is a *monoidal categorification* of the cluster algebra $\mathcal{A}_\ell$ [22]. We illustrate the conjecture with simple examples.

Example 4.3 We take $\mathfrak{g} = \mathfrak{sl}_4$, of type A_3, and we choose $\ell = 1$. The isomorphism ι_1 is displayed in Fig. 2, which shows the image of the initial seed of $\mathcal{A}_1$. For instance, $\iota_1(z_{2,-1}) = \left[W^{(2)}_{1,q^0}\right]$. The 3 frozen variables are marked with a box.

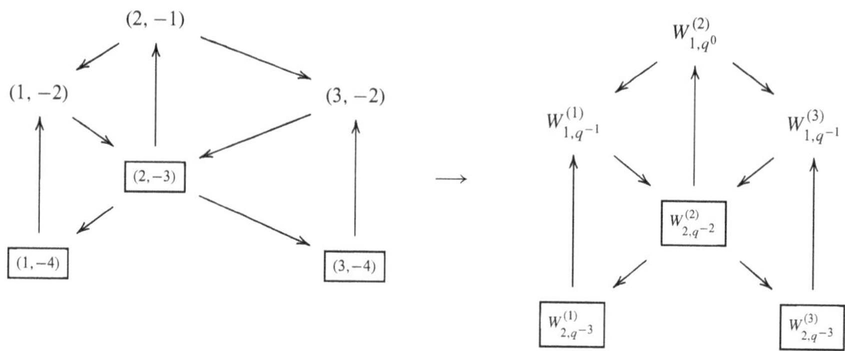

Fig. 2 The isomorphism ι_ℓ in type A_3 for $\ell = 1$

The cluster variable $z^*_{2,-1}$ obtained by mutating $z_{2,-1}$ in this initial seed is given by the exchange relation:

$$z_{2,-1}z^*_{2,-1} = z_{2,-3} + z_{1,-2}z_{3,-2},$$

which translates under ι_1 into the T-system equation:

$$\left[W^{(2)}_{1,q^0}\right]\left[W^{(2)}_{1,q^{-2}}\right] = \left[W^{(2)}_{2,q^{-2}}\right] + \left[W^{(1)}_{1,q^{-1}}\right]\left[W^{(3)}_{1,q^{-1}}\right].$$

Thus, $\iota_1(z^*_{2,-1}) = \left[W^{(2)}_{1,q^{-2}}\right]$.

In this case, $\mathcal{A}_1$ is a cluster algebra of finite type A_3 in the Fomin–Zelevinsky classification, and Conjecture 4.2 is proved [22]. Moreover, since $\mathcal{A}_1$ has finite type, the inclusions of Conjecture 4.2 are in fact bijections. The algebra $\mathcal{A}_1$ has 9 cluster variables plus 3 frozen variables. The prime simple modules of $\mathcal{C}_1$ corresponding to cluster variables are

$$\begin{aligned}
&W^{(1)}_{1,q^{-1}} = L((\varpi_1, q^{-1})),\quad W^{(2)}_{1,q^{-2}} = L((\varpi_2, q^{-2})),\ W^{(3)}_{1,q^{-1}} = L((\varpi_3, q^{-1})),\\
&W^{(1)}_{1,q^{-3}} = L((\varpi_1, q^{-3})),\quad W^{(2)}_{1,q^{0}} = L((\varpi_2, q^{0})),\quad W^{(3)}_{1,q^{-3}} = L((\varpi_3, q^{-3})),\\
&L((\varpi_1, q^{-3}) + (\varpi_2, q^0)),\ L((\varpi_1, q^{-3}) + (\varpi_2, q^0) + (\varpi_3, q^{-3})),\\
&\qquad\qquad\qquad\qquad\qquad\qquad L((\varpi_2, q^0) + (\varpi_3, q^{-3})).
\end{aligned}$$

There are 6 fundamental modules, and 2 minimal affinizations, but the 70-dimensional module $L((\varpi_1, q^{-3}) + (\varpi_2, q^0) + (\varpi_3, q^{-3}))$, which restricts to $L(\varpi_1 + \varpi_2 + \varpi_3) \oplus L(\varpi_2)$ as a $U_q(\mathfrak{sl}_4)$-module, is *not* a minimal affinization.

By [15], there is a bijection between the set of cluster variables of $\mathcal{A}_1$ and the set of almost positive roots of a root system of type A_3. A natural bijection was given in [22]:

$$
\begin{array}{ll}
L((\varpi_1, q^{-1})) & \leftrightarrow -\alpha_1 \\
L((\varpi_2, q^{-2})) & \leftrightarrow -\alpha_2 \\
L((\varpi_3, q^{-1})) & \leftrightarrow -\alpha_3 \\
L((\varpi_1, q^{-3})) & \leftrightarrow \alpha_1 \\
L((\varpi_2, q^{0})) & \leftrightarrow \alpha_2 \\
L((\varpi_3, q^{-3})) & \leftrightarrow \alpha_3 \\
L((\varpi_1, q^{-3}) + (\varpi_2, q^{0})) & \leftrightarrow \alpha_1 + \alpha_2 \\
L((\varpi_2, q^{0}) + (\varpi_3, q^{-3})) & \leftrightarrow \alpha_2 + \alpha_3 \\
L((\varpi_1, q^{-3}) + (\varpi_2, q^{0}) + (\varpi_3, q^{-3})) & \leftrightarrow \alpha_1 + \alpha_2 + \alpha_3
\end{array}
$$

Following [15] and using this bijection one can read the 14 clusters of $\mathcal{A}_1 \cong K_0(\mathcal{C}_1)$ from the associahedron of Fig. 3. Every cluster variable corresponds to a face, indicated by the attached almost positive root (the negative simple root $-\alpha_i$ labels the unique rear face parallel to the face labelled by α_i). Every cluster corresponds to a vertex and consists of the 3 faces adjacent to it. For instance, there is a cluster

$$\{\alpha_1, \alpha_3, -\alpha_2\} \equiv \{L((\varpi_1, q^{-3})),\ L((\varpi_3, q^{-3})),\ L((\varpi_2, q^{-2}))\}.$$

The neat final result is that every simple module of $\mathcal{C}_1$ is a tensor product of prime simple modules belonging to a single cluster, and of frozen simple modules (corresponding to the frozen variables of $\mathcal{A}_1$).

Example 4.4 We take $\mathfrak{g} = \mathfrak{so}_5$, of type B_2, and we choose $\ell = 2$. The isomorphism ι_2 is displayed in Fig. 4, which shows the image of the initial seed of $\mathcal{A}_2$. For instance, $\iota_2(z_{2,-2}) = \left[W^{(2)}_{1,q^{-1}}\right]$. The 3 frozen variables are marked with a box.

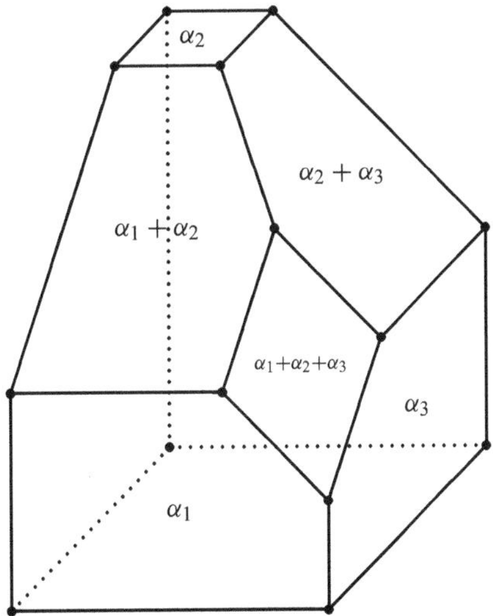

Fig. 3 The associahedron in type A_3

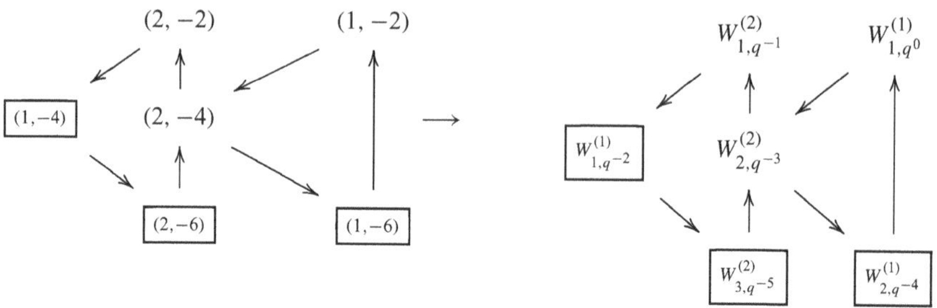

Fig. 4 The isomorphism ι_ℓ in type B_2 for $\ell = 2$

Again, $\mathcal{A}_2$ has finite cluster type A_3. So $\mathcal{C}_2$ has 12 prime objects, namely, the 6 Kirillov–Reshetikhin modules of the initial seed, together with

$$W^{(1)}_{1,q^{-4}},\ W^{(2)}_{1,q^{-3}},\ W^{(2)}_{1,q^{-5}},\ W^{(2)}_{2,q^{-5}},\ L((\varpi_1,q^0)+(\varpi_2,q^{-5})),$$
$$L((\varpi_1,q^0)+(\varpi_2,q^{-3})+(\varpi_2,q^{-5})).$$

A bijection between the highest loop-weights of the unfrozen primes and the almost positive roots of type A_3, allowing to determine the clusters using the associahedron as in Example 4.3, is for instance:

$$\begin{array}{ll}
(\varpi_1,q^{-4}) & \leftrightarrow -\alpha_1\\
(\varpi_2,q^{-3})+(\varpi_2,q^{-1}) & \leftrightarrow -\alpha_2\\
(\varpi_2,q^{-1}) & \leftrightarrow -\alpha_3\\
(\varpi_1,q^{0}) & \leftrightarrow \alpha_1\\
(\varpi_2,q^{-5}) & \leftrightarrow \alpha_2\\
(\varpi_2,q^{-3}) & \leftrightarrow \alpha_3\\
(\varpi_1,q^{0})+(\varpi_2,q^{-5}) & \leftrightarrow \alpha_1+\alpha_2\\
(\varpi_2,q^{-3})+(\varpi_2,q^{-5}) & \leftrightarrow \alpha_2+\alpha_3\\
(\varpi_1,q^{-0})+(\varpi_2,q^{-3})+(\varpi_2,q^{-5}) & \leftrightarrow \alpha_1+\alpha_2+\alpha_3.
\end{array}$$

Example 4.5 We take $\mathfrak{g}$ of type G_2, and we choose $\ell = 3$. The isomorphism ι_3 is displayed in Fig. 5, which shows the image of the initial seed of $\mathcal{A}_3$. There are 4 frozen variables. $\mathcal{A}_3$ has finite cluster type A_4. So $\mathcal{C}_3$ has 18 prime objects, namely, the 8 Kirillov–Reshetikhin modules of the initial seed, together with

$$W^{(1)}_{1,q^{-6}},\ W^{(2)}_{1,q^{-3}},\ W^{(2)}_{1,q^{-5}},\ W^{(2)}_{2,q^{-5}},\ W^{(2)}_{2,q^{-7}},\ W^{(2)}_{3,q^{-7}},\ W^{(2)}_{1,q^{-7}},$$
$$L((\varpi_1,q^0)+(\varpi_2,q^{-7})),\ L((\varpi_1,q^0)+(\varpi_2,q^{-5})+(\varpi_2,q^{-7})),$$
$$L((\varpi_1,q^0)+(\varpi_2,q^{-3})+(\varpi_2,q^{-5})+(\varpi_2,q^{-7})).$$

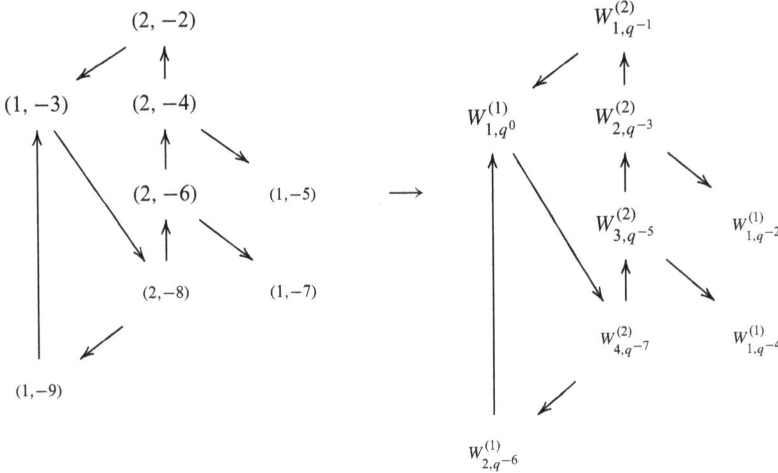

Fig. 5 The isomorphism ι_ℓ in type G_2 for $\ell = 3$

Remark 4.6

(i) There is a similar Conjecture in which the finite quiver Γ_ℓ is replaced by the semi-infinite quiver Γ^- with vertex set

$$V^- := \{(i, r) \in V | r + d_i \leq 0\},$$

and the category $\mathcal{C}_\ell$ is replaced by the category $\mathcal{C}^-$ of finite-dimensional $U_q(\widehat{\mathfrak{g}})$-modules whose composition factors are of the form $L(\widehat{\lambda})$, $\widehat{\lambda} \in \bigoplus_{(i,r)\in V^-} \mathbb{N}(\varpi_i, q^{r+d_i})$, see [23].

(ii) There is also a similar Conjecture in which the finite quiver Γ_ℓ is replaced by the doubly-infinite quiver Γ. In that case, the corresponding category is no longer a subcategory of the category of finite-dimensional $U_q(\widehat{\mathfrak{g}})$-modules. We have to consider a certain subcategory of the category $\mathcal{O}$ of (possibly infinite-dimensional) representations over a quantum Borel subalgebra $U_q(\widehat{\mathfrak{b}})$ of $U_q(\widehat{\mathfrak{g}})$, see [26]. The category $\mathcal{C}_\mathbb{Z}$ of Definition 3.1 can be regarded as a subcategory of this category of $U_q(\widehat{\mathfrak{b}})$-modules. The initial seed consists of the classes of prefundamental representations, which are simple infinite-dimensional modules of $U_q(\widehat{\mathfrak{b}})$ which cannot be extended to $U_q(\widehat{\mathfrak{g}})$-modules.

4.2 *What Is Known?*

4.2.1 Part (ii)

The difficult part of Conjecture 4.2 is (i). If (i) is known, then (ii) follows from a result of [17] which says that if a cluster algebra is a factorial ring, then every cluster variable is a prime element of this ring.

4.2.2 First Evidences

As explained in Sect. 1, when $\mathfrak{g} = \mathfrak{sl}_2$ Conjecture 4.2 follows from [3].

In [22], (i) was proved for type A_n and D_4 when $\ell = 1$. The proof was algebraic and combinatorial, and certain parts of the proof were more general.

In [38], Nakajima proved (i) for types A, D, E and $\ell = 1$, using the geometric approach to $U_q(\widehat{\mathfrak{g}})$ via quiver varieties. (An introduction to this proof is presented in [35].)

A variant of Conjecture 4.2 for type A_n and D_n when $\ell = 1$ was proved in [23]. This involves finite subquivers of Γ different from Γ_ℓ. Very recently Brito and Chari [1] generalized the results of [23] in type A using purely representation theoretic methods.

4.2.3 Proof in Simply-Laced Cases

In [40], Qin gave a proof of (i) for types A, D, E and arbitrary ℓ. The proof also relies on the geometric approach, and uses the t-deformation of $K_0(\mathcal{C}_\ell)$ introduced by Varagnolo–Vasserot and Nakajima in terms of quiver varieties.

4.2.4 Connection with Quiver Hecke Algebras

In type A, D, E, for $\ell = h/2 - 1$, where h is the Coxeter number supposed to be even, there is another approach as follows. In [24], we have shown that there is a ring isomorphism

$$i : \mathbb{C} \otimes_{\mathbb{Z}} K_0(\mathcal{C}_{h/2-1}) \longrightarrow \mathbb{C}[N],$$

where N is a maximal unipotent subgroup of a simple Lie group G with $\mathrm{Lie}(G) = \mathfrak{g}$. The ring of polynomial functions $\mathbb{C}[N]$ has a well-known cluster algebra structure, and the isomorphism i transports the cluster structure of $\mathcal{A}_{h/2-1} \cong \mathbb{C} \otimes_{\mathbb{Z}} K_0(\mathcal{C}_{h/2-1})$ to the cluster structure of $\mathbb{C}[N]$. We have shown that the isomorphism i maps the basis of $\mathbb{C} \otimes_{\mathbb{Z}} K_0(\mathcal{C}_{h/2-1})$ consisting of classes of simple objects to the dual canonical basis (or upper global basis) of $\mathbb{C}[N]$. Therefore to prove (i) in this case amounts to prove:

(i') The cluster monomials of $\mathbb{C}[N]$ form a subset of the dual canonical basis of $\mathbb{C}[N]$.

This was proved by Kang et al. [30]. They used the categorification of the dual canonical basis of $\mathbb{C}[N]$ by simple objects of a category $\mathcal{H}$ of graded modules over quiver Hecke algebras. This raises the question of a relation between the two categories $\mathcal{H}$ and $\mathcal{C}_{h/2-1}$. In [28], Kang, Kashiwara and Kim constructed a functor from $\mathcal{H}$ to $\mathcal{C}_{h/2-1}$ inducing the isomorphism i^{-1} at the level of Grothendieck rings. In type A this can be regarded as a variant of the quantum affine Schur–Weyl duality of Chari–Pressley and Ginzburg–Reshetikhin–Vasserot [6, 18]. Recently, Fujita [10, 11] proved that the KKK-functor is in fact an equivalence of categories.

4.2.5 Non-Simply-Laced Cases

For non-simply-laced types, let us start with the example of the category $\mathcal{C}_2$ in type B_2 discussed in Example 4.4. Comparing with the category $\mathcal{C}_1$ in type A_3 discussed in Example 4.3, we observe that not only the cluster algebras $\mathcal{A}_1$ in type A_3 and $\mathcal{A}_2$ in type B_2 have the same cluster type A_3 and the same numbers of frozen variables, but also the quivers in Figs. 3 and 4 are mutation-equivalent. This is illustrated in Fig. 6 with the mutation sequence at nodes $(3, -2)$, $(2, -1)$, $(1, -2)$. Arrows between frozen vertices may be omitted as this does not change the cluster algebra structure. Hence we get a distinguished isomorphism between Grothendieck rings

$$K_0(\mathcal{C}_{1,\, A_3}) \simeq K_0(\mathcal{C}_{2,\, B_2})$$

which is compatible with the cluster algebra structures. This last point is the most important since we know already that the two rings are isomorphic to the polynomial ring in 6 variables.

This example is the first instance of a family of isomorphisms of cluster algebras

$$K_0(\mathcal{C}_{h/2-1,\, A_{2n-1}}) \simeq K_0(\mathcal{C}'_{B_n})$$

obtained by the first author in a joint work with Hironori Oya [27]. These are distinguished isomorphisms between the Grothendieck rings of the type A_{2n-1} categories $\mathcal{C}_{h/2-1,\, A_{2n-1}}$ already mentioned in Sect. 4.2.4 above, and remarkable subcategories $\mathcal{C}'_{B_n}$ of finite-dimensional representations in type B_n. The proof is established at the level of quantum cluster algebras, in order to demonstrate a conjecture on quantum Grothendieck rings and related analogs of Kazhdan–Lusztig polynomials formulated in [20].

Using a completely different method based on functors from categories of representations of quiver Hecke algebras, Kashiwara–Kim–Oh and Kashiwara–Oh [31, 32] constructed isomorphisms of Grothendieck rings in types A_{2n-1}/B_n preserving the classes of simple modules. In fact, these match the distinguished isomorphisms obtained from cluster algebra structures in [27]. Hence the cluster algebra isomorphisms also preserve classes of simple modules. For instance, in the example above, the bijection between prime simples can be directly written from the bijection with almost positive roots in Examples 4.3 and 4.4. Consequently, combining these results, the analogue of Conjecture 4.2 for the subcategories $\mathcal{C}'_{B_n}$ holds.

It is expected that this approach will be extended to larger categories and to more general types.

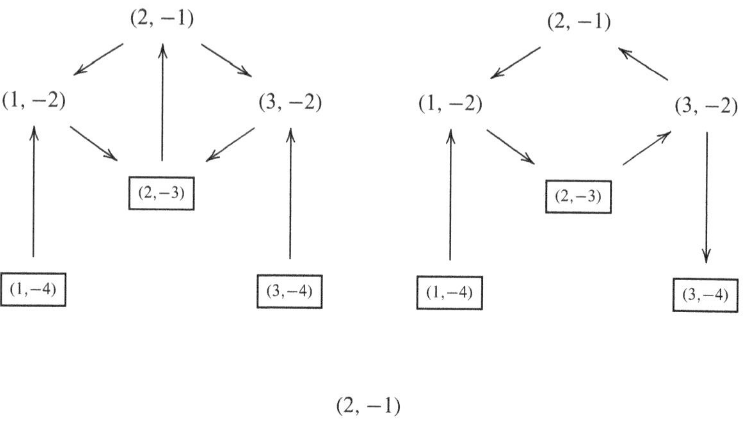

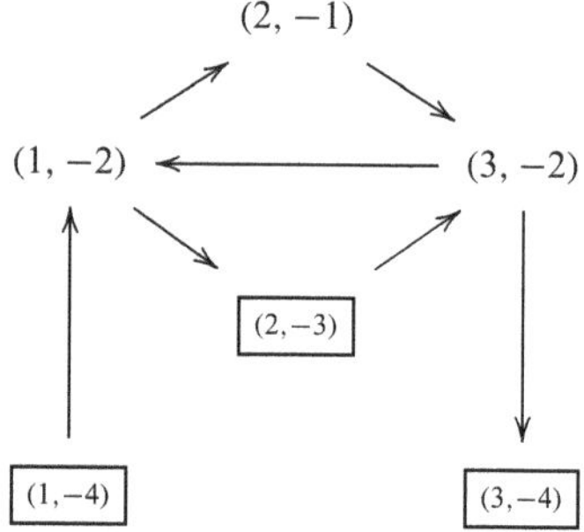

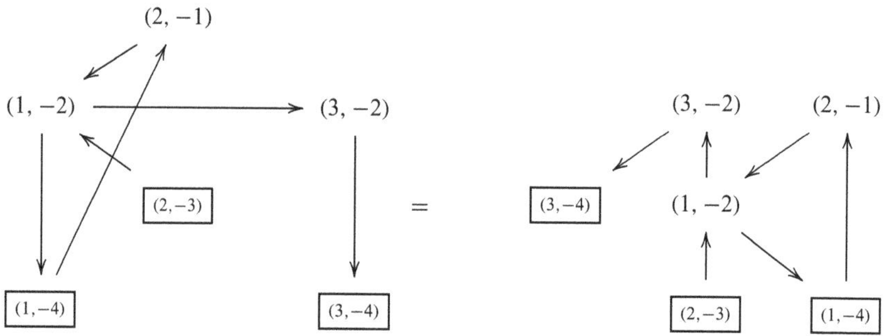

Fig. 6 Mutations : from $\mathcal{A}_1$ in type A_3 to $\mathcal{A}_2$ in type B_2

4.2.6 Real Modules

When the cluster algebra $\mathcal{A}_\ell$ is not of finite cluster type, cluster monomials do not span the vector space $\mathbb{C} \otimes \mathcal{A}_\ell$. This raises the question of describing the simple objects of $\mathcal{C}_\ell$ whose class in the Grothendieck ring is a cluster monomial.

Conjecture 4.7 ([22, 23]) The class of a simple object S of $\mathcal{C}_\ell$ in the Grothendieck ring is a cluster monomial if and only if S is a *real simple object*, that is, if and only if $S \otimes S$ is simple.

Let us assume that Conjecture 4.2 (i) holds. Then one direction of Conjecture 4.7 is obvious: the square of a cluster monomial is clearly a cluster monomial. The converse is wide open.

Real $U_q(\widehat{\mathfrak{g}})$-modules have interesting properties. For instance Kang, Kashiwara, Kim and Oh proved the following theorem, which was conjectured in type A in [34].

Theorem 4.8 ([29]) *If S_1 and S_2 are two simple $U_q(\widehat{\mathfrak{g}})$-modules, and one of them (at least) is real, then $S_1 \otimes S_2$ has a simple socle and a simple head. Moreover the socle and the head are isomorphic if and only if $S_1 \otimes S_2$ is simple.*

Classification of real simple modules (in terms of their highest loop-weight) is a difficult open problem. Recently Lapid and Minguez [36] classified in type A all real simples satisfying a certain regularity condition. Surprisingly, this classification is related to the classification of rationally smooth Schubert varieties in type A flag varieties.

5 Geometric Character Formulas

An important obstacle for proving Conjecture 4.2 in general is the absence of Nakajima's geometric theory in the non-symmetric cases B_n, C_n, F_4, G_2. It turns out that, applying the results of Derksen, Weyman and Zelevinsky [8, 9] to the cluster algebras $\mathcal{A}_\ell$, one can define projective varieties whose Euler characteristics calculate the q-characters of the *standard* $U_q(\widehat{\mathfrak{g}})$-modules in all types. These varieties can be seen as generalizations of the Nakajima graded varieties $\mathcal{L}^\bullet(V, W)$ for types A, D, E. We shall now review this theorem of [23].

5.1 *Quiver Grassmannians and F-Polynomials*

Let Q be a quiver with vertex set I. Let M be a representation of Q over the field $\mathbb{C}$ of complex numbers. Let $\mathbf{e} = (e_i)_{i \in I} \in \mathbb{N}^I$ be a dimension vector, and write $e := \sum_{i\in I} e_i$. The variety $\mathrm{Gr}(\mathbf{e}, M)$ is the closed subvariety of the Grassmannian $\mathrm{Gr}(e, M)$ of e-dimensional subspaces of M whose points parametrize the sub-representations of M with dimension vector $\mathbf{e}$. Thus, $\mathrm{Gr}(\mathbf{e}, M)$ is a projective complex variety, called a *quiver Grassmannian*.

Definition 5.1 The *F-polynomial* of the representation M of Q is

$$F_M(\mathbf{v}) := \sum_{\mathbf{e}\in\mathbb{N}^I} \chi(\mathrm{Gr}(\mathbf{e}, M))\mathbf{v}^{\mathbf{e}},$$

where $\mathbf{v} := (v_i)_{i\in I}$ is a sequence of commutative variables, $\mathbf{v}^{\mathbf{e}} := \prod_i v_i^{e_i}$, and $\chi(V)$ denotes the Euler characteristic of a complex projective variety V.

5.2 The Algebra A

Recall that in Sect. 3.1 we have associated with $U_q(\widehat{\mathfrak{g}})$ an infinite quiver Γ. Recall also the notation of Sect. 2.1 for the Cartan matrix. For every negative entry $c_{ij} < 0$ of the Cartan matrix and every $(i, r) \in V$, the graph Γ contains an oriented cycle $\gamma_{i,j,r}$:

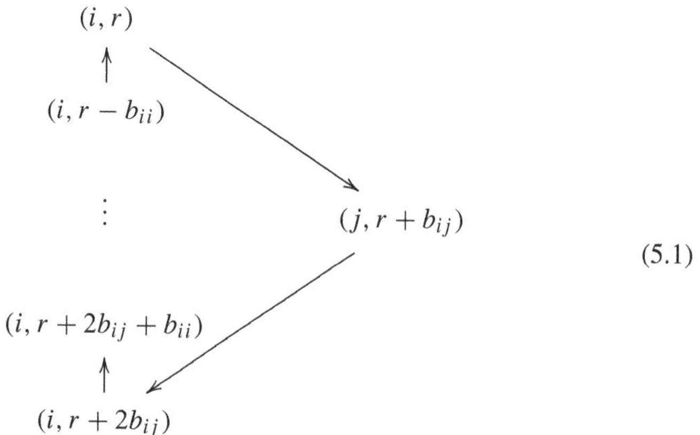

(5.1)

We define a *potential* S as the formal sum of all these oriented cycles $\gamma_{i,j,r}$ up to cyclic permutations, see [8, §3]. This is an infinite sum, but note that a given arrow of Γ can only occur in a finite number of summands. Hence all the cyclic derivatives of S, defined as in [8, Definition 3.1], are finite sums of paths in Γ. Let R be the list of all cyclic derivatives of S. Let J denote the two-sided ideal of the path algebra $\mathbb{C}\Gamma$ generated by R. Following [8], we now introduce

Definition 5.2 Let A be the infinite-dimensional $\mathbb{C}$-algebra $\mathbb{C}\Gamma/J$.

Example 5.3 Let $\mathfrak{g} = \mathfrak{sl}_3$, of type A_2. The quiver Γ is displayed in Fig. 7. We slightly abuse notation using the same letter α for every arrow of the form $(2, r) \to (1, r - 1)$, and similarly for β, γ, δ. All cycles $\gamma_{i,j,r}$ are of length 3, and are either of the form (γ, β, α) or of the form (β, δ, α). Therefore the potential is

$$S = \sum(\gamma, \beta, \alpha) + \sum(\beta, \delta, \alpha),$$

and the relations obtained by taking its cyclic derivatives are of the form:

$$\begin{aligned}
&\text{derivative with respect to } \alpha \rightsquigarrow \gamma\beta + \beta\delta = 0,\\
&\text{derivative with respect to } \beta \rightsquigarrow \alpha\gamma + \delta\alpha = 0,\\
&\text{derivative with respect to } \gamma \rightsquigarrow \quad \beta\alpha = 0,\\
&\text{derivative with respect to } \delta \rightsquigarrow \quad \alpha\beta = 0.
\end{aligned}$$

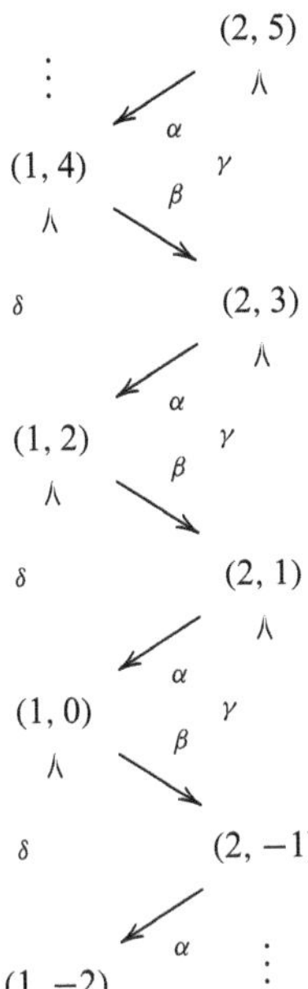

Fig. 7 The algebra A in type A_2

So A is the algebra defined by the quiver Γ of Fig. 7, subject to the above 4 families of relations.

5.3 Some A-Modules

The algebra A is infinite-dimensional. For every $(i, r) \in V$ there is a one-dimensional A-module $S_{(i,r)}$ supported on vertex (i, r). Let $I_{(i,r)}$ denote the injective hull of $S_{(i,r)}$, an infinite-dimensional indecomposable A-module.

Example 5.4 We continue Example 5.3. The injective module $I_{(1,2)}$ is represented in Fig. 8. Each vertex occurring in the picture carries a one-dimensional vector space, and all occurring arrows are nonzero. The β arrows are all zero and therefore not represented. The infinite socle series of this module is

$$\begin{gathered}\vdots \\ S_{(1,-2)} \oplus S_{(2,1)} \\ S_{(1,0)} \oplus S_{(2,3)} \\ S_{(1,2)}\end{gathered}$$

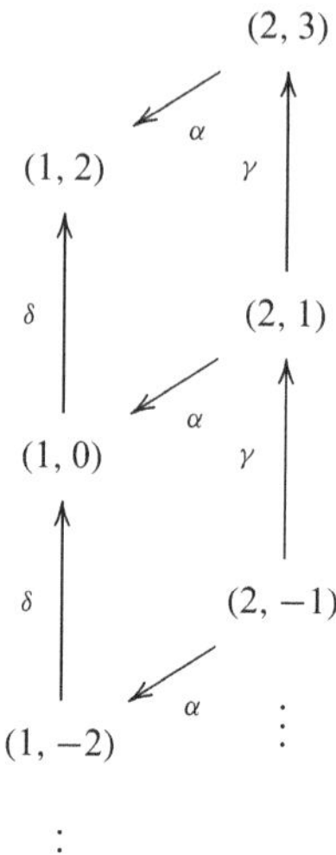

Fig. 8 The injective A-module $I_{(1,2)}$ in type A_2

Note that we have a short exact sequence of A-modules

$$0 \to K_{(1,2)} \to I_{(1,2)} \to I_{(1,0)} \to 0,$$

where $K_{(1,2)}$ is the two-dimensional module:

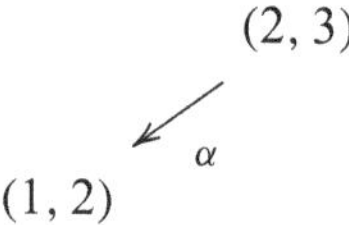

Proposition 5.5 *For every $(i,r) \in V$ there is a unique submodule $K_{(i,r)}$ of $I_{(i,r)}$ such that $I_{(i,r)}/K_{(i,r)}$ is isomorphic to $I_{(i,r-d_i)}$. The module $K_{(i,r)}$ is finite-dimensional.*

5.4 Geometric q-Character Formula

The next theorem gives a geometric description of the q-character of the fundamental $U_q(\widehat{\mathfrak{g}})$-module $L((\varpi_i, q^{r-d_i}))$ in terms of the F-polynomial of the A-module $K_{(i,r)}$. To state it we need some more notation. To every $(i,r) \in V$ we attach a commutative variable $z_{(i,r)}$, and we set

$$\widehat{y}_{(i,r)} := \prod_{(i,r)\to(j,s)} z_{(j,s)} \prod_{(k,l)\to(i,r)} z_{(k,l)}^{-1},$$

where the first product runs over all arrows of Γ starting at vertex (i, r), and the second product over all arrows of Γ ending in vertex (i, r).

Theorem 5.6 ([23]) *The F-polynomial $F_{K_{(i,r)}}(\widehat{\mathbf{y}})$ of the A-module $K_{(i,r)}$ evaluated in the variables $\widehat{y}_{(i,r)}$ can be expressed as a Laurent polynomial in the new variables*

$$Y_{j,q^{s-d_j}} := \frac{z_{(j,s-2d_j)}}{z_{(j,s)}}, \qquad ((j,s) \in V).$$

Then we have

$$\chi_q(L((\varpi_i, q^{r-d_i}))) = Y_{i,q^{r-d_i}} F_{K_{(i,r)}}(\widehat{\mathbf{y}}),$$

where in the right-hand side $F_{K_{(i,r)}}(\widehat{\mathbf{y}})$ is expressed in terms of the variables $Y_{j,q^{s-d_j}}$.

Example 5.7 We continue Example 5.4. The A-module $K_{(1,2)}$ has exactly three submodules: $\{0\}$, $S_{(1,2)}$, and $K_{(1,2)}$. So there are three nonempty quiver Grassmannians, each reduced to a single point, hence each having Euler characteristic 1. It follows that

$$F_{K_{(1,2)}}(\mathbf{v}) = 1 + v_{(1,2)} + v_{(1,2)}v_{(2,3)}.$$

On the other hand

$$\widehat{y}_{(1,2)} = \frac{z_{(2,1)}z_{(1,4)}}{z_{(1,0)}z_{(2,3)}}, \qquad \widehat{y}_{(2,3)} = \frac{z_{(1,2)}z_{(2,5)}}{z_{(1,4)}z_{(2,1)}}.$$

Hence

$$F_{K_{(1,2)}}(\mathbf{y}) = 1 + \frac{z_{(2,1)}z_{(1,4)}}{z_{(1,0)}z_{(2,3)}} + \frac{z_{(1,2)}z_{(2,5)}}{z_{(1,0)}z_{(2,3)}} = 1 + Y_{1,q}^{-1}Y_{1,q^3}^{-1}Y_{2,q^2} + Y_{1,q}^{-1}Y_{2,q^4}^{-1},$$

and

$$Y_{1,q}F_{K_{(1,2)}}(\mathbf{y}) = Y_{1,q} + Y_{1,q^3}^{-1}Y_{2,q^2} + Y_{2,q^4}^{-1} = \chi_q(L((\varpi_1, q))).$$

Example 5.8 We take $\mathfrak{g} = \mathfrak{so}_5$, of type B_2. The A-modules $K_{(1,0)}$ and $K_{(2,0)}$ are displayed in Fig. 9. In this case too, the nonempty quiver Grassmannians are reduced

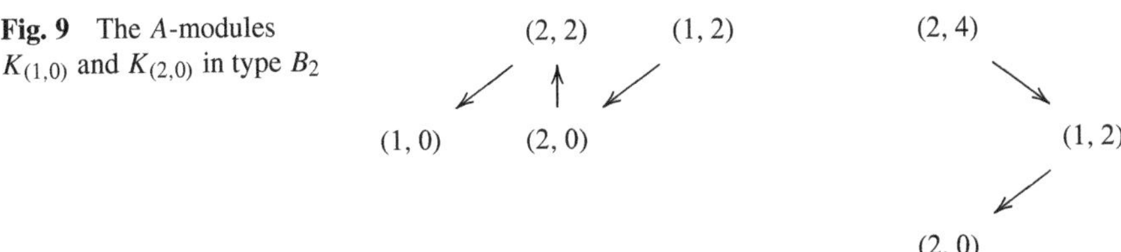

Fig. 9 The A-modules $K_{(1,0)}$ and $K_{(2,0)}$ in type B_2

to points, and we can calculate easily:

$$\chi_q(L((\varpi_1, q^{-2})) = Y_{1,q^{-2}}\Big(1 + \widehat{y}_{(1,0)} + \widehat{y}_{(1,0)}\widehat{y}_{(2,2)} + \widehat{y}_{(1,0)}\widehat{y}_{(2,2)}\widehat{y}_{(2,0)} + \widehat{y}_{(1,0)}\widehat{y}_{(2,2)}\widehat{y}_{(2,0)}\widehat{y}_{(1,2)}\Big),$$

$$\chi_q(L((\varpi_2, q^{-1})) = Y_{2,q^{-1}}\left(1 + \widehat{y}_{(2,0)} + \widehat{y}_{(2,0)}\widehat{y}_{(1,2)} + \widehat{y}_{(2,0)}\widehat{y}_{(1,2)}\widehat{y}_{(2,4)}\right).$$

To express these q-characters in terms of variables Y_{j,q^s} one uses the formulas:

$$\widehat{y}_{(1,r)} = \frac{z_{(1,r+4)}z_{(2,r-2)}}{z_{(1,r-4)}z_{(2,r+2)}} = \frac{Y_{2,q^{r-1}}Y_{2,q^{r+1}}}{Y_{1,q^{r+2}}Y_{1,q^{r-2}}},$$

$$\widehat{y}_{(2,r)} = \frac{z_{(1,r-2)}z_{(2,r+2)}}{z_{(1,r+2)}z_{(2,r-2)}} = \frac{Y_{1,q^r}}{Y_{2,q^{r-1}}Y_{2,q^{r+1}}}.$$

5.5 Comments on Theorem 5.6

5.5.1 Kirillov–Reshetikhin Modules

In [23] we give a similar q-character formula for every Kirillov–Reshetikhin module. One only needs to replace the A-modules $K_{(1,r)}$ by some larger finite-dimensional submodules of the injective modules $1I_{(i,r)}$.

5.5.2 Standard Modules

Classical properties of Euler characteristics imply that, given two finite-dimensional A-modules M and N, we have

$$F_{M\oplus N}(\mathbf{v}) = F_M(\mathbf{v})F_N(\mathbf{v}).$$

On the other hand, q-characters are multiplicative on tensor products. So Theorem 5.6 readily extends to tensor products of fundamental $U_q(\widehat{\mathfrak{g}})$-modules, that is, we have a similar geometric q-character formula for *standard* $U_q(\widehat{\mathfrak{g}})$-modules (or local Weyl modules), in which one uses quiver Grassmannians of direct sums of A-modules $K_{(i,r)}$.

5.5.3 Relation to Nakajima's Theory

If $\mathfrak{g}$ is of type A, D, E it follows from results of Lusztig [37] and Savage–Tingley [41] that the quiver Grassmannians

$$G(\mathbf{e}, \mathbf{a}) := \mathrm{Gr}\left(\mathbf{e}, \bigoplus_{(i,r)\in V} K_{(i,r)}^{\oplus a_{(i,r)}}\right)$$

are homeomorphic to Nakajima varieties $\mathcal{L}^\bullet(V, W)$, where the graded dimension of V is encoded in the dimension vector $\mathbf{e}$, and the graded dimension of W is given by the multiplicity vector $\mathbf{a} = (a_{(i,r)})$. One can therefore regard the varieties $G(\mathbf{e}, \mathbf{a})$ in non-simply-laced types B, C, F, G as natural candidates for replacing the graded Nakajima varieties $\mathcal{L}^\bullet(V, W)$.

5.5.4 Beyond KR-Modules and Standard Modules

By the Derksen–Weyman–Zelevinsky theory, every cluster monomial of $\mathcal{A}_\ell$ has an expression of the form

$$m = \mathbf{z}^{\mathbf{g}}\, F_K(\widehat{\mathbf{y}})$$

for an appropriate A-module K. So, if Conjecture 4.2 is true, all the simple $U_q(\widehat{\mathfrak{g}})$-modules corresponding to cluster monomials (all the real modules, if Conjecture 4.7 is true) have a similar geometric q-character formula in terms of quiver Grassmannians.

Acknowledgments D. Hernandez is supported in part by the European Research Council under the European Union's Framework Programme H2020 with ERC Grant Agreement number 647353 QAffine.

References

1. M. Brito, V. Chari, *Tensor products and q-characters of HL-modules and monoidal categorifications*, J. Éc. polytech. math. 6 (2019), 581–619.
2. V. Chari, *Braid group actions and tensor products*, Int. Math. Res. Not. (2002), no. 7, 357–382.
3. V. Chari, A. Pressley, *Quantum affine algebras*, Commun. Math. Phys. **142** (1991), 261–283.
4. V. Chari, A. Pressley, *A guide to quantum groups*. Cambridge University Press 1994.
5. V. Chari, A. Pressley, *Minimal affinizations of representations of quantum groups: the simply laced case*, J. Algebra **184** (1996), 1–30.
6. V. Chari, A. Pressley, *Quantum affine algebras and affine Hecke algebras*, Pacific J. Math. **174** (1996), 295–326.
7. V. Chari, A. Pressley, *Factorizations of representations of quantum affine algebras* in Modular Interfaces, (Riverside, Calif. 1995), AMS/IP Stud. Adv. Math. **4**, Amer. Math. Soc., Providence, 1997, 33–40.
8. H. Derksen, J. Weyman, A. Zelevinsky, *Quivers with potential and their representations I: Mutations*, Selecta Math., **14** (2008), 59–119.

9. H. Derksen, J. Weyman, A. Zelevinsky, *Quivers with potential and their representations II: Applications to cluster algebras*, J. Amer. Math. Soc. **23** (2010), 749–790.
10. R. Fujita, *Affine highest weight categories and quantum affine Schur-Weyl duality of Dynkin quiver types*, arXiv:1710.11288
11. R. Fujita, *Geometric realization of Dynkin quiver type quantum affine Schur-Weyl duality*, Int. Math. Res. Not. IMRN **22** (2020), 8353–8386.
12. E. Frenkel, E. Mukhin, *Combinatorics of q-characters of finite-dimensional representations of quantum affine algebras*, Comm. Math. Phys. **216** (2001), 23–57.
13. E. Frenkel, N. Reshetikhin, *The q-characters of representations of quantum affine algebras*, Recent developments in quantum affine algebras and related topics, Contemp. Math. **248** (1999), 163–205.
14. S. Fomin, A. Zelevinsky, *Cluster algebras I: Foundations*, J. Amer. Math. Soc. **15** (2002), 497–529.
15. S. Fomin, A. Zelevinsky, *Cluster algebras II: Finite type classification*, Invent. Math. 154 (2003), 63–121.
16. S. Fomin, A. Zelevinsky, *Cluster algebras: notes for the CDM-03 conference*, in Current developments in mathematics, 2003, 1–34, Int. Press, Somerville, MA, 2003.
17. C. Geiss, B. Leclerc, J. Schröer, *Factorial cluster algebras*, Doc. Math. **18** (2013), 249–274.
18. V. Ginzburg, N. Reshetikhin, E. Vasserot, *Quantum groups and flag varieties*, in Mathematical aspects of conformal and topological field theories and quantum groups (South Hadley, MA, 1992), 101–130, Contemp. Math., 175, Amer. Math. Soc., Providence, RI, 1994.
19. M. Gekhtman, M. Shapiro, A. Vainshtein, *Cluster algebras and Poisson geometry*, AMS Math. Survey and Monographs **167**, AMS 2010.
20. D. Hernandez, *Algebraic approach to q,t-characters*, Adv. Math. **187** (2004), 1–52.
21. D. Hernandez, *The Kirillov-Reshetikhin conjecture and solutions of T-systems*, J. Reine Angew. Math. **596** (2006), 63–87.
22. D. Hernandez, B. Leclerc, *Cluster algebras and quantum affine algebras*, Duke Math. J. **154** (2010), 265–341.
23. D. Hernandez, B. Leclerc, *Monoidal categorifications of cluster algebras of type A and D*, in Symmetries, integrable systems and representations, (K. Iohara, S. Morier-Genoud, B. Rémy, eds.), Springer proceedings in mathematics and statistics **40**, 2013, 175–193.
24. D. Hernandez, B. Leclerc, *Quantum Grothendieck rings and derived Hall algebras*, J. Reine Angew. Math. **701** (2015), 77–126.
25. D. Hernandez, B. Leclerc, *A cluster algebra approach to q-characters of Kirillov-Reshetikhin modules*, J. Eur. Math. Soc., **18** (2016), 1113-1159.
26. D. Hernandez, B. Leclerc, *Cluster algebras and category $\mathcal{O}$ for representations of Borel subalgebras of quantum affine algebras*, Algebra Number Theory **10** (2016), 2015–2052.
27. D. Hernandez, H. Oya, *Quantum Grothendieck ring isomorphisms, cluster algebras and Kazhdan-Lusztig algorithm*, Adv. Math. **374** (2019), 192–272.
28. S.-J. Kang, M. Kashiwara, M. Kim, *Symmetric quiver Hecke algebras and R-matrices of quantum affine algebras, II*, Duke Math. J. **164** (2015), 1549–1602.
29. S.-J. Kang, M. Kashiwara, M. Kim, S.-J. Oh, *Simplicity of heads and socles of tensor products*, Compos. Math. **151** (2015), 377–396.
30. S.-J. Kang, M. Kashiwara, M. Kim, S.-J. Oh, *Monoidal categorification of cluster algebras*, J. Amer. Math. Soc. **31** (2018), 349–426.
31. M. Kashiwara, M. Kim, S.-J. Oh, *Monoidal categories of modules over quantum affine algebras of type A and B*, Proc. London Math. Soc. **118** (2019) 43–77.
32. M. Kashiwara, S.-J. Oh, *Categorical relations between Langlands dual quantum affine algebras: doubly laced types,* to appear in J. Algebraic Combin. **49** (2019), 401–435. https://doi.org/10.1007/s10801-018-0829-z
33. A. Kuniba, T. Nakanishi, J. Suzuki, *Functional relations in solvable lattice models: I. Functional relations and representation theory*, Int. J. Mod. Phys. A**9** (1994), 5215–5266.

34. B. Leclerc, *Imaginary vectors in the dual canonical basis of $U_q(n)$*, Transform. Groups **8** (2003), 95–104.
35. B. Leclerc, *Quantum loop algebras, quiver varieties, and cluster algebras*, in Representations of Algebras and Related Topics, (A. Skowroński and K. Yamagata, eds.), European Math. Soc. Series of Congress Reports, 2011, 117–152.
36. E. Lapid, A. Minguez, *Geometric conditions for □-irreducibility of certain representations of the general linear group over a non-Archimedean local field*, Adv. Math. **339** (2018), 113–190.
37. G. Lusztig, *On quiver varieties*, Adv. Math. **136** (1998), 141–182.
38. H. Nakajima, *t-analogs of q-characters of Kirillov-Reshetikhin modules of quantum affine algebras*, Represent. Theory **7** (2003), 259–274.
39. H. Nakajima, *Quiver varieties and cluster algebras*, Kyoto J. Math. **51** (2011), 71–126.
40. Fan Qin, *Triangular bases in quantum cluster algebras and monoidal categorification conjectures*, Duke Math. J. **166** (2017), 2337–2442.
41. A. Savage, P. Tingley, *Quiver Grassmannians, quiver varieties and preprojective algebras*, Pacific J. Math. **251** (2011), 393–429.

Part II
Surveys

Work of Vyjayanthi Chari

Jacob Greenstein and David Hernandez

Abstract The goal of this survey is to describe the most important contributions of Vjyayanthi Chari in representation theory.

For almost four decades Vyjayanthi Chari's work in representation theory of affine Kac–Moody algebras, affine quantum groups, and current algebras has changed the face of these subjects.

Lie theory has been a central theme in mathematics since its origin in the late nineteenth century. In the second half of the twenty-first century, the introduction of Kac–Moody algebras (infinite-dimensional analogs of classical simple finite-dimensional Lie algebras) and of quantum groups (quantization of simple Lie algebras and Kac–Moody algebras) by Drinfeld and Jimbo increased considerably the richness of the field. These expansions led to many important developments and the discovery of new structures that play a fundamental role in Lie theory and in other areas of mathematics and physics. They are also the underlying algebraic structures for important models in statistical and quantum physics, such as the XXZ spin chain model (the quantum Heisenberg model) or the ice model. The corresponding representation theory is crucial for the understanding of the properties of the algebraic structure as well as for its applications.

Vyjayanthi Chari obtained several groundbreaking results in this field. She is one of the founders of large parts of the representation theory of infinite-dimensional Lie algebras and their quantum groups.

J. Greenstein (✉)
Department of Mathematics, University of California, Riverside, Riverside, CA, USA
e-mail: jacobg@ucr.edu

D. Hernandez
Université de Paris, Université of Paris Diderot, CNRS Institut de Mathématiques de Jussieu-Paris Rive Gauche UMR 7586, Paris, France
e-mail: david.hernandez@imj-prg.fr

J. Greenstein et al. (eds.), *Interactions of Quantum Affine Algebras with Cluster Algebras, Current Algebras and Categorification*, Progress in Mathematics 337,
https://doi.org/10.1007/978-3-030-63849-8_3

We will focus on some of her most influential results.

The fundamental contribution of Vyjayanthi Chari to the field already started in her PhD thesis (which yielded, in particular, two papers published in *Inventiones Mathematicae* [9, 10]). In [9], she extended Duflo's theorem [36] describing annihilators of Verma modules in the universal enveloping algebra of a semisimple Lie algebra to an arbitrary Kac–Moody algebra. A special case was settled in [19] by a different method. In the same paper, she proved that the center of the universal enveloping algebra of a Kac–Moody Lie algebra is "trivial," namely coincides with the universal enveloping algebra of the center of the Lie algebra itself. In the semisimple case, the situation is exactly the opposite: the Lie algebra's center is trivial, while the universal enveloping algebra has a large and quite nontrivial center.

In [10], Chari classified simple integrable modules over an affine Lie algebra. In particular, she showed that, apart from the highest and lowest weight modules, which exist for any Kac–Moody algebra, there is one more class, the *loop modules*, those existence stems from the "loop algebra" nature of affine Lie algebras. More precisely, an (untwisted) affine Lie algebra $\widehat{\mathfrak{g}}$ is obtained by attaching a derivation to the central extension of the Lie algebra $L\mathfrak{g}$ of maps from the unit circle to a simple finite-dimensional Lie algebra $\mathfrak{g}$. More algebraically, $L\mathfrak{g}$ is the tensor product of the algebra $\mathbb{C}[t, t^{-1}]$ of Laurent polynomials in t with $\mathfrak{g}$ with the Lie bracket defined by

$$[x \otimes f, y \otimes g] = [x, y] \otimes fg, \qquad x, y \in \mathfrak{g}, \quad f, g \in \mathbb{C}[t, t^{-1}]$$

and is referred to as the *loop algebra* of $\mathfrak{g}$. Several ideas that will be crucial in the quantum setting too already appeared in this work.

Then, Vyjayanthi Chari started a famous and very fruitful collaboration with Andrew Pressley. It began with [25–27] where explicit realizations of simple loop modules from [10] were obtained. Furthermore, Chari and Pressley initiated a new field, the systematic study of finite-dimensional representations of quantum affine algebras. The relations between these various algebraic structures (Lie algebras, affine algebras, quantum groups, and affine quantum groups) may be summed up in the following diagram:

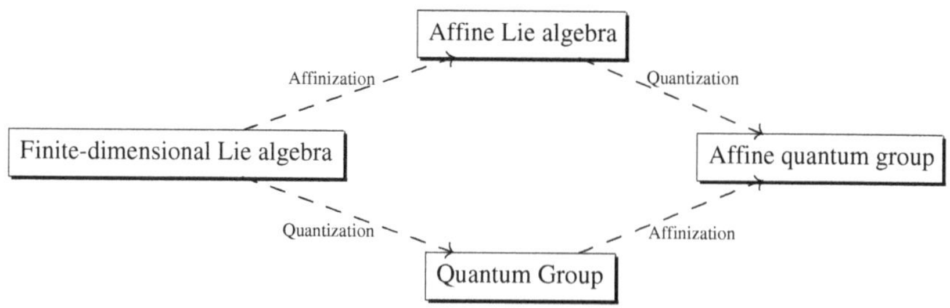

One of their most famous results is the classification of simple finite-dimensional representations of quantum versions of $L\mathfrak{g}$, called quantum affine algebras, in terms of "Drinfeld polynomials" [11, 29, 31]. Namely, let n be the rank of $\mathfrak{g}$. Then there is a bijection between the isomorphism classes of simple finite-dimensional representations of $U_q(L\mathfrak{g})$ and n-tuples of polynomials in one variable with constant term 1. This is a quantum affine analog of the classical parametrization of complex simple finite-dimensional representations of $\mathfrak{g}$ by integral dominant weights, that is, n-tuples of non-negative integers (in fact, the degree of the Drinfeld polynomials). This result of Chari and Pressley is familiar to everyone who ever worked with the representations of quantum affine algebras.

This collaboration led to several other fundamental results. In particular, the Schur–Weyl duality established in [32] between categories of representations of quantum affine algebras associated with $\mathfrak{sl}_{n+1}$ and affine Hecke algebras has a lot of applications, and echoes today is the recent work of Kang, Kashiwara, and Kim [38], which generalizes this construction to Schur–Weyl dualities between quiver Hecke algebras (Khovanov–Laura–Rouquier algebras [40, 42]) and representations of quantum affine algebras.

One very important feature of the tensor category of finite-dimensional representations of quantum affine algebras or Yangians is that it is not braided, but only generically braided. More precisely, for any finite-dimensional representations V, W of the Yangian, there is an isomorphism

$$\mathcal{R}_{V,W}(z) : V(z) \otimes W \to W \otimes V(z)$$

which is meromorphic in the formal parameter z. Here, the representation $V(z)$ is obtained from V by twisting with a natural "shift" automorphism of the Yangian corresponding to z. The operator $\mathcal{R}_{V,W}(z)$ may have poles when regarded as a function of z. In fact, in general,

$$V \otimes W \not\simeq W \otimes V.$$

This is one of the features that make the structure of the category of finite-dimensional representations of a Yangian so intricate. A fundamental problem is to localize the poles of these meromorphic operators. Chari and Pressley made a very important contribution on this problem in [28], but it is still open in general.

Another fundamental result of Chari and Pressley is the complete description of simple finite-dimensional representations V of the quantum affine algebra $\mathcal{U}_q(L\mathfrak{sl}_2)$ associated with the simple Lie algebra $\mathfrak{sl}_2$. This description is given in the form of a factorization into a tensor product of various evaluation representations W_i:

$$V \simeq W_1 \otimes W_2 \otimes \cdots \otimes W_N,$$

that is, simple representations of $\mathcal{U}_q(L\mathfrak{sl}_2)$ obtained from a simple representation of the underlying finite-type quantum group $\mathcal{U}_q(\mathfrak{sl}_2)$ through a Jimbo evaluation algebra morphism

$$\mathcal{U}_q(L\mathfrak{sl}_2) \to \mathcal{U}_q(\mathfrak{sl}_2).$$

This has remarkable consequences. For example, it implies that all simple finite-dimensional representations V of this algebra $\mathcal{U}_q(\widehat{\mathfrak{sl}_2})$ are real, in the sense of monoidal categorification of cluster algebras, that is, the tensor square of V is simple. Moreover, each such simple representation admits a unique factorization into prime representations, that is, so that each factor has no nontrivial tensor factorization, with a very precise combinatorial rule to describe the factors. Besides, there are very explicit relations between the classes of these prime representations in the Grothendieck ring of the category. This may be seen as a premise of the theory of cluster algebras and their monoidal categorifications. This also leads to the general notion of prime representations and factorization of representations [33].

In [34, 35], Chari and Pressley introduced a class of representations that play the role of Weyl modules in representation theory in positive characteristics. Together with J. Beck [1], they established important intrinsic properties of the algebraic structure of affine quantum groups and their canonical bases.

Chari made important advancements toward establishing the famous Kirillov–Reshetikhin conjecture. As before, let $\mathfrak{g}$ be a simple finite-dimensional complex Lie algebra. Since the category of finite-dimensional $U_q(\mathfrak{g})$-modules is semisimple, every finite-dimensional $U_q(L\mathfrak{g})$-module decomposes into a direct sum of simple finite-dimensional $U_q(\mathfrak{g})$-modules. Motivated by statistical physics, Kirillov and Reshetikhin conjectured in 1987 [41] that there exists a class of simple finite-dimensional $U_q(L\mathfrak{g})$-modules, whose tensor products decompose into a direct sum of simple modules over $U_q(\mathfrak{g})$ with multiplicities given in terms of combinatorial formulae, known as fermionic formulae. Such modules are called *Kirillov–Reshetikhin* modules. They are known to be prime in the sense that they are not isomorphic to tensor products of nontrivial $U_q(L\mathfrak{g})$-modules.

In [12], Chari used specializations of $U_q(L\mathfrak{g})$-modules at $q = 1$ to prove the Kirillov–Reshetikhin conjecture in classical types. In [13], she obtained a proof of an important theorem on the cyclicity of a tensor product

$$W_1 \otimes W_2 \otimes \cdots \otimes W_k$$

of Kirillov–Reshetikhin modules W_i (analogous statements were also proved simultaneously by Kashiwara and by Varagnolo and Vasserot using different methods). Under certain precise conditions on the order of the tensor factors, such a tensor product is generated by the tensor product of the highest weight vectors $w_i \in W_i$

$$w_1 \otimes \cdots \otimes w_k \in W_1 \otimes \cdots \otimes W_k.$$

An important idea of this work was to use the action of the braid group corresponding to $\mathfrak{g}$ on ℓ-weights. Furthermore, she introduced [11] an important class of representations called *minimal affinizations* of simple finite-dimensional $U_q(\mathfrak{g})$-modules. Roughly speaking, these are the "smallest," with respect to a certain partial order, modules over $U_q(L\mathfrak{g})$ whose $U_q(\mathfrak{g})$-module simple "top" has the given dominant integral weight. For example, a Kirillov–Reshetikhin module is the minimal affinization corresponding to a multiple of a fundamental weight. Minimal affinizations, also known as Chari modules, appeared later in the work of Hernandez and Leclerc [37] as classes of certain cluster variables in the context of monoidal categorification of cluster algebras.

It should be noted that quasi-classical limits of simple finite-dimensional modules over quantum affine algebras are indecomposable but usually not simple, and many important classes of such modules can be considered as graded modules over the current algebra $\mathfrak{g}[t] = \mathfrak{g} \otimes \mathbb{C}[t]$, which is, naturally, a Lie subalgebra of $L\mathfrak{g}$. So, it is only natural to study the category of such modules. This was another important direction of Chari's work (see [2–4, 15, 17, 18, 20, 21, 23, 24]) in recent years, in collaboration with her students Matthew Bennet, R. J. Dolbin, Nathan Manning, and Tim Ridenour, as well as Jacob Greenstein, Bogdan Ion, Apoorva Khare, Sergei Loktev, and Adriano Moura. In particular, different highest weight category structures were discovered, an analog of BGG reciprocity was established, tilting modules were constructed, and homological properties of various important families of indecomposable modules were understood. This line of investigation also led Vyjayanthi Chari and her students to study several questions pertaining to very classic aspects of Lie theory.

Vyjayanthi Chari obtained, with Adriano Moura, a description of blocks of the category of finite-dimensional representations of untwisted quantum affine algebras [22] (the result was extended very recently in [39] to twisted quantum affine algebras by another method). An analogous result for integrable level-zero modules was obtained in [16].

In [14], with Ghislain Fourier and Tanusree Khandal, she developed a categorical way of defining and studying Weyl modules.

In recent years, Vyjayanthi Chari pursued very active researches on the representations of current algebras [6, 7] as well as on the representations of quantum affine algebras and their relations to cluster algebras [5, 8].

Not only seminal papers of Vyjayanthi Chari served as a source of inspiration for many researchers. In addition, she and Andrew Pressley wrote a very influential book "A Guide to Quantum Groups" [30]. It is well-known and is used as a standard reference book in the field of quantum groups. Many students and researchers around the world, including ourselves, learned the theory of quantum groups using "the book." Vyjayanthi Chari mentored more than twenty graduate students and a large number of young researchers. Notably, most of her collaborative projects in recent years were with graduate students, post-docs, and junior faculty members.

References

1. J. Beck, V. Chari, and A. Pressley, *An algebraic characterization of the affine canonical basis*, Duke Math. J. **99** (1999), no. 3, 455–487, https://doi.org/10.1215/S0012-7094-99-09915-5.
2. M. Bennett, A. Berenstein, V. Chari, A. Khoroshkin, and S. Loktev, *Macdonald polynomials and BGG reciprocity for current algebras*, Selecta Math. (N.S.) **20** (2014), no. 2, 585–607, https://doi.org/10.1007/s00029-013-0141-7.

3. M. Bennett and V. Chari, *Tilting modules for the current algebra of a simple Lie algebra*, Recent developments in Lie algebras, groups and representation theory, Proc. Sympos. Pure Math., vol. 86, Amer. Math. Soc., Providence, RI, 2012, pp. 75–97, https://doi.org/10.1090/pspum/086/1411.
4. M. Bennett, V. Chari, and N. Manning, *BGG reciprocity for current algebras*, Adv. Math. **231** (2012), no. 1, 276–305, https://doi.org/10.1016/j.aim.2012.05.005.
5. R. Biswal, V. Chari, and D. Kus, *Demazure flags*, q-*Fibonacci polynomials and hypergeometric series*, Res. Math. Sci. **5** (2018), no. 1, Paper No. 12, 34, https://doi.org/10.1007/s40687-018-0129-1.
6. R. Biswal, V. Chari, P. Shereen, and J. Wand, *Macdonald Polynomials and level two Demazure modules for affine* $\mathfrak{sl}_{n+1}$, available at arXiv:1910.0548.
7. M. Brito and V. Chari, *Tensor products and* q-*characters of HL-modules and monoidal categorifications*, J. Éc. Polytech. Math. **6** (2019), 581–619, https://doi.org/10.5802/jep.101.
8. M. Brito and V. Chari,, *Resolutions and a Weyl Character formula for prime representations of quantum affine* $\mathfrak{sl}_{n+1}$, available at arXiv:1704.02520.
9. V. Chari, *Annihilators of Verma modules for Kac-Moody Lie algebras*, Invent. Math. **81** (1985), no. 1, 47–58, https://doi.org/10.1007/BF01388771.
10. V. Chari, *Integrable representations of a?ne Lie algebras*, Invent. Math. **85** (1986), no. 2, 317–335, https://doi.org/10.1007/BF01389093.
11. V. Chari, *Minimal affinizations of representations of quantum groups: the rank* 2 *case*, Publ. Res. Inst. Math. Sci. **31** (1995), no. 5, 873–911, https://doi.org/10.2977/prims/1195163722.
12. V. Chari, *On the fermionic formula and the Kirillov-Reshetikhin conjecture*, Internat. Math. Res. Notices **12** (2001), 629–654, https://doi.org/10.1155/S1073792801000332.
13. V. Chari, *Braid group actions and tensor products*, Int. Math. Res. Not. **7** (2002), 357–382, https://doi.org/10.1155/S107379280210612X.
14. V. Chari, G. Fourier, and T. Khandai, *A categorical approach to Weyl modules*, Transform. Groups **15** (2010), no. 3, 517–549, https://doi.org/10.1007/s00031-010-9090-9.
15. V. Chari and J. Greenstein, *Current algebras, highest weight categories and quivers*, Adv. Math. **216** (2007), no. 2, 811–840, https://doi.org/10.1016/j.aim.2007.06.006.
16. V. Chari and J. Greenstein, *Graded level zero integrable representations of affine Lie algebras*, Trans. Amer. Math. Soc. **360** (2008), no. 6, 2923–2940, https://doi.org/10.1090/S0002-9947-07-04394-2.
17. V. Chari and J. Greenstein, *A family of Koszul algebras arising from finite-dimensional representations of simple Lie algebras*, Adv. Math. **220** (2009), no. 4, 1193–1221, https://doi.org/10.1016/j.aim.2008.11.007.
18. V. Chari and J. Greenstein, *Minimal affinizations as projective objects*, J. Geom. Phys. **61** (2011), no. 3, 594–609, https://doi.org/10.1016/j.geomphys.2010.11.008.
19. V. Chari and S. Ilangovan, *On the Harish-Chandra homomorphism for infinite-dimensional Lie algebras*, J. Algebra **90** (1984), no. 2, 476–490, https://doi.org/10.1016/0021-8693(84)90185-6.
20. V. Chari and B. Ion, *BGG reciprocity for current algebras*, Compos. Math. **151** (2015), no. 7, 1265–1287, https://doi.org/10.1112/S0010437X14007908.
21. V. Chari and S. Loktev, *Weyl, Demazure and fusion modules for the current algebra of* $\mathfrak{sl}_{r+1}$, Adv. Math. **207** (2006), no. 2, 928–960, https://doi.org/10.1016/j.aim.2006.01.012.
22. V. Chari and A. Moura, *Characters and blocks for finite-dimensional representations of quantum affine algebras*, Int. Math. Res. Not. **5** (2005), 257–298, https://doi.org/10.1155/IMRN.2005.257.
23. V. Chari and A. Moura, *The restricted Kirillov-Reshetikhin modules for the current and twisted current algebras*, Comm. Math. Phys. **266** (2006), no. 2, 431–454, https://doi.org/10.1007/s00220-006-0032-2.
24. V. Chari, A. Moura, and C. Young, *Prime representations from a homological perspective*, Math. Z. **274** (2013), no. 1-2, 613–645, https://doi.org/10.1007/s00209-012-1088-7.

25. V. Chari and A. Pressley, *New unitary representations of loop groups*, Math. Ann. **275** (1986), no. 1, 87–104, https://doi.org/10.1007/BF01458586.
26. V. Chari and A. Pressley, *A new family of irreducible, integrable modules for affine Lie algebras*, Math. Ann. **277** (1987), no. 3, 543–562, https://doi.org/10.1007/BF01458331.
27. V. Chari and A. Pressley, *Integrable representations of twisted affine Lie algebras*, J. Algebra **113** (1988), no. 2, 438–464, https://doi.org/10.1016/0021-8693(88)90171-8.
28. V. Chari and A. Pressley, *Fundamental representations of Yangians and singularities of* R-*matrices*, J. Reine Angew. Math. **417** (1991), 87–128.
29. V. Chari and A. Pressley, *Quantum affine algebras*, Comm. Math. Phys. **142** (1991), no. 2, 261–283.
30. V. Chari and A. Pressley, *A guide to quantum groups*, Cambridge University Press, Cambridge, 1994.
31. V. Chari and A. Pressley, *Quantum affine algebras and their representations*, Representations of groups (Banff, AB, 1994), CMS Conf. Proc., vol. 16, Amer. Math. Soc., Providence, RI, 1995, pp. 59–78.
32. V. Chari and A. Pressley, *Yangians: their representations and characters*, Acta Appl. Math. **44** (1996), no. 1-2, 39–58, https://doi.org/10.1007/BF00116515. Representations of Lie groups, Lie algebras and their quantum analogues.
33. V. Chari and A. Pressley, *Quantum affine algebras and integrable quantum systems*, Quantum fields and quantum space time (Cargèse, 1996), NATO Adv. Sci. Inst. Ser. B Phys., vol. 364, Plenum, New York, 1997, pp. 245–263.
34. V. Chari and A. Pressley, *Weyl modules for classical and quantum affine algebras*, Represent. Theory **5** (2001), 191–223, https://doi.org/10.1090/S1088-4165-01-00115-7.
35. V. Chari and A. Pressley, *Integrable and Weyl modules for quantum affine* sl_2, Quantum groups and Lie theory (Durham, 1999), London Math. Soc. Lecture Note Ser., vol. 290, Cambridge Univ. Press, Cambridge, 2001, pp. 48–62.
36. M. Duflo, *Sur la classification des idéaux primitifs dans l'algèbre enveloppante d'une algèbre de Lie semi-simple*, Ann. of Math. (2) **105** (1977), no. 1, 107–120, https://doi.org/10.2307/1971027.
37. D. Hernandez and B. Leclerc, *Cluster algebras and quantum a?ne algebras*, Duke Math. J. **154** (2010), no. 2, 265–341, https://doi.org/10.1215/00127094-2010-040.
38. S.-J. Kang, M. Kashiwara, and M. Kim, *Symmetric quiver Hecke algebras and R-matrices of quantum affine algebras*, Invent. Math. **211** (2018), no. 2, 591–685, https://doi.org/10.1007/s00222-017-0754-0.
39. M. Kashiwara, M. Kim, S. Oh, and E. Park, *Monoidal categorification and quantum affine algebras*, Compos. Math. **156** (2020), no. 5, 1039–1077, https://doi.org/10.1112/s0010437x20007137.
40. M. Khovanov and A. D. Lauda, *A diagrammatic approach to categorification of quantum groups. I*, Represent. Theory **13** (2009), 309–347, https://doi.org/10.1090/S1088-4165-09-00346-X.
41. A. N. Kirillov and N. Yu. Reshetikhin, *Representations of Yangians and multiplicities of the inclusion of the irreducible components of the tensor product of representations of simple Lie algebras*, Zap. Nauchn. Sem. Leningrad. Otdel. Mat. Inst. Steklov. (LOMI) **160** (1987), no. Anal. Teor. Chisel i Teor. Funktsiĭ. 8, 211–221, 301, https://doi.org/10.1007/BF02342935.
42. R. Rouquier, *2-Kac-Moody algebras*, available at arXiv:0812.5023.

Steinberg Groups for Jordan Pairs: An Introduction with Open Problems

Erhard Neher

Dedicated to Vyjayanthi Chari on the occasion of her 60th birthday

Abstract The paper gives an introduction to Steinberg groups for Jordan pairs, a theory developed in the book recent book by Ottmar Loos and the author.

1 Introduction

The connection between Jordan structures (Jordan algebras, Jordan pairs) and Lie algebras and groups has a long and successful history, starting with the work of Chevalley–Schafer [3] and continued by Jacobson [8–10], Kantor [13–15], Koecher [17–20], Loos [24, 25, 27], Springer [40], Springer–Veldkamp [41] and Tits [46, 47].

The book [30] by Loos and the author is a further contribution to the theme "Groups and Jordan Structures". It contains a detailed study of Steinberg groups associated with certain types of Jordan pairs. These groups generalize the classical linear and unitary Steinberg groups of a ring by, roughly speaking, replacing associative coordinates with Jordan algebras or Jordan pairs. We are able to prove the basic results on Steinberg groups (central closedness, universal central extension in the stable case) in our setting, thereby recovering all previous results, except those on groups of type E_8, F_4 and G_2, and in addition deal with new types, not considered before. The main novelty, however, is our approach based on 3-graded root systems and Jordan pairs.

The present paper is an introduction to the theory developed in [30]. In Sect. 2 we describe the linear Steinberg group $\mathrm{St}(A)$ of a ring A from the point of view of Jordan pairs. This is motivation for Sect. 3 where we define the Steinberg group of

E. Neher (✉)
Department of Mathematics and Statistics, University of Ottawa, Ottawa, ON, Canada
e-mail: neher@uottawa.ca

J. Greenstein et al. (eds.), *Interactions of Quantum Affine Algebras with Cluster Algebras, Current Algebras and Categorification*, Progress in Mathematics 337,
https://doi.org/10.1007/978-3-030-63849-8_4

a root graded Jordan pair and state the main results of [30] regarding these groups. The final Sect. 4 discusses some open research problems in the area of Steinberg groups and Jordan pairs.

The paper does not assume any prior knowledge of linear Steinberg groups or Jordan pairs: all relevant definitions are given in the paper. We demonstrate their scope by many examples and refer the reader to [29] and [30] for most proofs. But we include the details of our discussion of the linear Steinberg group and the elementary linear group of a ring from the point of view of Jordan theory (Sects. 2.7–2.10 and 2.12, respectively). We also give all details of our description of the Tits–Kantor–Koecher algebra and the projective elementary group of a rectangular Jordan pair (Sects. 3.12 and 3.13).

Notation Throughout k is a unital commutative associative ring and A is a not necessarily commutative, but unital associative k-algebra. Its identity element and zero element are denoted by 1_A and 0_A, respectively. We will often simply write 1 for 1_A if A is clear from the context, and analogously for $0 \in A$. We use $A^\times$ to denote the invertible elements of A. If $k = \mathbb{Z}$, we will refer to A as a ring.

For non-empty sets I and J, we denote by $\mathrm{Mat}_{IJ}(A)$ the k-module of $I \times J$-matrices over A, i.e. maps $x : I \times J \to A$ with only finitely many values different from 0. As usual, we write a matrix in the form $x = (x_{ij})_{(i,j)\in I\times J}$. In case $I = J$ we abbreviate $\mathrm{Mat}_I(A) = \mathrm{Mat}_{IJ}(A)$. This is an associative k-algebra with respect to ordinary matrix multiplication, which is unital if and only if I is finite. We put $\mathrm{Mat}_n(A) = \mathrm{Mat}_I(A)$ if $|I| = n < \infty$. Here and in general $|I|$ denotes the cardinality of the set I. The identity element of $\mathrm{Mat}_n(A)$ is denoted by $\mathbf{1}_n$, and the group $\mathrm{Mat}_n(A)^\times$ by $\mathrm{GL}_n(A)$. The group commutator of elements g, h in a group G is $(\!(g, h)\!) = ghg^{-1}h^{-1}$.

2 Elementary Linear Groups and Their Steinberg Groups

In this section we give an introduction to elementary linear groups over a ring A (Sect. 2.1) and their associated Steinberg groups (Sect. 2.3). After a review of central extensions in Sect. 2.4, we state the Kervaire–Milnor–Steinberg Theorem (2.6), which says, for example, that the stable Steinberg group is the universal central extension of the stable elementary group. We also exhibit a new set of generators and relations for the Steinberg groups considered in this Sect. 2.7–2.9, which we take as axioms for a new Steinberg group defined in Sect. 2.10. The main result is Theorem 2.11: the classical and the new Steinberg groups are isomorphic.

2.1 Elementary Linear Groups

Let $n \in \mathbb{N}$, $n \geqslant 2$. As usual, $E_{ij} \in \mathrm{Mat}_n(A)$ is the $n \times n$-matrix with entry 1_A at the position (ij) and 0_A elsewhere. For $1 \leqslant i \neq j \leqslant n$ and $a \in A$, we put

$$\mathrm{e}_{ij}(a) = \mathbf{1}_n + aE_{ij}, \quad (a \in A).$$

The well-known multiplication rules $aE_{ij}\, bE_{kl} = \delta_{jk}\, abE_{il}$ for $a, b \in A$ imply

$$\mathrm{e}_{ij}(a)\,\mathrm{e}_{ij}(b) = \mathrm{e}_{ij}(a+b). \tag{E1}$$

Hence $\mathrm{e}_{ij}(a)\,\mathrm{e}_{ij}(-a) = \mathbf{1}_n = \mathrm{e}_{ij}(-a)\,\mathrm{e}_{ij}(a)$, and so $\mathrm{e}_{ij}(a) \in \mathrm{GL}_n(A)$. The *elementary linear group (of rank n)* is the subgroup

$$\mathrm{E}_n(A) = \big\langle \mathrm{e}_{ij}(a) : 1 \leqslant i \neq j \leqslant n, a \in A \big\rangle$$

of $\mathrm{Mat}_n(A)^\times$ generated by all $\mathrm{e}_{ij}(a)$.

One easily verifies two further relations of the $\mathrm{e}_{ij}(a)$:

$$(\!(\mathrm{e}_{ij}(a),\ \mathrm{e}_{kl}(b))\!) = \mathbf{1}_n \qquad (j \neq k, i \neq l), \tag{E2}$$

$$(\!(\mathrm{e}_{ij}(a),\ \mathrm{e}_{jl}(b))\!) = \mathrm{e}_{il}(ab) \qquad (i, j, l \neq). \tag{E3}$$

Taking the inverse of (E3) and using $(\!(g, h)\!)^{-1} = (\!(h, g)\!)$ yield the equivalent relation

$$(\!(\mathrm{e}_{ij}(a),\ \mathrm{e}_{ki}(b))\!) = \mathrm{e}_{kj}(-ba) \qquad (i, j, k \neq). \tag{E4}$$

We will also need an infinite variant of $\mathrm{Mat}_n(A)$ and the group $\mathrm{E}_n(A)$. Let

$$\mathrm{Mat}_{\mathbb{N}}(A)$$

be the set of all $\mathbb{N} \times \mathbb{N}$-matrices $x = (x_{ij})_{i,j\in\mathbb{N}}$ with entries from A, but with only finitely many $x_{ij} \neq 0$. The usual addition and multiplication of matrices are well-defined operations on $\mathrm{Mat}_{\mathbb{N}}(A)$ satisfying all axioms of a ring, except the existence of an identity element. To remedy this, let $\mathbf{1}_{\mathbb{N}} = \mathrm{diag}(1_A, 1_A, \ldots)$ be the diagonal matrix of size $\mathbb{N} \times \mathbb{N}$ with every diagonal entry being 1_A. Then,

$$\mathrm{Mat}_{\mathbb{N}}(A)_{\mathrm{ex}} := k\mathbf{1}_{\mathbb{N}} + \mathrm{Mat}_{\mathbb{N}}(A)$$

is a ring with the usual addition and multiplication of matrices. Its identity element is $\mathbf{1}_{\mathbb{N}}$ and its zero element is the zero matrix, see for example [5, 1.2B] where this ring is denoted by $\mathrm{Mat}_\infty(A)$ (its elements are the $\mathbb{N} \times \mathbb{N}$-matrices with entries from A that have only finitely many non-zero entries off the diagonal and whose diagonal elements become eventually constant).

We associate with $x \in \mathrm{Mat}_n(A)$ the matrix $\iota_n(x) \in \mathrm{Mat}_{\mathbb{N}}(A)_{\mathrm{ex}}$ by putting x in the upper-left corner and filling the diagonal outside x with 1_A:

$$\iota_n(x) = \begin{pmatrix} x & 0 \\ 0 & \mathrm{diag}(1_A, \dots) \end{pmatrix}.$$

Then ι_n maps invertible matrices in $\mathrm{Mat}_n(A)$ to invertible matrices of $\mathrm{Mat}_{\mathbb{N}}(A)_{\mathrm{ex}}$, in particular $\iota_n(\mathrm{e}_{ij}(a)) \in \mathrm{Mat}_{\mathbb{N}}(A)^{\times}_{\mathrm{ex}}$. Since $\iota_n(\mathrm{e}_{ij}(a)) = \iota_p(\mathrm{e}_{ij}(a))$ for $p \geqslant n$, we can take the maps ι_n as identification and view all $\mathrm{e}_{ij}(a)$, $i, j \in \mathbb{N}$ with $i \neq j$, as elements of $\mathrm{Mat}_{\mathbb{N}}(A)^{\times}_{\mathrm{ex}}$. The *(stable) elementary linear group* is the subgroup $\mathrm{E}(A)$ of $\mathrm{Mat}_{\mathbb{N}}(A)^{\times}_{\mathrm{ex}}$ generated by all the $\mathrm{e}_{ij}(a)$:

$$\mathrm{E}(A) = \langle \mathrm{e}_{ij}(a) : i, j \in \mathbb{N}, i \neq j \rangle.$$

It is immediate that the relations (E1)–(E4) also hold in $\mathrm{E}(A)$. The group $\mathrm{E}(A)$ is canonically isomorphic to the limit of the inductive system $(\mathrm{E}_n(A), \iota_{pn})$ where $\iota_{pn}\colon \mathrm{E}_n(A) \to \mathrm{E}_p(A)$ for $p \geqslant n$ is defined by taking the upper-left $(p \times p)$-corner of $\iota_n(x)$.

2.2 Why Is $\mathrm{E}_n(A)$ Important?

One reason is that $\mathrm{E}_n(F) = \mathrm{SL}_n(F)$ in case of $A = F$ is a field—in other words, every matrix of determinant 1 can be reduced to the identity matrix by elementary row and column reductions. The equality $\mathrm{E}_n(A) = \mathrm{SL}_n(A)$ even holds for a noncommutative local ring, for example a division ring, if one uses the Dieudonné determinant ([5, 2.2.2] or [38, 2.2.2–2.2.6]). If A is commutative, then obviously $\mathrm{E}_n(A) \subset \mathrm{SL}_n(A)$. Equality holds for example if A is a Euclidean ring [5, 1.2.11].

While all of this is interesting, the real interest in $\mathrm{E}_n(A)$ and $\mathrm{E}(A)$ stems from their connection to Steinberg groups of A and to the K-group $K_2(A)$, defined in (2.3.2).

2.3 The Steinberg Groups $\mathrm{St}_n(A)$ and $\mathrm{St}(A)$

We assume $n \in \mathbb{N}$, $n \geqslant 3$ (the case $n = 2$ is uninteresting since then the definitions below yield free products of A. The group $\mathrm{St}_2(A)$ is defined differently, see e.g. [33]; it will not play a role in this paper).

The *Steinberg group* $\mathrm{St}_n(A)$ is the group presented by

- generators $\mathrm{x}_{ij}(a)$, $1 \leqslant i \neq j \leqslant n$ and $a \in A$, and
- relations (E1)–(E3) of Sect. 2.1:

$$\mathrm{x}_{ij}(a)\,\mathrm{x}_{ij}(b) = \mathrm{x}_{ij}(a+b) \quad \text{for all } 1 \leqslant i \neq j \leqslant n \text{ and } a, b \in A,$$
$$(\!(\mathrm{x}_{ij}(a), \mathrm{x}_{kl}(b))\!) = 1 \quad \text{if } j \neq k \text{ and } l \neq i,$$
$$(\!(\mathrm{x}_{ij}(a), \mathrm{x}_{jl}(b))\!) = \mathrm{x}_{il}(ab) \quad \text{if } i, j, l \neq .$$

The *(stable) Steinberg group* $\mathrm{St}(A)$ is the group presented by

- generators $\mathrm{x}_{ij}(a)$, $i, j \in \mathbb{N}$ distinct, $a \in A$, and
- relations (E1)–(E3) for $i, j \in \mathbb{N}$.

Since the defining relations (E1)–(E3) hold in $\mathrm{E}_n(A)$ and $\mathrm{E}(A)$, we get surjective group homomorphisms

$$\wp_n\colon \mathrm{St}_n(A) \to \mathrm{E}_n(A) \qquad \text{and} \qquad \wp\colon \mathrm{St}(A) \to \mathrm{E}(A) \tag{2.3.1}$$

determined by $\mathrm{x}_{ij}(a) \mapsto \mathrm{e}_{ij}(a)$. The second K-group of A is then defined as

$$K_2(A) := \mathrm{Ker}(\wp). \tag{2.3.2}$$

This is an important but also mysterious group, even for fields. The reader can find more about this group in the classic [33], and in [5, Ch. 1], [32, Parts IV and V], [38, Ch. 4] or [49, III] (the list is incomplete).

To put all of this in a bigger picture, we make a short detour on central extensions of groups.

2.4 Central Extensions

Let G be a group. An *extension of* G is a surjective group homomorphism $p\colon E \to G$. An extension is called *central* if $\mathrm{Ker}(p)$ is contained in the centre of the group G. A central extension $q\colon X \to G$ is a *universal central extension* if for all central extensions $p\colon E \to G$ there exists a unique homomorphism $f\colon X \to E$ such that $q = p \circ f$:

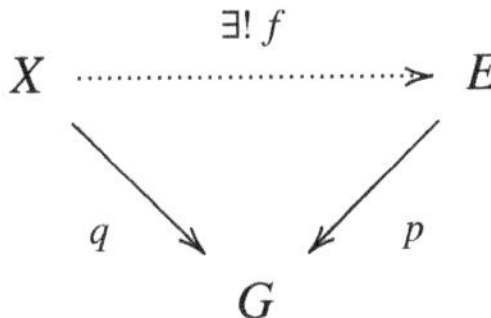

A group X is called *centrally closed* if $\mathrm{Id}_X\colon X \to X$ is a universal central extension. Thus, X is centrally closed if and only if every central extension $p\colon E \to X$ splits uniquely in the sense that there exists a unique group homomorphism $f\colon X \to E$ satisfying $p \circ f = \mathrm{Id}_X$. The concepts defined above are related by the following facts, proved for example in [5, 1.4C], [33, §5], [38, 4.1] or [43, §7].

(a) A group G has a universal central extension if and only if it is *perfect*, i.e. generated by all commutator $((g, h))$ of $g, h \in G$. In particular, a centrally closed group is perfect.
(b) For two universal central extensions of a group G, say $q: X \to G$ and $q': X' \to G$, there exists an isomorphism $f: X \to X'$ of groups, uniquely determined by the condition $q = q' \circ f$.
(c) Let $q: X \to G$ be a universal central extension, whence G is perfect by (a). Then X is centrally closed and thus also perfect, again by (a). Obviously, G is a central quotient of X. The following fact (d) says that every universal central extension of G is obtained as a central quotient of a centrally closed group.
(d) A surjective group homomorphism $q: X \to G$ is a universal central extension if and only if (i) X is centrally closed and (ii) $\mathrm{Ker}(q)$ is central.
(e) Let $f: X \to G$ and $g: G \to \bar{G}$ be central extensions. Then f is a universal central extension if and only if $g \circ f$ is a universal central extension.

To describe a universal central extension of a group G, we have, by (d) and (e), two approaches:

(I) Find successive central extensions $G_1 \to G_0 = G, \ldots, G_n \to G_{n-1}, \ldots$ until one of them, say $G_n \to G_{n-1}$, becomes universal, and then take the composition $G_n \to G$ of these central extensions, or
(II) find an extension $q: X \to G$ with X centrally closed and then find conditions for $\mathrm{Ker}(q)$ to be central.

Although (I) sees to be the more natural approach, we will actually follow (II). But first back to Steinberg groups.

In [42] Steinberg gave a very elegant description of the universal central extension of a perfect Chevalley group over a field. "Most" Chevalley groups are perfect by [43, Lemma 32]. In particular, for $n \geqslant 2$ and F a field, the group $\mathrm{SL}_n(F) = \mathrm{E}_n(F)$ (equality by Sect. 2.2) is a Chevalley group, and it is perfect if $n \geqslant 4$ or if $n = 3$ and $|F| \geqslant 3$ or if $n = 2$ and $|F| \geqslant 4$. A special case of Steinberg's result in [42] is the following theorem.

Theorem 2.5 ([42, 43, Thm. 10],[44, Thm. 1.1]) *Let $n \in \mathbb{N}$, $n \geqslant 2$, and let F be a field satisfying $|F| > 4$ if $n = 3$ and $|F| \notin \{2, 3, 4, 9\}$ if $n = 2$. Then the map $\wp_n : \mathrm{St}_n(F) \to \mathrm{E}_n(F)$ of (2.3.1) is a universal central extension.*

We have defined the maps $\wp_n : \mathrm{St}_n(A) \to \mathrm{E}_n(A)$ and $\wp : \mathrm{St}(A) \to \mathrm{E}(A)$ for any ring A. It is therefore natural to ask if Theorem 2.5 even holds for rings. An answer is given in the following Kervaire–Milnor–Steinberg Theorem.

Kervaire-Milnor-Steinberg Theorem 2.6 ([16, 33, 43]) *For an arbitrary ring A,*

(a) the group $\mathrm{St}_n(A)$, $n \geqslant 5$, is centrally closed.
(b) The map $\wp : \mathrm{St}(A) \to \mathrm{E}(A)$ is a universal central extension. □

An indication of the proof of *(a)* can be found in [5, 1.4.12] or [43, Cor. 1]. The attribution of part *(b)* of this theorem is somewhat complicated. Kervaire cites a preliminary version of [33], Milnor attributes it to Steinberg and Kervaire [33,

p. 43] and Steinberg says that *(b)*, proved in [43, Thm. 14], is "based in part on a letter from J. Milnor".

In view of *(a)*, the map $\wp_n\colon \mathrm{St}_n(A) \to \mathrm{E}_n(A)$, $n \geqslant 5$, is a universal central extension if and only if it is a central extension. It is known that this is not always the case, see [5, 4.2.20]. However, by Sect. 2.4(c), both $\mathrm{St}_n(A)$ and $\mathrm{St}(A)$ are centrally closed. It is this result that we will be concentrating on, following the strategy Sect. 2.4(II).

2.7 *Another Look at* $\mathrm{St}_n(A)$ *and* $\mathrm{St}(A)$*: Using Root Systems*

To treat $\mathrm{St}_n(A)$, $n \in \mathbb{N}$, $n \geqslant 3$ and $\mathrm{St}(A)$ at the same time, we use the subset $N \subset \mathbb{N}$, which is the finite integer interval $N = [1, n]$ or $N = \mathbb{N}$. We can then put

$$\mathrm{St}_N(A) = \begin{cases} \mathrm{St}_n(A) & \text{if } N = [1, n], \\ \mathrm{St}(A) & \text{if } N = \mathbb{N}. \end{cases}$$

By definition in Sect. 2.3, $\mathrm{St}_N(A)$ is generated by $\mathrm{x}_{ij}(a)$, $(i, j) \in N \times N$, $i \neq j$ and $a \in A$. We replace this indexing set by the root system $\dot{\mathrm{A}}_N$ (notation of Sect. 3.2), which we realize in the Euclidean space $X = \bigoplus_{i\in N} \mathbb{R}\varepsilon_i$ with basis $(\varepsilon_i)_{i\in N}$ and inner product $(\,|\,)$ defined by $(\varepsilon_i|\varepsilon_j) = \delta_{ij}$:

$$R = \dot{\mathrm{A}}_N = \{\varepsilon_i - \varepsilon_j : i, j \in N\}, \qquad R^\times = R \setminus \{0\}.$$

Thus $R^\times = \mathrm{A}_{n-1}$ for $N = [1, n]$ in the traditional notation, while for $N = \mathbb{N}$ we get an infinite locally finite root system—a concept that we will review later in Sect. 3.1. For the moment, it suffices to use the concretely given R above.

For $\alpha, \beta \in R$, one easily checks that $(\alpha|\beta) \in \{0, \pm1, \pm2\}$ with $(\alpha|\beta) = \pm2 \iff \alpha = \pm\beta$. To conveniently describe the remaining cases, we use the symbols

$$\alpha \perp \beta \iff (\alpha|\beta) = 0 \qquad \text{and} \qquad \alpha \text{ — } \beta \iff (\alpha|\beta) = 1. \tag{2.7.3}$$

A straightforward analysis of the indices shows for $\alpha = \varepsilon_i - \varepsilon_j$ and $\beta = \varepsilon_k - \varepsilon_l \in R$ that

$$\begin{aligned} \alpha \perp \beta \text{ or } \alpha \text{ — } \beta \quad &\iff \quad j \neq k \text{ and } l \neq i, \\ \alpha \text{ — } (-\beta) \quad &\iff \quad j = k, i, j, l \neq \text{ or } i = l, i, j, k \neq . \end{aligned} \tag{2.7.4}$$

Hence, putting

$$\mathrm{x}_\alpha(a) = \mathrm{x}_{ij}(a) \qquad \text{for } \alpha = \varepsilon_i - \varepsilon_j \in R^\times,$$

the relations (E1)–(E4) can be rewritten in terms of roots as follows. Let $\alpha, \beta \in R^\times$ and denote the relations corresponding to (Ei) by (ERi). Then the previous relations read

$$(\!(\mathrm{x}_\alpha(a),\ \mathrm{x}_\alpha(b))\!) = 1, \tag{ER1}$$

$$(\!(\mathrm{x}_\alpha(a),\ \mathrm{x}_\beta(b))\!) = 1, \qquad \text{if } \alpha \perp \beta \text{ or } \alpha \text{ --- } \beta, \tag{ER2}$$

$$(\!(\mathrm{x}_\alpha(a),\ \mathrm{x}_\beta(b))\!) = \mathrm{x}_{\alpha+\beta}(ab) \qquad \text{if } \alpha = \varepsilon_i - \varepsilon_j, \beta = \varepsilon_j - \varepsilon_l, i, j, l \neq, \tag{ER3}$$

$$(\!(\mathrm{x}_\alpha(a),\ \mathrm{x}_\beta(b))\!) = \mathrm{x}_{\alpha+\beta}(-ba) \qquad \text{if } \alpha = \varepsilon_i - \varepsilon_j, \beta = \varepsilon_k - \varepsilon_i, i, j, k \neq . \tag{ER4}$$

In particular, the two cases for α — $(-\beta)$ in (2.7.4) correspond precisely to the relations (ER3) and (ER4).

2.8 *Another Look at* $\mathrm{St}_N(A)$*: Fewer Generators*

We continue with N and R as in Sect. 2.7. In addition we choose a nontrivial partition

$$N = I \dot\cup J, \qquad \emptyset \neq I \neq N,$$

which we fix in the following. It induces a non-trivial partition

$$R = R_1 \dot\cup R_0 \dot\cup R_{-1}, \tag{2.8.5}$$

whose parts are

$$\begin{aligned} R_1 &= \{\varepsilon_i - \varepsilon_j : i \in I,\ j \in J\}, \\ R_{-1} &= \{\varepsilon_j - \varepsilon_i : i \in I, j \in J\} = -R_1, \\ R_0 &= \{\varepsilon_k - \varepsilon_l : (k,l) \in I \times I \text{ or } (k,l) \in J \times J\} = \dot{\mathrm{A}}_I \times \dot{\mathrm{A}}_J. \end{aligned}$$

The partition $R = R_1 \dot\cup R_0 \dot\cup R_{-1}$ will later be seen to be an example of a 3-grading of R, but we do not need this now. Observe that every $\mu \in R_0$ can be written (not uniquely) as $\mu = \alpha - \beta$ with α and $\beta \in R_1$ satisfying α — β. Indeed,

(i) if $\mu = \varepsilon_k - \varepsilon_l$ with $k, l \in I$, then $\mu = (\varepsilon_k - \varepsilon_j) - (\varepsilon_l - \varepsilon_j)$ for any $j \in J$, hence

$$\mathrm{x}_\mu(a) = \mathrm{x}_{kl}(a) = (\!(\mathrm{x}_{kj}(a),\ \mathrm{x}_{jl}(1))\!) = (\!(\mathrm{x}_\alpha(a),\ \mathrm{x}_{-\beta}(1))\!) \tag{2.8.6}$$

by (ER3) for $\alpha = \varepsilon_k - \varepsilon_j$ and $\beta = \varepsilon_l - \varepsilon_j \in R_1$, and

(ii) if $\mu = \varepsilon_k - \varepsilon_l$ with $k, l \in J$, then $\mu = (\varepsilon_i - \varepsilon_l) - (\varepsilon_i - \varepsilon_k)$ for any $i \in J$, hence

$$\mathrm{x}_\mu(a) = \mathrm{x}_{kl}(a) = ((\mathrm{x}_{il}(-a), \mathrm{x}_{ki}(1))) = ((\mathrm{x}_\alpha(-a), \mathrm{x}_{-\beta}(1))) \tag{2.8.7}$$

by (ER4) for $\alpha = \varepsilon_i - \varepsilon_l$ and $\beta = \varepsilon_i - \varepsilon_k \in R_1$.

Equations (2.8.6) and (2.8.7) show that $\mathrm{St}_N(A)$ is already generated by

$$\{\mathrm{x}_\alpha(a) : \alpha \in R_1 \cup R_{-1},\ a \in A\}. \tag{2.8.8}$$

2.9 *Another Look at* $\mathrm{St}_N(A)$*: Fewer Relations*

Our next goal is to rewrite some of the relations (ER1)–(ER4) in terms of the smaller generating set (2.8.8). Each of these relations depends on two roots $\xi, \tau \in R$. Because of $((g, h))^{-1} = ((h, g))$, we only need to consider the relations involving (ξ, τ) lying in one of the following subsets of $R \times R$:

$$R_1 \times R_1, \quad R_{-1} \times R_{-1}, \quad R_1 \times R_{-1}, \quad R_0 \times R_1, \quad R_0 \times R_{-1}, \quad R_0 \times R_0.$$

(a) Case $(\xi, \tau) = (\alpha, \beta) \in R_1 \times R_1$: Given $\alpha, \beta \in R_1$, exactly one of the relations $\alpha = \beta$, $\alpha — \beta$, $\alpha \perp \beta$ holds. Hence only (ER1) and (ER2) apply in this case and yield

$$((\mathrm{x}_\alpha(a), \mathrm{x}_\beta(b))) = 1 \qquad \text{for } \alpha, \beta \in R_1 \text{ and } a, b \in A. \tag{2.9.9}$$

It will now be more convenient to change notation (again) and put for $\alpha = \varepsilon_i - \varepsilon_j \in R_1$ and $u_\alpha = aE_\alpha^+$

$$E_\alpha^+ = E_{ij}, \qquad \mathrm{x}'_+(u_\alpha) = \mathrm{x}_\alpha(a) = \mathrm{x}_{ij}(a) \tag{2.9.10}$$

$$V_\alpha^+ = AE_\alpha^+, \qquad V^+ = \textstyle\bigoplus_{\alpha \in R_1} V_\alpha^+ = \bigoplus_{(ij) \in I \times J} AE_{ij}. \tag{2.9.11}$$

Because of (2.9.9), the map

$$\mathrm{x}'_+ : V^+ \longrightarrow \mathrm{St}_N(A), \quad u = \textstyle\sum_{\alpha \in R_1} u_\alpha \ \mapsto \ \prod_{\alpha \in R_1} \mathrm{x}'_+(u_\alpha) \tag{2.9.12}$$

is well-defined (independent of the chosen order for $\prod_{\alpha \in R_1}$) and satisfies

$$\mathrm{x}'_+(u + u') = \mathrm{x}'_+(u)\, \mathrm{x}'_+(u') \qquad \text{for } u, u' \in V^+. \tag{2.9.13}$$

It is clear that conversely (2.9.13) implies (2.9.9).

(b) Case $(\xi, \tau) = (-\alpha, -\beta) \in R_{-1} \times R_{-1}$: This case is analogous to Case (a). Given $\alpha = \varepsilon_i - \varepsilon_j \in R_1$ and $v_\alpha = aE_\alpha^-$, we define

$$E_\alpha^- = E_{ji}, \qquad \mathrm{x}_-(v_\alpha) = \mathrm{x}_{-\alpha}(-a) = \mathrm{x}_{ji}(-a),$$
$$V_\alpha^- = AE_\alpha^-, \qquad V^- = \bigoplus_{\alpha\in R_1} V_\alpha^- = \bigoplus_{(ji)\in J\times I} AE_{ji}. \tag{2.9.14}$$

(The minus sign in the definition of $\mathrm{x}_-(v_\alpha)$ is not significant but has been included so that the relations below are precisely those used later on. It avoids a minus sign in the formula (2.12.27).) As in Case (a), the relations (ER1)–(ER4) for $(-\alpha, -\beta) \in R_{-1} \times R_{-1}$ yield $(\!(\mathrm{x}'_-(v_\alpha), \mathrm{x}'_-(v'\beta))\!) = 1$ and thus give rise to a well-defined map

$$\mathrm{x}'_- : V^- \longrightarrow \mathrm{St}_N(A), \quad v = \sum_{\alpha\in R_1} a_\alpha E_\alpha^- \mapsto \prod_{\alpha\in R_1} \mathrm{x}'_-(-a_\alpha E_\alpha^-) \tag{2.9.15}$$

satisfying

$$\mathrm{x}'_-(v+v') = \mathrm{x}'_-(v)\,\mathrm{x}'_-(v') \qquad \text{for } v, v' \in V^-. \tag{2.9.16}$$

At this point, we obtain a new generating set of $\mathrm{St}_N(A)$

$$\mathrm{St}_N(A) = \big\langle \mathrm{x}'_+(V^+) \cup \mathrm{x}'_-(V^-)\big\rangle, \tag{2.9.17}$$

(c) Case $(\xi, \tau) = (\alpha, -\beta) \in R_1 \times R_{-1}$: From this case, we will only explicitly keep the relation (ER2), which in our new notation says

$$(\!(\mathrm{x}'_+(u), \mathrm{x}'_-(v))\!) = 1 \qquad \text{for } (u, v) \in V_\alpha^+ \times V_\beta^- \text{ with } \alpha \perp \beta. \tag{2.9.18}$$

In the following Case (d), we use the relations (ER3) and (ER4) for $(\alpha, -\beta) \in R_1 \times R_{-1}$ in double commutators.

(d) Case $(\xi, \tau) = (\mu, \gamma) \in R_0 \times R_1$: To deal with this case, we view the elements of V^+ as $I \times J$-matrices with only finitely many non-zero entries, as in (2.9.10). Similarly, elements in V^- are $J \times I$-matrices with finitely many non-zero entries. Matrix multiplication of matrices in $V^+ \times V^- \times V^+$ is then well-defined and yields the *Jordan triple product* $\{\cdots\}$, i.e. the map

$$\{\cdots\}\colon V^+ \times V^- \times V^+ \longrightarrow V^+, \quad (x, y, x) \mapsto \{x\,y\,z\} := xyz + zyx.$$

We claim that (ER2)–(ER4) imply

$$(\!(\,(\!(\mathrm{x}'_+(u_\alpha), \mathrm{x}'_-(v_\beta))\!), \mathrm{x}'_+(z_\gamma))\!) = \mathrm{x}'_+(-\{u_\alpha\, v_\beta\, z_\gamma\})$$
$$\text{for } \alpha, \beta, \gamma \in R_1 \text{ with } \alpha \,—\, \beta \text{ and all } u_\alpha \in V_\alpha^+, v_\beta \in V_\beta^-, z_\gamma \in V_\gamma^+. \tag{2.9.19}$$

We prove this by evaluating all possibilities for $\mu = \alpha - \beta$ with $\alpha, \beta \in R_1$ satisfying $\alpha \,—\, \beta$ and $\gamma = \varepsilon_r - \varepsilon_s \in R_1$. By Sects. 2.8(i) and 2.8(ii), there are two cases for such a representation of μ, discussed below as (I) and (II).

(I) $\alpha = \varepsilon_i - \varepsilon_j, \beta = \varepsilon_i - \varepsilon_l$ for $i \in I$ and $j, l \in J$ distinct. Thus $\mu = \alpha - \beta = \varepsilon_l - \varepsilon_j$. We let $u_\alpha = aE_{ij}, v_\beta = bE_{li}, z_\gamma = cE_{rs}$. Then, by (ER4)–(E4),

$$\begin{aligned}(\!(\,(\!(x'_+(u_\alpha),\, x'_-(v_\beta))\!),\, x'_+(z_\gamma))\!) &= (\!(\,(\!(x_{ij}(a),\, x_{li}(-b))\!),\, x_{rs}(c))\!) \\ &= (\!(x_{lj}(ba),\, x_{rs}(c))\!) =: A,\end{aligned}$$

$$\{u_\alpha\, v_\beta\, z_\gamma\} = \{aE_{ij}\, bE_{li}\, cE_{rs}\} = \delta_{sl}\, cbaE_{rj} =: B.$$

If $l = s$, then, again by (E4), $A = x_{rj}(-cba)$, while $B = cba\, E_{rj}$. Otherwise $l \neq s$, whence $A = 1$ by (E2) and clearly $B = 0$. This finishes the proof of (2.9.19) in case (I).

(II) $\alpha = \varepsilon_i - \varepsilon_j, \beta = \varepsilon_k - \varepsilon_j$ for distinct $i, k \in I$ and $j \in J$. This can be shown in the same way as (I).

To obtain a slightly simpler version of (2.9.19), we apply the commutator formula

$$(\!(g,\, h_1 h_2)\!) = (\!(g,\, h_1)\!) \cdot (\!(g,\, h_2)\!) \cdot (\!(\,(\!(h_2,\, g)\!),\, h_1)\!)$$

with $g = (\!(x'_+(u_\alpha),\, x'_-(v_\beta))\!)$, $h_1 = x'_+(z_\gamma)$ and $h_2 = x'_+(z_\delta)$ for arbitrary $\delta \in R_1$. We obtain $(\!(g,\, h_1 h_2)\!) = (\!(g,\, h_1)\!) \cdot (\!(g,\, h_2)\!)$, which allows us to rewrite (2.9.19) in the form

$$(\!(\,(\!(x'_+(u_\alpha),\, x'_-(v_\beta))\!),\, x'_+(z))\!) = x'_+(-\{u_\alpha\, v_\beta\, z\})$$
$$\text{for } \alpha, \beta \in R_1 \text{ with } \alpha \text{ — } \beta \text{ and arbitrary } u_\alpha \in V^+_\alpha,\, v_\beta \in V^-_\beta,\, z \in V^+. \tag{2.9.20}$$

(e) Case $(\xi, \tau) = (\mu, -\gamma) \in R_0 \times R_{-1}$. We proceed as in Case (d) and define the Jordan triple product

$$\{\cdots\}\colon V^- \times V^+ \times V^- \longrightarrow V^-, \quad (x, y, x) \mapsto \{x\, y\, z\} := xyz + zyx,$$

using matrix multiplication in the definition of $\{\cdots\}$. As in Case (d), one then proves the relation

$$(\!(\,(\!(x'_-(v_\alpha),\, x'_+(u_\beta))\!),\, x'_-(w))\!) = x'_-(-\{v_\alpha\, u_\beta\, w\})$$
$$\text{for } \alpha, \beta \in R_1 \text{ with } \alpha \text{ — } \beta \text{ and arbitrary } v_\alpha \in V^-_\alpha,\, u_\beta \in V^+_\beta,\, w \in V^-. \tag{2.9.21}$$

(f) Case $(\xi, \tau) \in R_0 \times R_0$: As we will see below, the relations involving these (ξ, τ) are not needed for presenting $\mathrm{St}_N(A)$.

2.10 *The Steinberg Group* $\mathrm{St}(\mathbb{M}_{IJ}(A), \mathfrak{R})$

We keep the setting of (Sects. 2.7–2.9). In Sect. 2.9 we defined a pair of matrix spaces

$$(V^+, V^-) = \big(\mathrm{Mat}_{IJ}(A), \, \mathrm{Mat}_{JI}(A)\big) =: \mathbb{M}_{IJ}(A),$$

and Jordan triple products

$$\{\cdots\}: V^\sigma \times V^{-\sigma} \times V^\sigma \to V^\sigma, \quad (x, y, z) \mapsto \{x\, y\, z\} = xyz + zyx$$

for $\sigma \in \{+, -\}$. In (2.9.11) and (2.9.14) we also introduced a family

$$\mathfrak{R} = (V_\alpha)_{\alpha \in R_1}, \quad V_\alpha = (V_\alpha^+, V_\alpha^-)$$

of pairs of subgroups with the property that $V^\sigma = \bigoplus_{\alpha \in R_1} V_\alpha^\sigma$. Furthermore, in (2.9.17) we found a new generating set for $\mathrm{St}_N(A)$, and we rewrote some of the relations defining $\mathrm{St}_N(A)$ in terms of this new generating set. It is then natural to define a new Steinberg group using these new generators and relations.

The *Steinberg group* $\mathrm{St}(\mathbb{M}_{IJ}(A), \mathfrak{R})$ is the group presented by

- generators $\mathrm{x}_+(u)$, $u \in V^+$ and $\mathrm{x}_-(v)$, $v \in V^-$, and
- the relations (2.9.13), (2.9.16), (2.9.18), (2.9.20) and (2.9.21). Taking $\sigma \in \{+, -\}$, these are

$$\mathrm{x}_\sigma(u + u') = \mathrm{x}_\sigma(u)\, \mathrm{x}_\sigma(u') \quad \text{for } u, u' \in V^\sigma, \tag{EJ1}$$

$$(\!(\mathrm{x}_+(u),\, \mathrm{x}_-(v))\!) = 1 \quad \text{for } (u, v) \in V_\alpha^+ \times V_\beta^-, \alpha \perp \beta, \tag{EJ2}$$

$$(\!(\,(\!(\mathrm{x}_\sigma(u),\, \mathrm{x}_{-\sigma}(v))\!),\, \mathrm{x}_-(z))\!) = \mathrm{x}_\sigma(-\{u\, v\, z\}) \tag{EJ3}$$

$$\text{for } u_\alpha \in V_\alpha^\sigma, v \in V_\beta^{-\sigma}, z \in V^\sigma \text{ with } \alpha \text{ --- } \beta.$$

(The letter "J" in (EJi) stands for "Jordan", to be explained in the next section.)

From the review above, it is clear that we have a surjective homomorphism of groups

$$\Phi\colon \mathrm{St}(\mathbb{M}_{IJ}(A), \mathfrak{R}) \to \mathrm{St}_N(A), \quad \mathrm{x}_\sigma(u) \mapsto \mathrm{x}'_\sigma(u) \tag{2.10.22}$$

where x'_σ is defined in (2.9.12) and (2.9.15). Moreover, composing Φ with the surjective group homomorphisms $\wp_N\colon \mathrm{St}_N(A) \to \mathrm{E}_N(A)$ of (2.3.1) yields another surjective group homomorphism

$$\begin{aligned} \mathfrak{p}_N\colon\ & \mathrm{St}(\mathbb{M}_{IJ}(A), \mathfrak{R}) \to \mathrm{E}_N(A), \\ & \mathrm{x}_+(aE_{ij}) \mapsto \mathrm{e}_{ij}(a), \quad \mathrm{x}_-(aE_{ji}) \mapsto \mathrm{e}_{ji}(-a) \end{aligned} \tag{2.10.23}$$

and hence a commutative diagram

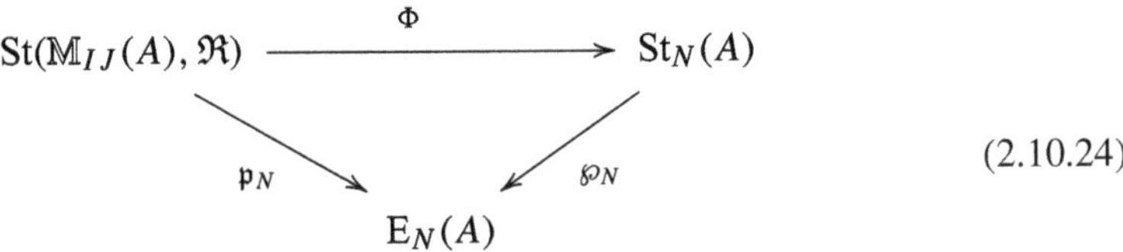

(2.10.24)

Theorem 2.11 ([30]) *The map* Φ *of* (2.10.22) *is an isomorphism of groups.*

In particular, $\mathrm{St}(\mathbb{M}_{IJ}(A), \mathfrak{R})$ *is centrally closed if* $|N| \geqslant 5$ *and* $\mathfrak{p}_N$ *is a universal central extension of* $\mathrm{E}(A)$ *if* $N = \mathbb{N}$.

Proof To prove bijectivity of Φ is bijective, a canonical approach is to show that the family of $\mathrm{x}_+(aE_{ij})$ and $\mathrm{x}_-(-bE_{ij}) \in \mathrm{St}(\mathbb{M}_{IJ}(A), \mathfrak{R})$ can be extended to a family of elements satisfying the defining relations (E1)–(E3) of $\mathrm{St}_N(A)$. As a consequence, this yields a group homomorphism $\Psi\colon \mathrm{St}_N(A) \to \mathrm{St}(\mathbb{M}_{IJ}(A), \mathfrak{R})$ such that $\Psi \circ \Phi$ and $\Phi \circ \Psi$ are the identity on the respective generators and therefore also on the corresponding groups. Another proof of the bijectivity of Φ is given in [30, 24.18], based on the interpretation of both groups as initial objects in an appropriate category of groups mapping onto $\mathrm{E}_N(A)$.

The second part of the theorem follows from the Kervaire–Milnor–Steinberg Theorem 2.6.

2.12 Another Look at $\mathrm{E}_N(A)$

It follows from the existence of the surjective group homomorphism $\mathfrak{p}_N$ of (2.10.23) that $\mathrm{E}_N(A)$ is generated by $\mathfrak{p}_N(\mathrm{x}_+(V^+) \cup \mathrm{x}_-(V^-))$ and that the matrices in this image satisfy the relations (EJ1)–(EJ3). It is instructive to verify this directly.

For $(u, v) \in \mathbb{M}_{IJ}(A)$, we define elements $\mathrm{e}_+(u)$ and $\mathrm{e}_-(v)$ of the ring $\mathrm{Mat}_N(A)_{\mathrm{ex}}$ of Sect. 2.1 by

$$\mathrm{e}_+(u) = \begin{pmatrix} \mathbf{1}_I & u \\ 0 & \mathbf{1}_J \end{pmatrix}, \qquad \mathrm{e}_-(v) = \begin{pmatrix} \mathbf{1}_I & 0 \\ -v & \mathbf{1}_J \end{pmatrix}. \tag{2.12.25}$$

Then clearly

$$\mathrm{e}_+(u + u') = \mathrm{e}_+(u)\,\mathrm{e}_+(u') \quad \text{and} \quad \mathrm{e}_-(v + v') = \mathrm{e}_-(v)\,\mathrm{e}_-(v'). \tag{2.12.26}$$

In particular, the matrices $\mathrm{e}_+(u)$ and $\mathrm{e}_-(v)$ are invertible with $\mathrm{e}_+(u)^{-1} = \mathrm{e}_+(-u)$ and $\mathrm{e}_-(v)^{-1} = \mathrm{e}_-(-v)$. Since $\mathrm{e}_+(aE_{ij}) = \mathrm{e}_{ij}(a)$ and $\mathrm{e}_-(vE_{ji}) = \mathrm{e}_{ji}(-v)$ for $(ij) \in I \times J$, Eqs. (2.12.26) also show that $\mathrm{e}_+(u) \in \mathrm{E}_N(A)$ and $\mathrm{e}_-(v) \in \mathrm{E}_N(A)$. By straightforward matrix multiplication, one obtains

$$((\mathrm{e}_+(u),\, \mathrm{e}_-(v))) = \begin{pmatrix} \mathbf{1}_I - uv + uvuv & uvu \\ vuv & \mathbf{1}_J + vu \end{pmatrix}. \tag{2.12.27}$$

In particular, taking (u, v) with $vu = 0$ or $uv = 0$, this proves

$$\begin{pmatrix} \mathbf{1}_I - uv & 0 \\ 0 & \mathbf{1}_J \end{pmatrix} \in \mathrm{E}_N(A) \quad \text{and} \quad \begin{pmatrix} \mathbf{1}_I & 0 \\ 0 & \mathbf{1}_J + vu \end{pmatrix} \in \mathrm{E}_N(A).$$

Specifying (u, v) even more, one then easily sees that all elementary matrices $\mathrm{e}_{kl}(a)$ with $(k, l) \in I \times I$ or $(k, l) \in J \times J$ lie in the subgroup of $\mathrm{E}_N(A)$ generated by $\mathrm{e}_+(V^+) \cup \mathrm{e}_-(V^-)$. Therefore, this subgroup equals $\mathrm{E}_N(A)$.

The relation (EJ1) is (2.12.26), and the relation (EJ2) follows from (2.12.27) since for $(u, v) \in V_\alpha^+ \times V_\alpha^-$ with $\alpha \perp \beta$ we have $uv = 0$ and $vu = 0$. In order to prove (EJ3) in case $\sigma = +$, let $(u, v) \in V_\alpha^+ \times V_\beta^-$ with $\alpha \perp \beta$ and let $z \in V^+$ arbitrary. Then $uvu = 0 = uvzvu$, $vuv = 0$ and $(\mathbf{1}_J + vu)^{-1} = \mathbf{1}_J - vu$. Hence, by (2.12.27),

$$\begin{aligned} (((\mathrm{e}_+(u),\, \mathrm{e}_-(v))),\, \mathrm{e}_+(z)) &= \left(\begin{pmatrix} \mathbf{1}_I - uv & 0 \\ 0 & (\mathbf{1}_J - vu)^{-1} \end{pmatrix}, \begin{pmatrix} \mathbf{1}_I & z \\ 0 & \mathbf{1}_J \end{pmatrix} \right) \\ &= \begin{pmatrix} \mathbf{1}_I & -z + (1 - uv)z(1 - vu) \\ 0 & \mathbf{1}_J \end{pmatrix} = \mathrm{e}_+(-\{u\, v\, z\}). \end{aligned}$$

The relation (EJ3) for $\sigma = -$ can be verified in the same way.

To put this example in the general setting of the following Sect. 3, we point out that the calculations above are not only valid for matrices of finite or countable size $|N|$ but also hold for N of arbitrary cardinality.

3 Generalizations

In this section we generalize the Steinberg groups considered in Sect. 2. The generalization has a combinatorial aspect, 3-graded root systems and an algebraic aspect, root graded Jordan pairs. They are presented in Sects. 3.1–3.3 and 3.4–3.6, respectively. We define the Steinberg group of a root graded Jordan pair (Sect. 3.7) and state the Jordan pair version of the Kervaire–Milnor–Steinberg Theorem (Sects. 3.8 and 3.11). Since the elementary linear group only makes sense for special Jordan pairs, we replace it by its central quotient, which can be defined for any Jordan pair: the projective elementary group $\mathrm{PE}(V)$ of a Jordan pair V defined in terms of the Tits–Kantor–Koecher algebra $\mathfrak{L}(V)$ (Sect. 3.10). We discuss $\mathfrak{L}(V)$ and $\mathrm{PE}(V)$ for the Jordan pair of rectangular matrices over a ring in Sects. 3.12 and 3.13.

3.1 Locally Finite Root Systems [29]

We use $\langle \cdot, \cdot \rangle$ to denote the canonical pairing between a real vector space X of arbitrary dimension and its dual space X^*, thus $\langle x, \varphi \rangle = \varphi(x)$ for $x \in X$ and $\varphi \in X^*$. If $\varphi \in X^*$ satisfies $\langle \alpha, \varphi \rangle = 2$, we define the *reflection* $s_{\alpha,\varphi} \in \mathrm{GL}(X)$ by

$$s_{\alpha,\varphi}(x) = x - \langle x, \varphi \rangle \alpha.$$

A *locally finite root system* is a pair (R, X) consisting of a real vector space X and a subset $R \subset X$ satisfying the axioms (i)–(iv) below:

(i) R spans X as a real vector space and $0 \in R$,
(ii) for every $\alpha \in R^\times = R \setminus \{0\}$, there exists $\alpha^\vee \in X^*$ such that $\langle \alpha, \alpha^\vee \rangle = 2$ and $s_{\alpha,\alpha^\vee}(R) = R$.
(iii) $\langle \alpha, \beta^\vee \rangle \in \mathbb{Z}$ for all $\alpha, \beta \in R^\times$.
(iv) R is locally finite in the sense that $R \cap Y$ is finite for every finite-dimensional subspace of X.

Locally finite root systems form a category **RS**, in which a morphism $f: (R, X) \to (S, Y)$ is an $\mathbb{R}$-linear map with $f(R) \subset S$. In this category, an isomorphism $f: (R, X) \to (S, Y)$ is a vector space isomorphism $f: X \to Y$ with $f(R) = S$. Such an isomorphism necessarily satisfies $\langle f(\alpha), f(\beta)^\vee \rangle = \langle \alpha, \beta^\vee \rangle$ for all $\alpha, \beta \in R^\times$.

Remarks, Facts and More Definitions

(a) The linear form $\alpha^\vee$ in (ii) is uniquely determined by the two conditions in (ii). Therefore, we simply write s_α instead of $s_{\alpha,\alpha^\vee}$ in the future.
(b) Our standard reference for locally finite root systems is [29]. As in [29], we will also here abbreviate the term "locally finite root system" by *root system*. Then a *finite root system* is a root system (R, X) with R a finite set, equivalently $\dim X < \infty$. Finite root systems are the root systems studied for example in [2, Ch. VI]. That [2] assumes that $0 \notin R$ does not pose any problem in applying the results developed there.
The real vector space X of a root system (R, X) is usually not important. We will therefore often just refer to R rather than to (R, X) as a root system.
(c) As in [29] and again in [30, §2], we assume here that $0 \in R$, which is more natural from a categorical point of view. In [30, §2] the real vector space X is replaced by a free abelian group X and condition (iv) becomes that $R \cap Y$ be finite for every finitely generated subgroup Y of X. With the obvious concept of a morphism, this defines a category of root systems over the integers, which is equivalent to the category **RS** [30, Prop. 2.9].
(d) A locally finite root system need not be *reduced* in the sense that $\mathbb{R}\alpha \cap R = \{\pm\alpha\}$ for every $\alpha \in R^\times$. The *rank* of a root system (R, X) is defined as the dimension of the real vector space X.
(e) The *direct sum* of a family $(R^{(j)}, X^{(j)})_{j \in J}$ of root systems is the pair

$$\left(\bigcup_{j\in J} R^{(j)},\ \bigoplus_{j\in J} X^{(j)}\right),$$

which is again a root system [29, 3.10], traditionally written as $R = \bigoplus_{j\in J} R^{(j)}$.
A non-empty root system is called *irreducible* if it is not isomorphic to a direct sum of two non-empty root systems. Every root system uniquely decomposes as a direct sum of irreducible root systems, called its *irreducible components* [29, 3.13].

(f) Every root system (R, X) admits an inner product $(\,|\,)\colon X \times X \to \mathbb{R}$, which is *invariant* in the sense that $(s_\alpha(x) \mid s_\alpha(y)) = (x \mid y)$ holds for all $\alpha \in R^\times$ and $x, y \in X$, equivalently

$$\langle \beta, \alpha^\vee \rangle = 2\,\frac{(\beta \mid \alpha)}{(\alpha \mid \alpha)} \qquad \text{for all } \alpha, \beta \in R^\times \tag{3.1.1}$$

[29, 4.2]. If R is irreducible, $(\,|\,)$ is unique up to a non-zero scalar. It follows that the definition of a root system given in [35] is equivalent to the definition above and that a finite reduced root system is the same as a "root system" in [7], again up to $0 \notin R$.

3.2 Classification of Root Systems

We first present, as examples, the *classical root systems* $\dot{\mathrm{A}}_I, \dots, \mathrm{BC}_I$. Let I be a set of cardinality $|I| \geqslant 2$, and let $X = \bigoplus_{i\in I} \mathbb{R}\varepsilon_i$ be the $\mathbb{R}$-vector space with basis $(\varepsilon_i)_{i\in I}$. Define

$$\dot{\mathrm{A}}_I = \{\varepsilon_i - \varepsilon_j : i, j \in I\}, \tag{3.2.2}$$

$$\mathrm{D}_I = \dot{\mathrm{A}}_I \cup \{\pm(\varepsilon_i + \varepsilon_j) : i \neq j\}, \tag{3.2.3}$$

$$\mathrm{B}_I = \mathrm{D}_I \cup \{\pm\varepsilon_i : i \in I\}, \tag{3.2.4}$$

$$\mathrm{C}_I = \mathrm{D}_I \cup \{\pm 2\varepsilon_i : i \in I\}, \tag{3.2.5}$$

$$\mathrm{BC}_I = \mathrm{B}_I \cup \mathrm{C}_I. \tag{3.2.6}$$

Then $\dot{\mathrm{A}}_I$ is a root system in $\dot{X} = \mathrm{Ker}(t)$ where $t \in X^*$ is defined by $t(\varepsilon_i) = 1$, $i \in I$. Its rank is therefore $|I| - 1$. The notation $\dot{\mathrm{A}}$ instead of the traditional A is meant to indicate this fact. Observe that $\dot{\mathrm{A}}_{\mathbb{N}}$ is the root system R of Sect. 2.7. All other sets $\mathrm{D}_I, \dots, \mathrm{BC}_I$ are root systems in X, whence of rank $|I|$. The root systems $\dot{\mathrm{A}}_I$, B_I, C_I and D_I are reduced, while BC_I is not.

The isomorphism class of a classical root system only depends on the cardinality of the set I. In particular, when I is finite of cardinality n, we will use the index

n instead of I. Thus, $\mathrm{D}_n = \mathrm{D}_{\{1,\dots,n\}}$ etc. Note $\dot{\mathrm{A}}_{n+1} = \dot{\mathrm{A}}_{\{0,1,\dots,n\}} = \mathrm{A}_n$ in the traditional notation.

The standard inner product $(\,|\,)$, defined by $(\varepsilon_i|\varepsilon_j) = \delta_{ij}$, is an invariant inner product in the sense of Sect. 3.1(f). With the exception of $\mathrm{D}_2 = \mathrm{A}_1 \oplus \mathrm{A}_1$, the root systems $\dot{\mathrm{A}}_I, \dots, \mathrm{BC}_I$ are irreducible. Apart from the low-rank isomorphisms $\mathrm{B}_2 \cong \mathrm{C}_2$, $\mathrm{D}_3 \cong \mathrm{A}_3$, they are pairwise non-isomorphic.

The classification of root systems [29, Thm. 8.4] says that *an irreducible root system is isomorphic either to a classical root system or to an exceptional finite root system.*

3.3 3-Graded Root Systems

A *3-grading* of a root system (R, X) is a partition $R = R_1 \dot\cup R_0 \dot\cup R_{-1}$ satisfying the following conditions (i)–(iii):

(i) $R_{-1} = -R_1$;
(ii) $(R_i + R_j) \cap R \subset R_{i+j}$ for $i, j \in \{1, 0, -1\}$, with the understanding that $R_k = \emptyset$ for $k \notin \{1, 0, -1\}$;
(iii) $R_1 + R_{-1} = R_0$, i.e. every root in R_0 is a difference of two roots in R_1.

In particular (ii) says that the sum of two roots in R_1 is never a root and $(R_1 + R_{-1}) \cap R \subset R_0$, a condition that is strengthened in (iii).

Since a 3-grading of a root system (R, X) is determined by the subset R_1 of R, we will denote a 3-graded root system by (R, R_1, X) or simply by (R, R_1). A *3-graded root system* is a root system equipped with a 3-grading. An *isomorphism* $f : (R, R_1, X) \to (S, S_1, Y)$ between 3-graded root systems is a vector space isomorphism $f : X \to Y$ satisfying $f(R_1) = S_1$, hence also $f(R_i) = S_i$ for $i \in \{1, 0, -1\}$, and is therefore an isomorphism $f : (R, X) \to (S, Y)$ of the underlying root systems. References for 3-graded root systems are [29, §17,§18], [30, Ch. IV] and [35].

Some Facts and Examples

(a) The decomposition (2.8.5) of the root system $R = \dot{\mathrm{A}}_{\mathbb{N}}$ is a 3-grading. The restrictions on N imposed in Sect. 2.7 are not necessary for the definition of a 3-graded root system, as we have seen in Sect. 3.2 for the root system $\dot{\mathrm{A}}_N$. Any non-trivial partition $N = I \dot\cup J$ induces a 3-grading of $\dot{\mathrm{A}}_N$ as in Sect. 2.8, denoted $\dot{\mathrm{A}}_N^I$. Thus, the 1-part of the 3-graded root system $\dot{\mathrm{A}}_N^I$ is

$$(\dot{\mathrm{A}}_N^I)_1 = \{\varepsilon_i - \varepsilon_j : i \in I, j \in N \setminus I\}. \tag{3.3.7}$$

Every 3-grading of $\dot{\mathrm{A}}_N$ is obtained in this way for a non-empty proper subset $I \subset N$.

(b) A 3-grading $\mathrm{B}_I^{\mathrm{qf}}$ of the root system B_I is obtained by choosing a distinguished element of I, say $0 \in I$, and putting $R_1 = \{\varepsilon_0\} \cup \{\varepsilon_0 \pm \varepsilon_i : 0 \neq i \in I\}$.

(c) The root system $R = \mathrm{C}_I$ has a 3-grading, denoted by $\mathrm{C}_I^{\mathrm{her}}$, whose 1-part is $R_1 = \{\varepsilon_i + \varepsilon_j : i, j \in I\}$. Note $R_0 = \{\varepsilon_i - \varepsilon_j : i, j \in I\} \cong \dot{\mathrm{A}}_I$.

(d) The root system D_I, $|I| \geqslant 4$, is a subsystem of B_I and C_I. The 3-gradings of these two root systems, defined in (b) and (c), induce 3-gradings $\mathrm{D}_I^{\mathrm{qf}}$ and $\mathrm{D}_I^{\mathrm{alt}}$ of D_I. The first of these is determined by $R_1 = \{\varepsilon_0 \pm \varepsilon_i : 0 \neq i \in I\}$ and the second by $R_1 = \{\varepsilon_i + \varepsilon_j : i, j \in I, i \neq j\}$. It is known that $\mathrm{D}_I^{\mathrm{qf}} \cong \mathrm{D}_I^{\mathrm{alt}}$ if $|I| = 4$, but $\mathrm{D}_I^{\mathrm{qf}} \not\cong \mathrm{D}_I^{\mathrm{alt}}$ if $|I| \geqslant 5$.

(e) The 3-gradings of a root system R are determined by the 3-gradings of its irreducible components $(R^{(j)})_{j \in J}$ as follows.
If (R, R_1) is a 3-grading, then $(R^{(j)}, R_1 \cap R^{(j)})$ is a 3-grading for every $j \in J$. Conversely, given 3-gradings $(R^{(j)}, R_1^{(j)})$ for every j, the set $R_1 = \bigcup_j R_1^{(j)}$ defines a 3-grading of R.
These easy observations reduce the classification of 3-graded root systems to the case of irreducible root systems. Their classification is given in [29, 17.8, 17.9]. It turns out that an irreducible root system has a 3-grading if and only if it is not isomorphic to E_8, F_4 and G_2. Some irreducible root systems have several non-isomorphic 3-gradings, such as $\dot{\mathrm{A}}_N$ or D_I. But every 3-grading of C_I is isomorphic to the 3-grading $\mathrm{C}_I^{\mathrm{her}}$ of (c).

(f) The relations $\perp$ and $—$ introduced in (2.7.3) in case $R = \dot{\mathrm{A}}_{\mathbb{N}}$ can be defined for any root system R without using an invariant inner product. For $\alpha, \beta \in R^\times$, we put

$$\begin{aligned} \alpha \perp \beta \quad &\Longleftrightarrow \quad \langle \alpha, \beta^\vee \rangle = 0, \quad \text{equivalently} \langle \beta, \alpha^\vee \rangle = 0, \\ \alpha — \beta \quad &\Longleftrightarrow \quad \langle \alpha, \beta^\vee \rangle = 1 = \langle \beta, \alpha^\vee \rangle, \\ \alpha \longrightarrow \beta \quad &\Longleftrightarrow \quad \langle \alpha, \beta^\vee \rangle = 2, \ \langle \beta, \alpha^\vee \rangle = 1. \end{aligned}$$

The formula (3.1.1) shows that the definitions of $\perp$ and $—$ above generalize (2.7.3). The relation $\longrightarrow$ occurs for example in the root system C_I: we have $2\varepsilon_i \longrightarrow \varepsilon_i + \varepsilon_j$ for $\neq j$. (In [29] the notations $\top$ and $\dashv$ are used in place of $—$ and $\longrightarrow$, respectively.)

(g) Given $\alpha, \beta \in R_1$, exactly one of these relations holds:

$$\alpha = \beta \quad \text{or} \quad \alpha \longrightarrow \beta \quad \text{or} \quad \alpha \leftarrow \beta \quad \text{or} \quad \alpha — \beta \quad \text{or} \quad \alpha \perp \beta. \tag{3.3.8}$$

Moreover, again for $\alpha, \beta \in R_1$,

$$2\alpha - \beta \in R \quad \Longleftrightarrow \quad 2\alpha - \beta \in R_1 \quad \Longleftrightarrow \quad \alpha \leftarrow \beta \text{ or } \alpha = \beta. \tag{3.3.9}$$

3.4 Jordan Pairs

This subsection contains a very short introduction to Jordan pairs over a commutative ring k, although for the purpose of this paper $k = \mathbb{Z}$ is completely sufficient. We will only present what is needed to understand this paper. A more detailed but still concise introduction to Jordan pairs is given in [30, §6]; the standard reference for Jordan pairs is [23].

We have already seen an example of a Jordan pair in Sect. 2.9: the rectangular matrix pair $\mathbb{M}_{IJ}(A) = (\mathrm{Mat}_{IJ}(A), \mathrm{Mat}_{JI}(A))$ equipped with the Jordan triple product $\{x\ y\ z\} = xyz+zyx$ for $(x, y, z) \in V^\sigma \times V^{-\sigma} \times V^\sigma$ and $\sigma = \pm$. The Jordan triple product is the linearization with respect to x of the expression $Q(x)y = xyx$, which did not play any role in Sect. 2, but which is the basic structure underlying Jordan pairs.

A *Jordan pair* is a pair $V = (V^+, V^-)$ of k-modules together with maps

$$Q^\sigma : V^\sigma \times V^{-\sigma} \to V^\sigma, \quad (x, y) \mapsto Q^\sigma(x)y, \qquad (\sigma = \pm),$$

which are quadratic in x and linear in y and which satisfy the identities (JP1)–(JP3) below in all base ring extensions. To define these identities, we will simplify the notation and omit σ, thus writing $Q(x)y$ or simply $Q_x y$. This does not lead to any confusion, as long as one takes care that the expressions make sense. Linearizing $Q_x y$ in x gives

$$Q_{x,z}y = Q(x, z)y = Q_{x+z}y - Q_x y - Q_z y,$$

which we use to define the *Jordan triple product*

$$\{\cdots\} : V^\sigma \times V^{-\sigma} \times V^\sigma \to V^\sigma, \quad (x, y, z) \mapsto \{x\ y\ z\} = Q_{x,z}y.$$

To improve readability, we will sometimes write $\{x,\ y,\ z\}$ instead of $\{x\ y\ z\}$. If K is a commutative associative unital k-algebra, we let $V_K^\sigma = V^\sigma \otimes_k K$ and observe that there exist unique extensions of the Q^σ to quadratic-linear maps $Q : V_K \times V_K \to V_K$.

The identities required to hold for $x, z \in V_K^\sigma$, $y, v \in V_K^{-\sigma}$, $\sigma \in \{+, -\}$ and any K as above are

$$\{x,\ y,\ Q_x v\} = Q_x\{y,\ x,\ v\}, \tag{JP1}$$

$$\{Q_x y,\ y,\ z\} = \{x,\ Q_y x,\ z\}, \tag{JP2}$$

$$Q_{Q_x y} v = Q_x\, Q_y\, Q_z v. \tag{JP3}$$

A *homomorphism* $f : V \to W$ of Jordan pairs is a pair $f = (f_+, f_-)$ of k-linear maps $f_\sigma : V^\sigma \to W^\sigma$ satisfying $f_\sigma(Q(x)y) = Q\big(f_\sigma(x)\big) f_{-\sigma}(y)$ for all $(x, y) \in V^\sigma \times V^{-\sigma}$ and $\sigma = \pm$.

Remarks and More Definitions

(a) Instead of requiring that (JP1)–(JP3) hold for all extensions K, one can demand that (JP1)–(JP3) as well as all their linearizations hold in V. For example, linearizing the identity (JP1) with respect to x gives the identity

$$\{z,\ y,\ Q_x v\} + \{x,\ y,\ Q_{x,z} v\} = Q_{x,z}\{y,\ x,\ v\} + Q_x\{y,\ z,\ v\}.$$

(b) If $V = (V^+, V^-)$ is a Jordan pair and $S = (S^+, S^-)$ is a pair of submodules of V satisfying $Q(S^\sigma)S^{-\sigma} \subset S^\sigma$ for $\sigma = \pm$, then S is a Jordan pair with the induced operations, called a *subpair* of V.

(c) An *idempotent* in a Jordan pair V is a pair $e = (e_+, e_-) \in V$ satisfying $Q(e_+)e_- = e_+$ and $Q(e_-)e_+ = e_-$. An idempotent e gives rise to the *Peirce decomposition* of V

$$V^\sigma = V_2^\sigma(e) \oplus V_1^\sigma(e) \oplus V_0^\sigma(e), \qquad \sigma = \pm,$$

where the *Peirce spaces* $V_i(e) = (V_i^+(e), V_i^-(e))$, $i = 0, 1, 2$, are given by

$$\begin{aligned} V_2^\sigma(e) &= \{x \in V^\sigma : Q(e_\sigma)Q(e_{-\sigma})x = x\},\\ V_1^\sigma(e) &= \{x \in V^\sigma : \{e_\sigma\ e_{-\sigma}\ x\} = x\},\\ V_0^\sigma(e) &= \{x \in V^\sigma : Q(e_\sigma)x = 0 = \{e_\sigma\ e_{-\sigma}\ x\}\}. \end{aligned}$$

The Peirce spaces $V_i^\pm = V_i^\pm(e)$ satisfy the multiplication rules

$$Q(V_i^\sigma)V_j^{-\sigma} \subset V_{2i-j}^\sigma, \qquad \{V_i^\sigma\ V_j^{-\sigma}\ V_l^\sigma\} \subset V_{i-j+l}^\sigma, \tag{3.4.10}$$

$$\{V_2^\sigma\ V_0^{-\sigma}\ V^\sigma\} = 0 = \{V_0^\sigma\ V_2^{-\sigma}\ V^\sigma\}, \tag{3.4.11}$$

where $i, j, l \in \{0, 1, 2\}$, with the understanding that $V_m^\sigma = 0$ if $m \notin \{0, 1, 2\}$. In particular, $V_i = V_i(e)$ are subpairs of V. If $2 \in k^\times$, we have $V_i^\sigma(e) = \{x \in V : \{e^\sigma\ e^{-\sigma}\ x\} = ix\}$ for $i = 0, 1, 2$.

3.5 Examples of Jordan Pairs

We now give concrete examples of Jordan pairs, to illustrate the abstract definition of Sect. 3.4.

(i) Any associative, not necessarily unital or commutative k-algebra A gives rise to a Jordan pair $V = (A, A)$ with respect to the operations $Q_x y = xyx$. Indeed, since a base ring extension of A is again associative, it suffices to verify the identities in V, where they follow from the following calculations:

$$\{x, y, Q_x v\} = xy(xvx) + (xvx)yx = x(yxv + vxy)x = Q_x\{y, v, x\},$$

$$\{Q_x y, y, z\} = (xyx)yz + zy(xyx) = x(yxy)z + z(yxy)x = \{x, Q_y x, z\},$$

$$Q_{Q_x y} v = (xyx)v(xyx) = x(y(xvx)y)x = Q_x Q_y Q_x v.$$

An idempotent c of the associative algebra A, defined by $c^2 = c$, induces an associative Peirce decomposition $A = A_{11} \oplus A_{10} \oplus A_{01} \oplus A_{00}$ with $A_{ij} = \{a \in A : ca = ia,\ ac = ja\}$. The pair $e = (c, c)$ is an idempotent of the Jordan pair V whose Peirce spaces are $V_2^\sigma(e) = A_{11}$, $V_1^\sigma(e) = A_{10} \oplus A_{01}$ and $V_0^\sigma(e) = A_{00}$. Not only $V_1(e)$ but also (A_{10}, A_{01}) and (A_{01}, A_{10}) are subpairs of V.

Not every idempotent of V has the form (c, c), c an idempotent of A. For example, if $u \in A^\times$ and c is an idempotent of A, then (uc, cu^{-1}) is an idempotent of V, which, however, has the same Peirce spaces as (c, c).

(ii) By (b) and (i), any pair $(S^+, S^-) \subset (A, A)$ of k-submodules closed under the operation $(x, y) \mapsto xyx$ is also a Jordan pair. Jordan pairs of this form are called *special*. Their Jordan triple product is

$$\{x\,y\,z\} = xyz + zyx \qquad (x, z \in S^\sigma, y \in S^{-\sigma}). \tag{3.5.12}$$

We next describe some important cases of special Jordan pairs.

(iii) Let I and J be non-empty sets. Let $N = I \dot{\cup} J'$ where J' is a set disjoint from I and in bijection with J under $j \mapsto j'$, and embed $\mathrm{Mat}_{IJ}(A)$ into the upper-right corner of the associative algebra $\mathrm{Mat}_N(A)$. Similarly, we identify $\mathrm{Mat}_{JI}(A)$ with the lower-left corner of $\mathrm{Mat}_N(A)$. Then,

$$\mathbb{M}_{IJ}(A) = \big(\mathrm{Mat}_{IJ}(A), \mathrm{Mat}_{JI}(A)\big)$$

is a subpair of the Jordan pair $\big(\mathrm{Mat}_N(A), \mathrm{Mat}_N(A)\big)$ and is therefore a Jordan pair, as claimed at the beginning of this subsection. If N is finite, $\mathbb{M}_{IJ}(A)$ is of type (A_{10}, A_{01}) for $A = \mathrm{Mat}_N(A)$ and the idempotent $c = \mathbf{1}_I$, see (i).

(iv) Let $a \mapsto a^J$ be an involution of the associative k-algebra A. Then $\mathrm{H}(A, J) = \{a \in A : a^J = a\}$ is closed under the Jordan pair product, whence $\mathbb{H}(A, J) = (\mathrm{H}(A, J), \mathrm{H}(A, J))$ is a Jordan pair.

More generally, extend J to an involution of the associative k-algebra $\mathrm{Mat}_I(A)$, $|I| \geqslant 2$, defined by $(x_{ij})^J = (x_{ji}^J)$ and again denoted by J. Then the *Hermitian matrix pair*

$$\mathbb{H}_I(A, J) = \big(\mathrm{H}(\mathrm{Mat}_I(A), J),\ \mathrm{H}(\mathrm{Mat}_I(A), J)\big)$$

is a special Jordan pair. In particular, taking $A = k$, $J = \mathrm{Id}_k$ and $I = \{1, \dots, n\}$, we get the *symmetric matrix pair* $\mathbb{H}_n(k) = (\mathrm{H}_n(k), \mathrm{H}_n(k))$.

(v) Let $\mathrm{Alt}_I(k)$ be the alternating $I \times I$-matrices over k, where $x = (x_{ij})$ is called alternating if $x_{ii} = 0 = x_{ij} + x_{ji}$ for $i, j \in I$. Then the *alternating matrix pair*

$\mathbb{A}_I(k) = (\mathrm{Alt}_I(k), \mathrm{Alt}_I(k))$ is a subpair of $\mathbb{M}_I(k)$, whence a special Jordan pair.

(vi) Let M be a k-module, and let $q: M \to k$ be a quadratic form with polar form b, defined by $b(x, y) = q(x + y) - q(x) - q(y)$. Then $\mathbb{J}(M, q) = (M, M)$ is a Jordan pair with quadratic operators $Q_x y = b(x, y)x - q(x)y$.

(vii) Let J be a unital quadratic Jordan algebra with quadratic operators U_x, $x \in J$ [11, 12]. Then (J, J) is a Jordan pair with quadratic maps $Q_x = U_x$. For example, if k is a field, the rectangular matrix pair $\mathbb{M}_{pp}(k)$ is of this form, but $\mathbb{M}_{pq}(k)$ for $p \neq q$ is not. Thus, there are "more" Jordan pairs than Jordan algebras.

Let $\mathcal{C}$ be an octonion k-algebra, see for example [41] in case k is a field or [31] in general, and let $J = \mathrm{H}_3(\mathcal{C})$ be the exceptional Jordan algebra of 3×3 matrices over $\mathcal{C}$, which are Hermitian with respect to the standard involution of $\mathcal{C}$. Then (J, J) is a Jordan pair, which is not special in the sense of (ii). Such Jordan pairs are called *exceptional*.

3.6 Root Graded Jordan Pairs

Let us first recast the Peirce decomposition Sect. 3.4(c) of an idempotent e in a Jordan pair V from the point of view of a grading.

We use the 3-graded root system $\mathrm{C}_I^{\mathrm{her}}$ of Sect. 3.3(c) with $I = \{0, 1\}$. Its 1-part is $R_1 = \{\varepsilon_i + \varepsilon_j : i, j \in \{0, 1\}\} = \{2\varepsilon_1, \varepsilon_1 + \varepsilon_0, 2\varepsilon_0\}$. Putting

$$V_\alpha^\sigma = V_{i+j}^\sigma(e) \qquad (\alpha = \varepsilon_i + \varepsilon_j \in R_1),$$

we have the decomposition $V^\sigma = \bigoplus_{\alpha \in R_1} V_\alpha^\sigma$, which satisfies

$$Q(V_\alpha^\sigma)V_\beta^{-\sigma} \subset V_{2\alpha-\beta}^\sigma, \qquad \{V_\alpha^\sigma\, V_\beta^{-\sigma}\, V_\gamma^\sigma\} \subset V_{\alpha-\beta+\gamma}^\sigma, \tag{RG1}$$

$$\{V_\alpha^\sigma\, V_\beta^{-\sigma}\, V^\sigma\} = 0 \qquad \text{if } \alpha \perp \beta. \tag{RG2}$$

Here $2\alpha - \beta$ and $\alpha - \beta + \gamma$ in (RG1) are calculated in $X = \mathbb{R} \cdot \varepsilon_0 \oplus \mathbb{R} \cdot \varepsilon_1 \cong \mathbb{R}^2$, and $V_{2\alpha-\beta}^\sigma = 0$ if $2\alpha - \beta \notin R_1$ or $V_{\alpha-\beta+\gamma}^\sigma = 0$ if $\alpha - \beta + \gamma \notin R_1$. We see that, apart from the actual definition of the Peirce spaces, the rules governing the Peirce decomposition can be completely described in terms of R_1. The following generalization is then natural.

Given a 3-graded root system (R, R_1) and a Jordan pair V, an *(R, R_1)-grading of V* is a decomposition $V^\sigma = \bigoplus_{\alpha \in R_1} V_\alpha^\sigma$, $\sigma = \pm$, satisfying (RG1) and (RG2). We will use $\mathfrak{R} = (V_\alpha)_{\alpha \in R_1}$ to denote such a grading. A *root graded Jordan pair* is a Jordan pair equipped with an (R, R_1)-grading for some 3-graded root system. In view of (3.3.9) we can make the first inclusion in (RG1) more precise:

$$\begin{aligned} &Q(V_\alpha^\sigma)V_\beta^{-\sigma} = 0 \quad \textit{unless } \alpha \longrightarrow \beta, \textit{ in which case} \\ &2\alpha - \beta \in R_1 \textit{ and } Q(V_\alpha^\sigma)V_\beta^{-\sigma} \subset V_{2\alpha-\beta}^\sigma. \end{aligned} \tag{3.6.13}$$

We call $\mathfrak{R}$ an *idempotent root grading* if there exist a subset $\Delta \subset R_1$ and a family $(e_\alpha)_{\alpha\in\Delta}$ of non-zero idempotents $e_\alpha \in V_\alpha$ such that V_β are given by

$$V_\beta = \bigcap_{\alpha\in\Delta} V_{\langle\beta,\alpha^\vee\rangle}(e_\alpha). \tag{3.6.14}$$

Observe that (3.6.14) makes sense since $\langle\alpha, \beta^\vee\rangle \in \{0, 1, 2\}$ by (3.3.8). Neither the idempotents nor the subset $\Delta \subset R_1$ is uniquely determined by an idempotent root grading, see for example (iii) below.

To avoid some technicalities, we will often assume that $\mathfrak{R}$ is a *fully idempotent root grading*, i.e. $\mathfrak{R}$ is idempotent with respect to a family of idempotents with $\Delta = R_1$. In the terminology of [34] this means that V is covered by the cog $(e_\alpha)_{\alpha\in R_1}$.

Examples

(i) Let $R = \mathrm{A}_1 = \{\alpha, -\alpha\}$ equipped with the 3-grading defined by $R_1 = \{\alpha\}$. An (R, R_1)-graded Jordan pair is simply a Jordan pair $V = (V^+, V^-)$ for which $V^\sigma = V_\alpha^\sigma$. This root grading is idempotent if and only if $V \cong (J, J)$ where J is a unital Jordan algebra. To see sufficiency in case J is a Jordan algebra with identity element 1_J, one uses $e_\alpha = (1_J, 1_J)$ and observes $(J, J) = V_2(e_\alpha)$.

(ii) Let $(R, R_1) = \mathrm{C}_2^{\mathrm{her}}$. We have seen above that the Peirce decomposition of an idempotent $e \in V$ can be viewed as a $\mathrm{C}_2^{\mathrm{her}}$-grading, which is idempotent with respect to $e = e_\alpha$, $\alpha = 2\varepsilon_1$. Thus here $\Delta = \{\alpha\}$.

(iii) Let $(R, R_1) = \dot{\mathrm{A}}_N^I$ be the 3-graded root system of Sect. 3.3(a). Put $J = N \setminus I$. An $\dot{\mathrm{A}}_I^N$-grading of a Jordan pair V is a decomposition

$$V = \bigoplus\nolimits_{(ij)\in I\times J} V_{(ij)}$$

such that for all (ij) and $(lm) \in I \times J$ and $\sigma = \pm$ we have, defining $V_{(ij)} = V_{\varepsilon_i-\varepsilon_j}$ for $\varepsilon_i - \varepsilon_j \in R_1$,

$$\begin{aligned} Q(V_{(ij)}^\sigma)V_{(ij)}^{-\sigma} &\subset V_{(ij)}^\sigma, & \{V_{(ij)}^\sigma\, V_{(ij)}^{-\sigma}\, V_{(im)}^\sigma\} &\subset V_{(im)}^\sigma, \\ \{V_{(ij)}^\sigma\, V_{(ij)}^{-\sigma}\, V_{(lj)}^\sigma\} &\subset V_{(lj)}^\sigma, & \{V_{(ij)}^\sigma\, V_{(lj)}^{-\sigma}\, V_{(lm)}^\sigma\} &\subset V_{(im)}^\sigma, \end{aligned}$$

and all other types of products vanish.

An example of an $\dot{\mathrm{A}}_I^N$-graded Jordan pair V is the rectangular matrix pair $\mathbb{M}_{IJ}(A)$ of an associative unital k-algebra $A \neq 0$, see Sect. 3.5(iii), with respect to the subpairs $V_{(ij)} = (AE_{ij}, AE_{ji})$. This $\dot{\mathrm{A}}_N^I$-grading of $\mathbb{M}_{IJ}(A)$ is fully idempotent with respect to the family $(e_\alpha)_{\alpha\in R_1}$, $e_\alpha = (a_{ij}E_{ij}, a_{ij}^{-1}E_{ji})$, where $\alpha = \varepsilon_i - \varepsilon_j$ and $a_{ij} \in A^\times$. It is also idempotent with respect to the

following smaller family: fix $i_0 \in I$ and $j_0 \in J$ and consider $(e_\alpha)_{\alpha\in\Delta}$ where $\Delta = \{\varepsilon_i - \varepsilon_{j_0} : i \in I\} \cup \{\varepsilon_{i_0} - \varepsilon_j : j \in J\}$.
Let $\mathfrak{a}, \mathfrak{b} \subset A$ be k-submodules with $\mathfrak{aba} \subset \mathfrak{a}$ and $\mathfrak{bab} \subset \mathfrak{b}$. Then $S = (\mathrm{Mat}_{IJ}(\mathfrak{a}), \mathrm{Mat}_{JI}(\mathfrak{b}))$ is a Jordan subpair of $\mathbb{M}_{IJ}(A)$. It inherits the $\dot{\mathrm{A}}_N^I$-grading from $\mathbb{M}_{IJ}(A)$ by putting $S_{(ij)} = (\mathfrak{a}E_{ij}, \mathfrak{b}E_{ji})$. This $\dot{\mathrm{A}}_N^I$-grading of S is in general not idempotent, e.g. if $\mathfrak{a}$ is a nil ideal.

(iv) The Hermitian matrix pair $V = \mathbb{H}_I(A, J)$ of Sect. 3.5(iv) has an idempotent $\mathrm{C}_I^{\mathrm{her}}$-grading. Indeed, define

$$h_{ij}(a) = \begin{cases} aE_{ij} + a^J E_{ji} & \text{if } a \in A \text{ and } i \neq j, \\ aE_{ii} & \text{if } i = j \text{ and } a \in \mathrm{H}(A, J). \end{cases}$$

Then $V = \bigoplus_{\alpha\in R_1} V_\alpha$ with

$$V_\alpha = V_{\varepsilon_i+\varepsilon_j} = \begin{cases} \big(h_{ij}(A), h_{ij}(A)\big), & \text{if } i \neq j \\ \big(h_{ii}(\mathrm{H}(A, J)), h_{ii}(\mathrm{H}(A, J))\big), & \text{if } i = j \end{cases}$$

is a $\mathrm{C}_I^{\mathrm{her}}$-grading of V. It is fully idempotent, for example with respect to the family $(e_\alpha)_{\alpha\in R_1}$ for which $e_\alpha = (h_{ij}(1), h_{ij}(1))$, $\alpha = \varepsilon_i + \varepsilon_j$.

(v) The remaining examples in Sect. 3.5 all have idempotent root gradings. The alternating matrix pair $\mathbb{A}_I(k)$ of Sect. 3.5(v) has an idempotent $\mathrm{D}_I^{\mathrm{alt}}$-grading [30, 23.24]. The Jordan pair $\mathbb{J}(M, q)$ associated with a quadratic form q in Sect. 3.5(v) has an idempotent $\mathrm{B}_I^{\mathrm{qf}}$ if q contains a hyperbolic plane, or even a $\mathrm{D}_I^{\mathrm{qf}}$-grading if q is hyperbolic [30, 23.25]. If $\mathcal{C}$ is a split octonion algebra in the sense of [41] or [31], the exceptional Jordan pair $(\mathrm{H}_3(\mathcal{C}), \mathrm{H}_3(\mathcal{C}))$ has an idempotent root grading with R of type E_7 [34, III, §3].

3.7 *The Steinberg Group* $\mathrm{St}(V, \mathfrak{R})$

Let (R, R_1) be a 3-graded root system, and let V be a Jordan pair with a root grading $\mathfrak{R} = (V_\alpha)_{\alpha\in R_1}$, not necessarily idempotent. The *Steinberg group* $\mathrm{St}(V, \mathfrak{R})$ is the group with the following presentation:

- The generators are $\mathrm{x}_+(u)$, $u \in V^+$, and $\mathrm{x}_-(v)$, $v \in V^-$.
  To formulate the relations, we first introduce, for $\alpha \neq \beta \in R_1$ and $(u, v) \in V_\alpha^+ \times V_\beta^-$, the element $\mathrm{b}(u, v)$ in the free group with the above generators by the equation

$$\mathrm{x}_+(u)\,\mathrm{x}_-(v) = \mathrm{x}_-(v + Q_v u)\,\mathrm{b}(u, v)\,\mathrm{x}_+(u + Q_u v). \tag{3.7.15}$$

- Then the relations are

$$\mathrm{x}_\sigma(u+u') = \mathrm{x}_\sigma(u)\,\mathrm{x}_\sigma(u') \qquad \text{for } u, u' \in V^\sigma, \tag{St1}$$

$$(\!(\mathrm{x}_+(u),\ \mathrm{x}_-(v))\!) = 1 \qquad \text{for } (u, v) \in V_\alpha^+ \times V_\beta^-, \alpha \perp \beta, \tag{St2}$$

$$\begin{cases} (\!(\mathrm{b}(u, v),\ \mathrm{x}_+(z))\!) = \mathrm{x}_+(-\{u\,v\,z\} + Q_u Q_v z), \\ (\!(\mathrm{b}(u, v)^{-1},\ \mathrm{x}_-(y))\!) = \mathrm{x}_-(-\{v\,u\,y\} + Q_v Q_u y) \end{cases} \tag{St3}$$

for all $(u, v) \in V_\alpha^+ \times V_\beta^-$ with $\alpha \neq \beta$ and all $(z, y) \in V$.

Remarks

(a) Let us have a closer look at the element $\mathrm{b}(u, v)$ in (3.7.15). By (3.3.8), the possibilities for α, β are $\alpha \perp \beta$, $\alpha \text{ — } \beta$, $\alpha \longrightarrow \beta$ and $\alpha \leftarrow \beta$ and by (3.6.13), $Q_v u = 0$ unless $\beta \longrightarrow \alpha$, and $Q_u v = 0$ unless $\alpha \longrightarrow \beta$. Therefore, by (St2),

$$\mathrm{b}(u, v) = \begin{cases} 1, & \text{if } \alpha \perp \beta, \\ (\!(\mathrm{x}_-(-v),\ \mathrm{x}_+(u))\!), & \text{if } \alpha \text{ — } \beta, \\ \mathrm{x}_-(-Q_v u)\ (\!(\mathrm{x}_-(-v),\ \mathrm{x}_+(u))\!), & \text{if } \alpha \longrightarrow \beta, \\ (\!(\mathrm{x}_-(-v),\ \mathrm{x}_+(u))\!)\ \mathrm{x}_+(-Q_u v), & \text{if } \alpha \leftarrow \beta. \end{cases} \tag{3.7.16}$$

In general, the factors $\mathrm{x}_-(-Q_v u)$ and $\mathrm{x}_+(-Q_u v)$ in the last two cases are not equal to 1.

The reader may be puzzled by the definition of $\mathrm{b}(u, v)$: why not take "$\mathrm{b}(u, v) = (\!(\mathrm{x}_-(-v), \mathrm{x}_+(u))\!)$"? We will give a justification for this in Sect. 3.11.

(b) We claim that (St3) follows from (St2) in case $\alpha \perp \beta$. Indeed, the left-hand sides of the two equations (St3) are 1 because $\mathrm{b}(u, v) = 1$, but also the right-hand sides are 1, since, say for $\sigma = +$, we have $\{u\,v\,z\} = 0$ by (RG2) and $Q_u Q_v z = 0$ by (3.7.17) below:

$$Q(V_\alpha^\sigma)Q(V_\beta^{-\sigma})V_\gamma^\sigma \neq 0 \implies \alpha = \beta \text{ or } \alpha \leftarrow \beta = \gamma \\ \text{or } \alpha \text{ — } \beta \leftarrow \gamma \perp \alpha, \tag{3.7.17}$$

which can be shown by repeated application of (3.3.9).

(c) Let $(V, \mathfrak{R}) = (\mathbb{M}_{IJ}(A), \mathfrak{R})$. Comparing the definition of the group $\mathrm{St}(\mathbb{M}_{IJ}(A), \mathfrak{R})$ with the one in Sect. 2.10, it is clear that the first two relations coincide: (EJ1) = (St1) and (EJ2) = (St2). We claim that the relations (EJ3) coincide with the two relations in (St3). Indeed, since we do not have a relation $\alpha \leftarrow \beta$ in $\dot{\mathrm{A}}_N^I$, it follows from (b) and the assumption $\alpha \neq \beta$ in (St3) that we only need to consider the case $\alpha \text{ — } \beta$. Then the map $V_\beta^- \to \mathrm{St}(V, \mathfrak{R})$, $v \mapsto (\!(\mathrm{x}_+(u),\ \mathrm{x}_-(v))\!)$ is homomorphism of groups, whence, by (3.7.16),

$$\mathrm{b}(u, v) = (\!(\mathrm{x}_-(-v),\ \mathrm{x}_+(u))\!) = (\!(\mathrm{x}_+(u),\ \mathrm{x}_-(-v))\!)^{-1} = (\!(\mathrm{x}_+(u),\ \mathrm{x}_-(v))\!).$$

Thus (EJ3) = (St3) for $\sigma = +$ because $Q_u Q_x z = 0$ by (3.7.17). The equality of the two relations for $\sigma = -$ can be established in the same way.

We can now state the generalization of part *(a)* of the Kervaire–Milnor–Steinberg Theorem 2.6 in the setting of this section.

3.8 Theorem A

Let (R, R_1) *be a 3-graded root system whose irreducible components all have rank* $\geqslant 5$, *and let* V *be a Jordan pair with a fully idempotent root grading* $\mathfrak{R}$. *Then the Steinberg group* $\mathrm{St}(V, \mathfrak{R})$ *is centrally closed.*

This theorem is one of the main results of [30]; its proof takes up all of Chapter VI of [30]. It is shown there in greater generality. First, it also true when R has connected components of rank 4, but not of type D_4. Moreover, it is not necessary to assume that the root grading $\mathfrak{R}$ is fully idempotent. For the irreducible components of type B_I or C_I, $|I| \geqslant 5$, one only needs idempotents $e_\alpha \in V_\alpha$ in case α is a long root in type B and α a short root in type C. This generality allows us to consider groups defined in terms of Hermitian matrices associated with form rings in the sense of [1].

3.9 Highlights of Our Approach

The novel aspect of our approach is the consistent use of the theory of 3-graded root systems and Jordan pairs, which introduces new methods in the theory of elementary and Steinberg groups. For example, instead of first dealing with the case of finite root systems and then taking a limit to get the stable (= infinite rank) case, we deal with both cases at the same time. Moreover, our approach avoids having to deal with concrete matrix realizations of the groups in question, as is traditionally done, see e.g. [1] or [5]. It allows for a concise description of the defining relations, independent of the types of root systems involved. Finally, as the discussion of the linear Steinberg group in Sects. 2.9–2.11 shows, we need fewer relations than in previous works, for example no relations involving two roots in R_0.

With the exception of groups defined in terms of root systems of type E_8, F_4 and G_2, which are not amenable to a Jordan approach, cf. Sect. 3.3(e), our Theorem Sect. 3.8 covers all types of Steinberg groups considered before. In addition, it also presents some new types, e.g. for elementary orthogonal groups. A detailed comparison of our Theorem Sect. 3.8 with previously known results is given in [30, 27.11].

At this point, it is natural to ask if there also exists a generalization of part *(b)* of Theorem 2.6, stating that the map $\wp\colon \mathrm{St}(A) \to \mathrm{E}(A)$ is a universal central extension. While the group $\mathrm{St}(V, \mathfrak{R})$ gives a satisfactory replacement for the linear

Steinberg group $\mathrm{St}(A)$, recasting the elementary linear group $\mathrm{E}(A)$ in the framework of Jordan pairs is limited to special Jordan pairs in the sense of Sect. 3.5(ii). While this can be done, see [26], we will instead replace the elementary group $\mathrm{E}(A)$ by the projective elementary group $\mathrm{PE}(V)$, see Sect. 3.10, that can be defined for any Jordan pair V. From the point of view of universal central extensions, this is harmless since, as we will see in Sect. 3.13, the group $\mathrm{PE}(V)$ is isomorphic to the central quotient $\mathrm{PE}(A) = \mathrm{E}(A)/\mathcal{Z}(\mathrm{E}(A))$ and universal central extensions of a group and its central quotients are essentially the same by Sect. 2.4(e).

3.10 The Tits–Kantor–Koecher Algebra and the Projective Elementary Group of a Jordan Pair

Let V be a Jordan pair, defined over a commutative ring k of scalars. It is fundamental (and well-known) that V gives rise to a $\mathbb{Z}$-graded Lie k-algebra

$$\mathfrak{L}(V) = \mathfrak{L}(V)_1 \oplus \mathfrak{L}(V)_0 \oplus \mathfrak{L}(V)_{-1}, \tag{3.10.18}$$

introduced at about the same time by Tits, Kantor and Koecher in [13–15, 17, 19, 46, 47] and called the *Tits–Kantor–Koecher algebra of* V. Various versions of $\mathfrak{L}(V)$ exist, but all agree that $\big(\mathfrak{L}(V)_1, \mathfrak{L}(V)_{-1}\big) = (V^+, V^-)$ as k-modules. For our purposes, the most appropriate choice for $\mathfrak{L}(V)_0$ is

$$\mathfrak{L}(V)_0 = k\zeta + \mathrm{Span}_k\{\delta(x, y) : (x, y) \in V\}, \tag{3.10.19}$$

where $\zeta = (\mathrm{Id}_{V^+}, \mathrm{Id}_{V^-})$ and $\delta(x, y) = (D(x, y), -D(y, x)) \in \mathrm{End}(V^+) \times \mathrm{End}(V^-)$, defined by $D(x, y)z = \{x\ y\ z\}$. We let $\mathfrak{gl}(V^\sigma)$ be the Lie algebra defined by $\mathrm{End}(V^\sigma)$ with the commutator as the Lie product. By definition, the Lie product of $\mathfrak{L}(V)$ is determined by the conditions that it be alternating, that $\mathfrak{L}(V)_0$ be a subalgebra of the Lie algebra $\mathfrak{gl}(V^+) \times \mathfrak{gl}(V^-)$ and that

$$[V^\sigma, V^\sigma] = 0, \qquad [D, z] = D_\sigma(z), \qquad [x, y] = -\delta(x, y)$$

for $D = (D_+, D_-) \in \mathfrak{L}(V)_0$, $z \in V^\sigma$ and $(x, y) \in V$. It follows from the identity (JP15) in [23]

$$[D(x, y),\ D(u, v)] = D(\{x\ y\ u\},\ v) - D(u,\ \{y\ x\ v\})$$

that $\mathfrak{L}(V)_0$ is indeed a subalgebra. As a k-Lie algebra, $\mathfrak{L}(V)$ is generated by ζ, V^+ and V^-, and it has trivial centre.

For a Jordan pair V with a fully idempotent root grading $\mathfrak{R}$, a description of the derived algebra $[\mathfrak{L}(V), \mathfrak{L}(V)]$ is given in [36]. The Tits–Kantor–Koecher algebra of a special Jordan pair is described in [26, §2]. We will work out $\mathfrak{L}(V)$ for a rectangular matrix pair in Sect. 3.12.

An automorphism f of V gives rise to an automorphism $\mathfrak{L}(f)$ of $\mathfrak{L}$, defined by

$$x \oplus D \oplus y \quad \mapsto \quad f_+(x) \oplus (f \circ D \circ f^{-1}) \oplus f_-(y).$$

The map $f \mapsto \mathfrak{L}(f)$ is an embedding of the automorphism group $\mathrm{Aut}(V)$ of V into the automorphism group of $\mathfrak{L}(V)$.

Any $(x, y) \in V$ gives rise to automorphisms $\exp_+(x)$ and $\exp_-(y)$ of $\mathfrak{L}(V)$, defined in terms of the decomposition (3.10.18) by the formal 3×3-matrices

$$\exp_+(x) = \begin{pmatrix} 1 & \operatorname{ad} x & Q_x \\ 0 & 1 & \operatorname{ad} x \\ 0 & 0 & 1 \end{pmatrix}, \qquad \exp_-(y) = \begin{pmatrix} 1 & 0 & 0 \\ \operatorname{ad} y & 1 & 0 \\ Q_y & \operatorname{ad} y & 1 \end{pmatrix}. \tag{3.10.20}$$

The map $\exp_\sigma$, $\sigma = \pm$, is an injective homomorphism from the abelian group $(V^\sigma, +)$ to the automorphism group of $\mathfrak{L}(V)$, whose image is denoted by U^σ. The *projective elementary group of* V is the subgroup $\mathrm{PE}(V)$ of $\mathrm{Aut}\big(\mathfrak{L}(V)\big)$ generated by $U^+ \cup U^-$, introduced in [26] and studied further in [30, §7, §8].

We have now explained all the concepts used in the generalization of part *(b)* of the Kervaire–Milnor–Steinberg Theorem 2.6.

3.11 Theorem B

Let (R, R_1) be a 3-graded root system, and let V be a Jordan pair with a root grading $\mathfrak{R} = (V_\alpha)_{\alpha \in R_1}$.

(a) *There exists a group homomorphism $\pi\colon \mathrm{St}(V, \mathfrak{R}) \to \mathrm{PE}(V)$, uniquely determined by*

$$\pi\big(\mathrm{x}_\sigma(u)\big) = \exp_\sigma(u), \qquad (u \in V^\sigma).$$

(b) *If all irreducible components of R have infinite rank and $\mathfrak{R}$ is fully idempotent with respect to a family $(e_\alpha)_{\alpha \in R_1}$, the homomorphism π is a universal central extension.*

Theorem B is established in [30]. Part (a) follows from [30, Cor. 21.12]. By Fact Sect. 2.4(d) and Theorem Sect. 3.8, the proof of (b) boils down to showing that $\operatorname{Ker} \pi$ is central, which we do in [30, Cor. 27.6]. As for Theorem Sect. 3.8, it is not necessary to assume that $\mathfrak{R}$ is fully idempotent.

In the setting of (a) let $(u, v) \in V_\alpha^+ \times V_\beta^-$ with $\alpha \neq \beta$ and let $\mathrm{b}(u, v)$ be the element of $\mathrm{St}(V, \mathfrak{R})$ defined in (3.7.15). Then $\pi(\mathrm{b}(u, v)) = \mathfrak{L}(f)$ for some $f \in \mathrm{Aut}(V)$ (for the experts: f is the inner automorphism $(B(u, v), B(v, u)^{-1})$ of [23,

3.9]). That $\pi(\mathrm{b}(u,v)) \in \mathfrak{L}\big(\operatorname{Aut}(V)\big) \subset \operatorname{Aut}(\mathfrak{L}(V))$ is the motivation for the perhaps surprising definition of $\mathrm{b}(u,v)$.

We finish this section by describing $\mathfrak{L}(V)$ and $\mathrm{PE}(V)$ for $V = \mathbb{M}_{IJ}(A)$.

3.12 *The Tits–Kantor–Koecher Algebra of a Rectangular Matrix Pair*

Let $V = \mathbb{M}_{IJ}(A) = \big(\mathrm{Mat}_{IJ}(A), \mathrm{Mat}_{JI}(A)\big)$ be the rectangular matrix pair of Sect. 3.5(iii). In this subsection we present a model for the Tits–Kantor–Koecher algebra $\mathfrak{L} = \mathfrak{L}(V)$ in terms of elementary Matrices, which will be used in Sect. 3.13 to link the elementary group of V and the abstractly defined group $\mathrm{PE}(V)$.

Let $\mathbf{1}_I = \mathrm{diag}(1_A, \dots)$ be the diagonal matrix of size $I \times I$, define $\mathbf{1}_J$ analogously and let $\mathfrak{A}$ be the unital associative k-algebra

$$\mathfrak{A} = \mathfrak{A}(V) = \begin{pmatrix} k\,\mathbf{1}_I + \mathrm{Mat}_I(A) & \mathrm{Mat}_{IJ}(A) \\ \mathrm{Mat}_{JI}(A) & k\,\mathbf{1}_J + \mathrm{Mat}_J(A) \end{pmatrix}$$

$$= \begin{pmatrix} \mathrm{Mat}_I(A)_{\mathrm{ex}} & \mathrm{Mat}_{IJ}(A) \\ \mathrm{Mat}_{JI}(A) & \mathrm{Mat}_J(A)_{\mathrm{ex}} \end{pmatrix}$$

whose operations are given by matrix addition and matrix multiplication. In particular,

$$e_1 = \begin{pmatrix} \mathbf{1}_I & 0 \\ 0 & 0 \end{pmatrix} \qquad \text{and} \qquad e_2 = \begin{pmatrix} 0 & 0 \\ 0 & \mathbf{1}_J \end{pmatrix}$$

are orthogonal idempotents of $\mathfrak{A}$. We consider $\mathfrak{A}$ rather than its subalgebra $\mathrm{Mat}_N(A)_{\mathrm{ex}}$, $N = I \dot\cup J$, since this will allow us to model the element ζ of (3.10.19).

The Peirce decomposition of $\mathfrak{A}$ with respect to the idempotent e_1 is

$$\mathfrak{A}_{11} = \begin{pmatrix} \mathrm{Mat}_I(A)_{\mathrm{ex}} & 0 \\ 0 & 0 \end{pmatrix}, \qquad \mathfrak{A}_{10} = \begin{pmatrix} 0 & \mathrm{Mat}_{IJ}(A) \\ 0 & 0 \end{pmatrix},$$

$$\mathfrak{A}_{01} = \begin{pmatrix} 0 & 0 \\ \mathrm{Mat}_{JI}(A) & 0 \end{pmatrix}, \qquad \mathfrak{A}_{00} = \begin{pmatrix} 0 & 0 \\ 0 & \mathrm{Mat}_J(A)_{\mathrm{ex}} \end{pmatrix}.$$

Let $\mathfrak{A}^{(-)}$ be the Lie algebra associated with $\mathfrak{A}$. Thus, $\mathfrak{A}^{(-)}$ is defined on the k-module underlying $\mathfrak{A}$ and its Lie algebra product is $[x,y] = xy - yx$ for $x, y \in \mathfrak{A}$. The Lie algebra $\mathfrak{A}^{(-)}$ is $\mathbb{Z}$-graded, $\mathfrak{A}^{(-)} = \bigoplus_{n\in\mathbb{Z}} \mathfrak{A}_n^{(-)}$ with

$$\mathfrak{A}_1^{(-)} = \mathfrak{A}_{10}, \quad \mathfrak{A}_0^{(-)} = \mathfrak{A}_{11} \oplus \mathfrak{A}_{00}, \quad \mathfrak{A}_{-1}^{(-)} = \mathfrak{A}_{01}$$

and $\mathfrak{A}_n^{(-)} = 0$ for $n \notin \{1, 0, -1\}$. We define $\mathfrak{e} = \mathfrak{e}(V)$ as the subalgebra of $\mathfrak{A}^-$ generated by e_1, e_2 and

$$\mathfrak{e}_1 = \begin{pmatrix} 0 & \mathrm{Mat}_{IJ}(A) \\ 0 & 0 \end{pmatrix} = \mathfrak{A}_1^{(-)} \quad \text{and} \quad \mathfrak{e}_{-1} = \begin{pmatrix} 0 & 0 \\ \mathrm{Mat}_{JI}(A) & 0 \end{pmatrix} = \mathfrak{A}_{-1}^{(-)}.$$

Put $\mathfrak{e}_0 = k\, e_1 + k\, e_2 + [\mathfrak{e}_1,\ \mathfrak{e}_{-1}]$ and $\mathfrak{e}_i = 0$ for $i \notin \{-1, 0, 1\}$. Then $\mathfrak{e} = \bigoplus_{i \in \mathbb{Z}} \mathfrak{e}_i$ is a $\mathbb{Z}$-graded Lie algebra.

We now relate $\mathfrak{e}$ to the Tits–Kantor–Koecher algebra $\mathfrak{L} = \mathfrak{L}(V)$ of V. First, for $\mathsf{a} = \left(\begin{smallmatrix} a & 0 \\ 0 & d \end{smallmatrix}\right) \in \mathfrak{e}_0$ define $\Delta(\mathsf{a}) = (\Delta(\mathsf{a})_+,\ \Delta(\mathsf{a})_-) \in \mathrm{End}_k(V^+) \times \mathrm{End}_k(V^-)$ by

$$\Delta(\mathsf{a})_+(u) = au - ud, \qquad \Delta(\mathsf{a})_-(v) = dv - va,$$

so that

$$\left[\mathsf{a}, \begin{pmatrix} 0 & b \\ c & 0 \end{pmatrix}\right] = \begin{pmatrix} 0 & au - ud \\ dv - va & 0 \end{pmatrix} = \begin{pmatrix} 0 & \Delta_+(\mathsf{a})(u) \\ \Delta_-(\mathsf{a})(v) & 0 \end{pmatrix}. \tag{3.12.21}$$

We claim: *the map*

$$\Psi : \mathfrak{e} \to \mathfrak{L}, \qquad \begin{pmatrix} a & b \\ c & d \end{pmatrix} \mapsto b \oplus \Delta(a, d) \oplus (-c)$$

is a surjective Lie algebra homomorphism whose kernel is $\mathfrak{z}(e)$, *the centre of* $\mathfrak{e}$, *and thus induces an isomorphism*

$$\mathfrak{e}/\mathfrak{z}(\mathfrak{e}) \cong \mathfrak{L} \tag{3.12.22}$$

of Lie algebras [26, 2.6], [30, 7.2]. Indeed, Ψ is surjective since $\Delta(e_1) = \zeta = -\Delta(e_2)$ and for $(x, y) \in V$

$$\Delta\left(\left[\left(\begin{smallmatrix} 0 & x \\ 0 & 0 \end{smallmatrix}\right), \left(\begin{smallmatrix} 0 & 0 \\ y & 0 \end{smallmatrix}\right)\right]\right) = \Delta\left(\begin{smallmatrix} xy & 0 \\ 0 & -yx \end{smallmatrix}\right) = \delta(x, y) \tag{3.12.23}$$

by (3.5.12). To see that $\mathrm{Ker}\,\Psi = \mathfrak{z}(\mathfrak{e})$, observe for $\mathsf{m} = \left(\begin{smallmatrix} a & b \\ c & d \end{smallmatrix}\right) \in \mathfrak{A}$ that

$$[\mathsf{m}, e_1] = 0 \iff b = 0 = c \iff [\mathsf{m}, e_2] = 0, \tag{3.12.24}$$

whence by (3.12.21),

$$\begin{aligned} \mathsf{m} \in \mathrm{Ker}\,\Psi \quad &\iff \quad b = 0 = c,\ \Delta(a, d) = 0,\ \left(\begin{smallmatrix} a & 0 \\ 0 & d \end{smallmatrix}\right) \in \mathfrak{e}_0, \\ &\iff \quad [\mathsf{m}, e_1] = 0 = [\mathsf{m}, e_2],\ [\mathsf{m},\ \mathfrak{e}_1] = 0 = [\mathsf{m},\ \mathfrak{e}_{-1}],\ \mathsf{m} \in \mathfrak{e}, \\ &\iff \quad \mathsf{m} \in \mathfrak{z}(\mathfrak{e}) \end{aligned}$$

because $e_1, e_2, \mathfrak{e}_1$ and $\mathfrak{e}_{-1}$ generate $\mathfrak{e}$ as Lie algebra. Finally, since both $\mathfrak{e}$ and $\mathfrak{L}$ are $\mathbb{Z}$-graded and Ψ preserves this grading, Ψ is a Lie algebra homomorphism as soon as Ψ preserves products of type $[\mathsf{x}_\mathsf{i}, \mathsf{y}_\mathsf{j}]$ for $(i, j) = (0, \pm 1), (1, -1)$ and $(0, 0)$. For $(0, \pm 1)$ and $(1, -1)$, this follows from (3.12.21) and (3.12.23), respectively; the case $(0, 0)$ is left to the reader.

In the remainder of this subsection we will give a more precise description of $\mathfrak{e}$ and its centre, see (3.12.25) and (3.12.27). Let $[A, A] = \operatorname{Span}\{ab - ba : a, b \in A\}$, the derived algebra of the Lie algebra A^-, and let

$$\mathfrak{sl}_N(A) := \{x = (x_{kl}) \in \operatorname{Mat}_N(A) \colon \textstyle\sum_{n \in N} x_{nn} \in [A, A]\}.$$

From $[aE_{kl}, bE_{rs}] = ab\delta_{lr}E_{ks} - ba\delta_{ks}E_{rl}$, one then gets

$$\begin{aligned} &\mathfrak{e}_1 \oplus [\mathfrak{e}_1, \mathfrak{e}_{-1}] \oplus \mathfrak{e}_{-1} = \mathfrak{sl}_N(A), \qquad \text{whence} \\ &k\, e_1 + k\, e_2 + \mathfrak{sl}_N(A) = \mathfrak{e}. \end{aligned} \tag{3.12.25}$$

The description of the centre $\mathfrak{z}(\mathfrak{e})$ depends on the cardinality of N because

$$\mathfrak{A}_0^{(-)} \cap A\mathbf{1}_N = \begin{cases} A\,\mathbf{1}_N & \text{if } |N| < \infty, \\ k\,\mathbf{1}_N & \text{if } |N| = \infty. \end{cases}$$

Denoting by $\mathrm{Z}(A) = \{z \in A : [z, A] = 0\}$ the centre of A (= centre of $A^{(-)}$), a straightforward calculation shows

$$\big\{x \in \mathfrak{A}_0^{(-)} : [x, \mathfrak{e}_1] = 0 = [x, \mathfrak{e}_{-1}]\big\} = \mathfrak{A}_0^{(-)} \cap \mathrm{Z}(A)\mathbf{1}_N = \begin{cases} \mathrm{Z}(A)\mathbf{1}_N, & |N| < \infty, \\ k\mathbf{1}_N, & |N| = \infty. \end{cases} \tag{3.12.26}$$

Since $\mathfrak{e}$ is generated by $e_1, e_2, \mathfrak{e}_1$ and $\mathfrak{e}_{-1}$, (3.12.24) and (3.12.26) imply

$$\mathfrak{z}(\mathfrak{e}) = \mathfrak{e}_0 \cap (\mathrm{Z}(A)\,\mathbf{1}_N) = \begin{cases} \mathfrak{e}_0 \cap (\mathrm{Z}(A)\mathbf{1}_N), & |N| < \infty, \\ k\mathbf{1}_N, & |N| = \infty. \end{cases} \tag{3.12.27}$$

For example, if $A = k$ is a field of characteristic 0 and $|N| = n$ is finite, we get $\mathfrak{e} = \mathfrak{gl}_n(k)$, $\mathfrak{z}(\mathfrak{e}) = k\mathbf{1}_n$ and $\mathfrak{L} \cong \mathfrak{sl}_n(k)$.

3.13 *The Projective Elementary Group of a Rectangular Matrix Pair*

We use the notation of Sect. 3.12 and let $V = \mathbb{M}_{IJ}(A)$. The goal of this subsection is to show that the group $\mathrm{PE}(V)$ is isomorphic to a central quotient of the elementary group $\mathrm{E}(A)$. We put

$$\mathrm{E}(V) = \mathrm{E}_N(A)$$

and call it the *elementary group of* V. Since by Sect. 2.12 the group $\mathrm{E}_N(A)$ is generated by $\mathrm{e}_+(V^+) \cup \mathrm{e}_-(V^-)$, this agrees with the definition of the elementary group of an arbitrary special Jordan pair in [26, §2] or [30, 6.2]. We will identify the Tits–Kantor–Koecher algebra $\mathfrak{L}(V) = \mathfrak{L}$ with $\mathfrak{e}/\mathfrak{z}(\mathfrak{e})$ via the isomorphism (3.12.22) induced by $\Psi\colon \mathfrak{e} \to \mathfrak{L}$.

Any $g \in \mathfrak{A}^\times$ gives rise to an automorphism $\operatorname{Ad} g$ of $\mathfrak{A}^-$ defined by $(\operatorname{Ad} g)(x) = gxg^{-1}$. If $\operatorname{Ad} g$ stabilizes the subalgebra $\mathfrak{e}$ of $\mathfrak{A}^-$, it also stabilizes $\mathfrak{z}(\mathfrak{e})$ and therefore descends to an automorphism $\overline{\operatorname{Ad}}(g)$ of $\mathfrak{e}/\mathfrak{z}(\mathfrak{e}) = \mathfrak{L}$ satisfying $\Psi \circ (\operatorname{Ad} g|_{\mathfrak{e}}) = (\overline{\operatorname{Ad}} g) \circ \Psi$. The map

$$\overline{\operatorname{Ad}}\colon \{g \in \mathfrak{A}^\times : (\operatorname{Ad} g)(\mathfrak{e}) = \mathfrak{e}\} \to \operatorname{Aut}(\mathfrak{L}), \qquad g \mapsto \overline{\operatorname{Ad}} g$$

is a group homomorphism. For $g = \mathrm{e}_+(x) = \left(\begin{smallmatrix} 1 & x \\ 0 & 1 \end{smallmatrix}\right) \in \mathfrak{A}^\times$ as in (2.12.25), the automorphism $\operatorname{Ad} \mathrm{e}_+(x)$ acts as follows:

$$\begin{aligned}
\big(\operatorname{Ad} \mathrm{e}_+(x)\big) \begin{pmatrix} 0 & b \\ 0 & 0 \end{pmatrix} &= \begin{pmatrix} 0 & b \\ 0 & 0 \end{pmatrix}, \\
\big(\operatorname{Ad} \mathrm{e}_+(x)\big) \begin{pmatrix} a & 0 \\ 0 & d \end{pmatrix} &= \begin{pmatrix} a & -ax + xd \\ 0 & d \end{pmatrix} \\
&= \begin{pmatrix} a & 0 \\ 0 & d \end{pmatrix} + \Big[\begin{pmatrix} 0 & x \\ 0 & 0 \end{pmatrix}, \begin{pmatrix} a & 0 \\ 0 & d \end{pmatrix} \Big], \\
\big(\operatorname{Ad} \mathrm{e}_+(x)\big) \begin{pmatrix} 0 & 0 \\ -c & 0 \end{pmatrix} &= \begin{pmatrix} -xc & xcx \\ -c & cx \end{pmatrix} \\
&= \begin{pmatrix} 0 & 0 \\ -c & 0 \end{pmatrix} + \Big[\begin{pmatrix} 0 & x \\ 0 & 0 \end{pmatrix}, \begin{pmatrix} 0 & 0 \\ -c & 0 \end{pmatrix} \Big] + \begin{pmatrix} 0 & Q_x c \\ 0 & 0 \end{pmatrix}.
\end{aligned}$$

These equations show that the automorphism $\operatorname{Ad} \mathrm{e}_+(x)$ stabilizes $\mathfrak{e}$ and, by comparison with (3.10.20), that $\overline{\operatorname{Ad}} \mathrm{e}_+(x) = \exp_+(x)$. One proves in the same way that $\operatorname{Ad} \mathrm{e}_-(y)$, $y \in V^-$, stabilizes $\mathfrak{e}$ and that $\overline{\operatorname{Ad}} \mathrm{e}_-(y) = \exp_-(y)$. Since $\mathrm{PE}(V)$ is generated by $\exp_+(V^+) \cup \exp_-(V^-)$, the homomorphism $\overline{\operatorname{Ad}}$ restricts to a surjective group homomorphism

$$\overline{\operatorname{Ad}}_{\mathrm{E}}\colon \mathrm{E}(V) \to \mathrm{PE}(V), \quad g \mapsto \overline{\operatorname{Ad}} g.$$

We claim that its kernel is the centre $\mathrm{Z}(\mathrm{E}(V))$ of $\mathrm{E}(V)$:

$$\mathrm{Z}(\mathrm{E}(V)) = \operatorname{Ker} \operatorname{Ad}|_{\mathrm{E}(V)} = \operatorname{Ker} \overline{\operatorname{Ad}}_{\mathrm{E}}, \qquad \text{whence} \tag{3.13.28}$$

$$\mathrm{E}(V)/\mathrm{Z}(\mathrm{E}(V)) \cong \mathrm{PE}(V). \tag{3.13.29}$$

Proof of (3.13.28): Clearly, $\mathrm{Z}(\mathrm{E}(V)) = \mathrm{Ker}\,\mathrm{Ad}\,|_{\mathrm{E}(V)} \subset \mathrm{Ker}\,\overline{\mathrm{Ad}}_{\mathrm{E}}$, so it remains to show that $g \in \mathrm{Ker}\,\overline{\mathrm{Ad}}_{\mathrm{E}}$ is central in $\mathrm{E}(V)$. Let $g = \left(\begin{smallmatrix} a & b \\ c & d \end{smallmatrix}\right)$ and $g^{-1} = \left(\begin{smallmatrix} a' & b' \\ c' & d' \end{smallmatrix}\right)$. Then,

$$g \begin{pmatrix} \mathbf{1}_I & 0 \\ 0 & 0 \end{pmatrix} g^{-1} = \begin{pmatrix} aa' & ab' \\ ca' & cb' \end{pmatrix}, \tag{3.13.30}$$

$$g \begin{pmatrix} 0 & 0 \\ 0 & \mathbf{1_J} \end{pmatrix} g^{-1} = \begin{pmatrix} bc' & bd' \\ dc' & dd' \end{pmatrix}, \tag{3.13.31}$$

$$g^{-1} \begin{pmatrix} 0 & x \\ 0 & 0 \end{pmatrix} g = \begin{pmatrix} a'xc & a'xd \\ c'xc & c'xd \end{pmatrix}, \tag{3.13.32}$$

$$g^{-1} \begin{pmatrix} 0 & 0 \\ y & 0 \end{pmatrix} g = \begin{pmatrix} b'ya & b'yb \\ d'ya & d'yp \end{pmatrix}. \tag{3.13.33}$$

Since $(\mathrm{Ad}\, g)\,(\mathsf{m}) \equiv \mathsf{m} \equiv (\mathrm{Ad}\, g^{-1})(\mathsf{m}) \mod \mathfrak{z}(\mathfrak{e})$ for all $\mathsf{m} \in \mathfrak{e}$ and since $\mathfrak{z}(\mathfrak{e})$ is diagonal by (3.12.27), it follows from (3.13.30), (3.13.31) and (3.13.33) that

$$ab' = 0 = ca' = bd', \qquad d'ya = y \text{ for } y \in V^-.$$

Applied to $c \in V^-$, this proves $cb' = (d'ca) \cdot b' = d'c \cdot ab' = 0$ and then $bc' = b \cdot (d'c'a) = bd' \cdot c'a = 0$. From $\mathbf{1}_N = gg^{-1}$, we obtain

$$\begin{pmatrix} \mathbf{1}_I & 0 \\ 0 & \mathbf{1}_J \end{pmatrix} = \begin{pmatrix} aa' + bc' & * \\ * & * \end{pmatrix} = \begin{pmatrix} aa' & * \\ * & * \end{pmatrix}.$$

Together with the already established equations, this shows, using (3.13.30), that $(\mathrm{Ad}\, g)(e_1) = e_1$. Because $g \in \mathfrak{A}$, we get $b = 0 = c$ from (3.12.24). Thus also $b' = 0 = c'$, so that (3.13.32) and (3.13.33) prove that $\mathrm{Ad}\, g$ fixes $\mathfrak{e}_{\pm 1}$. Since $\mathrm{e}_\pm(V^\pm) = \mathbf{1}_N + \mathfrak{e}_{\pm 1}$, we see that $\mathrm{Ad}\, g$ fixes the generators $\mathrm{e}_\pm(V^\pm)$ of $\mathrm{E}(V)$, i.e. g is central.

A similar result holds for any special Jordan pair V: there always exists a surjective group homomorphism from the elementary group $\mathrm{E}(V)$ (which we have not defined) onto the projective elementary group $\mathrm{PE}(V)$, whose kernel is central, but not necessarily the centre of $\mathrm{E}(V)$, see [26, Thm. 2.8].

4 Some Open Problems

We describe some open problems for Steinberg and projective elementary groups of Jordan pairs. Our list is very much limited by the author's taste and knowledge. This section requires some expertise in Jordan pairs.

4.1 The Normal Subgroup Structure of PE(V)

The problem is quite easily stated: *Given a Jordan pair V, describe all normal subgroups of* PE(V). As stated, this may be too general. We therefore discuss some special cases.

(a) In view of the results of [27] it is natural to ask: when is PE(V) a perfect group, when is it simple? Indeed, [27, Thm. 2.6] says that, for a nondegenerate Jordan pair V with dcc on principal ideals, PE(V) is a perfect group if and only if V has no simple factors isomorphic to $(\mathbb{F}_2, \mathbb{F}_2)$, $(\mathbb{F}_3, \mathbb{F}_3)$ or $\big(\mathrm{H}_2(\mathbb{F}_2), \mathrm{H}_2(\mathbb{F}_2)\big)$. Here $\mathbb{F}_q$ is the field with q elements. Also, by [27, Thm. 2.8], PE(V) is a simple (abstract) group if and only if V is simple and not isomorphic to $(\mathbb{F}_2, \mathbb{F}_2)$, $(\mathbb{F}_3, \mathbb{F}_3)$ or $\big(\mathrm{H}_2(\mathbb{F}_2), \mathrm{H}_2(\mathbb{F}_2)\big)$ [27, Thm. 2.8]. That the exceptional cases have to be excluded in these two theorems is evident from the isomorphisms PE$(\mathbb{F}_2, \mathbb{F}_2) \cong \mathfrak{S}_3$ (the symmetric group on three letters), PE$(\mathbb{F}_3, \mathbb{F}_3) \cong \mathfrak{A}_4$ (the alternating group on four letters), and PE$\big(\mathrm{H}_2(\mathbb{F}_2), \mathrm{H}_3(\mathbb{F}_2)\big) \cong \mathfrak{S}_6$. Thus, the problem is to find out if these two theorems of [27] hold for more general Jordan pairs. It follows from (b) that is natural to assume simplicity of V for the second theorem.

(b) Every ideal I of the Jordan pair V gives rise to a normal subgroup of PE(V). Indeed, one can show [30, 7.5] that the canonical map can: $V \to V/I$ induces a surjective group homomorphism PE(can): PE$(V) \to$ PE(V/I). We let PE(V, I) be its kernel:

$$1 \longrightarrow \mathrm{PE}(V, I) \longrightarrow \mathrm{PE}(V) \xrightarrow{\mathrm{PE(can)}} \mathrm{PE}(V/I) \longrightarrow 1$$

Problem: describe PE(V, I) by generators and relations. For elementary linear groups over rings, this is a standard result, see for example [6, Lemma 3]. The paper [4] shows that even in case $\mathrm{SL}_2(A)$ one needs methods from Jordan algebras.

4.2 Central Closedness of St$(V, \mathfrak{R})$ *in Low Ranks*

We have excluded low rank cases in Theorem Sect. 3.8 for the simple reason that it is not true without further assumptions in low ranks. We discuss $2 \leqslant \operatorname{rank} R \leqslant 4$ in (a) and $\operatorname{rank} R = 1$ in (b).

(a) One knows [30, 27.11] that St$(V, \mathfrak{R})$ is a classical linear or unitary Steinberg group. Let us first consider the case that V is defined over a field F and that $\dim_F V_\alpha = 1$ for all $\alpha \in R_1$. Then [44, Thm. 1.1] applies and yields that St$(V, \mathfrak{R})$ is not centrally closed if and only if (R, R_1) and F satisfy one of the following conditions.

(R, R_1)	A_2^1	A_3^1 or A_3^2	C_2^{her}	B_3^{qf}	C_3^{her}	D_4^{alt}
$\lvert F\rvert$	2, 4	2	2	3	2	2

(4.2.1)

In this table we use the abbreviation $A_2^1 = A_N^I$ for $|N| = 2$, $|I| = 1$ and analogously for A_3^1, …, D_4^{alt}. The cases $R = A_4, B_4, C_4$ do not appear in the table because in these cases $\text{St}(V, \mathfrak{R})$ is centrally closed, as mentioned in Sect. 3.8.

Still assuming that V is defined over a field F, it is natural to replace the assumption $\dim_F V_\alpha = 1$ by the requirement that the fully idempotent root grading $\mathfrak{R}$ of V is a *division grading* in the sense that all root spaces V_α are Jordan division pairs, which means that for every non-zero $x \in V_\alpha^\sigma$ the endomorphism $Q(x)|V_\alpha^{-\sigma}$ is invertible. Preliminary investigations lead us to conjecture:

(C) *If* $\text{St}(V, \mathfrak{R})$ *is not centrally closed, then* $\dim_F V_\alpha = 1$ *for all* $\alpha \in R_1$ *and* F *satisfies the restrictions of table* (4.2.1).

(b) $R = A_1$: As in (a) we assume that $\mathfrak{R}$ is a division grading, i.e. V is a division pair and is therefore isomorphic to the Jordan pair (J, J) of a division Jordan algebra J. By [30, 9.13], this is equivalent to $\text{PE}(V)$ being a rank one group in the sense of [28]. Since the grading is trivial, $\text{St}(V, \mathfrak{R})$ is the free product of the abelian groups V^+ and V^-, which is not perfect in general, a necessary condition for a group to be centrally closed (Sect. 2.4(a)). Following the example of Chevalley groups [43], it seems more promising to consider the group $\text{St}(J)$ defined by the following presentation:

- generators $\mathrm{x}_\sigma(a)$, $a \in J$, $\sigma = \pm$ and putting

$$\mathrm{w}_b = \mathrm{x}_-(b^{-1})\,\mathrm{x}_+(b)\,\mathrm{x}_-(b^{-1})$$

  for $0 \neq b \in J$,
- relations

$$\mathrm{x}_\sigma(a+b) = \mathrm{x}_\sigma(a)\,\mathrm{x}_\sigma(b) \text{ for } a, b \in J \text{ and}$$

$$\mathrm{w}_b\,\mathrm{x}_-(a)\,\mathrm{w}_b^{-1} = \mathrm{x}_+\big(U(b)a\big) \text{ for all } a \in J \text{ and all } 0 \neq b \in J.$$

We remark that $\text{St}(J)$ is the Steinberg group $\text{St}(V, \mathcal{S})$ of [30, 13.1], where $\mathcal{S}$ is the set of all non-zero idempotents of V. By [30, 13.6], $\text{St}(J)$ is the classical Steinberg group $\text{St}(A)$ in case $V = (A, A)$ and A an associative division algebra.

To motivate our conjecture in this case, let us first consider the special case $J = \mathbb{F}_q$. Since by Sect. 4.1(a)) the group $\text{PE}(V)$ is not perfect in case $J = \mathbb{F}_q$, $q = 2, 3$, these cases have to be excluded. Moreover, by [44, Th. 1.1], $\text{St}(J)$ is not centrally closed in case $V = (\mathbb{F}_q, \mathbb{F}_q)$ and $q \in \{4, 9\}$, but these values of q are the only exceptions for $V = (F, F)$, F a field. This leads us to ask:

(Q) *Is* St(J) *centrally closed whenever* $J \neq \mathbb{F}_q$ *with* $q \in \{2, 3, 4, 9\}$ *?*

There exists an example of an associative unital $\mathbb{F}_5$-algebra A for which St(A) is not centrally closed [45, Ex. 4], but A is not a division algebra.

4.3 Centrality of Ker(π)

Let $\pi\colon \mathrm{St}(V, \mathfrak{R}) \to \mathrm{PE}(V)$ be the homomorphism of Theorem Sect. 3.11(a). For simplicity, let us assume that R is irreducible. If R has infinite rank, part (b) of Sect. 3.11 says that π is a universal central extension. The problem here is: *find sufficient conditions for* Ker(π) *to be central if R has finite rank.*

Some special cases are known. For example, if V is split in the sense of [37], centrality of Ker(π) is established in [21, 22, 48] and [39] for rank $\geqslant 3$. The quoted papers all use the same method, pioneered by [48], namely a "basis-free presentation of St$(V, \mathfrak{R})$". *Can the method of* [48] *be generalized to treat* St$(V, \mathfrak{R})$, V *split root graded, in a case-free manner?*

For a slightly different type of Steinberg group and a unit regular V, centrality of Ker(π) is shown in [27, Th. 1.12].

Acknowledgments The author thanks Ottmar Loos for many helpful comments on an earlier version of this paper.

The author acknowledges with thanks partial support by the Natural Sciences and Engineering Research Council of Canada (NSERC) through a Discovery grant.

References

1. A. Bak, *K-theory of forms*, Ann. of Math. Studies, vol. 98, Princeton University Press, 1981.
2. N. Bourbaki, *Groupes et algèbres de Lie, chapitres 4–6*, Masson, Paris, 1981.
3. C. Chevalley and R. D. Schafer, *The exceptional simple Lie algebras* F_4 *and* E_6, Proc. Nat. Acad. Sci. U.S.A. **36** (1950), 137–141.
4. D. L. Costa and G. E. Keller, *The* $E(2, A)$ *sections of* SL$(2, A)$, Ann. of Math. (2) **134** (1991), no. 1, 159–188.
5. A. J. Hahn and O. T. OMeara, *The classical groups and K-theory*, Grundlehren, vol. **291**, Springer-Verlag, 1989.
6. R. Hazrat, N. Vavilov, and Z. Zhang, *The commutators of classical groups*, Zap. Nauchn. Sem. S.-Peterburg. Otdel. Mat. Inst. Steklov. (POMI) **443** (2016), no. Voprosy Teorii Predstavleniĭ Algebr i Grupp. 29, 151–221.
7. J. Humphreys, *Introduction to Lie algebras and Representation Theory*, Graduate Texts in Mathematics vol. **9**, Springer-Verlag, New York, 1972.
8. N. Jacobson, *Some groups of transformations defined by Jordan algebras. I*, J. Reine Angew. Math. **201** (1959), 178–195.
9. ——, *Some groups of transformations defined by Jordan algebras. II. Groups of type F4*, J. Reine Angew. Math. **204** (1960), 74–98.
10. ——, *Some groups of transformations defined by Jordan algebras. III*, J. Reine Angew. Math. **207** (1961), 61–85.

11. ——, *Lectures on quadratic Jordan algebras*, Tata Institute of Fundamental Research, 1969.
12. ——, *Structure theory of Jordan algebras*, Lecture Notes in Math., vol. **5**, The University of Arkansas, 1981.
13. I. L. Kantor, *Classification of irreducible transitive differential groups*, Dokl. Akad. Nauk SSSR **158** (1964), 1271–1274.
14. ——, *Non-linear groups of transformations defined by general norms of Jordan algebras*, Dokl. Akad. Nauk SSSR **172** (1967), 779–782.
15. ——, *Certain generalizations of Jordan algebras*, Trudy Sem. Vektor. Tenzor. Anal. **16** (1972), 407–499.
16. M. Kervaire, *Multiplicateurs de Schur et K-théorie*, Essays on Topology and Related Topics (Mémoires dédiés à Georges de Rham), Springer, New York, 1970, pp. 212–225.
17. M. Koecher, *Imbedding of Jordan algebras into Lie algebras. I*, Amer. J. Math. **89** (1967), 787–816.
18. ——, *Über eine Gruppe von rationalen Abbildungen*, Invent. Math. **3** (1967), 136–171.
19. ——, *Imbedding of Jordan algebras into Lie algebras. II*, Amer. J. Math. **90** (1968), 476–510.
20. ——, *Gruppen und Lie-Algebren von rationalen Funktionen*, Math. Z. **109** (1969), 349–392.
21. A. Lavrenov, *Another presentation for symplectic Steinberg groups*, J. Pure Appl. Algebra **219** (2015), no. 9, 3755–3780.
22. A. Lavrenov and S. Sinchuk, *On centrality of even orthogonal* K_2, J. Pure Appl. Algebra **221** (2017), no. 5, 1134–1145.
23. O. Loos, *Jordan pairs*, Springer-Verlag, Berline, 1975, Lecture Notes in Mathematics, vol **460**.
24. ——, *Homogeneous algebraic varieties defined by Jordan pairs*, Mh. Math. **86** (1978), 107–129.
25. ——, *On algebraic groups defined by Jordan pairs*, Nagoya Math. J. **74** (1979), 23–66.
26. ——, *Elementary groups and stability for Jordan pairs*, K-Theory **9** (1995), 77–116.
27. ——, *Steinberg groups and simplicity of elementary groups defined by Jordan pairs*, J. Algebra **186** (1996), no. 1, 207–234.
28. ——, *Rank one groups and division pairs*, Bull. Belg. Math. Soc. Simon Stevin **21** (2014), no. 3, 489–521.
29. O. Loos and E. Neher, *Locally finite root systems*, Mem. Amer. Math. Soc. **171** (2004), no. 811.
30. ——, *Steinberg groups for Jordan pairs*, Progress in Mathematics, vol. 332, Birkhäuser Springer, 2019
31. O. Loos, H. P. Petersson and M. L. Racine, *Inner derivations of alternative algebras over commutative rings*, Algebra & Number Theory, **2**, (2008), 927–968.
32. B. Magurn, *An algebraic introduction to K-Theory*, Encyclopedia of Mathematics and Its Applications **87**, Cambridge University Press 2002.
33. J. Milnor, *Introduction to algebraic K-theory*, Ann. of Math. Studies, vol. 72, Princeton University Press, 1971.
34. E. Neher, *Jordan triple systems by the grid approach*, Lecture Notes in Math., vol. **1280**, Springer-Verlag, 1987.
35. ——, *Systèmes de racines 3-gradués*, C. R. Acad. Sci. Paris Sér. I **310** (1990), 687–690.
36. ——, *Lie algebras graded by 3-graded root systems and Jordan pairs covered by a grid*, Amer. J. Math **118** (1996), 439–491.
37. ——, *Polynomial identities and nonidentities of split Jordan pairs*, J. Algebra **211** (1999, 206–224.
38. J. Rosenberg, *Algebraic K-theory and its applications*, Graduate Texts in Mathematics, vol. **147**, Springer-Verlag, New York, 1994.
39. S. Sinchuk, *On centrality of* K_2 *for Chevalley groups of type* E_ℓ, J. Pure Appl. Algebra **220** (2016), 857–875.
40. T. A. Springer, *Jordan algebras and algebraic groups*, Springer-Verlag, New York, 1973, Ergebnisse der Mathematik und ihrer Grenzgebiete, vol **75**.
41. T. A. Springer and F. D. Veldkamp, *Octonions, Jordan algebras and exceptional groups*. Springer Monographs in Mathematics. Springer-Verlag, Berlin, 2000

42. R. Steinberg, *Générateurs, relations, et revêtements de groupes algébriques*, Colloq. Théorie des Groupes Algébriques (Bruxelles), 1962, 113–127.
43. ——, *Lectures on Chevalley groups*, Yale University Lecture Notes, New Haven, Conn., 1967.
44. ——, *Generators, relations and coverings of algebraic groups II*, J. Algebra **71** (1981), 527–543.
45. J. R. Strooker, *The fundamental group of the general linear group*, J. Algebra **48** (1977), 477–508.
46. J. Tits, *Une classe d'algèbres de Lie en relation avec les algèbres de Jordan*, Nederl. Akad. Wetensch. Proc. Ser. A **65** = Indag. Math. **24** (1962), 530–535.
47. J. Tits, *Algèbres alternatives, algèbres de Jordan et algèbres de Lie exceptionnelles. I. Construction*, Nederl. Akad. Wetensch. Proc. Ser. A 69 = Indag. Math. **28** (1966), 223–237.
48. W. van der Kallen, *Another presentation for Steinberg groups*, Nederl. Akad. Wetensch. Proc. Ser. A **80** = Indag. Math. **39** (1977), 304–312.
49. C. Weibel, *The K-book: an introduction to algebraic K-theory*, Graduate Studies in Mathematics **145**, AMS 2013.

On the Hecke-Algebraic Approach for General Linear Groups Over a p-Adic Field

Maxim Gurevich

Abstract We survey some of the known links between the representation theories of p-adic groups, affine Hecke algebras and quantum affine algebras in type A, and their connection to quantum groups. Some recent applications are reviewed.

1 Introduction

Representation theory often deals with families of objects and associated categories that are defined by combinatorial Lie-theoretic data, such as roots systems and their Dynkin diagrams. Roughly speaking, a choice of datum defines a Lie algebra, a p-adic group, a Hecke algebra, a quantum group, etc. Each of these will then give rise to categories of representations of its own particular nature.

The aim of this survey is to present certain known similarities that arise between different settings, when the root data is of type A. In particular, the focus will be biased towards known and potential applications of such links in the study of the smooth representation theory of p-adic groups.

We consider groups of the form $G_n = GL_n(D)$, where D is a division algebra over a non-Archimedean local field F. The category of complex smooth representations of G_n is decomposed into smaller categories known as Bernstein blocks. Each block can then be shown to be equivalent to a category of modules of an appropriately defined affine Hecke algebra. These are associative algebras defined by concrete generators and relations (of type A).

This is the topic of Sect. 2, which presents in some detail an approach for these results, seen in the works of Bernstein and others. We put special emphasis on relatively recent results of Heiermann. We discuss how the Zelevinski classification of irreducible representations of p-adic groups behaves under transition to affine

M. Gurevich (✉)
Department of Mathematics, Technion, Haifa, Israel
e-mail: maxg@technion.ac.il

J. Greenstein et al. (eds.), *Interactions of Quantum Affine Algebras with Cluster Algebras, Current Algebras and Categorification*, Progress in Mathematics 337,
https://doi.org/10.1007/978-3-030-63849-8_5

Hecke algebras. Finally, we discuss in less detail an alternative approach for same result using type-theory.

The next two sections survey the links, in type A, between affine Hecke algebras and the domain of quantum algebra.

Section 3 discusses how the positive part of the quantized algebra $U_v(\mathfrak{sl}_N)$ can naturally be viewed as a deformed Grothendieck ring of representations of affine Hecke algebras. The involvement of Kazhdan–Lusztig polynomials of finite symmetric groups may be viewed as a link which connects between the geometries that govern the two settings.

The last section recalls the Chari–Pressley version of the classical Schur–Weyl duality. While the classical functors relate the representation theory of symmetric groups with that of Lie groups (or algebras) of type A, our interest is in the quantum affine variation. Namely, these functors relate representations of affine Hecke algebras to representations of quantum affine algebras. We survey some recent developments related to this link.

All topics covered here are well-known to experts and each one of them surely received more detailed expositions than the one provided by this survey. My hope is that most of such sources were referenced within the text for the convenience of the reader. Nevertheless, the specific choice of topics, with a particular stress on the representation theory of p-adic groups and the Zelevinski classification, is also aimed at promoting the benefits of our suggested point of view.

The survey is based on several series of seminar talks and informal discussions I held during 2018 in the National University of Singapore, University of Vienna and Tel-Aviv University, in addition to the enlightening conference celebrating Vyjayanthi Chari's 60th birthday in Washington, DC. I would like to thank the participants of these events and others who provided much valued feedback to my understanding and presentation of these topics. Among those are Valentin Buciumas, Ryo Fujita, Wee Teck Gan, Zahi Hazan, David Hernandez, Erez Lapid, Bernard Leclerc, Anton Mellit, Alberto Mínguez, Eugene Mukhin, David Soudry, and Eric Vasserot.

2 From p-Adic Groups to Affine Hecke Algebras

Let F be a non-Archimedean local field. Let $\mathbf{G}$ be a connected reductive group defined over F and $G = \mathbf{G}(F)$ be its group of F-points. We will refer to such G as a *reductive p-adic group*.

The topology on F naturally equips G with a local compact totally disconnected topology.

A *smooth representation* of G is a complex vector space on which G acts linearly, in such manner that the stabilizer of each vector is an open subgroup.

We will write $\mathcal{M}(G)$ for the Abelian category of smooth representations of G. (Morphisms are defined as intertwiners between representations.)

Such categories are the source of much mathematical interest. In this section, we will survey parts of their basic structure, largely following the renowned lecture notes of J. Bernstein [5].

We note that the theory of smooth representations of p-adic groups, and the categories to which it gives rise, is in a highly mature state, spanning links to various mathematical fields and particularly motivated by its involvement in automorphic representations. This survey will not attempt to give an inclusive overview of the theory, but rather focus on specific aspects. (See, for example [48], for a more detailed review.)

2.1 Bernstein Decomposition

The category $\mathcal{M}(G)$ can be decomposed into smaller Abelian categories named *Bernstein blocks* (sometimes we will abbreviate to "blocks"), each of which is amenable to convenient descriptions.

A recurring ingredient in the study of p-adic groups is the pair of functors of (normalized) parabolic induction and restriction (the latter more commonly known as the Jacquet functor).

A parabolic subgroup $P < G$ is one which is given as $P = \mathbf{P}(F)$, for a F-subgroup $\mathbf{P} < \mathbf{G}$ with projective quotient $\mathbf{G}/\mathbf{P}$. Parabolic subgroups admit Levi decompositions $P = MN$, where $N < P$ is the unipotent radical and $M < P$ is a (non-unique) reductive subgroup normalizing N in P.

Each parabolic subgroup with a given Levi decomposition $P = MN$ gives rise to functors

$$i_P^G : \mathcal{M}(M) \to \mathcal{M}(G), \qquad r_P^G : \mathcal{M}(G) \to \mathcal{M}(M),$$

which are adjoint in the sense that

$$\mathrm{Hom}_G(\pi, i_P^G(\sigma)) \cong \mathrm{Hom}_M(r_P^G(\pi), \sigma)$$

holds, for every $\pi \in \mathcal{M}(G)$ and $\sigma \in \mathcal{M}(M)$. Moreover, both functors send irreducible objects to finite-length objects.

Let $\mathrm{Irr}(G)$ denote the collection (of isomorphism classes) of irreducible representations in $\mathcal{M}(G)$ (i.e. not having proper sub-objects).

Let $\Psi(G)$ denote the Abelian group of *unramified characters* of G, that is, continuous homomorphisms $G \to \mathbb{C}^\times$, which are trivial on the subgroup $G_0 < G$ generated by all compact subgroups.

Note, that the group $\Psi(G)$ naturally acts on $\mathrm{Irr}(G)$ by tensoring a representation with a character.

A special subclass $\mathrm{Irr}_c(G) \subseteq \mathrm{Irr}(G)$ consists of irreducible representations ρ, which satisfy $r_P^G(\rho) = 0$, for every parabolic subgroup $P < G$. These are called (irreducible) *supercuspidal* representations. The $\Psi(G)$-orbit of $\rho \in \mathrm{Irr}_c(G)$ is called the *inertia class* of ρ.

The basic example of a Bernstein block is defined by a choice of $\rho \in \mathrm{Irr}_c(G)$. Namely, the supercuspidal Bernstein block $\mathcal{M}_{[\rho]} = \mathcal{M}(G)_{[G,\rho]}$ is defined to be the full subcategory of $\mathcal{M}(G)$ consisting of representations whose irreducible subquotients all belong to the inertia class of ρ.

Now, a general Bernstein block can be defined using parabolic induction. Let $P = MN < G$ be a parabolic subgroup with Levi decomposition, and let $\sigma \in \mathrm{Irr}_c(M)$ be a choice of supercuspidal representation. Then, $\mathcal{M}(G)_{[M,\sigma]}$ is the full subcategory of $\mathcal{M}(G)$ consisting of representations whose irreducible subquotients all appear in $i_P^G(\mathcal{M}_{[\sigma]})$ (that is, the image under i_P^G of the given supercuspidal block in $\mathcal{M}(M)$).

Let $\mathfrak{B}(G)$ denote the collection of Bernstein blocks as defined above. For M and σ as above, we will write $[M,\sigma] \in \mathfrak{B}(G)$ to index the block $\mathcal{M}(G)_{[M,\sigma]}$.

Note, that the data (M,σ) used to define a block is not unique. In fact, the resulting equivalence relation $[M,\sigma] = [M',\sigma']$ of data which defines the same block has a clear description. The equality holds only when M and M' are conjugate subgroups inside G, in such a way that $M = gM'g^{-1}$ for an element $g \in G$ which intertwines the actions of σ' and $\tilde{\sigma}$, where $\tilde{\sigma}$ is a M-representation in the inertia class of σ.

We are ready to formulate the Bernstein decomposition theorem, which states that $\mathcal{M}(G)$ is a product of the blocks defined above. Yet, prior to that, let us recall the meaning of such a statement.

By decomposing an Abelian category $\mathcal{M}$ we mean finding a family of full Abelian subcategories $(\mathcal{M}_i)_{i\in I}$, such that for every object M of $\mathcal{M}$, there are objects M_i of $\mathcal{M}_i$ with $M \cong \oplus_{i\in I} M_i$.

When such a family could be found, we would write $\mathcal{M} = \prod_{i\in I} \mathcal{M}_i$. In particular, we have

$$\mathrm{Hom}_{\mathcal{M}}(M,N) = \oplus_{i\in I} \mathrm{Hom}_{\mathcal{M}_i}(M_i,N_i) ,$$

for all objects $M = \oplus M_i$, $N = \oplus_i N_i$ in $\mathcal{M}$.

Another clear consequence is that $\mathrm{Irr}(\mathcal{M}) = \bigsqcup_{i\in I} \mathrm{Irr}(\mathcal{M}_i)$, where Irr stands for the collection of isomorphism classes of the irreducible objects in a category.

Theorem 2.1.1 (Bernstein Decomposition) *For a reductive p-adic group G, we have*

$$\mathcal{M}(G) = \prod_{[M,\sigma]\in\mathfrak{B}(G)} \mathcal{M}(G)_{[M,\sigma]} .$$

2.1.1 Equivalences of Categories

Let us now treat each block $\mathcal{M} = \mathcal{M}(G)_{[M,\sigma]}$ separately.

For an object Π in $\mathcal{M}$, let

$$\mathcal{H}(\Pi) := \mathrm{Hom}_{\mathcal{M}}(\Pi, \Pi)$$

denote the resulting complex associative algebra. Note, that for every representation M in $\mathcal{M}$, the vector space $\mathrm{Hom}_{\mathcal{M}}(\Pi, M)$ is naturally a right module over $\mathcal{H}(\Pi)$ by composition.

Hence, we can define a functor

$$\begin{aligned} \mathcal{F}_{\Pi} : \mathcal{M} &\to \mathcal{H}(\Pi) - \mathrm{mod} , \\ M &\mapsto \mathrm{Hom}_{\mathcal{M}}(\Pi, M) . \end{aligned}$$

By definition, $\mathcal{F}_{\Pi}$ is an exact functor, when Π is a projective object. The object Π is called a *generator* for $\mathcal{M}$, if $\mathcal{F}_{\Pi}$ is faithful. In other words, Π is a projective generator, if it is projective and $\mathcal{F}_{\Pi}(M) \neq 0$, for every object M in $\mathcal{M}$.

Lemma 2.1.1 *If Π is a projective generator for $\mathcal{M}$ which is a finitely generated representation, then $\mathcal{F}_{\Pi}$ is an equivalence of categories.*

The question now becomes whether we can find finitely generated projective generators Π for $\mathcal{M}$, for which the algebra $\mathcal{H}(\Pi)$ and the module category $\mathcal{H}(\Pi)$ – mod are well-understood.

Indeed, for supercuspidal blocks, this task is relatively easy using the basic theory of p-adic groups. For $\rho \in \mathrm{Irr}_c(G)$, let us write $\Pi(\rho) = \mathrm{ind}_{G_0}^{G}(\rho|_{G_0})$, where ind stands for induction with compact support (Recall that $G_0 < G$ is the subgroup generated by all compact subgroups). It is easy to verify that $\Pi(\rho)$ belongs to $\mathcal{M}_{[\rho]}$.

For any representation π in $\mathcal{M}_{[\rho]}$, the semisimple behavior of supercuspidal representations when restricted to G_0, implies by Frobenius reciprocity that

$$\mathcal{F}_{\Pi(\rho)}(\pi) \cong \mathrm{Hom}_{G_0}(\rho, \pi) \neq 0 .$$

Similarly, one can see that $\mathcal{F}_{\Pi(\rho)}$ is exact. Hence, $\Pi(\rho)$ is (one choice of) a finitely generated projective generator for $\mathcal{M}_{[\rho]}$.

Note, that by taking a G_0-irreducible subrepresentation $\tau < \rho|_{G_0}$, we could have just as easily shown that

$$\Pi'(\rho) = \mathrm{ind}_{G_0}^{G}(\tau) \tag{2.1}$$

is also a finitely generated projective generator for $\mathcal{M}_{[\rho]}$.

In fact, the algebra $\mathcal{H}(\Pi'(\rho))$ will become more relevant for our needs (although by our discussion is still Morita-equivalent to the algebra $\mathcal{H}(\Pi(\rho))$).

As for the case of a general block, we have the following pivotal theorem of Bernstein.

Theorem 2.1.2 *Let $P = MN < G$ be a parabolic subgroup with a Levi decomposition. For $\sigma \in \mathrm{Irr}_c(M)$, let Π be a finitely generated projective generator for $\mathcal{M}(M)_{[\sigma]}$.*

Then, $i_P^G(\Pi)$ is a finitely generated projective generator for the block $\mathcal{M}(G)_{[M,\sigma]}$.

The difficult part in the proof of the above theorem is showing that $i_P^G(\Pi)$ is in fact a projective object. This is obtained through finding a *right*-adjoint functor to parabolic induction (the so-called second-adjointness).

Summing up, we see that each Bernstein block $\mathcal{M}(G)_{[M,\sigma]}$ may be studied through the equivalent category of modules over the complex associative algebra

$$\mathcal{H}(i_P^G(\Pi)) ,$$

where Π is an (easily constructed) choice of finitely generated projective generator for the supercuspidal block $\mathcal{M}(M)_{[\sigma]}$.

In some groups of classical type, the work of Heiermann [30] computed these algebras explicitly and showed that their structure is closely related to that of *affine Hecke algebras*. We will make this statement more concrete for groups related to $\mathbf{GL}_n$ in the next section.

2.1.2 Parabolic Induction

We mention a result of Roche [49] which says that the parabolic induction functor (in fact, a similar statement is true for the Jacquet functor) can be transferred through the equivalences of the form $\mathcal{F}_\Pi$ into an intrinsic operation of algebraic module induction.

Namely, for a representation Π in $\mathcal{M}(M)$, the algebra $\mathcal{H}(\Pi)$ is embedded as a sub-algebra of $\mathcal{H}(i_P^G(\Pi))$ due to functoriality of i_P^G. Thus, a natural module induction functor

$$I_{\Pi,P} : \mathcal{H}(\Pi) - \mathrm{mod} \;\to\; \mathcal{H}(i_P^G(\Pi)) - \mathrm{mod}$$

exists.

Theorem 2.1.3 *For a finitely generated projective generator Π in $\mathcal{M}_{[\sigma]}$, the commutation relation*

$$I_{\Pi,P} \circ \mathcal{F}_\Pi = \mathcal{F}_{i_P^G(\Pi)} \circ i_P^G$$

holds.

2.2 The Case of Type A

Suppose now that $\mathbf{G}$ is either $\mathbf{GL}_m$ or one of its inner forms. In other words, $G = GL_n(D)$, the group of $n \times n$ invertible matrices with entries in a division algebra D over F (with $n^2 \cdot \dim_F D = m^2$). The basic case is when $\mathbf{G}$ is F-split, that is, $D = F$.

Let us write $G_n = GL_n(D)$, for a fixed D. We will often want to consider the sequence of groups $\{G_n\}_{n=1}^{\infty}$ and their smooth representations taken together.

For a given n, let $\alpha = (n_1, \ldots, n_r)$ be a composition of n. We denote by M_α the subgroup of G_n isomorphic to $G_{n_1} \times \cdots \times G_{n_r}$ consisting of matrices which are diagonal by blocks of size $n_1, \ldots, n_r$ and by P_α the subgroup of G_n generated by M_α and the upper unitriangular matrices. A standard parabolic subgroup of G_n is a subgroup of the form P_α and its standard Levi subgroup is M_α.

In particular, we can put the group $M = G_{n_1} \times G_{n_2}$ as a standard Levi subgroup of $G_{n_1+n_2}$, associated with the standard parabolic subgroup $P = P_{(n_1,n_2)}$. Then, for representations π_1, π_2 in $\mathcal{M}(G_{n_1})$, $\mathcal{M}(G_{n_1})$, respectively, we define the *Bernstein–Zelevinski product* [14]

$$\pi_1 \times \pi_2 := i_P^{G_{n_1+n_2}} (\pi_1 \otimes \pi_2) \ ,$$

as a representation in $\mathcal{M}(G_{n_1+n_2})$. This product structure is fundamental in the understanding of the representation theory of the groups $\{G_n\}_n$.

Every character $\chi \in \Psi(G_n)$ is of the form $\chi = \nu^s := |\,\mathrm{Nrd}\,|_F^s$, where $\mathrm{Nrd} : G_n \to F^\times$ is the reduced norm of the algebra $M_n(D)$, $|\cdot|_F$ is the absolute value of F, and $s \in \mathbb{C}$.

For every $\rho \in \mathrm{Irr}_c$, there exists a unique positive $t_\rho \in \mathbb{R}$, for which $\rho \times \nu^{t_\rho}\rho$ is reducible (see Appendix A of [41]). Let us write $\nu_\rho^s = \nu^{st_\rho}$. For the case of $D = F$, we have $t_\rho = 1$ and $\nu_\rho = \nu$, for all $\rho \in \mathrm{Irr}$.

For reasons of conjugation, the Bernstein blocks of $\mathcal{M}(G_n)$ are exhausted by representations of standard Levi subgroups induced through standard parabolic subgroups.

Let $\sigma \in \mathrm{Irr}_c(M_\alpha)$ be given. It is easily verified that an element $g \in G_n$ can be found, which normalizes M_α and permutes its factors so that

$$\sigma^g \cong \rho_1^1 \otimes \cdots \otimes \rho_1^{l_1} \otimes \cdots \otimes \rho_k^1 \otimes \cdots \otimes \rho_k^{l_k} \ ,$$

with $\{\rho_i^j\}$ irreducible supercuspidal representations satisfying that $\rho_i^1, \ldots, \rho_i^{l_i}$ all lie in the same $\Psi(G_{n_i'})$-orbit $\mathcal{O}_i$, for every $1 \leqslant i \leqslant k$, and $\mathcal{O}_{i_1} \neq \mathcal{O}_{i_2}$ for $i_1 \neq i_2$.

Hence, $[M_\alpha, \sigma] = [M_\alpha, \sigma^g] \in \mathfrak{B}(G_n)$. Moreover, since the block $[M_\alpha, \sigma]$ is determined up to the $\Psi(M_\alpha)$-orbit of σ, we may assume that

$$\sigma = \rho_1 \otimes \cdots \otimes \rho_1 \otimes \cdots \otimes \rho_k \otimes \cdots \otimes \rho_k \ , \tag{2.2}$$

for a given collection $\{\rho_i \in \mathrm{Irr}_c(G_{n'_i})\}$, with distinct i_1, i_2 satisfying that either $n'_{i_1} \neq n'_{i_2}$ or ρ_{i_1}, ρ_{i_2} not lying in the same $\Psi(G_{n'_{i_1}})$-orbit.

Let us write $M_\alpha < M(\sigma) < G_n$ for the standard Levi subgroup given as

$$M(\sigma) = G_{n'_1 l_1} \times \cdots \times G_{n'_k l_k} ,$$

and $P(\sigma) < G_n$ for the corresponding standard parabolic subgroup. Then, $[M_\alpha, \sigma]$ also defines a Bernstein block for the group $M(\sigma)$.

Proposition 2.2.1 *The parabolic induction functor*

$$\begin{aligned} i_{P(\sigma)}^{G_n} : \mathcal{M}\left(G_{n'_1 l_1} \times \cdots \times G_{n'_k l_k}\right)_{[M,\sigma]} &\to \mathcal{M}(G_n)_{[M_\alpha,\sigma]} \\ \pi_1 \otimes \cdots \otimes \pi_k &\mapsto \pi_1 \times \cdots \times \pi_k \end{aligned}$$

is an equivalence of categories.

When $k = 1$ (i.e. $M(\sigma) = G_n$), we will call $\mathcal{M}(G_n)_{[M_\alpha,\sigma]}$ a *simple block.*

Any given $\rho \in \mathrm{Irr}_c(G_d)$ and an integer $r \geqslant 1$ define a simple block $\Theta(\rho, r) \in \mathfrak{B}(G_{rd})$ by taking the Levi subgroup $M_\alpha < G_{rd}$ with $\alpha = (d, \ldots, d)$ and setting

$$\Theta(\rho, r) = [M_\alpha, \rho \otimes \cdots \otimes \rho] .$$

In light of Proposition 2.2.1, it makes sense to denote a general block of $\mathfrak{B}(G_n)$ which is given as in (2.2) by

$$[M, \sigma] = \Theta(\rho_1, l_1) \times \cdots \times \Theta(\rho_k, l_k) .$$

2.2.1 Hecke Algebras

The complex (finite) *Hecke algebra* for the symmetric group $\mathfrak{S}_n$ and a parameter $q \in \mathbb{C}^\times$ is defined to be the complex associative algebra $H_f(n, q)$ generated by $T_1, \ldots, T_{n-1}$, subject to the relations

$$\begin{aligned} &T_i T_{i+1} T_i = T_{i+1} T_i T_{i+1}, && \forall 1 \leqslant i \leqslant n-2 \\ &(T_i - q)(T_i + 1) = 0, && \forall 1 \leqslant i \leqslant n-1 \\ &T_i T_j = T_j T_i, && \forall |j - i| > 1 . \end{aligned} \tag{2.3}$$

Note, that for $q = 1$, $H_f(n, 1)$ becomes nothing but the complex group algebra of $\mathfrak{S}_n$. Thus, the finite Hecke algebra can be viewed as a deformation of the Weyl group of GL_n.

We will define the complex *affine Hecke algebra* for GL_n and a parameter $q \in \mathbb{C}^\times$ to be the vector space

$$H(n,q) = \mathbb{C}[y_1^{\pm 1},\ldots,y_n^{\pm 1}] \otimes H_f(n,q)\,,$$

with the associative algebra structure given by both subalgebras $\mathbb{C}[y_1^{\pm 1},\ldots,y_n^{\pm 1}]$ (the commutative ring of Laurent polynomials) and $H_f(n,q)$ and the commutation relations

$$\begin{aligned} T_i y_i T_i &= q y_{i+1}, \ \forall 1 \leqslant i \leqslant n-1 \\ T_i y_j &= y_j T_i, \quad \ \forall j \neq i, i+1\,, \end{aligned}$$

between them.

Theorem 2.2.1 ([30]) *Let $\Pi'(\sigma)$ be the projective generator as defined in* (2.1) *for the supercuspidal block $\mathcal{M}_{[\sigma]}$, with σ as in* (2.2).

Then, the algebra $\mathcal{H}(i_P^G(\Pi'(\sigma)))$ is isomorphic to

$$H\left(n_1', q_F^{t_{\rho_1} o(\rho_1)}\right) \otimes \cdots \otimes H\left(n_k', q_F^{t_{\rho_k} o(\rho_k)}\right)\,,$$

where q_F is the cardinality of the finite residue field of F and $o(\rho_i)$ is the cardinality of the stabilizer subgroup of ρ_i in $\Psi(G_{n_i'})$.

In particular, each simple block $\Theta(\rho, r)$ is equivalent as an Abelian category to the category of right modules over an affine Hecke algebra for GL_r.

Note, that the isomorphisms in the theorem above are compatible with parabolic induction in the following sense.

Suppose that $\Theta(\rho, r_1), \Theta(\rho, r_2)$ are simple blocks for representations of the groups G_{n_1}, G_{n_2}, respectively. Let Π_1, Π_2 be the corresponding generators described by Theorem 2.2.1, that is, $\Pi_i = i_{P_i}^{G_{n_i}}(\Pi'(\sigma_i))$, $i = 1, 2$, for suitable supercuspidal representations σ_1, σ_2. Then, $\mathcal{H}(\Pi_i) \cong H(r_i, q_F^{o(\rho)})$.

Recall again that the group $M = G_{n_1} \times G_{n_2}$ appears as a standard Levi subgroup of $G_{n_1+n_2}$. Parabolic induction then gives an embedding

$$\mathcal{H}(\Pi_1) \otimes \mathcal{H}(\Pi_2) \to \mathcal{H}(\Pi'(\sigma_1) \times \Pi'(\sigma_2)) \tag{2.4}$$

of algebras, for the suitable parabolic subgroup P. Yet, since clearly $\Pi'(\sigma_1 \otimes \sigma_2) \cong \Pi'(\sigma_1) \otimes \Pi'(\sigma_2)$ holds and the Bernstein block defined by $\sigma_1 \otimes \sigma_2$ in $\mathcal{M}(G_{n_1+n_2})$ is nothing but $\Theta(\rho, r_1 + r_2)$, Theorem 2.2.1 implies that the embedding in (2.4) may be described as

$$H(r_1, q_F^{t_\rho o(\rho)}) \otimes H(r_2, q_F^{t_\rho o(\rho)}) \to H(r_1 + r_2, q_F^{t_\rho o(\rho)})\,. \tag{2.5}$$

The explication of Heiermann's isomorphisms shows that the above map is in fact the natural map we obtain by sending generators to generators in our definition of affine Hecke algebras.

In similarity to the Bernstein–Zelevinski product for group representations, for modules π_1, π_2 over the algebras $H(r_1, q_F^{t_\rho o(\rho)})$, $H(r_2, q_F^{t_\rho o(\rho)})$, respectively, we define $\pi_1 \times \pi_2$ to be the $H(r_1 + r_2, q_F^{t_\rho o(\rho)})$-module resulting from the module-theoretic operation of induction of $\pi_1 \otimes \pi_2$ through the algebra embedding in (2.5).

Thus, by means of Theorem 2.1.3, we see that the category equivalences resulting from Theorem 2.2.1 intertwine the Bernstein–Zelevinski product of group representations with the one for affine Hecke algebras.

2.3 *Classification of Irreducible Representations*

We would like to discuss the meaning of the category equivalences implied by Theorem 2.2.1 on the level of irreducible representations.

For an Abelian category $\mathcal{M}$, let $\mathrm{Irr}(\mathcal{M})$ denote its collection of isomorphism classes of irreducible objects in $\mathcal{M}$.

We will write

$$\mathrm{Irr} = \{0\} \cup \bigsqcup_{n=1}^{\infty} \mathrm{Irr}(\mathcal{M}(G_n)), \qquad \mathrm{Irr}_c = \bigsqcup_{n=1}^{\infty} \mathrm{Irr}_c(\mathcal{M}(G_n)) .$$

Given $\rho \in \mathrm{Irr}_c$, we will write

$$\mathrm{Irr}_{[\rho]} = \{0\} \cup \bigsqcup_{r=1}^{\infty} \mathrm{Irr}(\Theta(\rho, r)) .$$

It is a consequence of Proposition 2.2.1 that there is a natural identification

$$\mathrm{Irr} = \prod_{[\rho]}{}' \mathrm{Irr}_{[\rho]} , \tag{2.6}$$

where the product goes over all inertia classes in Irr_c. Here, the product (of sets) is taken in a restricted manner, that is, all but finitely many factors should be zero.

An early cornerstone achievement of the representation theory of p-adic was the *Zelevinski classification* [57] of Irr, for the case of $D = F$, in combinatorial terms. Through a long list of works involving several approaches (see Appendix A of [41] for a thorough discussion), it was later established that the classification of Irr for a general division algebra D remains essentially similar.

Let us present it here in a way which is compatible with the decomposition of (2.6).

A *segment* $[a, b]$ is a formal object given by two numbers $a, b \in \mathbb{C}$ whose difference $b - a$ is a non-negative integer. We will write Seg for the collection of segments.

We will write Mult for the collection of *multisegments*, that is, all sets with prescribed multiplicities of elements of Seg. It is convenient to think of Mult as $\mathbb{Z}_{\geqslant 0}(\mathrm{Seg})$, that is, the monoid of functions $\mathrm{Seg} \to \mathbb{Z}_{\geqslant 0}$ with finite support. This view also allows to write multisegments additively, e.g.

$$[1, 1] + [1.3, 2.3] + [1.3, 2.3] + [i - 5, i - 11] \in \mathrm{Mult} .$$

Given $\Delta = [a, b] \in \mathrm{Seg}$ and $\rho \in \mathrm{Irr}_c$, the segment representation $\Delta_\rho \in \mathrm{Irr}_{[\rho]}$ is defined as the unique irreducible subrepresentation of

$$\nu_\rho^a \rho \times \nu_\rho^{a+1} \rho \times \cdots \times \nu_\rho^b \rho .$$

Note, that for $s \in \mathbb{C}$ with $\nu_\rho^s \rho \cong \rho$, the segment $\tilde{\Delta} = [a + s, b + s]$ would give $\tilde{\Delta}_\rho \cong \Delta_\rho$. In fact, such numbers s must form a discrete subgroup of the imaginary complex axis.

Let $\mathrm{Seg}_{[\rho]}$ be the quotient of the collection Seg by relations of the form $\Delta \sim \tilde{\Delta}$, which simply identifies segments up to a certain twist on the imaginary axis. Similarly, we write $\mathrm{Mult}_{[\rho]} = \mathbb{Z}_{\geqslant 0}(\mathrm{Seg}_{[\rho]})$.

Theorem 2.3.1 *For every choice of $\rho \in \mathrm{Irr}_c$, there is a bijection*

$$Z_\rho : \mathrm{Mult}_{[\rho]} \to \mathrm{Irr}_{[\rho]} ,$$

so that for each $\mathfrak{m} = \sum_{i=1}^t \Delta^i \in \mathrm{Mult}$, $Z_\rho(\mathfrak{m})$ is the unique irreducible subrepresentation of

$$\Delta_\rho^1 \times \cdots \times \Delta_\rho^t ,$$

where the segments defining $\mathfrak{m}$ are ordered in a prescribed manner.

The prescribed ordering of segments in the theorem can in fact be taken as any ordering which *avoids* a situation in which $\Delta_\rho^i = [a_i, b_i]$ and $\Delta_\rho^j = [a_j, b_j]$ exist, for which $i < j$, $a_j - a_i$, $b_j - b_i$ are positive integers, and $a_j \leqslant b_j + 1$.

Let us consider the family of categories $\mathcal{M}(n, q)$ of right modules over the algebra $H(n, q)$. For $q \in \mathbb{C}^\times$ we will write

$$\mathrm{Irr}_{\mathrm{Hecke}}(q) = \{0\} \cup \bigsqcup_{n=1}^{\infty} \mathrm{Irr}\, \mathcal{M}(n, q) .$$

For every choice of $\rho \in \mathrm{Irr}_c$ and $r \geqslant 1$, Theorem 2.2.1 supplies an identification of $\mathrm{Irr}\,\Theta(\rho, r)$ with $\mathrm{Irr}\,\mathcal{M}(r, q_F^{t_\rho o(\rho)})$. Thus, we obtain an identification

$$T_\rho : \mathrm{Irr}_{[\rho]} \cong \mathrm{Irr}_{\mathrm{Hecke}}(q_F^{t_\rho o(\rho)}) \tag{2.7}$$

of sets, which depends on the choice of ρ.

Let us vary the field F and the algebra D for the sake of current discussion. Suppose that $\rho_i \in \mathrm{Irr}_c(GL_{d_i}(D_i)), i = 1, 2$ are two representations with $q_{F_1}^{t_{\rho_1} o(\rho_1)} = q_{F_2}^{t_{\rho_2} o(\rho_2)}$. Then, $T_{\rho_1,\rho_2} = T_{\rho_2}^{-1} \circ T_{\rho_1}$ is a bijection from $\mathrm{Irr}_{[\rho_1]}$ to $\mathrm{Irr}_{[\rho_2]}$.

Proposition 2.3.1 ([27, Proposition 3.2]) *The assumption* $q_{F_1}^{t_{\rho_1} o(\rho_1)} = q_{F_2}^{t_{\rho_2} o(\rho_2)}$ *implies the existence of a bijection* $\phi_{\rho_1,\rho_2} : \mathrm{Seg}_{[\rho_1]} \to \mathrm{Seg}_{[\rho_2]}$, *which naturally extends to a bijection* $\phi_{\rho_1,\rho_2} : \mathrm{Mult}_{[\rho_1]} \to \mathrm{Mult}_{[\rho_2]}$.

The resulting map is compatible with the correspondence T_{ρ_1,ρ_2}, *that is,*

$$Z_{\rho_2} \circ \phi_{\rho_1,\rho_2} = T_{\rho_1,\rho_2} \circ Z_{\rho_1} .$$

Let us mention that a direct classification (in the spirit of the Zelevinski classification) of $\mathrm{Irr}_{\mathrm{Hecke}}(q)$, for any q which is not a root of unity, was conducted by Rogawski [50], using purely Hecke-algebraic techniques.

2.4 Alternative Type-Theory Approach

The relation between Bernstein blocks of $\mathcal{M}(G_n)$ and module categories over affine Hecke algebras, in fact, originated earlier than the result described in Theorem 2.2.1. Such relation was a major outcome of an approach called *type-theory* for the study of $\mathcal{M}(G_n)$ (or, p-adic groups in general).

Roughly speaking, type-theory studies compact subgroups of $J < G$ of a p-adic group G and irreducible representations σ of such J, so that for every $\pi \in \mathrm{Irr}(\mathcal{M}(G))$, $\mathrm{Hom}_J(\sigma, \pi|_J) \neq 0$ holds, if and only if, π belongs to a given Bernstein block $\mathcal{M}(G)_{\mathfrak{s}}$. Such pair (J, σ) is called a type for the block $\mathcal{M}(G)_{\mathfrak{s}}$.

By Frobenius reciprocity, $\mathrm{Hom}_J(\sigma, \pi|_J) \cong \mathrm{Hom}_G(\mathrm{ind}_J^G \sigma, \pi)$. Thus, when (J, σ) is a type for a given Bernstein block, $\Pi = \mathrm{ind}_J^G \sigma$ becomes a finitely generated projective generator for that block, and by Lemma 2.1.1 the block is equivalent to the category of modules over $\mathcal{H}(\Pi)$.

An early motivation for such theory was a discovery [13, 16] of a type for the *principle block* of $\mathcal{M}(G)$, for any connected reductive G. This is the block produced by taking the trivial representation of a minimal Levi subgroup of G, as a supercuspidal representation. In other words, the principle block is the one which is generated by *unramified principal series*. The type constructed for the principal block was of the form (I, trv), where $I < G$ is a compact subgroup which may be referred to as the *Iwahori subgroup* for certain cases, while trv is its trivial representation.

Using these results, the principle block $\mathcal{M}(G)$ could be realized (see, for example, [9]) as the category of modules of the algebra

$$\mathcal{H}(\Pi) \cong C_c^\infty(I \setminus G/I) , \tag{2.8}$$

that is, the algebra (under a natural convolution product) of smooth functions on G with compact support, that are bi-invariant under the subgroup I.

For the case of $G = GL_n(F)$, the next development in this line of research came in the seminal works of Bushnell–Kutzko [8, 10].

Theorem 2.4.1 ([8]) *For every simple block $\Theta(\rho, r)$ of the category $\mathcal{M}(GL_n(F))$, there is a type (J, σ).*

Moreover, there is a subgroup $C < GL_n(F)$, such that $C \cong GL_r(E)$, where E/F is a field extension of degree n/r, $C \cap J$ is a Iwahori subgroup of C, and

$$\mathcal{H}(\mathrm{ind}_J^{GL_n(F)}(\sigma)) \cong C_c^\infty(C \cap J \setminus C / C \cap J)\ .$$

Thus, it was shown that all simple blocks for the groups $GL_n(F)$ are equivalent to module categories over algebras of the form (2.8), up to a change of field. In the subsequent work [10], it was further established that all blocks in $\mathcal{M}(GL_n(F))$ are in fact governed by tensor products of algebras of the form (2.8).

The relation to affine Hecke algebras (that is, the class of algebras which specializes in the case of GL_n to the definition of the previous section) from this point of view is intriguing on its own. Algebras of the form (2.8) were initially studied by Iwahori–Matsumoto [37]. The fact that they can be presented as affine Hecke algebras (generalizing our definition for GL_n) was a non-trivial discovery attributed to Bernstein and Zelevinski and presented in a paper of Lusztig [44].

In that respect, we mention Bezrukavnikov's work [6], which sheds more light from a categorical perspective on this non-trivial isomorphism.

When taking under account, for the case of $G = GL_n(F)$, the identification of (2.8) with affine Hecke algebras, we see that Theorem 2.4.1 (together with its extensions) and Theorem 2.2.1 give essentially the same result using two distinct approaches.

Finally, the work of Bushnell–Kutzko received a successful extension for the general case of $G_n = GL_n(D)$ in the works of Sécherre and Stevens [52, 53]. Their results showed again that all blocks admit types which produce generators whose endomorphism algebras are isomorphic to the algebras of (2.8). In this manner, we see that the type-theory approach provides a full alternative (and earlier) proof of Theorem 2.2.1.

3 From Affine Hecke Algebras to Quantum Groups

As mentioned before, one of the key challenges in the study of the representation theory of the groups G_n, or the algebras $H(n, q)$, is the decomposition of representations of the form $\pi_1 \times \pi_2$, where $\pi_1, \pi_2 \in \mathrm{Irr}$, into their irreducible constituents. One tool which is available to us in this task is realizing the problem as a specialization of a decomposition problem arising in the domain of quantum groups. We would like to overview these relations. A large part of the following discussion will closely follow the expository treatment in [43].

The discussion of the previous section shows that it is enough to treat such problems in the setting of $\mathcal{M}(n,q)$, that is, the module categories of affine Hecke algebras.

We will fix a parameter $0 \neq q \in \mathbb{C}$ which is not a root of unity (indeed, when coming from group representations, q will always be a positive power of the prime power q_F). Let R_n be the complex Grothendieck group of the category $\mathcal{M}_{fl}(n,q)$ consisting of finite-length objects in $\mathcal{M}(n,q)$, that is, the complex vector space with a formal basis $\{[\pi]\}_{\pi\in \mathrm{Irr}\,\mathcal{M}(n,q)}$. Then, for each object σ in $\mathcal{M}_{fl}(n,q)$, the element $[\sigma] \in R_n$ stands for the formal sum given by the Jordan-Hölder series of σ.

Naturally, the Bernstein–Zelevinski product equips $R := \oplus_{n=0}^{\infty} R_n$ with an associative algebra structure (here, $R_0 := \mathbb{C}\cdot 1$ serves as the ring identity element). In fact, the algebra R turns out to be commutative.

Recall, that for $\rho \in \mathrm{Irr}_c(G_d)$ (assuming $q = q_F^{t_\rho o(\rho)}$), $\mathrm{Irr}_{[\rho]}$ may be identified (2.7) with the basis of simple modules in R. In particular, we can index this basis by $\mathrm{Mult}_{[\rho]}$ through the map Z_ρ of Theorem 2.3.1.

Let us make a further basic reduction in the analysis of the product structure on R. For $s \in \mathbb{C}$, let $\mathrm{Mult}^s = \mathrm{Mult}^s_{[\rho]} \subseteq \mathrm{Mult}_{[\rho]}$ be the collection of multisegments containing only segments of the form $[a+s, b+s]$, for $a, b \in \mathbb{Z}$. Let $R^s \subseteq R$ be space spanned by $Z_\rho(\mathrm{Mult}^s)$.

For each $s \in \mathbb{C}$, R^s is a subalgebra of R, and we have

$$R \cong \otimes_{\nu^s\in\Psi(G_d)/\{\nu^k\,:\,k\in\mathbb{Z}\}} R^s \ .$$

Furthermore, R^r and R^s are isomorphic rings, for all $r, s \in \mathbb{C}$, as a result of the shift functor $\pi \mapsto \nu^{s-r}\pi$.

In particular, it is enough to study the product structure of the algebra R^0, whose basis is parameterized by Mult^0, that is, multisets of segments $[a, b]$, with integer a, b. We write Seg^0 for these segments, i.e., $\mathrm{Mult}^0 = \mathbb{N}(\mathrm{Seg}^0)$.

3.1 A Quantization

Let us recall a different Lie-theoretic setting in which Mult^0 naturally parameterizes a basis. Let $\mathfrak{n}$ be a maximal nilpotent sub-algebra of the simple Lie algebra $\mathfrak{sl}_n$. Its basis, given by the set of positive roots of $\mathfrak{sl}_n$, can be indexed by segments $[a, b]$, with integers $1 \leqslant a \leqslant b \leqslant n-1$.

Hence, by the PBW theorem, the universal enveloping algebra $U(\mathfrak{n}) = U(\mathfrak{sl}_n)^+$ comes with a (non-canonical) basis parameterized by multisets of segments of the form above. In fact, to make things more convenient we may take the infinite-dimensional Lie algebra $\mathfrak{sl}_\infty$, with $\mathfrak{n}$ being the algebra of strictly uppertriangular bi-infinite matrices. Again, through a PBW theorem, $U(\mathfrak{sl}_\infty)^+$ becomes equipped with a basis indexed by Mult^0.

In order to see the precise links between the algebras R^0 and $U(\mathfrak{sl}_\infty)^+$, we need to *quantize* the setting.

Let $U_v = U_v(\mathfrak{sl}_\infty)^+$ be the $\mathbb{C}(v)$-algebra (Here $\mathbb{C}(v)$ is the complex field of rational function in a formal variable v) given by the countable set of generators $\{E_i\}_{i\in\mathbb{Z}}$, subject to the (so-called quantum Serre) relations

$$\begin{cases} E_iE_j = E_jE_i, & |i-j|>1 \\ E_i^2E_j - (v+v^{-1})E_iE_jE_i + E_jE_i^2 = 0 & |i-j|=1 \end{cases}.$$

In analogy to the classical PBW situation, each segment $[a,b] \in \mathrm{Seg}^0$ may give rise to an element $E^*([a,b]) \in U_v$ in the algebra generated by $\{E_a, E_{a+1}, \ldots, E_b\}$. The collection $\mathcal{E}^* = \{E^*(\mathfrak{m})\}_{\mathfrak{m}\in\mathrm{Mult}^0}$, where $E^*(\mathfrak{m}) = E^*(\Delta^1)\cdot\ldots\cdot E^*(\Delta^k)$, for a prescribed (in a similar sense to the prescription in Theorem 2.3.1) ordering $\mathfrak{m} = \sum_{i=1}^k \Delta^i$, is a (dual-PBW) basis for U_v.

While U_v has a natural specialization at $v=1$ to the classical algebra $U(\mathfrak{sl}_\infty)^+$, it has yet another specialization to our commutative algebra of interest R^0. Next, we will give an overview of the phenomenon.

3.2 Standard Modules and Underlying Geometry

Recall, that for $\mathfrak{m} = \sum_{i=1}^k \Delta^i \in \mathrm{Mult}_{[\rho]}$, $Z_\rho(\mathfrak{m})$ was defined as an irreducible subrepresentation of $\zeta_\rho(\mathfrak{m}) := \prod_{i=1}^k \Delta^i_\rho$, for a prescribed ordering of the segments in $\mathfrak{m}$. We will call representations of the form $\zeta_\rho(\mathfrak{m})$ *standard representations* (or modules).

It is known that the standard representations give a basis to the complex algebra R. In particular, $\{[\zeta_\rho(\mathfrak{m})]\}_{\mathfrak{m}\in\mathrm{Mult}^0}$ is a basis for R^0.

Note, that in the commutative algebra R the order of segments which defines $[\zeta(\mathfrak{m})]$ is immaterial. Moreover, it is evident that

$$[\zeta_\rho(\mathfrak{m})][\zeta_\rho(\mathfrak{n})] = [\zeta_\rho(\mathfrak{m}+\mathfrak{n})] \tag{3.1}$$

holds in R, for all $\mathfrak{m}, \mathfrak{n} \in \mathrm{Mult}_{[\rho]}$, even though as representations, we may often have $\zeta_\rho(\mathfrak{m}) \times \zeta_\rho(\mathfrak{n}) \not\cong \zeta_\rho(\mathfrak{m}+\mathfrak{n})$.

The last multiplicative property makes the basis of standard representations especially relevant to the study of the product structure on R^0. It is a simple linear algebra exercise to see that if we fully knew the coefficients of the transition matrix between the basis of irreducible representations to that of standard ones, then we would have a closed formula for the decomposition of $Z_\rho(\mathfrak{m}) \times Z_\rho(\mathfrak{n})$ into irreducible constituents.

Indeed, we would like to describe how the said coefficients may be computed by means of the Kazhdan–Lusztig theory.

First, let us reinterpret the collection Mult^0 in terms of quiver representations (see, for example, [45]). Let Q be the Dynkin quiver of type A_∞, that is, the directed graph whose vertices are given by $\mathbb{Z}$, with an arrow $i \to i+1$, for each $i \in \mathbb{Z}$. A (complex) representation of Q is a finite-dimensional graded complex vector space $V = \oplus_{i\in\mathbb{Z}} V_i$ together with linear maps $g_i : V_i \to V_{i+1}$. The representations of Q (with their natural morphisms) form an Abelian category $\mathcal{M}(Q)$.

A segment $[a, b] \in \mathrm{Seg}^0$ defines an indecomposable object $V_{[a,b]}$ of $\mathcal{M}(Q)$ by setting

$$\left(V_{[a,b]}\right)_i = \begin{cases} \mathbb{C} & a \leqslant i \leqslant b \\ 0 & i \notin [a,b] \end{cases},$$

and $g_i : V_i \cong V_{i+1}$, for all $a \leqslant i < b$.

It can be verified that all representations of Q are isomorphic to direct sums of objects of the form V_Δ, for $\Delta \in \mathrm{Seg}^0$. In other words, the objects of $\mathcal{M}(Q)$ are in a natural correspondence with the multisegments in Mult^0.

A geometric context can be added to this setting. Let us fix a dimension vector, that is, $\underline{d} = (d_i)_{i\in\mathbb{Z}}$, with $d_i \in \mathbb{Z}_{\geqslant 0}$, and $d_i = 0$, for all but finitely many i. We set the vector space

$$E_{\underline{d}} = \oplus_{i\in\mathbb{Z}} \mathrm{Hom}_{\mathbb{C}}(\mathbb{C}^{d_i}, \mathbb{C}^{d_{i+1}})\ .$$

Then, each representation V of Q, with $\dim V_i = d_i$, for all i, can be viewed as a point in $E_{\underline{d}}$. Moreover, the isomorphism classes of Q-representations with the fixed dimension vector $\underline{d}$ can be thought of as orbits of the group $\prod_{i\in\mathbb{Z}} GL_{d_i}(\mathbb{C})$ acting on the space $E_{\underline{d}}$.

For $\mathfrak{m} \in \mathrm{Mult}^0$, let $\mathcal{O}_{\mathfrak{m}}$ denote the orbit in $E_{\underline{d}(\mathfrak{m})}$, for a suitable $\underline{d}(\mathfrak{m})$ (ith coordinate of $\underline{d}(\mathfrak{m})$ is determined by the number of segments in $\mathfrak{m}$ that contain i), of the corresponding Q-representation isomorphism class.

Some of the results of [57] can be stated in the following geometric manner.

Proposition 3.2.1 *For* $\mathfrak{m} \in \mathrm{Mult}^0$ *and* $\rho \in \mathrm{Irr}_c$, *the irreducible constituents of the representation* $\zeta_\rho(\mathfrak{m})$ *are precisely those of the form* $Z_\rho(\mathfrak{n})$, *where* $\mathfrak{n} \in \mathrm{Mult}^0$ *is such that* $\mathcal{O}_{\mathfrak{m}} \subseteq \overline{\mathcal{O}_{\mathfrak{n}}}$.

The above proposition hints that the transition matrix between the bases of standard and irreducible representation is related to the geometry of the varieties $\{\mathcal{O}_{\mathfrak{m}}\}_{\mathfrak{m}\in\mathrm{Mult}^0}$. This might have lead Zelevinski [58] to formulate a Kazhdan–Lusztig-type conjecture for the setting in hand.

For any $\mathfrak{m}, \mathfrak{n} \in \mathrm{Mult}^0$ satisfying the inclusion $\mathcal{O}_{\mathfrak{m}} \subseteq \overline{\mathcal{O}_{\mathfrak{n}}}$, let us write $\mathcal{H}^i(\overline{\mathcal{O}_{\mathfrak{n}}})_{\mathfrak{m}}$ for the stalk at a point of $\mathcal{O}_{\mathfrak{m}}$ of the i-th intersection cohomology sheaf of the variety $\overline{\mathcal{O}_{\mathfrak{n}}}$. The intersection cohomology polynomial is then defined as

$$IC_{\mathfrak{m},\mathfrak{n}}(q) = \sum_{i\geqslant 0} q^i \dim \mathcal{H}^{2i}(\overline{\mathcal{O}_{\mathfrak{n}}})_{\mathfrak{m}} \,. \tag{3.2}$$

Theorem 3.2.1 *For any $\rho \in \mathrm{Irr}_c$ and $\mathfrak{m}, \mathfrak{n} \in \mathrm{Mult}^0$ such that $\mathcal{O}_{\mathfrak{m}} \subseteq \overline{\mathcal{O}_{\mathfrak{n}}}$, the multiplicity of the irreducible representation $Z_\rho(\mathfrak{n})$ in the Jordan-Hölder series of $\zeta_\rho(\mathfrak{m})$ is given by the value $IC_{\mathfrak{m},\mathfrak{n}}(1)$.*

This theorem was eventually proved in [17] using a realization of affine Hecke algebras as Grothendieck groups of equivariant coherent sheaves on Steinberg varieties (see further the remark following Theorem 3.4.1).

3.3 Dual Canonical Basis

Theorem 3.2.1 states that the geometry of the varieties associated with Q-representations governs some of the behavior of representations in $\mathcal{M}(n, q)$. Yet, it is also evident that when specializing the polynomials $IC_{\mathfrak{m},\mathfrak{n}}(q)$ at $q = 1$, some of the geometric information encoded in these varieties is lost. This additional information still bears meaning in our context of representation theory, which can be recovered through the algebra U_v.

Lusztig defined in [45] another $\mathbb{C}(v)$-basis for U_v, indexed again by Mult^0, which is named the *canonical basis*, due to some remarkable properties which go beyond the scope of this survey. For our purposes, we will deal with yet another related basis $\mathcal{B}^* = \{G^*(\mathfrak{m})\}_{\mathfrak{m}\in\mathrm{Mult}^0}$, which is *dual* to the canonical basis relative to a given scalar product on U_v.

It was then shown that the transition matrix between the dual-PBW basis $\mathcal{E}^*$ defined above to the dual canonical basis $\mathcal{B}^*$ is given precisely by the geometric data encoded in the polynomials $IC_{\mathfrak{m},\mathfrak{n}}$.

Theorem 3.3.1 ([45]) *For any $\mathfrak{m}, \mathfrak{n} \in \mathrm{Mult}^0$, let $c_{\mathfrak{m},\mathfrak{n}} \in \mathbb{C}(v)$ be the transition coefficient given by the equations*

$$E^*(\mathfrak{m}) = \sum_{\mathfrak{n}\in\mathrm{Mult}^0} c_{\mathfrak{m},\mathfrak{n}} G^*(\mathfrak{n}) \,.$$

Then, $c_{\mathfrak{m},\mathfrak{n}} = 0$, unless $\mathcal{O}_{\mathfrak{m}} \subseteq \overline{\mathcal{O}_{\mathfrak{n}}}$ is satisfied. In the latter case, we have

$$c_{\mathfrak{m},\mathfrak{n}}(v) = v^{\dim \mathcal{O}_{\mathfrak{n}} - \dim \mathcal{O}_{\mathfrak{m}}} IC_{\mathfrak{m},\mathfrak{n}}(v^{-2}) \,.$$

We see now that the above transition matrix inside the quantum group is specialized at $v = 1$ precisely to the transition matrix between the irreducible and standard bases of R^0. This statement is made more precise in the following proposition, whose proof is essentially the combination of Theorems 3.2.1 and 3.3.1.

Proposition 3.3.1 ([43]) *Let $\mathcal{L} \subseteq U_v$ be the $\mathbb{Z}[v, v^{-1}]$-ring (lattice) spanned by the basis $\mathcal{E}^*$.*

Let $U = \mathbb{C}_{v\to 1} \otimes_{\mathbb{Z}[v,v^{-1}]} \mathcal{L}$ be the complex algebra constructed by taking $\mathbb{C}_{v\to 1} = \mathbb{C}$ as a $\mathbb{Z}[v, v^{-1}]$-module in which v acts trivially.

Then, the linear map defined as

$$\begin{array}{rl} U & \to R^0 \, , \\ 1 \otimes E^*(\mathfrak{m}) & \mapsto \zeta_\rho(\mathfrak{m}) \, , \ \ \forall \mathfrak{m} \in \mathrm{Mult}^0 \end{array}$$

for any choice of $\rho \in \mathrm{Irr}_c$, is an isomorphism of algebras, which sends each dual canonical basis element $1 \otimes G^*(\mathfrak{m})$ to the irreducible representation $Z_\rho(\mathfrak{m})$.

The understanding that U_v serves as a quantization of the ring R^0 coming from representation categories of affine Hecke algebra and p-adic groups, gives a new perspective on the problem of decomposing the Bernstein–Zelevinski product.

Namely, for any $\mathfrak{m}, \mathfrak{n} \in \mathrm{Mult}^0$, we may decompose the product

$$G^*(\mathfrak{m})G^*(\mathfrak{n}) = \sum_{\mathfrak{p}\in \mathrm{Mult}^0} k^{\mathfrak{p}}_{\mathfrak{m},\mathfrak{n}} G^*(\mathfrak{p}) \tag{3.3}$$

in U_v, for some coefficients $k^{\mathfrak{p}}_{\mathfrak{m},\mathfrak{n}} \in \mathbb{Z}[v, v^{-1}]$. By Lusztig's positivity result [46, Theorem 14.4.13(b)], we actually have $k^{\mathfrak{p}}_{\mathfrak{m},\mathfrak{n}} \in \mathbb{Z}_{\geqslant 0}[v, v^{-1}]$.

Yet, by Proposition 3.3.1 we may specialize (3.3) at $v = 1$ to obtain the equality

$$[Z_\rho(\mathfrak{m}) \times Z_\rho(\mathfrak{n})] = \sum_{\mathfrak{p}\in \mathrm{Mult}^0} k^{\mathfrak{p}}_{\mathfrak{m},\mathfrak{n}}(1)[Z_\rho(\mathfrak{p})]$$

in the algebra R^0.

For example, multisegments $\mathfrak{m} \in \mathrm{Mult}^0$ of the form $\mathfrak{m} = [a_1, b_1]+\ldots+[a_k, b_k]$, for integers $a_1 < \ldots < a_k$ and $b_1 < \ldots < b_k$, are called *ladders*. In the context of quantum groups, the corresponding dual canonical basis elements were studied in [15] under the name of *quantum minors*.

It was proved in [28] that when $\mathfrak{m}, \mathfrak{n}$ are both ladders, the irreducible constituents of $Z_\rho(\mathfrak{m}) \times Z_\rho(\mathfrak{n})$ all appear with multiplicity 1. In other words, $k^{\mathfrak{p}}_{\mathfrak{m},\mathfrak{n}}(1) \in \{0, 1\}$, for all $\mathfrak{p} \in \mathrm{Mult}^0$.

Thus, for non-zero $k^{\mathfrak{p}}_{\mathfrak{m},\mathfrak{n}}$ with ladder $\mathfrak{m}, \mathfrak{n}$, we have $k^{\mathfrak{p}}_{\mathfrak{m},\mathfrak{n}}(v) = v^{d^{\mathfrak{p}}_{\mathfrak{m},\mathfrak{n}}}$, for a certain integer $d^{\mathfrak{p}}_{\mathfrak{m},\mathfrak{n}} \in \mathbb{Z}$. These newly obtained integer invariants shed additional light on the original decomposition problem (as was studied in [26]).

3.3.1 KLR-Algebras

Let us mention briefly the tight relation of this discussion to the representation theory of Khovanov–Lauda–Rouquier algebras (also known as quiver Hecke algebras). The ring R^0 was obtained by taking the Grothendieck ring resulting from representation categories of affine Hecke algebras. In much similarity, the quantum group U_v can be viewed as the Grothendieck ring of representations of corresponding KLR algebras. This is a process of *categorification* of quantum groups, which was achieved in [40] and [51].

Furthermore, it was shown in [56] that the dual canonical basis is categorified into the set of simple modules. This correspondence together with Proposition 3.3.1 immediately characterizes the representation categories of KLR algebras of type A as the "categorical quantization" of representations of affine Hecke algebras of type A. In fact, such a phenomenon can be better explained through the remarkable algebraic relations between the algebras involved discovered in [11] and [51].

The categorical equivalent of a quantization is a *grading* on the algebras. KLR algebras and their natural modules are ($\mathbb{Z}$-)graded, while the $\mathbb{Z}[v, v^{-1}]$ action on U_v, viewed as an action on a Grothendieck group, is a shift of grading on a module. Thus, the coefficients $k^{\mathfrak{p}}_{\mathfrak{m},\mathfrak{n}}$ of (3.3) give the multiplicities of distinct shifts of the same (isomorphism class of a) KLR-module.

3.4 *Kazhdan–Lusztig Polynomials*

Another interesting feature of the theory we discuss is that the intersection cohomology polynomials of (3.2) are nothing but another manifestation of the ubiquitous *Kazhdan–Lusztig polynomials* (specifically, those that are defined by symmetric groups). Therefore, the quantization process described in the previous sections through which we are exposed to the "full" polynomials, may be viewed as a link between the representation theory of affine Hecke algebras (and p-adic groups) and the Kazhdan–Lusztig theory.

Such link may be exploited in both ways. On one hand, Kazhdan–Lusztig polynomials are in principle computable invariants with vast literature on their algorithmic aspects (see [12]). This gives us the ability to compute representation-theoretic invariants, at least in low-rank cases, and gain insight into open questions related to the categories in question. On the other hand, it may occur that representation-theoretic results will supply more precise formulas and insight into the nature of Kazhdan–Lusztig polynomials.

In order to explain the precise links we mentioned, we will need to adopt a new point on view on multisegments.

A pair

$$\mathcal{A} = \begin{pmatrix} a_1 \dots a_k \\ b_1 \dots b_k \end{pmatrix}$$

of two sequences of integers satisfying $a_1 \leqslant \ldots \leqslant a_k$, $b_1 \geqslant \ldots \geqslant b_k$ and $a_i \leqslant b_{k+1-i} + 1$, for all $1 \leqslant i \leqslant k$, will be called a *bi-sequence* of length k.

A bi-sequence $\mathcal{A}$ as above defines two standard parabolic subgroups of the group S_k, in the following manner. Let $P_1(\mathcal{A}) < S_k$ (respectively, $P_2(\mathcal{A})$) be subgroup generated by transpositions $(i\; i+1)$, for which $b_i = b_{i+1}$ (respectively, $a_i = a_{i+1}$) holds.

We write $D(\mathcal{A}) = P_1(\mathcal{A}) \setminus S_k / P_2(\mathcal{A})$ for the set of double-cosets related to $\mathcal{A}$. For $\sigma \in D(\mathcal{A})$, we will write $\tilde{\sigma} \in S_k$ for the shortest representative of σ.

Let us define a permutation $\sigma_0 = \sigma_0(\mathcal{A})$ for each bi-sequence $\mathcal{A}$ by the following recursion. Given $\sigma_0^{-1}(k), \ldots, \sigma_0^{-1}(i+1)$, we set

$$\sigma_0^{-1}(i) = \max\{j \notin \sigma_0^{-1}(\{i+1, \ldots, k\}) \;:\; a_j \leqslant b_i + 1\}\,.$$

By [42, Section 6.1], for any permutation $\sigma \in S_k$, the inequality $\sigma_0 \leqslant \sigma$ holds in the Bruhat order of S_k, if and only if, $a_i \leqslant b_{\sigma(i)} + 1$, for all i.

Given a bi-sequence $\mathcal{A}$ of length k as above and a permutation $\sigma_0(\mathcal{A}) \leqslant \sigma \in S_k$, we can construct a multisegment

$$\mathfrak{m}_\sigma(\mathcal{A}) = \sum_{i=1}^{k} [a_i, b_{\sigma(i)}] \in \mathrm{Mult}^0\,,$$

by considering expressions of the form $[a, b]$, with $b < a$, as empty segments (These will only occur in the form of $b = a - 1$).

Note, that $\mathfrak{m}_{\sigma'}(\mathcal{A}) = \mathfrak{m}_\sigma(\mathcal{A})$, for any $\sigma' \in P_1(\mathcal{A})\sigma P_2(\mathcal{A})$. Thus, we can also write $\mathfrak{m}_{\overline{\sigma}}(\mathcal{A})$ by specifying a double-coset $\overline{\sigma} \in D(\mathcal{A})$.

In fact, it can be easily seen that every element of Mult^0 can be written in the form $\mathfrak{m}_\sigma(\mathcal{A})$, for some (non-unique) σ and $\mathcal{A}$.

Given a multisegment $\mathfrak{m} = \mathfrak{m}_\sigma(\mathcal{A}) \in \mathrm{Mult}$, for a bi-sequence $\mathcal{A}$ and (a double-coset of) a permutation $\sigma \in D(\mathcal{A})$, the closure of the corresponding variety $\mathcal{O}_\mathfrak{m}$, which is relevant to Proposition 3.2.1, decomposes as

$$\overline{\mathcal{O}_\mathfrak{m}} = \bigsqcup_{\substack{\omega \in D(\mathcal{A}),\\ \tilde{\sigma} \leqslant \tilde{\omega}}} \mathcal{O}_{\mathfrak{m}_\omega(\mathcal{A})}\,.$$

Let us recall that the Kazhdan–Lusztig polynomials (for the symmetric group S_k) are a collection of integer polynomials $\{P_{\sigma,\omega}(q)\}$ attached to each pair of permutations $\sigma \leqslant \omega$ in S_k. Briefly, the Hecke algebra $H_f(k, q)$ may be equipped with a standard basis and a canonical one. The Kazhdan–Lusztig polynomials essentially form the transition matrix between these bases (see, for example [36]).

Some monumental developments of mathematics were the discoveries that the same polynomials encode the singularities of Schubert varieties [39] and count multiplicities in representation categories of Lie algebras [2, 7].

For our discussion, the Kazhdan–Lusztig polynomials became relevant through Zelevinski's realization that the varieties $\overline{\mathcal{O}_\mathfrak{m}}$ are locally isomorphic to Schubert varieties [59]. Using further combinatorial work of Henderson [31], the following was established.

Theorem 3.4.1 *For any bi-sequence $\mathcal{A}$ and $\sigma, \omega \in D(\mathcal{A})$ with $\sigma_0(\mathcal{A}) \leqslant \tilde{\sigma} \leqslant \tilde{\omega}$, we have*

$$IC_{\mathfrak{m}_\omega(\mathcal{A}),\mathfrak{m}_\sigma(\mathcal{A})} = P_{\tilde{\omega}\omega_0,\tilde{\sigma}\omega_0} \,.$$

It is worth mentioning that another approach for establishing Theorems 3.2.1 and 3.4.1 is through application of the Arakawa–Suzuki functors [1, 54], which take us from the classically familiar category $\mathcal{O}(\mathfrak{sl}_k)$ to that of modules over *graded* affine Hecke algebras (in the sense of [44]). More on this approach can be read in [4] or the introduction to [31].

Note, that a particular consequence of Theorems 3.2.1 and 3.4.1 is that the multiplicity of a representation of the form $Z_\rho(\mathfrak{m}_\sigma(\mathcal{A}))$ inside the Jordan-Hölder series of a standard representation $\zeta_\rho(\mathfrak{m}_\omega(\mathcal{A}))$ does not depend on the bi-sequence $\mathcal{A}$. Moreover, the complexity of the computations involved in finding such multiplicities depends on the number of segments in the defining multisegment, rather than the rank of the group (which can be arbitrarily large).

Let us finish with brief examples of the kind of interplay between p-adic representation theory and Kazhdan–Lusztig theory that we described above.

In [42], the authors tackled the question of characterizing all representations $\pi \in \mathrm{Irr}$, which satisfy $\pi \times \pi \in \mathrm{Irr}$. These were called *square-irreducible* representations. The importance of this class becomes more evident in the quantum affine setting, after passing through the quantum affine Schur–Weyl duality of our next section. Namely, the class is expected to play a role in newly discovered cluster structures on relevant ring (see [35]).

We say that a bi-sequence $\mathcal{A}$ is *regular*, if $P_1(\mathcal{A})$ and $P_2(\mathcal{A})$ are trivial groups.

Theorem 3.4.2 *For a regular bi-sequence $\mathcal{A}$ and $\rho \in \mathrm{Irr}_c$, any representation of the form $Z_\rho(\mathfrak{m}_\omega(\mathcal{A})) \in \mathrm{Irr}$ is square-irreducible, if and only if, the Kazhdan–Lusztig polynomial $P_{\sigma_0,\omega}$ is trivial (i.e. $P_{\sigma_0,\omega} \equiv 1$), for $\sigma_0 = \sigma_0(\mathcal{A})$.*

Consequently, in [29], the representation-theoretic Theorem 3.4.2 was applied, using the quantization procedure discussed above, to obtain new identities that are satisfied by the *parabolic* analogues of Kazhdan–Lusztig polynomials.

4 Quantum Affine Algebras

A valuable tool in the study of representations of affine Hecke algebras of type A is the quantum affine version of the Schur–Weyl duality. While the classical duality involves functors which take representations of (finite) symmetric groups to representations of Lie groups of type A, the quantum affine analogue will take us from modules over the quantum (that is, deformed) affine version of the symmetric group, i.e. the algebra $H(n, q)$, to modules over the quantum version of the affine Lie algebra $\widehat{\mathfrak{sl}}_N$.

Before our discussion will commence, let us mention some lines of research which are *not* covered by the scope of this manuscript, but need to be mentioned for their close ties to our topic: One is the work of Brubaker-Buciumas-Bump-Friedberg [3] and its related papers. This is an intriguing study on quantum structures of Whittaker models of representations of G_n. Another topic is the affine Schur algebra [24, 47, 55] which can serve as another bridge between affine Hecke algebras and the quantum affine setting.

Let us first recall the meaning of the latter objects and the basics of the theory of their representations. We refer the reader to [19], [18, Chapter 12], [21] for a comprehensive study.

Each symmetric generalized Cartan matrix C defines a Kac–Moody algebra $\mathfrak{g}(C)$, a notion which generalizes semisimple Lie algebras. The theory of quantum groups allows for a proper definition of a quantized version $U_v(\mathfrak{g}(C))$ of the universal enveloping algebra of $\mathfrak{g}(C)$.

A particular family of Kac–Moody algebras is that of affine Lie algebras. To each simple Lie algebra $\mathfrak{g}$, one can attach its affine version $\widehat{\mathfrak{g}}$, into which the original algebra $\mathfrak{g}$ embeds.

The associative algebras resulting from quantizing the universal enveloping algebras of affine Lie algebras are known as the *quantum affine algebras*. Their structure is far better understood than the case of a general Kac–Moody algebra, while their representation theory is a rapidly developing subject, partly motivated by its high relevance for mathematical physics (see [33]).

Our case of interest will be $\widehat{\mathfrak{sl}}_N$. For a choice of non-root of unity $v \in \mathbb{C}^\times$, its quantum version, $\mathcal{A}_{N,v} := U_v(\widehat{\mathfrak{sl}}_N)$, is the complex associative algebras generated by a Laurent polynomial algebra $\mathbb{C}[k_0^{\pm 1}, k_1^{\pm}, \ldots, k_{N-1}^{\pm 1}]$ together with additional $2N$ generators $x_0^{\pm}, \ldots, x_{N-1}^{\pm}$, subject to the relations

$$\begin{cases} x_i^{\pm} x_j^{\pm} = x_j^{\pm} x_i^{\pm}, & |i-j| \neq 0, 1 \ (\mathrm{mod}\ N) \\ \left(x_i^{\pm}\right)^2 x_j^{\pm} - (v+v^{-1}) x_i^{\pm} x_j^{\pm} x_i^{\pm} + x_j^{\pm} \left(x_i^{\pm}\right)^2 = 0 & |i-j| = 1 \ (\mathrm{mod}\ N) \end{cases},$$

$$\begin{cases} k_j x_i^{\pm} = x_i^{\pm} k_j, & |i-j| \neq 0, 1 \ (\mathrm{mod}\ N) \\ k_j x_i^{\pm} k_j^{-1} = v^{\pm 2} x_i^{\pm}, & |i-j| = 0 \ (\mathrm{mod}\ N) \\ k_j x_i^{\pm} k_j^{-1} = v^{\mp 1} x_i^{\pm} & |i-j| = 1 \ (\mathrm{mod}\ N) \end{cases}, \quad x_i^{\pm} x_j^{\mp} - x_j^{\pm} x_i^{\mp} = \delta_{ij} \frac{k_i - k_i^{-1}}{v - v^{-1}}.$$

The defining presentation for $\mathcal{A}_{N,v}$ given above will actually not be material for our discussion. However, let us take note that the sub-algebra of $\mathcal{A}_{N,v}$ generated by $k_1, \ldots, k_{N-1}$ and $x_1^{\pm}, \ldots, x_{N-1}^{\pm}$ is precisely the quantum group $U_v(\mathfrak{sl}_N)$.

Instead, the alternative Drinfeld presentation of $\mathcal{A}_{N,v}$, whose details will be omitted here, identifies 3 subalgebras: U^0, which contains $k_1, \ldots, k_{N-1}$ and $U^{\pm}$, which contains, respectively, $x_1^{\pm}, \ldots, x_{N-1}^{\pm}$. These satisfy a triangular decomposition (as a vector space)

$$\mathcal{A}_{N,v} = U^- \otimes U^0 \otimes U^+ ,$$

whose behavior resembles the classical semisimple theory.

We are interested in the Abelian category $\mathcal{C}(N, v)$ of finite-dimensional modules over $\mathcal{A}_{N,v}$ of *type* 1. The last condition means that a module V has a weight decomposition of the form $V = \oplus_\mu V_\mu$, where

$$V_\mu = \{x \in V \,:\, k_i \cdot x = v^{\mu(i)} x,\ \forall i = 0, \dots, N-1\} ,$$

for $\mu : \{0, \dots, N-1\} \to \mathbb{Z}$.

This notion is in fact not too restrictive, since any finite-dimensional module over $\mathcal{A}_{N,v}$ is known to be of type 1, up to a suitable algebra automorphism.

For a module V in $\mathcal{C}(N, v)$, we say that a vector $x \in V$ is a *highest weight* vector, if x is an eigenvector for the sub-algebra U^0 and $U^+ \cdot x = 0$. The character of U^0 by which it acts on a highest weight vector x is called the ℓ-weight of x.

Given a $V \in \mathrm{Irr}(\mathcal{C}(N, v))$, it is known there is a unique, up to scalar, highest weight vector $x \in V$. The ℓ-weight P of x characterizes the isomorphism class of V. We can write in this case $V = V(P)$.

Let us write $\mathcal{D}_N$ for the subset of characters P of U^0 which give rise to $V(P) \in \mathrm{Irr}(\mathcal{C}(N, v))$. In other words, we obtain a bijection $\mathrm{Irr}(\mathcal{C}(N, v)) \leftrightarrow \mathcal{D}_N$.

The set $\mathcal{D}_N$ can now be described using what is known as *Drinfeld polynomials*. Each element $P \in \mathcal{D}_N$ is identified with a tuple $P = (P_1, \dots, P_{N-1})$ of monic complex polynomials with non-zero constant term. Moreover, each such tuple gives an element of $\mathcal{D}_N$.

Finally, recall that $\mathcal{C}(N, v)$ is a monoidal category, that is, modules can be tensored one against another to produce a new module, coming from the Hopf algebra structure of $\mathcal{A}_{N,v}$. The co-multiplication $\Delta : \mathcal{A}_{N,v} \to \mathcal{A}_{N,v} \otimes \mathcal{A}_{N,v}$ used to define such structure is given by

$$\Delta(k_i) = k_i \otimes k_i, \quad \Delta(x_i^+) = x_i^+ \otimes k_i + 1 \otimes x_i^+, \quad \Delta(x_i^-) = x_i^- \otimes 1 + k_i^{-1} \otimes x_i^- ,$$

on the generators used to define the algebra.

4.1 *Quantum Affine Schur–Weyl Duality*

The classical Schur–Weyl duality may be viewed as a functor of tensoring with a bimodule, which enjoys remarkable properties. In particular, it serves as a gateway between different settings of representation theory in type A.

To recall the classical setting, consider the natural action of $\mathfrak{sl}_N$ on the vector space $\mathbb{V} = \mathbb{C}^N$, as a left module over the enveloping algebra $U(\mathfrak{sl}_N)$. For any $n \geqslant 1$, the tensor space $M_{N,n} = \mathbb{V}^{\otimes n}$ is again a left module over $U(\mathfrak{sl}_N)$ (using the natural Hopf algebra structure).

Yet, the space $M_{N,n}$ also has a natural action of the symmetric group S_n by permuting its tensor components. These two actions obviously commute, making $M_{N,n}$ into a $(U(\mathfrak{sl}_N), \mathbb{C}[S_n])$-bimodule. The Schur–Weyl duality can then be stated in terms of properties of the functor $M_{N,n} \otimes_{\mathbb{C}[S_n]} -$, taking representations of S_n to representations of $\mathfrak{sl}_N$.

This situation was quantized in the work of Jimbo [38]. Suppose that $q \in \mathbb{C}^\times$ is a choice of non-root of unity. The Hecke algebra $H_f(n, q^2)$ then acts on $M_{N,n}$ through the following formulas. If $\{e_i\}_{i=1}^N$ is a choice of basis for $\mathbb{V}$, then

$$T_i(\cdots e_r \otimes e_s \cdots) = \begin{cases} q^2(\cdots e_r \otimes e_r \cdots) & r = s \\ q(\cdots e_s \otimes e_r \cdots) & r < s \\ q(\cdots e_r \otimes e_s \cdots) + (q^2 - 1)(\cdots e_r \otimes e_s \cdots) & r > s \end{cases},$$

where $T_i, i = 1, \ldots, N-1$, are the generators as in (2.3) and $\cdots x \otimes y \cdots$ denotes the i-th and $i+1$-th tensor components of pure tensors in $M_{N,n}$.

The space $\mathbb{V}$ is also equipped with a natural structure of a $U_q(\mathfrak{sl}_N)$-module. It conveniently happens that when tensoring $\mathbb{V}$ as a module to produce $M_{N,n}$, the resulting quantum group action commutes with the Hecke algebra action we described. Hence, $M_{N,n}$ becomes a $(U_q(\mathfrak{sl}_N), H_f(n, q^2))$-bimodule.

Theorem 4.1.1 *The functor $\mathcal{F}_{N,n}$ from the category of finite-dimensional modules over $H_f(n, q^2)$ to the category of finite-dimensional modules over $U_q(\mathfrak{sl}_N)$, given by*

$$\mathcal{F}_{N,n}(V) = M_{N,n} \otimes_{H_f(n,q^2)} V ,$$

is fully faithful, when $n \leqslant N-1$.

In other words, when $n \leqslant N$, the functor $\mathcal{F}_{N,n}$ injects $H_f(n, q^2)$-modules as a full subcategory of $U_q(\mathfrak{sl}_N)$-modules.

The next generalization step is the main interest of our discussion. The following quantum affine duality was precisely stated and proved by Chari–Pressley [20], while an intriguing geometric point of view was established by Ginzburg–Reshetikhin–Vasserot in [25]. See further the introduction of [22].

Theorem 4.1.2 *For a finite-dimensional module V in $\mathcal{M}(n, q^2)$, the space $\mathcal{F}_{N,n}(V)$, for the functor defined in Theorem 4.1.1, can be equipped with an additional structure of a (type 1) $\mathcal{A}_{N,q}$-module. The resulting functor*

$$\mathcal{F}^{\mathrm{aff}}_{N,n} : \mathcal{M}_{fl}(n, q^2) \to \mathcal{C}(N, q)$$

is fully faithful, when $n \leqslant N-1$.

Thus, the structure of the category $\mathcal{M}_{fl}(n, q^2)$, and as a consequence of Theorem 2.2.1, of any finite-length objects in a simple Bernstein block $\Theta(\rho, n)$ (when setting $q = q_F^{o(\rho)/2}$), can be studied as a full subcategory of $\mathcal{C}(N, q)$.

Moreover, the duality functors are monoidal [20, Proposition 4.7], in the sense that

$$\mathcal{F}^{\mathrm{aff}}_{N,n_1}(\pi_1) \otimes \mathcal{F}^{\mathrm{aff}}_{N,n_2}(\pi_2) \cong \mathcal{F}^{\mathrm{aff}}_{N,n_1+n_2}(\pi_1 \times \pi_2), \tag{4.1}$$

for any representations π_1, π_2 in $\mathcal{M}_{fl}(n_1, q^2)$, $\mathcal{M}_{fl}(n_2, q^2)$, respectively. Recall, that $\pi_1 \times \pi_2$ denotes the induction product which produces a representation in $\mathcal{M}_{fl}(n_1 + n_2, q^2)$ through the embedding (2.5).

Let us clarify the correspondence that we obtain on the level of irreducible representations. By Theorem 4.1.2, we see that for $n \leqslant N$, the functor $\mathcal{F}^{\mathrm{aff}}_{N,n}$ sends simple objects to simple objects, hence, gives an injection

$$\mathrm{Irr}(\mathcal{M}(n, q^2)) \hookrightarrow \mathrm{Irr}(\mathcal{C}(N, q)) \ . \tag{4.2}$$

Going back to the p-adic group setting, let us fix $\rho \in \mathrm{Irr}_c$ and $q = q_F^{o(\rho)/2}$. Then, by Theorem 2.3.1 and (2.7), we see that $\mathrm{Irr}(\mathcal{M}(n, q^2))$ may be identified with a subset of the collection of multisegments $\mathrm{Mult}_{[\rho]}$. On the other hand, we have seen that $\mathrm{Irr}(\mathcal{C}(N, q))$ has a combinatorial description in terms of the polynomials $\mathcal{D}_N$.

Chari–Pressley have identified the correspondence between subsets of combinatorial objects resulting from (4.2). Let us recall it here in an explicit manner.

For all $2 \leqslant N$, let $\iota_N : \mathcal{D}_N \to \mathrm{Mult}_{[\rho]}$ be the map given as follows. For a Drinfeld polynomial $P = (P_1, \ldots, P_{N-1}) \in \mathcal{D}_N$, let $a_{i,1}, \ldots, a_{i,\deg P_i} \in \mathbb{C}^\times$ be the set of roots of P_i (counted with multiplicities), for $i = 1, \ldots, N-1$. Choose numbers $s_{i,j} \in \mathbb{C}$, for which $a_{i,j} = q^{-2s_{i,j}}$ holds. Then,

$$\iota_N(P) = \sum_{i=1}^{N-1} \sum_{j=1}^{\deg P_i} \left[s_{i,j} - \frac{i-1}{2}, s_{i,j} + \frac{i-1}{2} \right] \in \mathrm{Mult}_{[\rho]} \tag{4.3}$$

is a well defined multisegment.

For $k(P) := \sum_{i=1}^{N-1} i \deg P_i$, we clearly obtain $Z_\rho(\iota_N(P)) \in \mathrm{Irr}(\Theta(\rho, k(P)))$.

Proposition 4.1.1 ([20, Theorem 7.6]) *For* $\mathfrak{m} \in \mathrm{Mult}_{[\rho]}$, *let us use the correspondence* T_ρ *of* (2.7) *to identify* $Z_\rho(\mathfrak{m})$ *as an element of* $\mathrm{Irr}\,\mathcal{M}(n, q^2)$, *for suitable* $n \geqslant 1$.

For a choice of $n \leqslant N-1$, *let* $P \in \mathcal{D}_N$ *be the Drinfeld polynomial which satisfies*

$$\mathcal{F}^{\mathrm{aff}}_{N,n}(Z_\rho(\mathfrak{m})) \cong V(P) \in \mathrm{Irr}(\mathcal{C}(N, q)) \ .$$

Then, $\mathfrak{m} = \iota_N(P)$.

Let us comment that the map ι_N, which is shown above to give the combinatorial counterpart of the quantum affine Schur–Weyl duality, is of a simple linear nature. Namely, the information given by a Drinfeld polynomial is the location of the root

$z \in \mathbb{C}$ of the i-th polynomial. Such root becomes a segment of length $i-1$ centered in a point determined by z.

4.2 Irreducibility and Cyclic Modules

One remarkable contribution of the quantum affine approach to the study of representations of affine Hecke algebras and p-adic groups is the following result on irreducibility. A celebrated recent result of Hernandez points on the following phenomenon.

Theorem 4.2.1 ([32]) *Suppose that $V_1, \ldots, V_k \in \mathrm{Irr}(\mathcal{C}(N, v))$ are modules, for which $V_i \otimes V_j$ is a simple module, for all $i \neq j$. Then, the module*

$$V_1 \otimes \cdots \otimes V_k$$

is simple.

The combination of Theorems 4.2.1, 4.1.2, the multiplicative property (4.1), and Theorem 2.2.1 immediately gives the following corollary about representations of p-adic groups.

Corollary 4.2.1 *Suppose that $\pi_1, \ldots, \pi_k \in \mathrm{Irr}$ are representations, for which $\pi_i \times \pi_j$ is irreducible, for all $i \neq j$. Then,*

$$\pi_1 \times \cdots \times \pi_k$$

is an irreducible representation.

Hernandez later proved in [34] a more subtle extension of Theorem 4.2.1, which again gave implications through the Schur–Weyl duality functors. This extension deals with a notion called *cyclic products.*

Suppose that $V_1, \ldots, V_k \in \mathrm{Irr}(\mathcal{C}(N, v))$ are given. Let $x_i \in V_i$, $1 \leqslant i \leqslant k$ be the corresponding highest weight vectors. We say that the product $V_1 \otimes \cdots \otimes V_k$ is cyclic (or, highest weight), if it is generated as a module by the vector $x_1 \otimes \cdots \otimes x_k$.

Theorem 4.2.2 *Suppose that $V_1, \ldots, V_k \in \mathrm{Irr}(\mathcal{C}(N, v))$ are modules, for which $V_i \otimes V_j$ is a cyclic product, for all $1 \leqslant i < j \leqslant k$. Then, the product*

$$V_1 \otimes \cdots \otimes V_k$$

is cyclic.

Though the last theorem may appear as a statement on the intrinsic structure of modules, the cyclicity property can in fact be stated in a more categorical language. This will allow to transfer Theorem 4.2.2 through the Schur–Weyl duality functors.

Let us recall that the set $\mathcal{D}_N$ comes with a natural monoid structure, through entry-wise multiplication of complex polynomials.

Now, suppose that V, W are modules in $\mathcal{C}(N, v)$. Given highest weight vectors $x_P \in V$, $x_Q \in W$ of respective ℓ-weights $P, Q \in \mathcal{D}_N$, the vector $x_P \otimes x_Q \in V \otimes W$ is also a highest weight vector, whose ℓ-weight is given by $P \cdot Q \in \mathcal{D}_N$. In particular, we always have the isomorphism class of $V(PQ)$ as an irreducible subquotient of $V \otimes W$.

Proposition 4.2.1 *Given irreducible modules $V(P^1), \ldots, V(P^k) \in \mathrm{Irr}(\mathcal{C}(N, v))$ that are parameterized by $P^1, \ldots, P^k \in \mathcal{D}_N$, the product $V(P^1) \otimes \cdots \otimes V(P^k)$ is cyclic, if and only if, it has a unique irreducible quotient whose isomorphism class is given by $V(P^1 \cdot \ldots \cdots P^k)$.*

Note that the maps ι_N from (4.3) clearly respect the additive structure of the monoids $\mathcal{D}_N$ and $\mathrm{Mult}_{[\rho]}$. Thus, in light of Propositions 4.1.1 and 4.2.1, we can apply the functors of Theorems 4.1.2 and 2.2.1 to reformulate Theorem 4.2.2 as follows.

Corollary 4.2.2 *For a choice of $\rho \in \mathrm{Irr}_c$, let $\pi_1 = Z_\rho(\mathfrak{m}_1), \ldots, \pi_k = Z_\rho(\mathfrak{m}_k) \in \mathrm{Irr}$ be given irreducible representations, parameterized by $\mathfrak{m}_1, \ldots, \mathfrak{m}_k \in \mathrm{Mult}_{[\rho]}$.*

If $\pi_i \times \pi_j$ has a unique irreducible quotient which is parameterized by $Z_\rho(\mathfrak{m}_i + \mathfrak{m}_j)$, for all $1 \leqslant i < j \leqslant k$, then the representation $\pi_1 \times \cdots \times \pi_k$ has a unique irreducible quotient which is parameterized by $Z_\rho(\mathfrak{m}_1 + \cdots + \mathfrak{m}_k)$.

Both Corollaries 4.2.1 and 4.2.2 are remarkable in the sense that no proof which uses purely p-adic (or purely Hecke-algebraic) methods, without relying on the quantum affine Schur–Weyl duality, was known for these statements, up until recently. In a joint work with Alberto Mínguez under preparation, we were able to construct such a "pure" proof, using the ideas of the quantum affine proofs of Hernandez.

Corollary 4.2.2 has in fact intriguing applications for natural questions in the representation theory of p-adic groups.

Suppose that $\pi \in \mathrm{Irr}(\mathcal{M}(G_n))$ is a given irreducible representation. The *restriction problem* asks how does π decompose when restricted to the smaller group G_{n-1} embedded inside G_n. In particular, one could ask how to describe the irreducible quotients of $\pi|_{G_{n-1}}$. A variant of the Gan–Gross–Prasad conjectures [23] attempts to give a partial answer to this question. Recently, in [27] we found that Corollary 4.2.2 was the key missing tool for settling a large part of these conjectures.

References

1. Tomoyuki Arakawa and Takeshi Suzuki. Duality between $\mathfrak{sl}_n(\mathbf{C})$ and the degenerate affine Hecke algebra. *J. Algebra*, 209(1):288–304, 1998.
2. Alexandre Beĭlinson and Joseph Bernstein. Localisation de $\mathfrak{g}$-modules. *C. R. Acad. Sci. Paris Sér. I Math.*, 292(1):15–18, 1981.

3. Ben Brubaker, Valentin Buciumas, Daniel Bump, and Solomon Friedberg. Hecke modules from metaplectic ice. *Selecta Mathematica, New Series*, 24(3):2523–2570, 2018.
4. Dan Barbasch and Dan Ciubotaru. Ladder representations of $GL_n(\mathbb{Q}_p)$. *Representations of Reductive Groups: In Honor of the 60th Birthday of David A. Vogan, Jr.*, 312:117, 2015.
5. Joseph Bernstein. Representations of p-adic groups. *Notes taken by K. Rumelhart of lectures by J. Bernstein at Harvard*, 1992.
6. Roman Bezrukavnikov. On two geometric realizations of an affine Hecke algebra. *Publ. Math. Inst. Hautes Études Sci.*, 123:1–67, 2016.
7. J.-L. Brylinski and M. Kashiwara. Kazhdan-Lusztig conjecture and holonomic systems. *Invent. Math.*, 64(3):387–410, 1981.
8. Colin J. Bushnell and Philip C. Kutzko. *The admissible dual of* GL(N) *via compact open subgroups*, volume 129 of *Annals of Mathematics Studies*. Princeton University Press, Princeton, NJ, 1993.
9. CJ Bushnell and PC Kutzko. Smooth representations of reductive p-adic groups: structure theory via types. *Proceedings of the London Mathematical Society*, 77(3):582634, 1998.
10. Colin J. Bushnell and Philip C. Kutzko. Semisimple types in GL_n. *Compositio Math.*, 119(1):53–97, 1999.
11. Jonathan Brundan and Alexander Kleshchev. Blocks of cyclotomic Hecke algebras and Khovanov-Lauda algebras. *Invent. Math.*, 178(3):451–484, 2009.
12. Sara Billey and V. Lakshmibai. *Singular loci of Schubert varieties*, volume 182 of *Progress in Mathematics*. Birkhäuser Boston, Inc., Boston, MA, 2000.
13. Armand Borel. Admissible representations of a semi-simple group over a local field with vectors fixed under an Iwahori subgroup. *Invent. Math.*, 35:233–259, 1976.
14. I. N. Bernstein and A. V. Zelevinsky. Induced representations of reductive p-adic groups. I. *Ann. Sci. École Norm. Sup. (4)*, 10(4):441–472, 1977.
15. Arkady Berenstein and Andrei Zelevinsky. String bases for quantum groups of type A_r. In *I. M. Gel′ fand Seminar*, volume 16 of *Adv. Soviet Math.*, pages 51–89. Amer. Math. Soc., Providence, RI, 1993.
16. W. Casselman. The unramified principal series of p-adic groups. I. The spherical function. *Compositio Math.*, 40(3):387–406, 1980.
17. Neil Chriss and Victor Ginzburg. *Representation theory and complex geometry*. Birkhäuser Boston, Inc., Boston, MA, 1997.
18. Vyjayanthi Chari and Andrew Pressley. *A guide to quantum groups*. Cambridge University Press, Cambridge, 1994.
19. Vyjayanthi Chari and Andrew Pressley. Quantum affine algebras and their representations. In *Representations of groups (Banff, AB, 1994)*, volume 16 of *CMS Conf. Proc.*, pages 59–78. Amer. Math. Soc., Providence, RI, 1995.
20. Vyjayanthi Chari and Andrew Pressley. Quantum affine algebras and affine Hecke algebras. *Pacific J. Math.*, 174(2):295–326, 1996.
21. Edward Frenkel and Nicolai Reshetikhin. The q-characters of representations of quantum affine algebras and deformations of W-algebras. In *Recent developments in quantum affine algebras and related topics (Raleigh, NC, 1998)*, volume 248 of *Contemp. Math.*, pages 163–205. Amer. Math. Soc., Providence, RI, 1999.
22. Ryo Fujita. Geometric realization of Dynkin quiver type quantum affine Schur–Weyl duality. *International Mathematics Research Notices*, 2019.
23. W. T. Gan, Benedict H. Gross, and Dipendra Prasad. Branching laws: The non-tempered case. *http://www.math.tifr.res.in/~dprasad/nongenericggp.pdf*.
24. R. M. Green. The affine q-Schur algebra. *J. Algebra*, 215(2):379–411, 1999.
25. Victor Ginzburg, Nicolai Reshetikhin, and Éric Vasserot. Quantum groups and flag varieties. In *Mathematical aspects of conformal and topological field theories and quantum groups (South Hadley, MA, 1992)*, volume 175 of *Contemp. Math.*, pages 101–130. Amer. Math. Soc., Providence, RI, 1994.
26. Maxim Gurevich. Quantum invariants for decomposition problems in type A rings of representations. *Journal of Combinatorial Theory, Series A*, 180: 105431, 2021.

27. Maxim Gurevich. On restriction of unitarizable representations of general linear groups and the non-generic local Gan-Gross-Prasad conjecture. to appear in *Journal of the European Mathematical Society arXiv preprint arXiv:1808.02640*, 2018.
28. Maxim Gurevich. Decomposition rules for the ring of representations of non-Archimedean GL_n. *International Mathematics Research Notices https://doi.org/10.1093/imrn/rnz006*, 2019.
29. Maxim Gurevich. An identity of parabolic Kazhdan-Lusztig polynomials arising from square-irreducible modules. *Journal of the Australian Mathematical Society https://doi.org/10.1017/S144678871900017X*, 2019.
30. Volker Heiermann. Opérateurs d'entrelacement et algèbres de Hecke avec paramètres d'un groupe réductif p-adique: le cas des groupes classiques. *Selecta Math. (N.S.)*, 17(3):713–756, 2011.
31. Anthony Henderson. Nilpotent orbits of linear and cyclic quivers and Kazhdan-Lusztig polynomials of type A. *Represent. Theory*, 11:95–121 (electronic), 2007.
32. David Hernandez. Simple tensor products. *Invent. Math.*, 181(3):649–675, 2010.
33. D. Hernandez. Avancées concernant les R-matrices et leurs applications (d'apres Maulik-Okounkov, Kang-Kashiwara-Kim-Oh...). *Astérisque*, 407:267–296, 2019.
34. David Hernandez. Cyclicity and R-matrices. *Selecta Mathematica*, 25(2):19, 2019.
35. David Hernandez and Bernard Leclerc. Quantum affine algebras and cluster algebras. *arXiv preprint arXiv:1902.01432*, 2019. (present volume)
36. James E. Humphreys. *Reflection groups and Coxeter groups*, volume 29 of *Cambridge Studies in Advanced Mathematics*. Cambridge University Press, Cambridge, 1990.
37. Nagayoshi Iwahori and Hideya Matsumoto. On some Bruhat decomposition and the structure of the hecke rings of p-adic Chevalley groups. *Publications Mathématiques de l'IHÉS*, 25:5–48, 1965.
38. Michio Jimbo. A q-analogue of $U(\mathfrak{gl}(N+1))$, Hecke algebra, and the Yang-Baxter equation. *Lett. Math. Phys.*, 11(3):247–252, 1986.
39. David Kazhdan and George Lusztig. Schubert varieties and Poincaré duality. In *Geometry of the Laplace operator (Proc. Sympos. Pure Math., Univ. Hawaii, Honolulu, Hawaii, 1979)*, Proc. Sympos. Pure Math., XXXVI, pages 185–203. Amer. Math. Soc., Providence, R.I., 1980.
40. Mikhail Khovanov and Aaron D. Lauda. A diagrammatic approach to categorification of quantum groups. I. *Represent. Theory*, 13:309–347, 2009.
41. Erez Lapid and Alberto Mínguez. On parabolic induction on inner forms of the general linear group over a non-Archimedean local field. *Selecta Math. (N.S.)*, 22(4):2347–2400, 2016.
42. Erez Lapid and Alberto Mínguez. Geometric conditions for □-irreducibility of certain representations of the general linear group over a non-Archimedean local field. *Adv. Math.*, 339:113–190, 2018.
43. Bernard Leclerc, Maxim Nazarov, and Jean-Yves Thibon. Induced representations of affine Hecke algebras and canonical bases of quantum groups. In *Studies in memory of Issai Schur (Chevaleret/Rehovot, 2000)*, volume 210 of *Progr. Math.*, pages 115–153. Birkhäuser Boston, Boston, MA, 2003.
44. George Lusztig. Affine Hecke algebras and their graded version. *J. Amer. Math. Soc.*, 2(3):599–635, 1989.
45. G. Lusztig. Canonical bases arising from quantized enveloping algebras. *J. Amer. Math. Soc.*, 3(2):447–498, 1990.
46. George Lusztig. *Introduction to quantum groups*, volume 110 of *Progress in Mathematics*. Birkhäuser Boston, Inc., Boston, MA, 1993.
47. Vanessa Miemietz and Catharina Stroppel. Affine quiver Schur algebras and p-adic GL_n. *Selecta Math. (N.S.)*, 25(2):Art. 32, 66, 2019.
48. David Renard. *Représentations des groupes réductifs p-adiques.* Cours Spécialisés. Société Mathématique de France, 2010.

49. Alan Roche. Parabolic induction and the Bernstein decomposition. *Compositio Math.*, 134(2):113–133, 2002.
50. J. D. Rogawski. On modules over the Hecke algebra of a p-adic group. *Invent. Math.*, 79(3):443–465, 1985.
51. Raphaël Rouquier. 2-Kac-Moody algebras. *arXiv preprint arXiv:0812.5023*, 2008.
52. Vincent Secherre. Représentations lisses de $\mathrm{GL}_m(D)$. III. Types simples. *Ann. Sci. École Norm. Sup. (4)*, 38(6):951–977, 2005.
53. Vincent Sécherre and Shaun Stevens. Smooth representations of $GL_m(D)$ VI: semisimple types. *Int. Math. Res. Not. IMRN*, (13):2994–3039, 2012.
54. Takeshi Suzuki. Rogawski's conjecture on the Jantzen filtration for the degenerate affine Hecke algebra of type A. *Represent. Theory*, 2:393–409 (electronic), 1998.
55. Marie-France Vignéras. Schur algebras of reductive p-adic groups. I. *Duke Math. J.*, 116(1):35–75, 2003.
56. M. Varagnolo and E. Vasserot. Canonical bases and KLR-algebras. *J. Reine Angew. Math.*, 659:67–100, 2011.
57. A. V. Zelevinsky. Induced representations of reductive p-adic groups. II. On irreducible representations of $\mathrm{GL}(n)$. *Ann. Sci. École Norm. Sup. (4)*, 13(2):165–210, 1980.
58. A. V. Zelevinskiĭ. The p-adic analogue of the Kazhdan-Lusztig conjecture. *Funktsional. Anal. i Prilozhen.*, 15(2):9–21, 96, 1981.
59. A. V. Zelevinskiĭ. Two remarks on graded nilpotent classes. *Uspekhi Mat. Nauk*, 40(1(241)):199–200, 1985.

Part III
Papers

Categorical Representations and Classical p-Adic Groups

Michela Varagnolo and Eric Vasserot

Abstract We characterize the categorical representation on finite dimensional modules of affine Hecke algebras of classical types (for arbitrary value of the parameters and of the characteristic). This reduces the classification of simple objects to computing a representation previously introduced by Enomoto and Kashiwara. We did this in some particular cases.

1 Introduction

The quiver-Hecke algebras introduced recently by Khovanov–Lauda and Rouquier are associated to any quiver without one-loops. When the quiver is of type A_∞, the quiver-Hecke algebra of rank n is known to be Morita equivalent to the affine Iwahori–Hecke algebra of type GL_n, and it categorifies some weight subspaces of the negative half of the quantum group of type A_∞. This is a fundamental result, which permits to relate the quantum group to the Grothendieck group of the affine Iwahori–Hecke algebras in characteristic zero, recovering some previous results of Ariki, etc. When the quiver is of type $A_d^{(1)}$, for a fixed positive integer d, one gets in a similar way the affine Hecke algebra of type GL_n at a d-th root of unity and the quantum group of the corresponding type. This yields in particular a link between modular representations of the affine Iwahori–Hecke algebra of type GL and the combinatorics of quantum groups of type $A^{(1)}$.

M. Varagnolo (✉)
CY Cergy Paris Université, UMR 8088 (CNRS), Cergy Pontoise, France
e-mail: michela.varagnolo@cyu.fr

E. Vasserot
Université de Paris, Paris, France

Institut Universitaire de France (IUF), Paris, France
e-mail: eric.vasserot@imj-prg.fr

J. Greenstein et al. (eds.), *Interactions of Quantum Affine Algebras with Cluster Algebras, Current Algebras and Categorification*, Progress in Mathematics 337,
https://doi.org/10.1007/978-3-030-63849-8_6

The Bernstein center theory relates the smooth representations of the group $\mathrm{GL}_n(F)$, where F is a local non-Archimedean field, over an algebraically closed field R of characteristic $\ell = 0$, to the representations of the affine Iwahori–Hecke algebras associated with standard Levi subgroups of GL_n over the same field. In particular, we get a relation between smooth finite length representations over R of the p-adic group and the combinatorics of an affine quantum group of type A_∞. Under this identification, the parabolic induction/restriction functors are related to Kashiwara's boson algebra, which acts on the positive half of the quantum group. The Bernstein center theory does not extend easily to the positive characteristic. Nevertheless, part of the identification above still holds in positive non-natural characteristic ℓ.

The goal of this paper is to study the cases of a positive non-natural characteristic ℓ and the case of classical groups. For a p-adic group of type $\mathrm{SO}(2n+1)$ or $\mathrm{SP}(2n)$, a new algebra has been introduced in [18] which plays a similar role to the quiver-Hecke algebra. In this note we review the main properties of this algebra and we explain its relation with the Bernstein center theory for classical types. Once again, the Grothendieck group of finite length smooth (unipotent) representations acquires an action of a version of Kashiwara's boson algebra. This case is more complicated than the GL_n one: the module we get is not known explicitly except in some particular cases.

Now, let us explain the plan of the paper.

The second section is a reminder on quiver-Hecke algebras. Then, we discuss the quiver-Hecke algebras of type C introduced in [18]. Our setting here is more general than in [18]. In this generality, the categorification theorem in loc. cit. do not hold anymore. In the appendix we explain the geometric realization of this version of the quiver-Hecke algebra of type C, generalizing the construction from [18].

The third section concerns representations of p-adic groups in characteristic $\ell = 0$. First, we give an introduction to the Bernstein center of general p-adic groups. Then, we consider the groups of types A, B, C. If $\ell = 0$, we equip the Grothendieck group of the category of unipotent representations with an action of a Boson algebra, using Bernstein's theory to reduce to the case of affine Hecke algebras. In type A, the resulting module is explicit described. In types B and C, this is unknown in general.

Finally, in the last section we consider the case of representations in positive non-natural characteristic. In this case there is no Bernstein theory to apply. However, it is still possible to define a categorical action of a boson algebra on the category of unipotent modules.

From now on R is a fixed algebraically closed field of characteristic ℓ and $\mathbb{A} = \mathbb{Q}[v, v^{-1}]$, $\mathbb{K} = \mathbb{Q}(v)$ where v is a formal variable.

2 Quiver-Hecke and Affine Hecke Algebras of Type C

Quiver-Hecke algebras of type C have been introduced in [18], with some restriction on the parameters. We will need a more general version here. In this section we recall briefly the definition of quiver-Hecke algebras of type C, in this more general version, and we recall the relation with affine Hecke algebras of type C.

2.1 Quiver-Hecke Algebras

2.1.1 Symmetric Quivers

Let (Q, D) be a *symmetric quiver*, i.e., a pair consisting of a quiver Q whose set of vertices is I and whose set of arrows is Ω, and an involution D which acts on the sets I and Ω so that it switches the source and the target of the arrows. Hence, we have $D(h)' = D(h'')$ for each $h \in \Omega$, where h' is the incoming vertex and h'' the outgoing one. We will assume that there are no 1-loops, i.e., no arrow joins a vertex to itself. For each elements i, j of I we set $c_{i,j} = \sharp\{h \in \Omega\,;\ h' = i,\ h'' = j\}$.

Example 2.1.1 Fix an element $q \in R \setminus \{0, \pm 1\}$ and a subset $I \subseteq R^\times$ which is stable by the multiplication by q and by the involution D given by $D(i) = i^{-1}$. Let $Q_{I,q} = (Q_{I,q}, D)$ be the symmetric quiver such that $\Omega = \{i \to j\,;\ j = iq\}$, so we have $c_{i,j} = \delta_{iq,j}$ for all i, j. Note that $Q_{R^\times,q}$ is the direct sum of quivers $Q_{I_z,q}$ such that $I_z = zq^{\mathbb{Z}} \bigcup z^{-1}q^{\mathbb{Z}}$ and either $z^2 \notin q^{\mathbb{Z}}$ or $z = \pm 1, \pm q^{1/2}$. Let $e \in \mathbb{N} \cup \{\infty\}$ be the multiplicative order of q. Fix a square root $q^{1/2}$ of q. For I_z with $z \in R^\times$ one of the following cases occurs:

- $z^2 \notin q^{\mathbb{Z}}$, then I_z does not contain any D-fixed point,
- $z = \pm 1$ and e is infinite or odd, then $I_z = zq^{\mathbb{Z}}$ admits a unique D-fixed point,
- $z = \pm q^{1/2}$ and e is infinite or even, then $I_z = zq^{\mathbb{Z}}$ does not contain any D-fixed point,
- $z = \pm 1$ and e is even, then $I_z = zq^{\mathbb{Z}}$ admits two D-fixed points.

We say that the symmetric quiver $Q_{I_z,q}$ is *without fixed points* if I_z does not contain any fixed point by the involution D.

2.1.2 The Quiver-Hecke Algebra of Type C

For a positive integer n, let $[n] = \{1 - n, \ldots, 0, 1, \ldots, n\}$. Fix a dimension vector $\lambda = \sum_i \lambda_i\, i$ in $\mathbb{N}I$. Set

$$I^{[n],D} = \{\nu = (\nu_{1-n}, \ldots, \nu_n) \in I^{[n]}\,;\ D(\nu_k) = \nu_{1-k},\ \forall k\}.$$

We will identify the sets I^n and $I^{[n],D}$ via the obvious bijection

$$(\nu_1, \ldots, \nu_n) \mapsto (D(\nu_n), \ldots, D(\nu_1), \nu_1, \ldots, \nu_n)$$

whenever this is more convenient. Write

$$\mathrm{A}(n) = R[x_1, \ldots, x_n], \quad \mathrm{P}(n) = \bigoplus_{\nu} \mathrm{A}(n)e(\nu), \quad \mathrm{RC}(0; \lambda) = R.$$

Let W_n be the subgroup of permutations w of $[n]$ such that $w(1-k) = 1 - w(k)$. For each $k = 0, 1, \ldots, n-1$ let $s_k \in \mathrm{W}_n$ be the permutation which flips the pairs $(k, k+1)$, $(1-k, -k)$ and fixes the other elements of $[n]$. The group W_n is a Coxeter group of type C_n with simple reflections $s_0, \ldots, s_{n-1}$. For each $\nu \in I^n$, $w \in \mathrm{W}_n$ we set

$$w\nu = (\nu_{w^{-1}(1-n)}, \ldots, \nu_{w^{-1}(n)}). \tag{2.1}$$

Moreover for each $w \in \mathrm{W}_n$ and $f \in \mathrm{A}(n)$ set

$$wf(x_1, \cdots, x_n) = f(x_{w(1)}, \cdots, x_{w(n)}). \tag{2.2}$$

Here, we write $x_{1-l} = -x_l$. Putting together (2.1) and (2.2) we get an action of W_n on $\mathrm{P}(n)$.

Definition 2.1 Let $\mathrm{RC}(n; \lambda)$ be the $\mathbb{Z}$-graded R-subalgebra of $\mathrm{End}_R(\mathrm{P}(n))$ generated by $e(\nu)$, x_l and τ_k with $\nu \in I^n$, $l = 1, \ldots, n$, $k = 0, 1, \ldots, n-1$ such that, if $k \neq 0$, we have

(a) $e(\nu) f e(\nu') = \delta_{\nu,\nu'} f e(\nu)$ for each $f \in \mathrm{A}(n)$,

(b) $\tau_0\, e(\nu) = \begin{cases} x_0^{\lambda_{\nu_1}}\, s_0\, e(\nu) & \text{if } \nu_0 \neq \nu_1, \\ x_0^{\lambda_{\nu_1}}\, \partial_0\, e(\nu) & \text{else,} \end{cases}$

(c) $\tau_k e(\nu) = \begin{cases} (x_k - x_{k+1})^{c_{\nu_k,\nu_{k+1}}}\, s_k\, e(\nu) & \text{if } \nu_k \neq \nu_{k+1}, \\ \partial_k\, e(\nu) & \text{else.} \end{cases}$

Here, we set $\partial_k = (x_k - x_{k+1})^{-1}(s_k - 1)$. The $\mathbb{Z}$-grading is given by

(d) $\deg(x_k e(\nu)) = 2$,

(e) $\deg(\tau_0 e(\nu)) = \begin{cases} \lambda_{\nu_0} + \lambda_{\nu_1} & \text{if } \nu_0 \neq \nu_1, \\ \lambda_{\nu_0} + \lambda_{\nu_1} - 2 & \text{else,} \end{cases}$

(f) $\deg(\tau_k e(\nu)) = \begin{cases} c_{\nu_k,\nu_{k+1}} + c_{\nu_{k+1},\nu_k} & \text{if } \nu_k \neq \nu_{k+1}, \\ -2 & \text{else.} \end{cases}$ □

The $\mathbb{Z}$-graded R-subalgebra $\mathrm{R}(n) \subseteq \mathrm{RC}(n;\lambda)$ generated by $e(\nu)$, x_l and τ_k with $\nu \in I^n$, $l = 1, \dots, n$, $k = 1, \dots, n-1$ is the quiver-Hecke algebra of rank n associated with the quiver Q. We may write $\mathrm{RC}(n;\lambda;Q,D)$ or $\mathrm{R}(n;Q)$ to stress the underlying quiver and we set

$$\mathrm{RC}(\lambda;Q,D) = \bigoplus_{n\geqslant 0} \mathrm{RC}(n;\lambda;Q,D), \qquad \mathrm{R}(Q) = \bigoplus_{n\geqslant 0} \mathrm{R}(n;Q).$$

For each tuple $\alpha \in \mathbb{N}I$ we may write $\mathrm{RC}(\alpha,\lambda;Q,D) = e(\alpha)\,\mathrm{RC}(n;\lambda;Q,D)e(\alpha)$, where $e(\alpha)$ is the sum of all tuples $e(\nu)$'s such that ν sums up to α.

Proposition 2.2 *The R-algebra $\mathrm{RC}(n;\lambda)$ is generated by $e(\nu)$, x_l, τ_h with $\nu \in I^n$, $l = 1, \dots, n$, $h = 0, \dots, n-1$ modulo the defining relations*

(a) $e(\nu)\,e(\nu') = \delta_{\nu,\nu'}e(\nu)$, $\tau_h e(\nu) = e(s_h\nu)\tau_h$, $x_l e(\nu) = e(\nu)x_l$,

(b) $x_l\, x_{l'} = x_l\, x_{l'}$,

(c) $(\tau_k x_l - x_{s_k(l)}\tau_k)e(\nu) = \begin{cases} -e(\nu) & \text{if } l = k,\ \nu_k = \nu_{k+1}, \\ e(\nu) & \text{if } l = k+1,\ \nu_k = \nu_{k+1}, \\ 0 & \text{else,} \end{cases}$

(d) $(\tau_0 x_l - x_{s_0(l)}\tau_0)e(\nu) = \begin{cases} x_0^{\lambda_{\nu_0}} e(\nu) & \text{if } l = 1,\ \nu_0 = \nu_1, \\ 0 & \text{else,} \end{cases}$

(e) $\tau_k^2 e(\nu) = Q_{\nu_k,\nu_{k+1}}(x_k, x_{k+1})e(\nu)$,

(f) $\tau_0^2 e(\nu) = \begin{cases} x_0^{\lambda_{\nu_0}} x_1^{\lambda_{\nu_1}} e(\nu) & \text{if } \nu_0 \neq \nu_1, \\ \partial_0(x_0^{\lambda_{\nu_0}})\tau_0 e(\nu) & \text{else,} \end{cases}$

(g) $\tau_h\tau_{h'} = \tau_{h'}\tau_h$ *if* $h \neq h' \pm 1$ *then*

(h) $(\tau_{k+1}\tau_k\tau_{k+1} - \tau_k\tau_{k+1}\tau_k)e(\nu) =$
$\delta_{\nu_k,\nu_{k+2}} \frac{Q_{\nu_{k+1},\nu_k}(x_{k+1},x_k) - Q_{\nu_{k+1},\nu_k}(x_{k+1},x_{k+2})}{x_k - x_{k+2}} e(\nu)$,

(i) *If* $\lambda_i = 0$ *or* 1 *for all* $i \in I$, $\big((\tau_1\tau_0)^2 - (\tau_0\tau_1)^2\big)e(\nu) =$

$$= \begin{cases} (-1)^{c_{\nu_0\nu_2}}(-x_2)^{\lambda_{\nu_2}} \frac{(x_1-x_2)^{a_{\nu_0\nu_2}} - (x_1+x_2)^{a_{\nu_0\nu_2}}}{2x_2}\,\tau_0 e(\nu), & \\ \qquad\qquad \nu_1^{-1} \neq \nu_1 \neq \nu_2 = \nu_2^{-1}, & \\ \frac{x_2^{\lambda_{\nu_1}}(-x_2)^{\lambda_{\nu_2}} - (-x_1)^{\lambda_{\nu_1}} x_1^{\lambda_{\nu_2}}}{x_1+x_2}\,\tau_1 e(\nu),\ \nu_2 = \nu_1^{-1} \neq \nu_1 & \\ 0, \qquad \text{otherwise.} & \end{cases}$$

Here we set $a_{i,j} = 2\delta_{i,j} - c_{i,j} - c_{j,i}$ *and* $Q_{i,j}(u,v) = (1-\delta_{i,j})(v-u)^{c_{i,j}}(u-v)^{c_{j,i}}$. *Further* $l' = 1, \dots, n$, $k = 1, \dots, n-1$ *and* $h' = 0, \dots, n-1$.

Proof Direct computation.

Remark 2.1.2 The relation (i) can be computed for any parameter λ. We will not need it. □

For each element $w \in W_n$ we fix a reduced decomposition $\dot{w} = s_{k_1}s_{k_2}\cdots s_{k_\ell}$ of w and we write $\tau_{\dot{w}} = \tau_{k_1}\tau_{k_2}\ldots\tau_{k_\ell}$.

Proposition 2.3

(a) $\mathrm{RC}(n;\lambda)$ *is free of finite rank over its center, which is equal to* $\mathrm{P}(n)^{W_n}$.
(b) $\mathrm{RC}(n;\lambda)$ *is free as a left* $\mathrm{P}(n)$*-module with basis* $\{\tau_{\dot{w}}\,;\ w \in W_n\}$.

Proof Part (a) is proved as in [18, prop. 7.7], part (b) as in [18, prop. 7.5].

2.1.3 Induction and Restriction

Denote by $\mathrm{Mod}(\mathrm{RC}(n;\lambda))$, $\mathrm{Proj}(\mathrm{RC}(n;\lambda))$, $\mathrm{mod}(\mathrm{RC}(n;\lambda))$ the categories of arbitrary, finitely generated projective, and finite dimensional $\mathbb{Z}$-graded left modules over the graded R-algebra $\mathrm{RC}(n;\lambda)$. Let $\mathrm{mod}_0(\mathrm{RC}(n;\lambda))$ be the category of all finite dimensional modules such that $x_1,\ldots,x_n$ are nilpotent. The morphisms are the degree preserving homomorphisms. Define

$$[\mathrm{Proj}(\mathrm{RC}(\lambda))] = \bigoplus_{n\in\mathbb{N}}[\mathrm{Proj}(\mathrm{RC}(n;\lambda))],$$

$$[\mathrm{mod}(\mathrm{RC}(\lambda))] = \bigoplus_{n\in\mathbb{N}}[\mathrm{mod}(\mathrm{RC}(n;\lambda))],$$

where $[\mathrm{Proj}(\mathrm{RC}(n;\lambda))]$ is the split Grothendieck group of the category $\mathrm{Proj}(\mathrm{RC}(n;\lambda))$ over $\mathbb{Q}$ while $[\mathrm{mod}(\mathrm{RC}(n;\lambda))]$ is the Grothendieck group of $\mathrm{mod}(\mathrm{RC}(n;\lambda))$ over $\mathbb{Q}$. We define the degree shift functors v by $vM = M\langle 1\rangle$ where $M\langle m\rangle_d = M_{d+1}$ and $M = \oplus_{d\in\mathbb{Z}}M_d$. Then, one can define $\mathbb{A}$-structures on $[\mathrm{Proj}(\mathrm{RC}(\lambda))]$ and $[\mathrm{mod}(\mathrm{RC}(\lambda))]$.

For each $i \in I$ and $n > 0$ let $\mathrm{RC}(n-1,i;\lambda) \subseteq \mathrm{RC}(n;\lambda)$ be the R-subalgebra generated by the elements $e(\nu)$, x_l, τ_k with $\nu_n = i$ and $k \neq n-1$. For a pair of vertices $i, j \in I$ we define the subalgebra $\mathrm{RC}(n-2,i,j;\lambda) \subseteq \mathrm{RC}(n;\lambda)$ in a similar way. Set $e(n-1,i) = \sum_\nu e(\nu)$, where the sum runs over all ν's as above. The inclusion $\mathrm{RC}(n-1,i;\lambda) \subseteq \mathrm{RC}(n;\lambda)$ and the graded R-algebra homomorphism $\mathrm{RC}(n-1,i;\lambda) \to \mathrm{RC}(n-1,i;\lambda)/(x_n) = \mathrm{RC}(n-1;\lambda)$ yield the functors

$$\begin{aligned} F_{i,n-1} &: \mathrm{mod}(\mathrm{RC}(n-1;\lambda)) \to \mathrm{mod}(\mathrm{RC}(n;\lambda)),\\ E_{i,n-1} &: \mathrm{mod}(\mathrm{RC}(n;\lambda)) \to \mathrm{mod}(\mathrm{RC}(n-1;\lambda)) \end{aligned} \tag{2.3}$$

given by

$$F_{i,n-1}(V) = \mathrm{RC}(n;\lambda) \otimes_{\mathrm{RC}(n-1,i;\lambda)} (V \otimes R[x_n])$$
$$= \mathrm{RC}(n;\lambda)e(n-1,i) \otimes_{\mathrm{RC}(n-1;\lambda)} V,$$
$$E_{i,n-1}(V) = \mathrm{RC}(n-1;\lambda) \otimes_{\mathrm{RC}(n-1,i;\lambda)} e(n-1,i)V.$$

Note that for a finitely generated $\mathrm{RC}(n;\lambda)$-module V the $\mathrm{RC}(n-1;\lambda)$-module obtained by taking the direct summand $e(n-1,i)V$ and restricting it to $\mathrm{RC}(n-1;\lambda)$ may not be finitely generated. The functors $E_i = \bigoplus_{n\geqslant 0} E_{i,n}$ and $F_i = \bigoplus_{n\geqslant 0} F_{i,n}$ yield $\mathbb{A}$-linear operators on $[\mathrm{Proj}(\mathrm{RC}(\lambda))]$.

2.1.4 The Mackey's Theorem

For each θ and each $a \in \mathbb{N}$, we write $\theta^{(a)} = \theta^a/[a]!$ for the v-version of the a-th divided power. Following [8] consider the following $\mathbb{K}$-algebra.

Definition 2.4 Let $B_v(Q,D)$ be the $\mathbb{K}$-algebra generated by e_i, f_i and invertible elements t_i for each $i \in I$, satisfying the following defining relations:

(a) $t_it_j = t_jt_i$ and $t_{D(i)} = t_i$,
(b) $t_ie_jt_i^{-1} = v^{a_{i,j}+a_{D(i),j}}e_j$,
(c) $t_if_jt_i^{-1} = v^{-a_{i,j}-a_{D(i),j}}f_j$,
(d) $e_if_j - v^{-a_{i,j}}f_je_i = \delta_{i,j} + \delta_{D(i),j}t_i$,
(e) $\sum_{a+b=1-a_{i,j}}(-1)^a\theta_i^{(a)}\theta_j\theta_i^{(b)} = 0$ if $i \neq j$ and $\theta_i = e_i$ or f_i. □

For each $i \in I$, the endofunctors E_i, F_i yield the linear endomorphisms $e_i = [E_i]$, $f_i = [F_i]$ of $[\mathrm{Proj}(\mathrm{RC}(\lambda))]$. We define an operator t_i on $[\mathrm{Proj}(\mathrm{RC}(\lambda))]$ by setting

$$t_i\,[P] = v^{\lambda_i+\lambda_{D(i)}-2\,\delta_{i,D(i)}-\alpha\cdot i}\,[P], \quad \forall P \in \mathrm{Proj}(\mathrm{RC}(\alpha;\lambda)),$$

where we write $\alpha\cdot\beta = \sum_{i,j\in I}\alpha_i\,\beta_j\,a_{i,j}$ for each elements $\alpha = \sum_i \alpha_i\,i$, $\beta = \sum_i \beta_i\,i$ of $\mathbb{N}I$.

Proposition 2.5 *The operators e_i, f_i, t_i with $i \in I$ define a representation of $B_v(Q,D)$ on* $[\mathrm{RC}(\lambda)\text{-proj}] \otimes_{\mathbb{A}} \mathbb{K}$. □

Proof We must check that the relations above are satisfied. They are obvious, except the relations (d) and (e). The relation (e) is proved as in [18] and goes back to [9]. To prove (d) it is enough to check the following version of the Mackey's theorem.

Lemma 2.6 *Fix $i,j \in I$. Let $\alpha,\beta \in \mathbb{N}I^D$ such that $\alpha + i + D(i) = \beta + j + D(j)$ and $2n = |\alpha| = |\beta|$. Set $\alpha' = \alpha - j - D(j)$. Then the $\mathbb{Z}$-graded* $(\mathrm{RC}(n,1;\lambda),\mathrm{RC}(n,1;\lambda))$*-bimodule $e(\alpha,i)\,\mathrm{RC}(n+1;\lambda)e(\beta,j)$ has a filtration by $\mathbb{Z}$-graded bimodules whose associated graded is isomorphic to*

(a) $\big(\mathrm{RC}(\alpha;\lambda)\otimes \mathrm{R}(i)\big)$
$\oplus\big(\mathrm{RC}(n,1;\lambda)e(\alpha',i,i)\otimes_{\mathrm{RC}(n-1,1,1;\lambda)} e(\alpha',i,i)\,\mathrm{RC}(n,1;\lambda)\big)\langle -2\rangle$
$\oplus\,\delta_{i,D(i)}\big(\mathrm{RC}(\alpha;\lambda)\otimes \mathrm{R}(i)\big)\langle -2-\alpha\cdot i\rangle$ *if* $j=i$,

(b) $\delta_{i,D(i)}\big(\mathrm{RC}(\alpha)\otimes \mathrm{R}(i)\big)$
$\oplus\big(\mathrm{RC}(n,1;\lambda)e(\alpha',D(i),i)\ \otimes_{\mathrm{RC}(n-1,1,1;\lambda)}\ e(\alpha',i,D(i))\,\mathrm{RC}(n,1;\lambda)\big)\langle -i\cdot D(i)\rangle\oplus\big(\mathrm{RC}(\alpha)\otimes \mathrm{R}(D(i))\big)\langle \lambda_i+\lambda_{i-1}-2\delta_{i,D(i)}-\alpha\cdot i\rangle$ *if* $j=D(i)$,

(c) $\big(\mathrm{RC}(n,1;\lambda)e(\alpha',j,i)\otimes_{\mathrm{RC}(n-1,1,1;\lambda)} e(\alpha',i,j)\,\mathrm{RC}(n,1;\lambda)\big)\langle -i\cdot j\rangle$ *if* $j\neq i,D(i)$.

Proof Set $\varepsilon_{n+1}=s_n\cdots s_1s_0s_1\cdots s_n$. Recall first that Proposition 2.3(b) yields

$$\mathrm{RC}(n+1;\lambda)=\bigoplus_{w\in \mathrm{W}_n}\mathrm{P}(n+1)\tau_{\dot w}\oplus\bigoplus_{w\in \mathrm{W}_n\,s_n\,\mathrm{W}_n}\mathrm{P}(n+1)\tau_{\dot w}\oplus\bigoplus_{w\in \mathrm{W}_n\,\varepsilon_{n+1}\,\mathrm{W}_n}\mathrm{P}(n+1)\tau_{\dot w}.$$

Next, for each $\nu,\nu'\in I^n$ and $u=e,s_n,\varepsilon_{n+1}$, the element

$$x_u=e(D(i),\nu,i)\,\tau_{\dot u}\,e(D(j),\nu',j)\in e(\alpha,i)\,\mathrm{RC}(n+1;\lambda)e(\beta,j)$$

is non-zero only if one of the following cases holds:

$$\begin{cases} i=j,\ \nu'=\nu,\ \deg(x_u)=0 & \text{if } u=e,\\ \nu'=(D(i),\nu_{2-n},\dots,\nu_{n-1},i),\ \deg(x_u)=-i\cdot j & \text{if } u=s_n,\\ i=D(j),\ \nu=\nu',\ \deg(x_u)=\lambda_i+\lambda_{D(i)}-2\delta_{i,D(i)}-\nu\cdot i & \text{if } u=\varepsilon_{n+1}.\end{cases}$$

The lemma follows.

We do not know what is the representation of $B_v(Q,D)$ on $[\mathrm{Proj}(\mathrm{RC}(\lambda))]\otimes_{\mathbb{A}}\mathbb{K}$ in general. More precisely, by [8, prop. 4.2], for each $\lambda\in\mathbb{N}I$ there is a unique $B_v(Q,D)$-module $V_v(\lambda)=V_v(\lambda;Q,D)$ up to isomorphism, which is generated by a non-zero vector ϕ_λ such that for each $i\in I$ we have

$$e_i\phi_\lambda=0,\quad t_i\phi_\lambda=v^{\lambda_i+\lambda_{D(i)}-2\delta_{i,D(i)}}\phi_\lambda,$$
$$\{x\in V_v(\lambda)\,;\ e_ix=0,\ \forall i\in I\}=\mathbb{K}\phi_\lambda.$$

The $B_v(Q,D)$-module $V_v(\lambda)$ is irreducible. If (Q,D) has no fixed points, then it is proved in [18, lem. 10.9, thm. 10.31] that the Enomoto–Kashiwara $B_v(Q,D)$-module $V_v(\lambda)$ is isomorphic to $[\mathrm{RC}(\lambda)\text{-proj}]\otimes_{\mathbb{A}}\mathbb{K}$. This is false in general, see Sect. 2.1.5. However, we have the following conjecture.

Conjecture 2.7 If $\lambda=0$, then $V_v(\lambda)\simeq[\mathrm{Proj}(\mathrm{RC}(\lambda))]\otimes_{\mathbb{A}}\mathbb{K}$. □

Remark 2.1.3 If $(Q,D)=\bigsqcup_k(Q^{(k)},D^{(k)})$ (a possibly infinite sum), then we have an obvious decomposition $\lambda=\sum_k\lambda^{(k)}$ as a finite sum of dimension vectors in $\mathbb{N}I^{(k)}$ such that

$$B_v(Q, D) = \otimes'_k B_v(Q^{(k)}, D^{(k)}), \quad V_v(\lambda; Q, D) = \otimes'_k V_v(\lambda^{(k)}; Q^{(k)}, D^{(k)}),$$

where $\otimes'$ is the restricted tensor product. Similarly, we have

$$\mathrm{RC}(n; \lambda; Q, D) = \bigoplus_{n=\sum_k n^{(k)}} \bigotimes_k \mathrm{RC}(n^{(k)}; \lambda^{(k)}; Q^{(k)}, D^{(k)}).$$

A similar decomposition holds for the quiver-Hecke algebra $\mathrm{R}(n; Q)$.

2.1.5 Example: The Case of Rank 1 or 2

Assume that $(Q, D) = Q_{I,q}$ is as in Example 2.1.1 with q of infinite order, and that the dimension vector λ is such that $\sum_i \lambda_i \leqslant 1$. Using Remark 2.1.3 and the results of [18], we may assume that $I = I_1$. The following holds.

Proposition 2.8 *If the element $\alpha \in \mathbb{N}I$ does not contain the vertex $1 \in I$, then we have*

$$V_v(\lambda)_{\lambda-\alpha} \simeq [\mathrm{Proj}(\mathrm{RC}(\alpha; \lambda))] \otimes_{\mathbb{A}} \mathbb{K}. \tag{2.4}$$

Proof Since $\alpha \in \mathbb{N}I$ does not contain the vertex $1 \in I$, the results in [18, §10.20] imply that for any simple module V and any $i \neq 1$, the induced modules $E_i(V)$, $F_i(V)$ defined in (2.3) have a simple head if non-zero. This permits to define analogues of Kashiwara's crystal operators on the set of the isomorphism classes of simple modules of the algebras $\mathrm{RC}(\alpha; \lambda)$ for all such α's. Then, the claim is proved as in [18, lem. 10.9, thm. 10.31].

Note that if V is a simple $\mathrm{RC}(\alpha; \lambda)$-module, then the induced module $F_1(V)$ may have a non-simple head. For example, if $\alpha = q + q^{-1}$ and $\lambda = 0$, then $F_1 F_q(R) = P_1 \oplus P_2$ where P_1, P_2 are indecomposable modules such that

$$F_q F_1(R) = v^{-1} P_1 \oplus v P_1, \quad F_{q^{-1}} F_1(R) = v^{-1} P_2 \oplus v P_2.$$

The module $F_q(R)$ is indecomposable. Let S, S_1, S_2 be the tops of $F_q(R)$, P_1, P_2. We have $F_1 S = S_1 \oplus S_2$.

If the rank n is 1 or 2, to prove the conjecture we are reduced to check (2.4) with $\alpha = 1+1, \alpha = 1+1+q+q^{-1}$, and $\alpha = 1+1+i+i^{-1}$ with $i \neq q, q^{-1}$. A direct computation yields the following.

- If $\alpha = 1 + 1$, then $\dim V_v(\lambda)_{\lambda-\alpha} = 1$, and $\dim[\mathrm{Proj}(\mathrm{RC}(\alpha; \lambda))] = 1$ if $\lambda_1 = 0$ and 2 if $\lambda_1 = 1$.
- If $\alpha = 1 + 1 + i + i^{-1}$ and $\lambda = 0$, then $V_v(\lambda)_{\lambda-\alpha}$ and $[\mathrm{Proj}(\mathrm{RC}(\alpha; \lambda))] \otimes_{\mathbb{A}} \mathbb{K}$ are one-dimensional. They are spanned by the element $f_i f_1(\phi_\lambda)$ and by the class of the module $F_i F_1(R)$, respectively.

- If $\alpha = 1+1+q+q^{-1}$ and $\lambda = 0$, then $V_v(\lambda)_{\lambda-\alpha}$ and $[\mathrm{Proj}(\mathrm{RC}(\alpha;\lambda))] \otimes_{\mathbb{A}} \mathbb{K}$ are 2 dimensional. They are spanned by $\{f_q f_1(\phi_\lambda),\ f_{q^{-1}} f_1(\phi_\lambda)\}$ and by the classes of the modules $F_q F_1(R)$, $F_{q^{-1}} F_1(R)$, respectively.

2.2 Affine Hecke Algebras

Given integers n, and elements q, q_1, q_0 in $R^\times$, let $\mathrm{H}(n, q, q_1, q_0)$ be the affine Hecke algebra of type C_n with parameters q_1 and q_0, which is associated with the root datum $(X_n, Y_n, R_n, R_n^\vee, \Pi_n)$ such that

- $X_n = Y_n = \mathbb{Z}^n$, with (e_i), $(e_i^\vee)$ the canonical bases of X_n and Y_n,
- $R_n = \{\pm e_i \pm e_j\,;\ 1 \leqslant i < j \leqslant n\} \cup \{\pm e_i\,;\ 1 \leqslant i \leqslant n\}$,
- $R_n^\vee = \{\pm e_i^\vee \pm e_j^\vee\,;\ 1 \leqslant i < j \leqslant n\} \cup \{\pm 2e_i^\vee\,;\ 1 \leqslant i \leqslant n\}$,
- $\Pi_n = \{\alpha_0, \alpha_1, \dots, \alpha_{n-1}\}$ with $\alpha_i = e_i - e_{i+1}$ if $i = 1, \dots, n-1$ and $\alpha_0 = -e_1$.

Let W_n be the corresponding Weyl group and $s_i \in \mathrm{W}_n$ be the reflexion with respect to α_i. The R-algebra $\mathrm{H}(n, q, q_1, q_0)$ is generated by elements T_i with $i = 0, 1, \dots, n-1$ and commuting invertible elements X_α with $\alpha \in X_n$ modulo the following relations:

- $X_\alpha X_\beta = X_{\alpha+\beta}$,
- $T_i T_j = T_j T_i$ if $|i - j| \neq 1$,
- $T_i T_j T_i = T_j T_i T_j$ if $|i - j| = 1$ and $i, j \neq 0$,
- $(T_i T_j)^2 = (T_j T_i)^2$ if $|i - j| = 1$ and $i = 0$ or $j = 0$,
- $(T_i + 1)(T_i - q) = 0$ if $i = 1, \dots, n-1$,
- $(T_0 + 1)(T_0 - q_1 q_0) = 0$,
- $X_\alpha T_i - T_i X_{s_i(\alpha)} = (q-1)(X_\alpha - X_{s_i(\alpha)}) \,/\, (1 - X_{-\alpha_i})$ if $i = 1, \dots, n-1$,
- $X_\alpha T_0 - T_0 X_{s_0(\alpha)} = \big((q_1 q_0 - 1) + X_{-\alpha_0}(q_1 - q_0)\big)(X_\alpha - X_{s_0(\alpha)}) \,/\, (1 - X_{-2\alpha_0})$.

We recover the presentation used in [18, Appendix A] by setting $p = q^{1/2}$, $X_i = X_{-e_i}$, $T_0 = q_1 T_0$, and $T_i = p T_i$ for $i \neq 0$. The center of $\mathrm{H}(N, q, q_1, q_0)$ is the subalgebra $\mathbb{Z}(n) = R[X_1^{\pm 1}, \dots, X_n^{\pm 1}]^{\mathrm{W}_n}$. The affine Hecke algebra of type GL_n is the subalgebra $\mathrm{H}(n, q) \subseteq \mathrm{H}(n, q, q_1, q_0)$ generated by $T_1, \dots, T_{n-1}$ and X_α with $\alpha \in X_n$. Let $\mathrm{mod}(\mathrm{H}(n, q, q_1, q_0))$ and $\mathrm{mod}(\mathrm{H}(n, q))$ be the categories of left modules over $\mathrm{H}(n, q, q_1, q_0)$ and $\mathrm{H}(n, q)$ which are finite dimensional over R.

Set $(Q, D) = Q_{I,q}$ with $I \subseteq R^\times$ and $q \in R \setminus \{0, \pm 1\}$. Fix any $q_1, q_0 \in R^\times$, and set $\lambda = \sum_{i \in I} \lambda_i\, i$ with

$$\lambda_i = \delta_{i,-q_0} + \delta_{i,q_1}. \tag{2.5}$$

Note that, since $-q_0$, q_1 may not belong to I, the parameter λ may be 0. The action of the elements X_k is locally finite with eigenvalues in $R^\times$ on any module in $\mathrm{mod}(\mathrm{H}(n, q, q_1, q_0))$. Let $\mathrm{mod}_I(\mathrm{H}(n, q, q_1, q_0))$ be the subcategory consisting of all modules such that the spectrum of X_l belongs to I for each $k = 1, \ldots, n$. We have the following, compare with [18, thm. A.4].

Proposition 2.9 *We have an equivalence of R-linear abelian categories*

$$\mathrm{mod}_0(\mathrm{RC}(n; \lambda; Q_{I_q})) \to \mathrm{mod}_I(\mathrm{H}(n, q, q_1, q_0))$$

which respects the dimension of the modules. □

Proof For each $k = 1, \ldots, n$ and $\nu \in I^{\alpha,D}$ we set

- $\varphi_0\, e(\nu) = \begin{cases} x_0^{-\lambda_{\nu_1}}\ \tau_0\, e(\nu) & \text{if } \nu_1^2 \neq 1, \\ (1 + 2x_0^{1-\lambda_{\nu_1}}\ \tau_0)\, e(\nu) & \text{else,} \end{cases}$
- $\varphi_k e(\nu) = \begin{cases} (x_k - x_{k+1})^{-\delta_{\nu_{k+1}, q\nu_k}}\ \tau_k\, e(\nu) & \text{if } \nu_k \neq \nu_{k+1}, \\ \big((x_k - x_{k+1})\tau_k + 1\big)\, e(\nu) & \text{else.} \end{cases}$

Similarly, for each $k = 1, \ldots, n$ we write

- $\phi_0 - 1 = \frac{X_1^2 - 1}{(X_1 + q_0)(X_1 - q_1)}(T_0 - q_1 q_0)$,
- $\phi_k - 1 = \frac{X_k - X_{k+1}}{qX_k - X_{k+1}}\,(T_k - q)$.

We must check that the assignment

$$\varphi_0 \mapsto \phi_0, \quad \varphi_k \mapsto \phi_k, \quad x_k \mapsto \nu_k^{-1} X_k - 1$$

yields an algebra isomorphism

$$\mathrm{RC}(n; \lambda) \otimes_{\mathrm{A}(n)} R[[x_1, \ldots, x_n]] \to \bigoplus_{\nu \in I^n} \mathrm{H}(n, q, q_1, q_0) \otimes_{\mathrm{Z}(n)} R[[X_1 - \nu_1, \ldots, X_n - \nu_n]].$$

The pushforward by this isomorphism is the equivalence in the proposition.

To prove the isomorphism, we first extend the representation of $\mathrm{RC}(n)$ on $\mathrm{P}(n)$ to a representation on $\bigoplus_\nu \mathrm{Frac}(\mathrm{A}(n))\, e(\nu)$. Then, the operators φ_0, φ_k introduced above act as s_0, s_k. Next, we consider the (faithful) representation of $\mathrm{H}(n, q, q_1, q_0)$ on $R(X_1, \ldots, X_n)$ such that the operators ϕ_0, ϕ_k introduced above act as s_0, s_k. Then, under the identification $x_k = \nu_k^{-1} X_k - 1$ for $k = 1, \ldots, n$, the claim follows from the following relations:

- if $\nu_{k+1} = q\nu_k$:

$$(T_k - q) = \frac{1}{q^{-1} - 1 + q^{-1}x_k - x_{k+1}}\tau_k e(\nu)$$

$$-\frac{x_k - x_{k+1}}{q^{-1} - 1 + q^{-1}x_k - x_{k+1}} e(\nu),$$

- if $\nu_{k+1} \neq \nu_k, q\nu_k$:

$$(T_k - q) = \frac{q\nu_k + q\nu_k x_k - \nu_{k+1} - \nu_{k+1}x_{k+1}}{\nu_k + \nu_k x_k - \nu_{k+1} - \nu_{k+1}x_{k+1}} (\tau_k - 1)e(\nu),$$

- if $\nu_k = \nu_{k+1}$:

$$(T_k - q) = (q - 1 + qx_k - x_{k+1})\tau_k e(\nu),$$

- if $\nu_1^2 \neq 1$:

$$(T_0 - q_1 q_0) = \frac{(1 + x_1 + \nu_1^{-1} q_0)(1 + x_1 - \nu_1^{-1} q_1)}{(1 + x_1)^2 - \nu_1^{-2}} (x_0^{-\lambda_{\nu_1}} \tau_0 - 1)e(\nu),$$

- if $\nu_1^2 = 1$:

$$(T_0 - q_1 q_0) = \frac{(1 + x_1 + \nu_1 q_0)(1 + x_1 - \nu_1 q_1)}{(1 + x_1)^2 - 1} 2x_0^{1-\lambda_{\nu_1}} \tau_0 e(\nu).$$

Note that, to get the required isomorphism, one of the four assertions below must be true:

- $\lambda_{\nu_1} = 0$,
- $\nu_1 = -q_0 \neq q_1$ and $\lambda_{\nu_1} = 1$,
- $\nu_1 = q_1 \neq -q_0$ and $\lambda_{\nu_1} = 1$,
- $\nu_1 = q_1 = -q_0$ and $\lambda_{\nu_1} = 2$,

which is ensured by (2.5).

Remark 2.2.1 A similar Morita equivalence holds for usual quiver-Hecke algebras, see [2, 16].

3 Reminder on Representations of p-Adic Groups

3.1 Generalities

Let F be a local non-Archimedean field of characteristic zero and let $v_F : F \to \mathbb{Z} \cup \infty$ be its non- Archimedean evaluation normalized in such a way that the image coincides with $\mathbb{Z} \cup \infty$. Let $\mathcal{O} = \{x \in F; v_F(x) \geqslant 0\}$ be the ring of integers of F and let $\mathbb{F}_q$ denote the residue field of characteristic p which we assume different from

2. Let ϖ be a uniformizer of F and $x \mapsto |x| = q^{-v_F(x)}$ be the normalized absolute value.

Let $\mathbb{G}$ be a connected reductive group over F. We will abbreviate $\mathrm{G} = \mathbb{G}(F)$ and $\bar{\mathrm{G}} = \mathrm{G}(q) = \mathbb{G}(\mathbb{F}_q)$. Let R be an algebraically closed field of characteristic $\ell \neq p$. We will write $\otimes = \otimes_R$.

Let $\mathrm{Mod}(R\mathrm{G})$ be the category of all smooth $R\mathrm{G}$-modules and $\mathrm{mod}(R\mathrm{G})$ be the subcategory of finite length modules. Both are Abelian categories. By a parabolic subgroup $\mathrm{P} \subseteq \mathrm{G}$ we will mean the set of F-points of a parabolic subgroup $\mathbb{P} \subseteq \mathbb{G}$ defined over F. If M is a Levi subgroup of P, we denote by $i^{\mathrm{G}}_{\mathrm{M}\subseteq\mathrm{P}}$ and $r^{\mathrm{G}}_{\mathrm{M}\subseteq\mathrm{P}}$ the normalized parabolic induction and restriction functors. Both functors are exact [19, §II.2.1]. They depend on the choice of the parabolic subgroup P. If M and P are both standard we will abbreviate $i^{\mathrm{G}}_{\mathrm{M}\subseteq\mathrm{P}} = i^{\mathrm{G}}_{\mathrm{M}}$ and $r^{\mathrm{G}}_{\mathrm{M}\subseteq\mathrm{P}} = r^{\mathrm{G}}_{\mathrm{M}}$. Let ind denote the compact induction.

Let $\mathrm{Irr}(R\mathrm{G})$ be the set of isomorphism classes of irreducible modules in $\mathrm{Mod}(R\mathrm{G})$ and $\mathrm{Cusp}(R\mathrm{G})$, $\mathrm{SCusp}(R\mathrm{G})$ be the subsets of *cuspidal* and *supercuspidal* ones. Recall that a smooth irreducible $R\mathrm{G}$-module is cuspidal if it is killed by all proper Jacquet functors, i.e., all proper parabolic restriction functors, and that it is supercuspidal if it does not occur as a subquotient of a properly parabolically induced module.

3.2 The Bernstein Center

3.2.1 Bernstein's Blocks

Let $X^*(\mathrm{G})$ be the group of characters of $\mathbb{G}$ defined over F and $^\circ\mathrm{G} \subseteq \mathrm{G}$ be the open, closed, normal subgroup given by

$$^\circ\mathrm{G} = \{g \in \mathrm{G};\ \chi(g) \in \mathcal{O}^\times\ ,\ \forall\chi \in X^*(\mathrm{G})\}.$$

Let $\Psi(R\mathrm{G})$ be the group of unramified R-valued characters of G, i.e., the set of all group homomorphisms $\mathrm{G} \to R^\times$ which factorize through $^\circ\mathrm{G}$. The quotient $\mathrm{G}/{}^\circ\mathrm{G}$ is a free abelian group of finite type and $\Psi(R\mathrm{G})$ is the group of R-points of the algebraic torus $\mathrm{Hom}_{\mathbb{Z}}(\mathrm{G}/{}^\circ\mathrm{G}, \mathbb{G}_{m,R})$. Let B_{G} be the R-algebra of regular functions on this torus. The *universal unramified character* ψ_{G} of G is the obvious representation of $R\mathrm{G}$ in B_{G}.

Given a representation π of $R\mathrm{G}$, let $\pi^\vee$ denote its contragredient module, and for any unramified character $\psi \in \Psi(R\mathrm{G})$ we will write $\pi\,\psi = \pi \otimes \psi$ for the corresponding twisted representation.

Two pairs (M, π), (M', π') consisting of a Levi subgroup of G and irreducible cuspidal representations π, π' of $R\mathrm{M}$, $R\mathrm{M}'$ are *inertially equivalent* if there is an element $g \in \mathrm{G}$ and an unramified character $\psi \in \Psi(R\mathrm{M}')$ such that $\mathrm{M}' = g\mathrm{M}g^{-1}$ and the g-conjugate ${}^g\pi = \pi(g \bullet g^{-1})$ is isomorphic to $\pi'\psi$. Let $[\mathrm{M}, \pi]$ be the inertial class of (M, π). Denote by $\mathcal{B}(R\mathrm{G})$ the Bernstein spectrum of G that is the set of all inertial classes of G and by $\mathcal{C}(R\mathrm{G}) \subseteq \mathcal{B}(R\mathrm{G})$ the subset of inertial equivalence classes of cuspidal RG-modules, i.e., we consider only cuspidal pairs such that the Levi is equal to G.

If $\tau \in \mathrm{Irr}(R\mathrm{G})$, then there is always a parabolic subgroup $\mathrm{P} \subseteq \mathrm{G}$ with a Levi subgroup $\mathrm{M} \subseteq \mathrm{P}$ such that τ is isomorphic to a submodule of $i^{\mathrm{G}}_{\mathrm{M}\subseteq\mathrm{P}}(\pi)$ for some cuspidal RM-module π. The G-conjugacy class of (M, π) is uniquely determined by τ and is called the *cuspidal support* of τ. Given an inertial class $\mathfrak{s}$, let $\mathrm{Mod}_{\mathfrak{s}}(R\mathrm{G}) \subseteq \mathrm{Mod}(R\mathrm{G})$ be the full subcategory generated by the irreducible modules whose cuspidal support belongs to $\mathfrak{s}$.

If $\ell = 0$, the theory of the Bernstein center gives a decomposition

$$\mathrm{Mod}(R\mathrm{G}) = \prod_{\mathfrak{s}\in\mathcal{B}(R\mathrm{G})} \mathrm{Mod}_{\mathfrak{s}}(R\mathrm{G}), \quad \mathrm{Irr}(R\mathrm{G}) = \bigsqcup_{\mathfrak{s}\in\mathcal{B}(R\mathrm{G})} \mathrm{Irr}_{\mathfrak{s}}(RG),$$

where each category $\mathrm{Mod}_{\mathfrak{s}}(R\mathrm{G})$ is indecomposable and $\mathrm{Irr}_{\mathfrak{s}}(R\mathrm{G}) = \mathrm{Irr}(R\mathrm{G}) \cap \mathrm{Mod}_{\mathfrak{s}}(R\mathrm{G})$.

Let $\mathfrak{s} = [\mathrm{M}, \pi] \in \mathcal{B}(R\mathrm{G})$ be an inertial class. Let $\mathcal{O}_\pi = \{\pi\psi\,;\ \psi \in \Psi(R\mathrm{M})\}$ be the $\Psi(R\mathrm{M})$-orbit of π in $\mathrm{Cusp}(R\mathrm{M})$. We view $\mathcal{O}_\pi$ as the algebraic torus given by the quotient of the torus $\Psi(R\mathrm{M})$ by the subgroup $\{\psi \in \Psi(R\mathrm{M})\,;\ \pi\psi \simeq \pi\}$. The function ring B_π of $\mathcal{O}_\pi$ has a natural structure of an RM-module which is a submodule of B_M. Let $\psi_\pi \subseteq \psi_M$ be the restriction of the universal unramified character ψ_M of M to the submodule $B_\pi \subseteq B_M$. We will consider the representation $\tilde{\pi}$ of RM given by

$$\tilde{\pi} = \pi\psi_\pi.$$

The RG-module $i^{\mathrm{G}}_{\mathrm{M}\subseteq\mathrm{P}}(\tilde{\pi})$ is a small progenerator of $\mathrm{Mod}_{\mathfrak{s}}(R\mathrm{G})$. Set

$$\mathrm{H}_{\mathfrak{s}} = \mathrm{End}_{R\mathrm{G}}(i^{\mathrm{G}}_{\mathrm{M}\subseteq\mathrm{P}}(\tilde{\pi}))^{\mathrm{op}}.$$

Then, we have an equivalence of Abelian categories

$$\mathrm{Mod}_{\mathfrak{s}}(R\mathrm{G}) \to \mathrm{Mod}(\mathrm{H}_{\mathfrak{s}}), \quad \sigma \mapsto \mathrm{Hom}_{R\mathrm{G}}\big(i^{\mathrm{G}}_{\mathrm{M}\subseteq\mathrm{P}}(\tilde{\pi}), \sigma\big). \tag{3.1}$$

We will write $\mathrm{mod}_{\mathfrak{s}}(R\mathrm{G}) = \mathrm{mod}(R\mathrm{G}) \cap \mathrm{Mod}_{\mathfrak{s}}(R\mathrm{G})$.

3.2.2 Bushnell–Kutzko Types

Let $\ell = 0$. Let $\mathrm{K} \subseteq \mathrm{G}$ be an open compact subgroup and let ρ be an irreducible representation of K. To such a couple we associate an algebra

$$\mathrm{H}_\rho = \mathrm{End}_{R\mathrm{G}}(ind^{\mathrm{G}}_{\mathrm{K}}(\rho))^{\mathrm{op}}$$

and a functor

$$\mathrm{Mod}(RG) \to \mathrm{Mod}(\mathrm{H}_\rho), \quad \sigma \mapsto \mathrm{Hom}_{R\mathrm{K}}(\rho, \sigma|_K) \simeq \mathrm{Hom}_{RG}\left(ind_{\mathrm{K}}^{\mathrm{G}}(\rho), \sigma\right). \tag{3.2}$$

The couple (K, ρ) is a *type* if the above functor induces an equivalence of categories

$$\mathrm{Mod}_\rho(RG) \to \mathrm{Mod}(\mathrm{H}_\rho),$$

where $\mathrm{Mod}_\rho(RG)$ is the subcategory of $\mathrm{Mod}(RG)$ of representations generated by their ρ-isotypic subspace, see [3]. For $\mathfrak{s} \in \mathcal{B}(RG)$ the type (K, ρ) is a $\mathfrak{s}$-type if any smooth irreducible representation with non-trivial ρ-isotypic subspace has cuspidal support in $\mathfrak{s}$. In that case $\mathrm{Mod}_\rho(RG) = \mathrm{Mod}_\mathfrak{s}(RG)$ as subcategories of $\mathrm{Mod}(RG)$.

3.3 *Level 0 and Unipotent Modules*

Given a parahoric subgroup $\mathrm{K} \subseteq \mathrm{G}$, let $\bar{\mathrm{K}}$ be the quotient of K by its pro-p-nilpotent radical K^+. It is the group of $\mathbb{F}_q$-points of a reductive connected group. An RG-module π is *of level 0* if it is generated by the subspace $\sum_{\mathrm{K}} \pi^{\mathrm{K}^+}$ where K runs over the set of all parahoric subgroups in G. Any subquotient of a level 0 module or any extension of level 0 modules is again a level 0 module, see, e.g., [19, §II.5.7–8]. Let $\mathrm{Irr}_0(\mathrm{G})$ be the set of isomorphism classes of level 0 irreducible modules, and let $\mathrm{mod}_0(RG) \subseteq \mathrm{mod}(RG)$ be the subcategory of level 0 modules of finite length. Set

$$\mathrm{Cusp}_0(RG) = \mathrm{Irr}_0(RG) \cap \mathrm{Cusp}(RG),$$
$$\mathrm{SCusp}_0(RG) = \mathrm{Irr}_0(RG) \cap \mathrm{SCusp}(RG).$$

Assume that the center of $\mathbb{G}$ is connected if $\ell \neq 0$. We call an irreducible cuspidal RG-module *unipotent* if it is isomorphic to $ind_{N_{\mathrm{G}}(\mathrm{K})}^{\mathrm{G}}(\sigma)$, where $\mathrm{K} \subseteq \mathrm{G}$ is a maximal standard parahoric subgroup of G, and σ is an $N_{\mathrm{G}}(\mathrm{K})$-module whose restriction to K is the inflation of an irreducible cuspidal unipotent module $\bar{\sigma}$ of $\bar{\mathrm{K}}$ in Lusztig's sense, i.e., $\bar{\sigma}$ is a constituent of the decomposition of a unipotent module over a field of characteristic zero. Let $\mathrm{Irr}_u(RG)$ be the set of isomorphism classes of unipotent irreducible modules, by which we means all modules isomorphic to an irreducible subquotient of a parabolically induced module from a cuspidal unipotent pair in G. Let $\mathrm{Cusp}_u(RG) = \mathrm{Cusp}(RG) \cap \mathrm{Irr}_u(RG)$ be the set of isomorphism classes of irreducible cuspidal unipotent RG-modules, and let $\mathrm{SCusp}_u(RG) = \mathrm{SCusp}(RG) \cap \mathrm{Irr}_u(RG)$. Finally, let $\mathrm{mod}_u(RG) \subseteq \mathrm{mod}(RG)$ be the Serre subcategory generated by $\mathrm{Irr}_u(RG)$. We have $\mathrm{mod}_u(RG) \subseteq \mathrm{mod}_0(RG)$.

Proposition 3.1 *The functors $r_{\mathrm{M}\subseteq\mathrm{P}}^{\mathrm{G}}$ and $i_{\mathrm{M}\subseteq\mathrm{P}}^{\mathrm{G}}$ preserve the categories of finite length, unipotent or level 0 modules.* □

Proof The first claim is [19, §II.5.13], the last one is [19, §II.5.12], see also [6, prop. 6.3]. The parabolic induction preserves the category of unipotent modules by definition. The parabolic restriction either by the geometric lemma.

Remark 3.3.2

(a) If the characteristic ℓ of R is 0, then $\mathrm{Irr}_u(RG)$ is the set of the irreducible unipotent modules considered by Lusztig in [12, thm. 6.11]. See [15].
(b) If $G = \mathrm{GL}_n$ and the characteristic ℓ of R is positive, then $\mathrm{Irr}_u(RG)$ is the set of the irreducible unipotent modules considered in [20], i.e., it is the set of all irreducible modules with supercuspidal support inertially equivalent to the trivial character of a minimal Levi subgroup, by [7, cor. 1.2]. □

4 Categorical Representations and Classical p-Adic Groups

4.1 Generalities on Representations of p-Adic Classical Groups

4.1.1 The General Linear Group

Given any integer $N \geqslant 1$ and any element $i \in R^\times$, let η_i denote the unramified character of GL_N such that $\eta_i(g) = i^{v_F(\det(g))}$ for all $g \in \mathrm{GL}_N$. So, we have $\Psi(R\,\mathrm{GL}_N) = \{\eta_i\,;\, i \in R^\times\}$. We will write $\eta = \eta_{q^{-1}}$.

Given a tuple $\mu = (N_1, N_2, \dots, N_a)$ of positive integers summing to N, we consider the standard Levi subgroup $\mathrm{GL}_\mu = \mathrm{GL}_{N_1} \times \cdots \times \mathrm{GL}_{N_a}$ of GL_N. Then, given an $R\,\mathrm{GL}_{N_b}$-module π_b for each $b = 1, 2, \dots, a$ we write $\pi_1 \times \pi_2 \times \cdots \times \pi_a$ for the $R\,\mathrm{GL}_N$-module given by the normalized parabolic induction of $\pi_1 \boxtimes \pi_2 \boxtimes \cdots \boxtimes \pi_a$.

For each isomorphism class $\rho \in \mathrm{Cusp}(R\,\mathrm{GL}_N)$ its inertial class $\mathfrak{c}_\rho \in \mathcal{C}(R\,\mathrm{GL}_N)$ is canonically identified with the $\Psi(R\,\mathrm{GL}_N)$-orbit $\mathcal{O}_\rho = \{\rho\,\eta_i\,;\, i \in R^\times\}$ in $\mathrm{Cusp}(R\,\mathrm{GL}_N)$. We will also view it as an $R^\times$-orbit. Following [14], we associate to ρ an unramified character $\eta_\rho \in \Psi(R\,\mathrm{GL}_N)$ and integers $N_\rho, e_\rho, q_\rho \geqslant 0$ such that

- $N_\rho = N$,
- $q_\rho \in q^{\mathbb{Z}}$ is such that for all $d > 0$ there is an R-algebra isomorphism $\mathrm{H}(d, q_\rho) \simeq \mathrm{End}_{R\,\mathrm{GL}_{dN}}(\tilde{\rho}^{\times d})$,
- e_ρ is the order of q_ρ in $R^\times$ (it may be infinite),
- $\eta_\rho = \eta_{q_\rho^{-1}}$, hence, by [13, §4.5], for all $\rho' \in \mathrm{Cusp}(R\,\mathrm{GL}_{N'})$ we have

$$\rho \times \rho' \text{ reducible} \iff N' = N \text{ and } \rho' = \rho\,\eta_\rho^{\pm 1}. \tag{4.1}$$

We abbreviate $\mathfrak{c} = \mathfrak{c}_\rho = \{\rho\eta_i\,;\, i \in R^\times\}$. The parameters $\eta_\rho, e_\rho, N_\rho, q_\rho$ do not depend on the choice of ρ in the class $\mathfrak{c}$. So we write $\eta_\mathfrak{c} = \eta_\rho$, $e_\mathfrak{c} = e_\rho$, $N_\mathfrak{c} = N_\rho$

and $q_{\mathfrak{c}} = q_\rho$. It is known that

$$\sharp\{\rho\,(\eta_{\mathfrak{c}})^i\ ;\ i \in \mathbb{Z}\} = e_{\mathfrak{c}}, \tag{4.2}$$

see, e.g., [14, lem. 5.3].

Let $Q_{\mathfrak{c}} = (\mathfrak{c}, \Omega_{\mathfrak{c}})$ be the quiver with vertex set $\mathfrak{c}$ and set of arrows

$$\Omega_{\mathfrak{c}} = \{\rho \to \rho\,\eta_{\mathfrak{c}}\ ;\ \rho \in \mathfrak{c}\}.$$

By (4.2), it is the disjoint union of an infinite number of copies of quivers of affine type $A^{(1)}_{e_{\mathfrak{c}}-1}$. Here $A^{(1)}_0$ holds for the Jordan quiver and $A^{(1)}_\infty$ for A_∞. If $\flat = \emptyset, u, 0$ we write

$$\mathrm{Cusp}_\flat(R\,\mathrm{GL}) = \bigsqcup_{N\geqslant 1} \mathrm{Cusp}_\flat(R\,\mathrm{GL}_N),$$

$$\mathrm{SCusp}_\flat(R\,\mathrm{GL}) = \bigsqcup_{N\geqslant 1} \mathrm{SCusp}_\flat(R\,\mathrm{GL}_N),$$

and we define the quiver $Q_{R\,\mathrm{GL}}$ by

$$Q_{R\,\mathrm{GL}} = (\mathrm{Cusp}(R\,\mathrm{GL}), \Omega_{R\,\mathrm{GL}}) = \bigsqcup_{N\geqslant 0}\ \bigsqcup_{\mathfrak{c}_N \in \mathcal{C}(R\,\mathrm{GL}_N)} Q_{\mathfrak{c}_N}. \tag{4.3}$$

The incidence matrix of the quiver $Q_{R\,\mathrm{GL}}$ is $C = (c_{\rho,\rho'})$ with $c_{\rho,\rho'} = \sharp\{\rho \to \rho' \in \Omega_{R\,\mathrm{GL}}\}$. Let $A = 2\mathrm{Id} - C - C^{\mathrm{T}}$ be the corresponding Cartan matrix. Consider the involution on the set $\mathrm{Cusp}(R\,\mathrm{GL})$ given by $\pi \mapsto \pi^\vee$. It yields an involution D of $Q_{R\,\mathrm{GL}}$ such that the pair $(Q_{R\,\mathrm{GL}}, D)$ is symmetric. Considering only the unipotent modules, we get a full symmetric subquiver $(Q_{R\,\mathrm{GL};u}, D)$ whose set of vertices is $\mathrm{Cusp}_u(R\,\mathrm{GL})$.

Example 4.1.1 If $\rho \in \mathrm{Cusp}(R\,\mathrm{GL}_N)$, then we have $q_\rho = q^N$, see [14, prop. 5.4]. In particular, if $\rho \in \mathrm{Cusp}_u(R\,\mathrm{GL}_N)$ and e is the order of q in $R^\times$, then we have $N \in \{1, e\ell^r\ ;\ r \in \mathbb{N}\}$ and we deduce that $q_\rho = q$ or 1. □

Example 4.1.2 Fix a supercuspidal representation $\rho \in \mathrm{SCusp}(R\,\mathrm{GL}_N)$. For each integer $n \geqslant 0$ let $\mathrm{St}(\rho, n)$ be the irreducible subquotient of $\rho \times \rho\eta_\rho \times \cdots \times \rho\eta_\rho^{n-1}$ defined in [14, thm. 6.14]. We will call it a *generalized Steinberg representation.* Recall that [14, prop. 6.4]

$$\mathrm{St}(\rho, n) \in \mathrm{Cusp}(R\,\mathrm{GL}_{nN}) \iff n = 1 \text{ or } n \in e_\rho\,\ell^{\mathbb{N}},$$

$$\mathrm{St}(\rho, n) \in \mathrm{SCusp}(R\,\mathrm{GL}_{nN}) \iff n = 1.$$

Any cuspidal representation of $R\,\mathrm{GL}_N$ is of this form. Further, we have $\mathrm{SCusp}_u(R\,\mathrm{GL}_N) = \emptyset$ if $N > 1$ and $\mathrm{SCusp}_u(R\,\mathrm{GL}_1) = \{\eta_i\ ;\ i \in R^\times\}$. □

4.1.2 Classical Groups

We set either $\mathrm{G}_n = \mathrm{SO}_{2n+1}$ for all $n > 0$, or $\mathrm{G}_n = \mathrm{SP}_{2n}$ for all $n > 0$. Set also $\mathrm{G}_0 = \{1\}$. Let $\mathrm{T}_n \subseteq \mathrm{B}_n \subseteq \mathrm{G}_n$ be the usual maximal torus and Borel subgroup. A parabolic subgroup of G_n containing T_n is called semistandard. If it also contains B_n it is called standard. A semistandard (resp. standard) parabolic subgroup has a unique Levi subgroup containing T_n. Such Levi subgroups are called semistandard (resp. standard). Given $n = m + N$ and a composition $\mu = (N_1, N_2, \ldots, N_a)$ of N we consider the standard Levi subgroup $\mathrm{G}_{m,\mu} = \mathrm{G}_m \times \mathrm{GL}_\mu$ in G_n and the corresponding standard parabolic subgroup $\mathrm{P}_{m,\mu} \subseteq \mathrm{G}_n$ and opposite semistandard parabolic subgroup $\mathrm{P}^-_{m,\mu} \subseteq \mathrm{G}_n$. We abbreviate

$$i^{\mathrm{G}_n}_{\mathrm{G}_{m,\mu}} = i^{\mathrm{G}_n}_{\mathrm{G}_{m,\mu} \subseteq \mathrm{P}_{m,\mu}}, \qquad r^{\mathrm{G}_n}_{\mathrm{G}_{m,\mu}} = r^{\mathrm{G}_n}_{\mathrm{G}_{m,\mu} \subseteq \mathrm{P}_{m,\mu}}, \qquad \bar{r}^{\mathrm{G}_n}_{\mathrm{G}_{m,\mu}} = r^{\mathrm{G}_n}_{\mathrm{G}_{m,\mu} \subseteq \mathrm{P}^-_{m,\mu}}.$$

We have an adjoint triple of functors $(r^{\mathrm{G}_n}_{\mathrm{G}_{m,\mu}}, i^{\mathrm{G}_n}_{\mathrm{G}_{m,\mu}}, \bar{r}^{\mathrm{G}_n}_{\mathrm{G}_{m,\mu}})$. The 2nd adjunction $(i^{\mathrm{G}_n}_{\mathrm{G}_{m,\mu}}, \bar{r}^{\mathrm{G}_n}_{\mathrm{G}_{m,\mu}})$ is called the *Bernstein adjunction*. It is known to hold for all classical groups. If $\flat = \emptyset, u, 0$, we write

$$\mathrm{mod}_\flat(R\mathrm{G}) = \bigoplus_{n \geqslant 0} \mathrm{mod}_\flat(R\mathrm{G}_n),$$

$$\mathcal{C}(R\mathrm{G}) = \bigsqcup_{n \geqslant 1} \mathcal{C}(R\mathrm{G}_n),$$

$$\mathcal{B}(R\mathrm{G}) = \bigsqcup_{n \geqslant 1} \mathcal{B}(R\mathrm{G}_n).$$

4.2 *Categorical Representations in Characteristic Zero*

In this section we assume that the characteristic ℓ of R is zero. In particular, any cuspidal module is also supercuspidal.

For each integers $m, N \geqslant 0$, given representants of some inertial classes $\mathfrak{c}_m \in \mathcal{C}(R\mathrm{G}_m)$ and $\mathfrak{s}_N \in \mathcal{B}(R\,\mathrm{GL}_N)$, the external tensor product gives a cuspidal module of some Levi subgroup $\mathrm{G}_{m,\mu}$ of G_{m+N} for some composition μ of N. Let $\mathfrak{c}_m \times \mathfrak{s}_N$ denote the corresponding inertial class of cuspidal pairs. Now, we fix m and we allow N to vary. Define

$$\mathrm{mod}_{\mathfrak{c}_m \times \bullet}(R\mathrm{G}) = \bigoplus_{N \geqslant 0} \bigoplus_{\mathfrak{s}_N \in \mathcal{B}(R\,\mathrm{GL}_N)} \mathrm{mod}_{\mathfrak{c}_m \times \mathfrak{s}_N}(R\mathrm{G}_{m+N}).$$

Then, we have

$$\mathrm{mod}(RG) = \bigoplus_{\mathfrak{c} \in \mathcal{C}(RG)} \mathrm{mod}_{\mathfrak{c} \times \bullet}(RG).$$

4.2.1 Categorical Action for GL

For each integer $n \geqslant 0$ set $G_n = GL_n$. Let $\mathfrak{s} = [G_\mu, \sigma] \in \mathcal{B}(RG_n)$ be an inertial class. We can assume, up to multiplication by some unramified characters in $\Psi(RG_\mu)$, that μ is a partition of n such that

$$\begin{aligned} \mu &= (\underbrace{m_1, \dots, m_1}_{l_1}, \dots, \underbrace{m_a, \dots, m_a}_{l_a}), \\ \sigma &= \underbrace{\sigma_1 \boxtimes \dots \boxtimes \sigma_1}_{l_1} \boxtimes \dots \boxtimes \underbrace{\sigma_a \boxtimes \dots \boxtimes \sigma_a}_{l_a}, \end{aligned} \tag{4.4}$$

with $\sigma_i \in \mathrm{Cusp}(RG_{m_i})$ and $\mathcal{O}_{\sigma_i} \neq \mathcal{O}_{\sigma_j}$ for $i \neq j$. Then it is known [4, 11] that

$$\mathrm{H}_{\mathfrak{s}} \simeq \bigotimes_{i=1}^{a} \mathrm{H}(l_i, q^{m_i}). \tag{4.5}$$

Definition 4.1 Let $Q = (I, \Omega)$ be a quiver. Let $B(Q)$ be the $\mathbb{Q}$-algebra generated by $\{e_i, f_i, h_i\,;\ i \in I\}$, satisfying the following defining relations for all $i, j \in I$:

(a) $h_i h_j = h_j h_i$
(b) $[h_i, e_j] = a_{i,j} e_j$, $[h_i, f_j] = -a_{i,j} f_j$,
(c) $[e_i, f_j] = \delta_{i,j}$,
(d) $\sum_{a+b=1-a_{i,j}} (-1)^a \theta_i^{(a)} \theta_j \theta_i^{(b)} = 0$ if $i \neq j$ and $\theta_i = e_i$ or f_i.

Here, we write $\theta_i^{(a)}$ for the a-th divided power.

For each inertial class $\mathfrak{c} \in \mathcal{C}(RG)$ the simply laced quiver $Q_\mathfrak{c} = (\mathfrak{c}, \Omega_\mathfrak{c})$ is the disjoint union of an infinite number of copies of quivers of type A_∞ by (4.2), because $\ell = 0$. More precisely, we have $Q_\mathfrak{c} \simeq Q_{R^\times, \eta_\mathfrak{c}}$ and the decomposition of the quiver $Q_\mathfrak{c}$ is given by the decomposition of $R^\times$ into $(\eta_\mathfrak{c})^\mathbb{Z}$-orbits.

Now, consider the quiver $Q = Q_{R\mathrm{GL}}$ in (4.3) and let U_Q^- be the universal enveloping algebra of the negative part of the Kac–Moody algebra of type Q over $\mathbb{Q}$. The restricted dual of U_Q^- is isomorphic to the coordinate algebra $\mathbb{Q}[N_Q]$ of a pro-unipotent group N_Q. It is also a $B(Q)$-module by the Ariki theorem, see [8]. For each $n \geqslant 1$ we have an isomorphism of $\mathbb{Q}$-vector spaces

$$\bigoplus_{\mathfrak{s} \in \mathcal{B}(RG_n)} [\mathrm{mod}(\mathrm{H}_\mathfrak{s})] \simeq \bigoplus_{\mathfrak{s} \in \mathcal{B}(RG_n)} [\mathrm{mod}_0(\bigotimes_{i=1}^{a} \mathrm{R}(l_i;\, Q_{R^\times, q^{m_i}}))]$$

$$\simeq [\mathrm{mod}_0(\mathrm{R}(n; Q))],$$

where a, l_i, m_i depend on $\mathfrak{s}$ and are as in (4.4). The first isomorphism follows from (4.5) and the Morita equivalence proved in [2, 16], see Remark 2.2.1. The second one follows from the decomposition in (4.3). Further, by the Morita equivalence and [8], the $\mathbb{Q}$-algebra $B(Q)$ acts on $\bigoplus_{n \geqslant 0} [\mathrm{mod}_0(\mathrm{R}(n; Q))]$, hence on $\bigoplus_{\mathfrak{s} \in \mathcal{B}(RG)} [\mathrm{mod}(\mathrm{H}_{\mathfrak{s}})]$, and we have an isomorphism of $B(Q)$-modules

$$\mathbb{Q}[N_Q] \simeq \bigoplus_{\mathfrak{s} \in \mathcal{B}(RG)} [\mathrm{mod}(\mathrm{H}_{\mathfrak{s}})].$$

Therefore, from (3.1) we deduce the following.

Proposition 4.2 *The Grothendieck group* $[\mathrm{mod}(RG)]$ *over* $\mathbb{Q}$ *is equipped with an action of the algebra* $B(Q)$ *and the resulting module is isomorphic to* $\mathbb{Q}[N_Q]$.

The goal of the next section is to prove an analogous statement for classical groups.

4.2.2 Categorical Action for SO

Set $\mathrm{G}_n = \mathrm{SO}_{2n+1}$ for all $n > 0$. We restrict ourselves to the subcategory $\mathrm{mod}_u(RG) \subseteq \mathrm{mod}(RG)$ to simplify. The group G_n is adjoint, hence $Z(\mathrm{G}_n) = \{1\}$ and ${}^\circ\mathrm{G}_n = \mathrm{G}_n$. The classification of the unipotent cuspidal pairs of G_n is given by the theory of types as in [12, 21]. More precisely, let $\Theta \subseteq \mathbb{N} \times 2\mathbb{N} \times \{\pm\}$ be the set of all triples $\theta = (r', r'', \varepsilon)$ such that $\varepsilon = +$ if $r'' = 0$. Write $n_\theta = r'(r'+1) + (r'')^2$ and $\Theta_{\leqslant n} = \{\theta \in \Theta\,;\ n_\theta \leqslant n\}$. Then, the Bernstein decomposition gives

$$\mathrm{mod}_u(R\mathrm{G}_n) = \bigoplus_{\theta \in \Theta_{\leqslant n}} \mathrm{mod}_{\mathfrak{s}_\theta}(R\mathrm{G}_n), \tag{4.6}$$

where $\mathfrak{s}_\theta \in \mathcal{B}(R\mathrm{G}_n)$ is the inertial class of the cuspidal pair $(\mathrm{G}_{n_\theta, 1^{n-n_\theta}}, \rho_\theta)$ such that the representation ρ_θ is compactly induced from the normalizer of a maximal parahoric subgroup of $\mathrm{G}_{n_\theta, 1^{n-n_\theta}}$. Given integers N, a, b with $N > 0$ and a triple $\theta \in \Theta_{\leqslant n}$ we write $a_\theta = \sup(2r'+1, 2r'')$, $b_\theta = \inf(2r'+1, 2r'')$ and

$$\mathrm{H}(N, q, a, b) = \mathrm{H}(N, q, q^{\frac{a+b}{2}}, q^{\frac{a-b}{2}}),$$
$$\mathrm{H}_\theta(N) = \mathrm{H}(N, q, a_\theta, b_\theta).$$

Let $\tilde{\rho}_\theta$ be the representation defined as in Sect. 3.2.1, with $\pi = \rho_\theta$.

Lemma 4.3 *For each* $\theta \in \Theta_{\leqslant n}$ *we have*

(a) *an* R*-algebra isomorphism* $\mathrm{H}_{\mathfrak{s}_\theta} \simeq \mathrm{H}_\theta(n - n_\theta)$,
(b) *an equivalence of categories*

$$\mathrm{mod}_{\mathfrak{s}_\theta}(R\mathrm{G}_n) \to \mathrm{mod}(\mathrm{H}_\theta(n - n_\theta)), \quad \sigma \mapsto \mathrm{Hom}_{R\mathrm{G}_n}\big(i_{\mathrm{G}_{n_\theta,1^{n-n_\theta}}}^{\mathrm{G}_n}(\tilde{\rho}_\theta)\,,\,\sigma\big). \tag{4.7}$$

Proof The Hecke algebra

$$\mathrm{H}_{\mathfrak{s}_\theta} = \mathrm{End}_{R\mathrm{G}_n}(i_{\mathrm{G}_{n_\theta,1^{n-n_\theta}}}^{\mathrm{G}_n}(\tilde{\rho}_\theta))^{\mathrm{op}}$$

is computed explicitly in [11, thm. 7.7]. There, it is proved that $\mathrm{H}_{\mathfrak{s}_\theta}$ is the semi-direct product of an affine Hecke algebra and a finite group. Since the cuspidal module ρ_θ is unipotent, this finite group is trivial. The parameters of the affine Hecke algebra can be computed from the Langlands parameter of ρ_θ as in [10], from which we deduce the isomorphism between the two algebras. The second assertion follows now from (3.1).

The discussion above yields a bijection $\mathcal{C}(R\mathrm{G}) \simeq \Theta$ and an equivalence of categories

$$\mathrm{mod}_u(R\mathrm{G}) \simeq \bigoplus_{\theta \in \Theta} \mathrm{mod}(\mathrm{H}_\theta), \tag{4.8}$$

where $\mathrm{mod}(\mathrm{H}_\theta) = \bigoplus_{N \geqslant 0} \mathrm{mod}(\mathrm{H}_\theta(N))$.

For each $i \in I$ and $N > 1$ let $\mathrm{RC}(N-1, i; \lambda) \subseteq \mathrm{RC}(N; \lambda)$ be the R-subalgebra defined in Sect. 2.1.3. We consider the functors

$$\begin{aligned} F_{i,N-1} &: \mathrm{Mod}(\mathrm{RC}(N-1; \lambda)) \to \mathrm{Mod}(\mathrm{RC}(N; \lambda)), \\ E_{i,N-1} &: \mathrm{Mod}(\mathrm{RC}(N; \lambda)) \to \mathrm{Mod}(\mathrm{RC}(N-1; \lambda)) \end{aligned} \tag{4.9}$$

given by

$$\begin{aligned} F_{i,N-1}(V) &= \mathrm{RC}(N; \lambda) \otimes_{\mathrm{RC}(N-1,i;\lambda)} (V \otimes \mathbb{C}[x_N]/(x_N)), \\ E_{i,N-1}(V) &= \mathrm{Hom}_{\mathrm{RC}(N-1,i;\lambda)}(\mathrm{RC}(N-1, i; \lambda)\,,\, V). \end{aligned}$$

The functors $E_i = \bigoplus_{N \geqslant 0} E_{i,N}$ and $F_i = \bigoplus_{N \geqslant 0} F_{i,N}$ are exact and preserve the category $\mathrm{mod}_0(\mathrm{RC}(\lambda))$. Consider the linear endomorphisms $e_i = [E_i]$, $f_i = [F_i]$ on $[\mathrm{mod}_0(\mathrm{RC}(\lambda))]$.

Now, specializing $v = 1$ in Definition 2.4, we consider the following algebra.

Definition 4.4 Let (Q, D) be any symmetric quiver. Let $B(Q, D)$ be the $\mathbb{Q}$-algebra generated by $\{e_i, f_i, h_i\,;\, i \in I\}$, satisfying the following defining relations for all $i, j \in I$:

(a) $h_i h_j = h_j h_i$ and $h_{D(i)} = h_i$,
(b) $[h_i, e_j] = (a_{i,j} + a_{D(i),j})e_j$, $[h_i, f_j] = (-a_{i,j} - a_{D(i),j})f_j$,
(c) $[e_i, f_j] = \delta_{i,j} + \delta_{D(i),j}$,
(d) $\sum_{a+b=1-a_{i,j}} (-1)^a \theta_i^{(a)} \theta_j \theta_i^{(b)} = 0$ if $i \neq j$ and $\theta_i = e_i$ or f_i.

One proves as in [8, prop. 4.2] that there is a unique (up to isomorphism) $B(Q, D)$-module $V(\lambda) = V(\lambda, Q, D)$ which is generated by a non-zero vector ϕ_λ such that for $i \in I$ we have

$$e_i\phi_\lambda = 0,\;\; h_i\phi_\lambda = (\lambda_i + \lambda_{D(i)} - 2\delta_{i,D(i)})\phi_\lambda,\;\; \{x \in V(\lambda)\,;\, e_j x = 0,\ \forall j\} = \mathbb{Q}\phi_\lambda.$$

Further, the $B(Q, D)$-module $V(\lambda)$ is irreducible. Let $V(\lambda)^\vee$ be the dual of $V(\lambda)$. It is equipped with the $B(Q, D)$-action obtained by composing the transpose of the operators e_i, f_i, h_i with the anti-homomorphism of $B(Q, D)$ such that $e_i, f_i, h_i \mapsto f_i, e_i, h_i$. The following follows from Proposition 2.5.

Proposition 4.5 *There are linear endomorphisms $\{h_i\,;\, i \in I\}$ of the vector space $[\mathrm{mod}_0(\mathrm{RC}(\lambda))]$ such that e_i, f_i, h_i define a $B(Q, D)$-module structure on $[\mathrm{mod}_0(\mathrm{RC}(\lambda))]$. If (Q, D) has no fixed points, the resulting $B(Q, D)$-module is isomorphic to $V(\lambda)^\vee$.*

Now, we consider the symmetric quiver

$$(Q, D) = (Q_{R\,\mathrm{GL};\,u}, D).$$

Since $\ell = 0$, we have $\mathrm{Cusp}_u(R\,\mathrm{GL}) = \Psi(R\,\mathrm{GL}_1) = \{\eta_i\,;\, i \in R^\times\}$. So there is a unique inertial class $\mathfrak{c}$ of unipotent cuspidal representations and $\eta_\mathfrak{c} = \eta$, yielding an isomorphism of symmetric quivers $(Q, D) \simeq Q_{R^\times, q}$. Given integers a, b, we consider the algebras $\mathrm{RC}(N, a, b) = \mathrm{RC}(N; \lambda; Q, D)$ and $B(Q, D)$, where the dimension vector $\lambda = \sum_i \lambda_i\, i$ is given by

$$\lambda_i = \delta_{i,-q^{(a-b)/2}} + \delta_{i,q^{(a+b)/2}}. \tag{4.10}$$

There is an equivalence of categories

$$\mathrm{mod}_0(\mathrm{RC}(N, a, b)) \to \mathrm{mod}(\mathrm{H}(N, q, a, b)) \tag{4.11}$$

which preserves the dimension of the modules. If $a = a_\theta$, $b = b_\theta$ we write $\lambda = \lambda_\theta$. Combining (4.8) and (4.11) we get an equivalence of categories

$$\mathrm{mod}_u(RG) \simeq \bigoplus_{\theta \in \Theta} \mathrm{mod}_0(\mathrm{RC}(\lambda_\theta)).$$

This equivalence together with Proposition 4.5 gives the following

Corollary 4.6 *For* $(Q, D) = (Q_{R\,\mathrm{GL};\,u}, D)$*, the Grothendieck group* $[\mathrm{mod}_u(\mathrm{G})]$ *over* $\mathbb{Q}$ *is equipped with an action of the algebra* $B(Q, D)$.

We do not know how to describe explicitly the resulting $B(Q, D)$-module (cf. Sect. 2.1.4). Now, we turn to the positive characteristic.

4.3 Categorical Representations in non-Natural Characteristic

If $\ell \neq 0$ we do not have the Bernstein decomposition. Nevertheless, we can define a categorical representation on the category of finite length modules. We will assume that $\mathrm{G}_n = \mathrm{SO}_{2n+1}$ for all $n > 0$ and $\mathrm{G}_0 = \{1\}$.

For each cuspidal representation $\rho \in \mathrm{Cusp}(R\,\mathrm{GL}_N)$ we define the exact endofunctor F_ρ of $\mathrm{mod}(R\mathrm{G})$ by setting

$$F_\rho(\pi) = i^{\mathrm{G}_n}_{\mathrm{G}_{m,N}}(\pi \boxtimes \rho) \tag{4.12}$$

for each m, n with $n = m+N$ and each $\pi \in \mathrm{mod}(R\mathrm{G}_m)$. Next, if ρ is supercuspidal of level 0, we define an exact endofunctor E_ρ of $\mathrm{mod}(R\mathrm{G})$ as follows. There is a unique supercuspidal representation $\bar{\rho}$ of $R\,\mathrm{GL}_N(q)$ and a unique lifting $\tilde{\rho}$ to the subgroup $\varpi^{\mathbb{Z}}\,\mathrm{GL}_N(\mathcal{O}) \subseteq \mathrm{GL}_N$, such that ρ is isomorphic to the compactly induced module $ind^{\mathrm{GL}_N}_{\varpi^{\mathbb{Z}}\,\mathrm{GL}_N(\mathcal{O})}(\tilde{\rho})$. Let $P_{\bar{\rho}}$ be the projective cover of $\bar{\rho}$, z be the central character of ρ, and $P_{\tilde{\rho}}$ be the lifting to $\varpi^{\mathbb{Z}}\,\mathrm{GL}_N(\mathcal{O})$ of $z \otimes P_{\bar{\rho}}$. For each representation $\pi \in \mathrm{mod}(R\mathrm{G}_n)$, we set

$$E'_\rho(\pi) = \mathrm{Hom}_{GL_N(\mathcal{O})}\big(P_{\tilde{\rho}}\,,\, \bar{r}^{\mathrm{G}_n}_{\mathrm{G}_{m,N}}(\pi)\big) \in \mathrm{Mod}(R\mathrm{G}_m).$$

The functor E'_ρ is exact, because the $\mathrm{GL}_N(\mathcal{O})$-module $P_{\tilde{\rho}}$ is projective. Since the $R\mathrm{G}_{m,N}$-module $\bar{r}^{\mathrm{G}_n}_{\mathrm{G}_{m,N}}(\pi)$ is smooth of finite length, the $\varpi^{\mathbb{Z}}$-action on $E'_\rho(\pi)$ is locally finite. We define

$$E_\rho(\pi) = \{v \in E'_\rho(\pi)\,;\, \exists d > 0 \text{ s.t. } (\varpi v - z(\varpi)v)^d = 0\}. \tag{4.13}$$

Proposition 4.7 *Let* $\rho, \sigma \in \mathrm{SCusp}_u(R\,\mathrm{GL})$.

(a) F_ρ, E_ρ *preserve the unipotent modules and are exact, yielding linear endomorphisms* f_ρ, e_ρ *of the rational Grothendieck group* $[\mathrm{mod}_u(R\mathrm{G})]$.
(b) *Assume that* $q-1$ *is invertible modulo* ℓ*. Then, the following relations hold:*

   (1) $[e_\sigma\,,\, e_\rho] = 0$ *and* $[f_\sigma\,,\, f_\rho] = 0$ *if* $a_{\sigma,\rho} = 0$.
   (2) $[e_\sigma\,,\, [e_\sigma\,,\, e_\rho]] = [f_\sigma\,,\, [f_\sigma\,,\, f_\rho]] = 0$ *if* $a_{\sigma,\rho} = -1$.
   (3) $[\,e_\sigma\,,\, f_\rho\,] = \delta_{\sigma,\rho} + \delta_{\sigma^\vee,\rho}$ *for all* σ, ρ. □

Proof Note that, since $\rho, \sigma \in \mathrm{SCusp}_u(R\,\mathrm{GL})$, they are both of the form η_i for some $i \in R^\times$. Part (a) follows from Proposition 3.1, except the fact that $E_\rho(\pi)$ is of finite length, which is left to the reader. To prove (b), relations (1) and (2), let us first observe that, for any representation $\pi \in \mathrm{mod}(R\mathrm{G}_{m+1})$, the module $E_{\eta_i}(\pi)$ is the set of all elements of

$$\mathrm{Hom}_{R(\mathbb{F}_q)^\times}\big(P\,,\,\bar{r}^{\mathrm{G}_{m+1}}_{\mathrm{G}_{m,1}}(\pi)^{1+\varpi\mathcal{O}}\big)$$

killed by a power of $\varpi - i$, where P is a projective cover of the trivial representation of $R(\mathbb{F}_q)^\times$ and $1+\varpi\mathcal{O} \subseteq \mathrm{GL}_1(\mathcal{O})$ is a subgroup of $\mathrm{G}_{m,1} = \mathrm{G}_m \times \mathrm{GL}_1$ in the obvious way. Since $q-1$ is invertible modulo ℓ, the module P is trivial, so we have

$$\mathrm{Hom}_{R(\mathbb{F}_q)^\times}\big(P\,,\,\bar{r}^{\mathrm{G}_{m+1}}_{\mathrm{G}_{m,1}}(\pi)^{1+\varpi\mathcal{O}}\big) = \bar{r}^{\mathrm{G}_{m+1}}_{\mathrm{G}_{m,1}}(\pi)^{\mathcal{O}^\times}.$$

We deduce that the functor E_ρ is right adjoint to F_ρ. Therefore, it is enough to check the relations for the operators f_ρ, f_σ. To do that, by the transitivity of induction, we must check that in the Grothendieck group $[\mathrm{mod}(R\,\mathrm{GL})]$, we have the following relations:

$$[\sigma \times \rho] = [\rho \times \sigma] \quad \text{if } a_{\sigma\,,\,\rho} = 0,$$
$$[\sigma \times \sigma \times \rho] + [\rho \times \sigma \times \sigma] = 2\,[\sigma \times \rho \times \sigma] \quad \text{if } a_{\sigma\,,\,\rho} = -1.$$

We can assume that R is the residue field of a local ring A with fraction field K of characteristic zero. Then the $R\,\mathrm{GL}_1$-modules ρ, σ lift to $A\,\mathrm{GL}_1$-modules and, by base change from A to K, we are reduced to the case where the characteristic ℓ of R is zero. In this case the relations follow from the Bernstein equivalence and the Ariki theorem, [1].

Finally, to prove (3), we compute the class $e_\sigma f_\rho(\pi)$ in the Grothendieck group $[\mathrm{mod}(R\mathrm{G})]$ for any $\sigma \in \mathrm{SCusp}(R\,\mathrm{GL}_{N_\sigma})$, $\rho \in \mathrm{Cusp}(R\,\mathrm{GL}_{N_\rho})$ and $\pi \in \mathrm{mod}(R\mathrm{G}_{m_\rho})$. Set $n = m_\rho + N_\rho = m_\sigma + N_\sigma$. The $R\mathrm{G}_{m_\sigma}$-module $E_\sigma F_\rho(\pi)$ is a generalized eigenspace of ϖ acting on the $R\mathrm{G}_{m_\sigma}$-module

$$E'_\sigma F_\rho(\pi) = \mathrm{Hom}_{\mathrm{GL}_{N_\sigma}(\mathcal{O})}\big(P_{\tilde{\sigma}}\,,\,\bar{r}^{\mathrm{G}_n}_{\mathrm{G}_{m_\sigma,N_\sigma}}\, i^{\mathrm{G}_n}_{\mathrm{G}_{m_\rho,N_\rho}}(\pi \boxtimes \rho)\big).$$

Hence, since the functor $E'_\sigma F_\rho$ is exact, the class $e_\sigma f_\rho(\pi)$ depends only on the class of the module

$$\bar{r}^{\mathrm{G}_n}_{\mathrm{G}_{m_\sigma,N_\sigma}}\, i^{\mathrm{G}_n}_{\mathrm{G}_{m_\rho,N_\rho}}(\pi \boxtimes \rho)$$

in the Grothendieck group $[\mathrm{mod}(R\mathrm{G}_{m_\sigma,N_\sigma})]$, which coincides with the class of the module

$$r^{\mathrm{G}_n}_{\mathrm{G}_{m_\sigma,N_\sigma}}\, i^{\mathrm{G}_n}_{\mathrm{G}_{m_\rho,N_\rho}}(\pi \boxtimes \rho).$$

Now, the geometric lemma, see, e.g., [19, §II.2.18], implies that this $R\mathrm{G}_{m_\sigma,N_\sigma}$-module is filtered and the sections of this filtration are

$$i_{w^{-1}(\mathrm{L})}^{\mathrm{G}_{m_\sigma,N_\sigma}} \circ w \circ r_{\mathrm{L}}^{\mathrm{G}_{m_\rho,N_\rho}}(\pi \boxtimes \rho),$$

where $\mathrm{L} = w(\mathrm{G}_{m_\rho,N_\rho}) \cap \mathrm{G}_{m_\sigma,N_\sigma}$ and w runs over the set of minimal length representatives in W_n of the double cosets in $\mathrm{W}_{m_\sigma,N_\sigma} \backslash \mathrm{W}_n \,/\, \mathrm{W}_{m_\rho,N_\rho}$. So w runs over the subset

$$\{w_{k,l'}\,;\ k, l' \geqslant 0\,, k + l' \leqslant N_\rho\,, l' \leqslant m_\sigma\} \subseteq \mathrm{W}_n$$

such that $w_{k,l'}$ is the unique element w as above satisfying the following conditions:

$$\sharp\,(w[m_\rho + 1, n] \cap [1 - n, -m_\sigma]) = k,$$
$$\sharp\,(w[m_\rho + 1, n] \cap [1, m_\sigma]) = l'.$$

The Weyl group of L is given by

$$\begin{aligned} \mathrm{W}_{\mathrm{L}} &= w(\mathrm{W}_{m_\rho\,,\,N_\rho}) \cap \mathrm{W}_{m_\sigma\,,\,N_\sigma} \\ &= \mathrm{W}_{m_\sigma - l'\,,\,l'\,,\,k\,,\,N_\sigma - N_\rho + l'\,,\,N_\rho - k - l'}\,. \end{aligned}$$

Now, by cuspidality of ρ, only three cases are possible:

- $k = 0$, $l' = N_\rho$: then, we have $\mathrm{W}_{\mathrm{L}} = \mathrm{W}_{m_\sigma - N_\rho\,,\,N_\rho\,,\,N_\sigma}$ and

$$w(h) = \begin{cases} h - N_\sigma, & \forall h \in [1 + m_\rho, n], \\ h + N_\rho, & \forall h \in [1 + m_\sigma - N_\rho, m_\rho], \\ h, & \forall h \in [1, m_\sigma - N_\rho]. \end{cases}$$

- $k = N_\rho$, $l' = 0$: then, we have $e_\sigma f_\rho(\pi) = 0$ unless $N_\sigma = N_\rho$ because σ is supercuspidal. Write $N = N_\sigma = N_\rho$ and $m = m_\sigma = m_\rho$. We have $\mathrm{W}_{\mathrm{L}} = \mathrm{W}_{m\,,\,N}$ and

$$w(h) = \begin{cases} h - m - n, & \forall h \in [1 + m, n], \\ h, & \forall h \in [1, m]. \end{cases}$$

- $k = l' = 0$: then, we have $e_\sigma f_\rho(\pi) = 0$ unless $N_\rho = N_\sigma$, because σ is supercuspidal. Write $N = N_\sigma = N_\rho$ and $m = m_\sigma = m_\rho$. We have $\mathrm{W}_{\mathrm{L}} = \mathrm{W}_{m\,,\,N}$ and $w = e$.

Further, according to Bernstein and Kazhdan, whenever $\tau \in \mathrm{Irr}(\mathrm{GL}_N)$, the representation $g \mapsto \tau((g^{\mathrm{T}})^{-1})$ is isomorphic to $\tau^\vee$. We deduce that

$$e_\sigma f_\rho(\pi) = f_\rho e_\sigma(\pi) + \delta_{\sigma^\vee, \rho} \cdot \pi + \delta_{\sigma, \rho} \cdot \pi.$$

Remark 4.3.1

(a) If $\mathrm{G}_n = \mathrm{GL}_n$ for all $n > 0$, then all claims in Proposition 4.7 hold, except the last relation which is replaced by the relation $[e_{[\sigma]}, f_{[\rho]}] = \delta_{[\sigma], [\rho]}$. The proof is similar to the case of type B/C and is left to the reader.

(b) For any representations $\pi \in \mathrm{mod}(R\mathrm{G}_m)$, $\tau \in \mathrm{mod}(R\,\mathrm{GL}_N)$ we write $F_\tau(\pi) = i_{\mathrm{G}_{m,N}}^{\mathrm{G}_n}(\pi \boxtimes \tau)$. The transitivity of the parabolic induction and restriction yields isomorphisms of functors $F_\rho F_\sigma \simeq F_{\rho\times\sigma}$. From (4.1) we deduce that if $a_{\sigma, \rho} = 0$, then $\rho \times \sigma$ and $\sigma \times \rho$ are both irreducible. Thus, by [14, prop. 2.2], they are isomorphic, hence the relation $F_\rho F_\sigma \simeq F_\sigma F_\rho$ holds in this more general setting. We expect Proposition 4.7 to be again true with $\rho, \sigma \in \mathrm{Cusp}_u(R\,\mathrm{GL})$. □

Consider the symmetric quiver given by

$$(Q, D) = (\mathrm{SCusp}_u(R\,\mathrm{GL}),\ \Omega_{R\,\mathrm{GL}},\ D),$$

compare (4.3). Thus, we have $(Q, D) = Q_{R^\times, q}$.

Conjecture 4.8 Assume that $q - 1$ is invertible modulo ℓ. The endomorphisms e_ρ, f_ρ with $\rho, \sigma \in \mathrm{SCusp}_u(R\,\mathrm{GL})$ extend to a $B(Q, D)$-module structure on $[\mathrm{mod}_u(R\mathrm{G})]$. □

Note that, for $\mathrm{G} = \mathrm{SP}(8)$, some irreducible G-module may occur as a subquotient of the induced module of two non-inertially equivalent supercuspidal pairs, see [7].

Remark 4.3.2 In the case $\mathrm{G}_n = \mathrm{SO}_{2n+1}$ and $\ell = 0$, composing (4.8) with (4.11) we get a chain of equivalences of categories

$$\mathrm{mod}_u(R\mathrm{G}) \xrightarrow{A} \bigoplus_{\theta\in\Theta} \mathrm{mod}(\mathrm{H}_\theta) \xrightarrow{B} \bigoplus_{\theta\in\Theta} \mathrm{mod}_0(\mathrm{RC}(\lambda_\theta)),$$

which yields an isomorphism of Grothendieck groups. We claim that the operators E_{η_i}, F_{η_i} on the left hand side coincide with the operators E_i, F_i on the right hand side. Hence, the linear operators e_{η_i}, f_{η_i} yield a representation of $B(Q, D)$ on $[\mathrm{mod}_u(R\mathrm{G})]$ where $(Q, D) = Q_{R^\times, q}$. Compare with Proposition 4.5.

To prove the claim, write

$$\mathrm{mod}_u(R\mathrm{G}) = \bigoplus_{n\geqslant 0} \bigoplus_{\theta\in\Theta_{\leqslant n}} \mathrm{mod}_{\mathfrak{s}_\theta}(R\mathrm{G}_n).$$

The equivalence A is the sum over all $\theta \in \Theta$ and all integers $N = n - n_\theta \geqslant 0$ of the functors

$$\mathrm{mod}_{\mathfrak{s}_\theta}(RG_n) \to \mathrm{mod}(\mathrm{H}_\theta(N)),\ \sigma \mapsto \mathrm{Hom}_{RG_n}\big(i^{G_n}_{G_{n_\theta,1^N}}(\tilde{\rho}_\theta)\,,\,\sigma\big),$$

considered in (4.7). Now, recall that $F_{\eta_i} = i^{G_{n+1}}_{G_{n,1}}(\bullet \boxtimes \eta_i)$ and that E_{η_i} is a direct summand of the functor $\bar{r}^{G_{n+1}}_{G_{n,1}}(\bullet)^{\mathcal{O}^\times}$. Further, the universal unramified character ψ_{GL_1} is the representation of $R\,\mathrm{GL}_1$ in the Laurent polynomial ring $R[x, x^{-1}]$ such that an element $g \in \mathrm{GL}_1$ acts by multiplication by the monomial $x^{v_F(\det(g))}$. For each modules $\pi \in \mathrm{mod}(RG_n)$, $\sigma \in \mathrm{mod}(RG_{n+1})$, the second adjunction yields the isomorphism

$$\mathrm{Hom}_{RG_{n+1}}\big(i^{G_{n+1}}_{G_{n,1}}(\pi \boxtimes \psi_{\mathrm{GL}_1})\,,\,\sigma\big) \simeq \mathrm{Hom}_{RG_n}\big(\pi\,,\,\bar{r}^{G_{n+1}}_{G_{n,1}}(\sigma)^{\mathcal{O}^\times}\big).$$

Setting $\pi = i^{G_n}_{G_{n_\theta,1^N}}(\tilde{\rho}_\theta)$ and summing over all θ, N, we get

$$\mathrm{Hom}_{RG_{n+1}}\big(i^{G_{n+1}}_{G_{n_\theta,1^{N+1}}}(\tilde{\rho}_\theta \boxtimes \psi_{\mathrm{GL}_1})\,,\,\sigma\big) \simeq A\big(\bar{r}^{G_{n+1}}_{G_{n,1}}(\sigma)^{\mathcal{O}^\times}\big).$$

Thus, the equivalence A intertwines the functor $\bar{r}^{G_{n+1}}_{G_{n,1}}(\bullet)^{\mathcal{O}^\times}$ with the restriction of affine Hecke algebra modules

$$\mathrm{Res}^{N+1}_{N,1} : \mathrm{Mod}(\mathrm{H}_\theta(N+1)) \to \mathrm{Mod}(\mathrm{H}_\theta(N) \otimes R[X_{N+1}, X_{N+1}^{-1}]).$$

Since $\eta_i = \psi_{\mathrm{GL}_1}/(x-i)$, it also intertwines E_{η_i} with the generalized eigenspace of $(X_{N+1} - i)$

$$E_{i,N} \subseteq \mathrm{Res}^{N+1}_{N,1}\,. \tag{4.14}$$

Since (4.7) is an equivalence, the unicity of the left adjoint implies that the equivalence A intertwines the functor $i^{G_{n+1}}_{G_{n,1}}(\bullet \boxtimes \psi_{\mathrm{GL}_1})$ with the induction of affine Hecke algebra modules

$$\mathrm{Ind}^{N+1}_{N,1} : \mathrm{Mod}(\mathrm{H}_\theta(N) \otimes R[X_{N+1}, X_{N+1}^{-1}]) \to \mathrm{Mod}(\mathrm{H}_\theta(N+1)).$$

Hence, it also intertwines F_{η_i} with the functor

$$F_{i,N} = \mathrm{Ind}^{N+1}_{N,1}(\bullet \boxtimes R[X_{N+1}, X_{N+1}^{-1}]/(X_{N+1} - i)). \tag{4.15}$$

Finally, the equivalence B intertwines the functors $E_{i,N}$, $F_{i,N}$ in (4.14), (4.15) with the eponymous functors in (4.9), see [18]. Now, the proof follows from Proposition 4.5. □

5 Appendix: Geometric Realization of the Quiver-Hecke Algebra

We give here a geometric realization of quiver-Hecke algebras of type C based on a complex flag variety of type C. Note that in [18] we used a flag variety of type D. In this section we set $Q = Q_{I,q}$ with $I \subseteq \mathbb{C}^\times$ and $q \in \mathbb{C} \setminus \{0, 1\}$. To simplify, we will also set $\ell = 0$.

5.1 *The Convolution Algebra*

5.1.1 Preliminaries

A *symplectic I-graded vector space* V is a pair consisting of a finite dimensional I-graded complex vector space $V = \bigoplus_{i \in I} V_i$ and a symplectic form ω such that $\omega(V_i, V_j) = 0$ whenever $D(i) \neq j$. Then, the dimension vector α of V belongs to the set

$$\mathbb{N}I^D = \{\sum_i \alpha_i\, i\,;\ \alpha_{D(i)} = \alpha_i,\ \forall i\}.$$

Two symplectic I-graded vector spaces with the same dimension vector are isomorphic. So we will fix once for all a symplectic I-graded vector space for each dimension vector α in $\mathbb{N}I^D$. Let us call it V, or V_α if some confusion is possible. Let $2n$ be the dimension of V. We will say that α is of *length* n.

An *I-graded symplectic basis of* V is a basis $(e_k\,;\ k \in [n])$ consisting of homogenous elements such that

$$\omega(e_k, e_l) = \delta_{l,1-k}, \quad \forall k, l \in [n] \text{ with } k > 0.$$

We identify $\wedge^2 V$ with the space of alternating forms of V via the isomorphism $V \simeq V^*$ defined by ω. So, the symplectic form ω is identified with the element $\sum_{k=1}^n e_k \wedge e_{1-k}$. Let $(\bullet)^*$ be the transpose relatively to the symplectic form ω. Any element of $\wedge^2 V$ is of the form

$$f(v, v') = \omega(u(v), v') = \omega(v, u(v')), \quad v, v' \in V$$

for some unique $u \in \mathrm{End}(V)$ such that $u^* = u$. Let $\mathrm{E}_\alpha \subseteq \wedge^2 V$ be the set of representations of the quiver Q on V such that $(x_h)^* = x_{D(h)}$ for all arrows h. The elements of E_α are called the *symmetric representations* of the quiver Q in V relatively to ω.

Fix an I-graded vector space $W = \bigoplus_{i \in I} W_i$ with dimension vector λ. Set

$$\mathrm{E}_{\alpha,\lambda} = \{(x, y) \in \mathrm{E}_\alpha \times \mathrm{Hom}_I(W, V)\,;\ (x_h)^* = x_{D(h)}\}.$$

We will abbreviate $\mathrm{GL}_n = \mathrm{GL}(V)$ and $\mathrm{G}_n = \mathrm{SP}(V)$. Set

$$\mathrm{GL}_\alpha = \prod_{i \in I} GL(V_i) \subseteq \mathrm{GL}_n,$$

$$\mathrm{G}_\alpha = \{(g_i) \in \mathrm{GL}_\alpha\,;\ (g_i)^* = (g_{D(i)})^{-1}\} \subseteq \mathrm{G}_n.$$

The group G_α acts by conjugation on $\mathrm{E}_{\alpha,\lambda}$.

5.1.2 The Convolution Algebra

Let

$$I^{\alpha,D} = \{\nu \in I^n\,;\ \sum_{k \in [n]} \nu_k = \alpha\}.$$

Let $(\bullet)^\perp$ denote the orthogonal for the symplectic form ω. A flag $\phi = (0 = \phi_{-n} \subseteq \cdots \subseteq \phi_n = V)$ is *isotropic of dimension* ν if $\dim(\phi_k/\phi_{k-1}) = \nu_k$ and $(\phi_k)^\perp = \phi_{-k}$ for all k. Let F_ν be the variety of all isotropic flags of dimension ν. Put $F_\alpha = \bigsqcup_{\nu \in I^{\alpha,D}} F_\nu$. Set

$$\tilde{F}_\nu = \{(\phi, x, y) \in F_\nu \times \mathrm{E}_{\alpha,\lambda}\,;\ x(\phi_k) \subseteq \phi_{k-1}\,,\ y(W) \subseteq \phi_0\}, \quad \tilde{F}_\alpha = \bigsqcup_{\nu \in I^{\alpha,D}} \tilde{F}_\nu. \tag{5.1}$$

For $\nu, \nu' \in I^{\alpha,D}$ we define the Steinberg-type variety

$$\begin{aligned} Z_{\nu,\nu'} &= \tilde{F}_\nu \times_{\mathrm{E}_{\alpha,\lambda}} \tilde{F}_{\nu'} \\ &= \{(\phi, \phi', x, y) \in F_\nu \times F_{\nu'} \times \mathrm{E}_{\alpha,\lambda}\,;\ (\phi, x, y) \in \tilde{F}_\nu,\ (\phi', x, y) \in \tilde{F}_{\nu'}\}, \end{aligned}$$

and let $Z_\alpha = \bigsqcup_{\nu,\nu' \in I^{\alpha,D}} Z_{\nu,\nu'}$. We define

$$\mathcal{RC}(\nu, \nu'; \lambda) = H_*^{\mathrm{G}_\alpha}(Z_{\nu,\nu'})\langle -2\dim \tilde{F}_\nu \rangle, \quad \mathcal{RC}(\alpha; \lambda) = \bigoplus_{\nu,\nu' \in I^{\alpha,D}} \mathcal{RC}(\nu, \nu'; \lambda),$$

$$\mathcal{P}(\nu) = H^*_{\mathrm{G}_\alpha}(\tilde{F}_\nu), \quad \mathcal{P}(\alpha) = \bigoplus_{\nu \in I^{\alpha,D}} \mathcal{P}(\nu).$$

Then $\mathcal{RC}(\alpha; \lambda)$ becomes an associative graded algebra with the convolution product

$$\star : H_p^{G_\alpha}(Z_{\nu,\nu'}) \otimes H_q^{G_\alpha}(Z_{\nu',\nu''}) \to H^{G_\alpha}_{p+q+2\dim \tilde{F}_{\nu'}}(Z_{\nu,\nu''})$$

defined as in [5, 17]. Moreover, $\mathrm{P}(\alpha)$ has the graded $\mathrm{RC}(\alpha;\lambda)$-module structure arising from the convolution product

$$\star : H_p^{G_\alpha}(Z_{\nu,\nu'}) \otimes H^q_{G_\alpha}(\tilde{F}_{\nu'}) \to H_p^{G_\alpha}(Z_{\nu,\nu''}) \times H^q_{G_\alpha}(Z_{\nu,\nu'}) \to H^{G_\alpha}_{p+q}(Z_{\nu,\nu'})$$
$$\to H^{G_\alpha}_{p+q}(\tilde{F}_\nu) \simeq H^{p+q+2\dim \tilde{F}_\nu}_{G_\alpha}(\tilde{F}_\nu).$$

Here, the first arrow is the pullback by the projection $Z_{\nu,\nu'} \to \tilde{F}_{\nu'}$, the second one is the Multiplication, and the last arrow is the pushforward by the proper map $Z_{\nu,\nu'} \to \tilde{F}_\nu$.

5.2 *The Theorem*

5.2.1 Definition of the Generators

Fix an I-graded symplectic basis $(e_k\,;\,k \in [n])$ of V. For each $k \in [n]$ let $\nu_k \in I$ be such that $e_k \in V_{\nu_k}$. Then, the tuple $\nu_\bullet = (\nu_k\,;\,k \in [n])$ belongs to $I^{\alpha,D}$. Let $D_k \subseteq V$ be the line spanned by the basis vector e_k. Note that $D_k \subseteq V_{\nu_k}$ for all k. The *canonical flag* associated with the basis $(e_k\,;\,k \in [n])$ is the flag $\phi_\bullet \in F_{\nu_\bullet}$ such that

$$\phi_\bullet = (0 \subseteq D_{1-n} \subseteq D_{1-n} \oplus D_{2-n} \subseteq \cdots \subseteq \bigoplus_{k\in[n]} D_k = V).$$

Let $T \subseteq G_\alpha \subseteq G_n$ be the maximal torus which fixes the lines $D_{1-n}, \ldots, D_n$. The Weyl group of the pair (G_n, T) is identified with the Coxeter group W_n by setting $w\phi_\bullet = \phi_w$ with

$$\phi_w = (0 \subseteq D_{w(1-n)} \subseteq D_{w(1-n)} \oplus D_{w(2-n)} \subseteq \cdots \subseteq \bigoplus_{-n<k\leqslant n} D_{w(k)} = V).$$

For each $\nu \in I^{\alpha,D}$, $w \in \mathrm{W}_n$ we set $\nu_w = w^{-1}\nu_\bullet = (\nu_{w,k}\,;\,k \in [n])$. We have $\phi_w \in F_{\nu_w}$, $\phi_e = \phi_\bullet$, where $e \in \mathrm{W}_n$ is the unit.

For each $w \in \mathrm{W}_n$ and $\nu, \nu' \in I^{\alpha,D}$ we define

$$\mathcal{O}^w_\alpha = \mathrm{G}_n \cdot (\phi_e, \phi_w) \cap (F_\alpha \times F_\alpha), \quad \bar{\mathcal{O}}^w_\alpha = \text{the closure of } \mathcal{O}^w_\alpha \text{ in } F_\alpha \times F_\alpha,$$
$$\mathcal{O}^w_{\nu,\nu'} = \mathcal{O}^w_\alpha \cap (F_\nu \times F_{\nu'}), \quad \bar{\mathcal{O}}^w_{\nu,\nu'} = \bar{\mathcal{O}}^w_\alpha \cap (F_\nu \times F_{\nu'}).$$

Note that we have $F_\alpha \times F_\alpha = \bigsqcup_{w\in\mathrm{W}_n} \mathcal{O}^w_\alpha$. Consider the G_α-equivariant morphism

$$q : Z_{\nu,\nu'} \to F_\nu \times F_{\nu'}, \quad (\phi, \phi', x, y) \mapsto (\phi, \phi').$$

For each $w \in \mathrm{W}_n$ and $\nu, \nu' \in I^{\alpha,D}$ we define

$$Z_\alpha^w = \overline{q^{-1}(\mathcal{O}_\alpha^w)}, \quad Z_\alpha^{\leqslant w} = \bigcup_{w' \leqslant w} Z_\alpha^{w'},$$

$$Z_{\nu,\nu'}^w = Z_\alpha^w \cap Z_{\nu,\nu'}, \quad Z_{\nu,\nu'}^{\leqslant w} = Z_\alpha^{\leqslant w} \cap Z_{\nu,\nu'},$$

where $\leqslant$ is the Bruhat order on W_n. We define

$$\mathcal{RC}(\nu, \nu'; \lambda)^{\leqslant w} = H_*^{\mathrm{G}_\alpha}(Z_{\nu,\nu'}^{\leqslant w})\langle -2 \dim \tilde{F}_\nu \rangle,$$

$$\mathcal{RC}(\alpha; \lambda)^{\leqslant w} = \bigoplus_{\nu,\nu' \in I^{\alpha,D}} \mathcal{RC}(\nu, \nu'; \lambda)^{\leqslant w}.$$

The fundamental class $[Z_\alpha^w] = \sum_{\nu,\nu' \in I^{\alpha,D}} [Z_{\nu,\nu'}^w]$ can be viewed as an element $\tau_w \in \mathcal{RC}(\alpha; \lambda)^{\leqslant w}$ or $\mathcal{RC}(\alpha; \lambda)$. If $w = s_k$ for some $k = 0, 1, \ldots, n-1$ and $\nu' = s_k\nu$, we abbreviate

$$\tau_k e(\nu) = \tau_{s_k} e(\nu) = [Z_{s_k\nu,\nu}^{s_k}] \in \mathcal{RC}(\alpha; \lambda)^{\leqslant s_k}. \tag{5.2}$$

For each $\nu \in I^{\alpha,D}$ and $k \in [n]$, let $\mathcal{L}_k(\nu)$ be the G_α-equivariant line bundle on F_ν assigning ϕ_k/ϕ_{k-1} to the flag ϕ. Let $\chi_k(\nu) \in H^2_{\mathrm{G}_\alpha}(\tilde{F}_\nu)$ be the 1-st Chern class of the pullback of $\mathcal{L}_k(\nu)$ to $\tilde{F}_\nu$. We have an isomorphism

$$\mathcal{P}(\nu) = H^*_{\mathrm{G}_\alpha}(\tilde{F}_\nu) = R[\chi_1(\nu), \ldots, \chi_n(\nu)] \simeq \mathrm{P}(\nu).$$

Since $Z_{\nu,\nu}^{\leqslant e} \simeq \tilde{F}_\nu$, the isomorphism above yields

$$\mathcal{RC}(\alpha; \lambda)^{\leqslant e} = H_*^{\mathrm{G}_\alpha}(Z_{\nu,\nu}^{\leqslant e})\langle -2 \dim \tilde{F}_\nu \rangle \simeq H^*_{\mathrm{G}_\alpha}(\tilde{F}_\nu) = R[\chi_1(\nu), \ldots, \chi_n(\nu)].$$

Let us denote the images of $\chi_k(\nu)$, 1 under this isomorphism by

$$x_k e(\nu), e(\nu) \in \mathcal{RC}(\alpha; \lambda)^{\leqslant e}. \tag{5.3}$$

5.2.2 The Theorem

We can now compute the algebra $\mathcal{RC}(\alpha; \lambda)$ and its representation in $\mathcal{P}(\alpha) \simeq \mathrm{P}(\alpha)$. The proof is essentially the same as in [17, 18].

Theorem 5.1

(a) *The representation of* $\mathcal{RC}(\alpha;\lambda)$ *on* $\mathrm{P}(\alpha)$ *is as in Definition 2.1.*
(b) $\mathcal{RC}(\alpha;\lambda)$ *is generated by* x_l, τ_k *and* $e(\nu)$ *with* $\nu \in I^{\alpha,D}$, $l = 1,\dots,n$, $k = 0,1,\dots,n-1$.
(c) *The representation of* $\mathcal{RC}(\alpha;\lambda)$ *on* $\mathrm{P}(\alpha)$ *is faithful.*

Corollary 5.2 *There is a* $\mathbb{Z}$*-graded* R*-algebra isomorphism* $\mathcal{RC}(n) \simeq \mathrm{RC}(n)$ *with takes the generators* $e(\nu)$, x_l, τ_k *to themseves and which interwines the representation of* $\mathrm{P}(n)$.

References

1. S. Ariki, *On the decomposition numbers of the Hecke algebra of* $G(m,1,n)$, J. Math. Kyoto Univ., **36** (4) (1996), 789–808.
2. J. Brundan, A. Kleshchev, *Blocks of cyclotomic Hecke algebras and Khovanov–Lauda algebras*, Invent. Math. **178** (2009), 451–484.
3. C. J. Bushnell, P. C. Kutzko, *Smooth representations of reductive p-adic groups: structure theory via types*, Proc. London Math. Soc. **7** (1998), 582–634.
4. C. J. Bushnell, P. C. Kutzko, *Semisimple types in* GL_n, Compositio Math. **119** (1999), 53–97.
5. N. Chriss, V. Ginzburg, Representation theory and complex geometry. Reprint of the 1997 edition. Modern Birkhäuser Classics. Birkhäuser Boston, Inc., Boston, MA, 2010.
6. J. F. Dat, Finitude pour les représentations lisses de groupes p-adiques, J. Inst. Math. Jussieu 8 (2009), 261–333.
7. J. F. Dat, Simple subquotients of big parabolically induced representations of p-adic groups, preprint 2017.
8. N. Enomoto, M. Kashiwara, *Symmetric Crystals and LLTA type conjectures for the affine Hecke algebras of type B*. Combinatorial representation theory and related topics, RIMS Kôkyûroku Bessatsu, B8, Res. Inst. Math. Sci. (RIMS), Kyoto, (2008), 1–20.
9. M. Khovanov, A. D. Lauda, *A diagrammatic approach to categorification of quantum groups I*, Represent. Theory **13** (2009), 309–347.
10. V. Heiermann, *Paramètres de Langlands et algèbres d'entrelacement*, Internat. Math. Res. Notices **9** (2010), 1607–1623.
11. V. Heiermann, *Opérateurs d'entrelacement et algèbres de Hecke avec paramètres d'un groupe réductif p-adique: le cas des groupes classiques*, Selecta Math. **17** (2011), 713–756.
12. G. Lusztig, *Classification of unipotent representations of simples p-adic groups*, Internat. Math. Res. Notices **11** (1995), 517–589.
13. A. Mínguez and V. Sécherre, *Types modulo* ℓ *pour les formes intérieures de* GL_n *sur un corps local non archimédien*, Proc. London Math. Soc. (3) **109** (2014), 823–891.
14. A. Mínguez and V. Sécherre, *Représentations lisses modulo* ℓ *de* $GL_m(D)$, Duke Math. J. **163** (2014), 795–887.
15. A. Moy and G. Prasad, *Jacquet functors an unrefined minimal K-types*, Comment. Math. Helvetici **71** (1996), 98–121.
16. R. Rouquier, *2-Kac-Moody algebras*, ArXiv:0812.5023.
17. M. Varagnolo and E. Vasserot, *Canonical Bases and KLR-Algebras*, J. Reine Angew. Math. **659** (2011), 67–100.
18. M. Varagnolo and E. Vasserot, *Canonical Bases and affine Hecke algebras of type B*, Invent. Math. **185** (2011), 593–693.

19. M. F. Vigneras, Représentations ℓ-modulaires d'un groupe réductif p-adique avec $\ell \neq p$, Progress in Mathematics **137**, Birkhauser, (1996).
20. M. F. Vigneras, *Schur algebras of reductive p-adic groups I*, Duke Math. J. **116** (2003), 35–75.
21. J. L. Waldspurger, *Représentations de réduction unipotente pour SO(2N+1): quelques conséquences d'un article de Lusztig*, Contributions to automorphic forms, geometry, and number theory, Johns Hopkins Univ. Press, Baltimore, MD, (2004), 803–910.

Formulae of $\imath$-Divided Powers in $U_q(\mathfrak{sl}_2)$, II

Weiqiang Wang and Collin Berman

Dedicated to Vyjayanthi Chari for her 60th birthday with admiration

Abstract The coideal subalgebra of the quantum $\mathfrak{sl}_2$ is a polynomial algebra in a generator t which depends on a parameter κ. The existence of the $\imath$-canonical basis (also known as the $\imath$-divided powers) for the coideal subalgebra of the quantum $\mathfrak{sl}_2$ were established by Bao and Wang. We establish closed formulae for the $\imath$-divided powers as polynomials in t and also in terms of Chevalley generators of the quantum $\mathfrak{sl}_2$ when the parameter κ is an arbitrary q-integer. The formulae were known earlier when $\kappa = 0, 1$.

1 Introduction

A highlight in the theory of Drinfeld-Jimbo quantum groups is the construction of canonical bases by Lusztig and Kashiwara; cf. [8]. Let $(\mathbf{U}, \mathbf{U}^\imath)$ be a quantum symmetric pair of finite type [7] (also cf. [1]). A general theory of ($\imath$-)canonical bases arising from quantum symmetric pairs of finite type was developed recently by Bao and the first author in [3], where $\imath$-canonical bases on based $\mathbf{U}$-modules and on the modified form of $\mathbf{U}^\imath$ were constructed.

One notable feature of the definition of quantum symmetric pairs is its dependence on parameters; see [1, 3, 7] where various conditions on parameters are imposed at different levels of generalities for various constructions. In [4] it is shown that the unequal parameters of type B Hecke algebras correspond under the $\imath$-Schur duality to certain specializations of parameters in the type AIII/AIV quantum symmetric pairs.

W. Wang (✉) · C. Berman
Department of Mathematics, University of Virginia, Charlottesville, VA, USA
e-mail: ww9c@virginia.edu; cmb5nh@virginia.edu

J. Greenstein et al. (eds.), *Interactions of Quantum Affine Algebras with Cluster Algebras, Current Algebras and Categorification*, Progress in Mathematics 337,
https://doi.org/10.1007/978-3-030-63849-8_7

For the remainder of the paper we let $\mathbf{U}$ be the quantum group of $\mathfrak{sl}_2$ over $\mathbb{Q}(q)$ with generators $E, F, K^{\pm 1}$. Let $\mathbf{U}^\imath$ be the $\mathbb{Q}(q)$-subalgebra of $\mathbf{U}$ (with a parameter κ), which is a polynomial algebra in t, where

$$t = F + q^{-1}EK^{-1} + \kappa K^{-1}.$$

Then $\mathbf{U}^\imath$ is a coideal subalgebra, and $(\mathbf{U}, \mathbf{U}^\imath)$ is an example of quantum symmetric pairs; cf. [6]. For the consideration of $\imath$-canonical bases (which will be referred to as $\imath$-divided powers from now on) the parameter κ is taken to be a bar invariant element in $\mathcal{A} = \mathbb{Z}[q, q^{-1}]$; cf. [3].

In contrast to the usual quantum $\mathfrak{sl}_2$ case where the divided powers are simple to define, finding explicit formulae for the $\imath$-divided powers is a highly nontrivial problem. Closed formulae for the $\imath$-divided powers in $\mathbf{U}^\imath$ were known earlier only in two distinguished cases when $\kappa = 0$ or 1 (cf. [5]); this verified a conjecture in [2] when $\kappa = 1$.

The goal of this paper is to establish closed formulae for the $\imath$-divided powers in $\mathbf{U}^\imath$ as polynomials in t and also viewed as elements in $\mathbf{U}$ (via the embedding of $\mathbf{U}^\imath$ to $\mathbf{U}$), when κ is an arbitrary q-integer which is clearly bar invariant. For arbitrary $\overline{\kappa} = \kappa \in \mathcal{A}$, we are able to present a closed formula only for the *second* $\imath$-divided power (note the first $\imath$-divided power is simply t itself).

In contrast to the quantum group setting, there are *two* $\mathcal{A}$-forms, ${}_\mathcal{A}\mathbf{U}^\imath_{\text{ev}}$ and ${}_\mathcal{A}\mathbf{U}^\imath_{\text{odd}}$, for $\mathbf{U}^\imath$ corresponding to the parities {ev, odd} of highest weights of finite-dimensional simple $\mathbf{U}$-modules [2, 3]. As a very special case of a main theorem in [3], ${}_\mathcal{A}\mathbf{U}^\imath_{\text{ev}}$ (and respectively, ${}_\mathcal{A}\mathbf{U}^\imath_{\text{odd}}$) admits $\imath$-canonical bases (= $\imath$-divided powers) for an arbitrary parameter $\overline{\kappa} = \kappa \in \mathcal{A}$, which are invariant with respect to a bar map (which fixes t and hence is not a restriction of Lusztig's bar map on $\mathbf{U}$ to $\mathbf{U}^\imath$) and satisfy an asymptotic compatibility with the $\imath$-canonical bases on finite-dimensional simple $\mathbf{U}$-modules.

Computations by hand and by Mathematica have led us to make an ansatz for the formulae for the $\imath$-divided powers as polynomials in t when κ is an arbitrary q-integer. In further discussions we need to separate the cases when κ is an even or odd q-integer, and let us restrict ourselves to the case for the q-integer κ being even in the remainder of the Introduction. Our ansatz is that the $\imath$-divided powers $\mathfrak{t}^{(n)}_{\text{ev}}$ in ${}_\mathcal{A}\mathbf{U}^\imath_{\text{ev}}$ and $\mathfrak{t}^{(n)}_{\text{odd}}$ in ${}_\mathcal{A}\mathbf{U}^\imath_{\text{odd}}$ are given as follows: for $a \in \mathbb{N}$,

$$\mathfrak{t}^{(n)}_{\text{ev}} = \begin{cases} \frac{t}{[2a]!}(t-[-2a+2])(t-[-2a+4])\cdots(t-[2a-4])(t-[2a-2]), \\ \hfill \text{if } n = 2a, \\ \\ \frac{1}{[2a+1]!}(t-[-2a])(t-[-2a+2])\cdots(t-[2a-2])(t-[2a]), \\ \hfill \text{if } n = 2a+1. \end{cases} \tag{1.1}$$

$$
t_{\mathrm{odd}}^{(n)} = \begin{cases} \frac{1}{[2a]!}(t-[-2a+1])(t-[-2a+3])\cdots(t-[2a-3])(t-[2a-1]), & \text{if } n=2a, \\ \frac{t}{[2a+1]!}(t-[-2a+1])(t-[-2a+3])\cdots(t-[2a-3])(t-[2a-1]), & \text{if } n=2a+1. \end{cases} \tag{1.2}
$$

That is, the $\imath$-divided power formulas as polynomials in t for κ being even q-integers are the same as for $\kappa = 0$ [2, 5].

To verify the above formulae (1.1)–(1.2) are indeed the $\imath$-canonical bases as defined and established in [3], we need to verify 2 properties: (1) these polynomials in t lie in the corresponding $\mathcal{A}$-forms of $\mathbf{U}^\imath$; (2) they satisfy the asymptotic compatibility with $\imath$-canonical bases on finite-dimensional simple $\mathbf{U}$-modules.

To that end, we find the expansions for the polynomials in t defined in (1.1)–(1.2) in terms of $E, F, K^{\pm 1}$ of $\mathbf{U}$ (via the embedding $\mathbf{U}^\imath \to \mathbf{U}$); they are given by explicit triple sum formulae. Once these explicit formulae are found, they are verified by lengthy inductions. In the formulation of the expansion formulae, a sequence of degree n polynomials $p^{(n)}(x) = p_n(x)/[n]!$ which are defined recursively arise naturally; see Sect. 2.4. The presence of $p^{(n)}(x)$ makes the formulation of the expansion formula and its proof much more difficult than the distinguished cases when $\kappa = 0$ or 1 treated in [5].

The polynomials $p^{(n)}(x)$ satisfy a crucial integrality property in the sense that $p^{(n)}(\kappa) \in \mathcal{A}$ which leads to the integral property (1) above. To prove such an integrality we express these polynomials in terms of another sequence of polynomials (which are a reincarnation of the $\imath$-divided powers viewed as polynomials) with some q^2-binomial coefficients. Note that $p_0(0) = 1$ and $p_n(0) = 0$ for $n \geqslant 1$, our triple sum expansion formula reduces at $\kappa = 0$ to a double sum formula in [5].

When applying the expansion formulas for the $\imath$-divided powers to the highest weight vectors of the finite-dimensional simple $\mathbf{U}$-modules, we establish the asymptotic compatibility property (2) above explicitly. Note this form of compatibility cannot be made as strong as the compatibility between $\imath$-divided powers on $\mathbf{U}^\imath$ and simple $\mathbf{U}$-modules when $\kappa = 0$ or 1 (as expected in [3]).

With Huanchen Bao's help, we compute the closed formulae for the second $\imath$-divided power for an arbitrary parameter $\overline{\kappa} = \kappa \in \mathcal{A}$; see Appendix 6. It is an interesting open problem to find closed formulae for higher $\imath$-divided power with such an arbitrary parameter κ.

The paper is organized as follows. There are 4 sections depending on the parities of the weights and of the parameter κ.

In Sect. 2, we recall the basics of the quantum symmetric pair $(\mathbf{U}, \mathbf{U}^\imath)$. Throughout the section we take κ to be an arbitrary even q-integer. We establish a key integral property of the polynomials $p^{(n)}(x)$, and use it to formulate and establish the expansion formula for the $\imath$-divided powers $t_{\mathrm{ev}}^{(n)}$ in ${}_\mathcal{A}\mathbf{U}^\imath_{\mathrm{ev}}$. We prove that $\{t_{\mathrm{ev}}^{(n)} | n \geqslant$

$0\}$ form an $\imath$-canonical basis for ${}_\mathcal{A}\mathbf{U}^\imath_{\rm ev}$ by showing $\mathrm{t}^{(n)}_{\rm ev} v^+_{2\lambda}$ is an $\imath$-canonical basis element on the finite-dimensional simple $\mathbf{U}$-modules $L(2\lambda)$, for integers $\lambda \gg n$.

In Sect. 3, letting κ be an arbitrary even q-integer, we establish the expansion formula for the $\imath$-divided powers $\mathrm{t}^{(n)}_{\rm odd}$ in ${}_\mathcal{A}\mathbf{U}^\imath_{\rm odd}$. We prove that $\{\mathrm{t}^{(n)}_{\rm odd} | n \geqslant 0\}$ form an $\imath$-canonical basis for ${}_\mathcal{A}\mathbf{U}^\imath_{\rm odd}$ by showing $\mathrm{t}^{(n)}_{\rm odd} v^+_{2\lambda+1}$ is an $\imath$-canonical basis element on the finite-dimensional simple $\mathbf{U}$-modules $L(2\lambda+1)$, for integers $\lambda \gg n$.

In Sect. 4, we take κ to be an arbitrary odd q-integer. We establish a key integral property of another sequence of polynomials $\mathfrak{p}^{(n)}(x)$, and use it to establish the expansion formula for the $\imath$-divided powers $t^{(n)}_{\rm ev}$ in ${}_\mathcal{A}\mathbf{U}^\imath_{\rm ev}$. We show that $\{t^{(n)}_{\rm ev} | n \geqslant 0\}$ form an $\imath$-canonical basis for ${}_\mathcal{A}\mathbf{U}^\imath_{\rm ev}$ and $t^{(n)}_{\rm ev} v^+_{2\lambda}$ is an $\imath$-canonical basis element on the finite-dimensional simple $\mathbf{U}$-modules $L(2\lambda)$, for integers $\lambda \gg n$.

In Sect. 5, we take κ to be an arbitrary odd q-integer. We establish the expansion formula for the $\imath$-divided powers $t^{(n)}_{\rm odd}$ in ${}_\mathcal{A}\mathbf{U}^\imath_{\rm odd}$. We show that $\{t^{(n)}_{\rm odd} | n \geqslant 0\}$ form an $\imath$-canonical basis for ${}_\mathcal{A}\mathbf{U}^\imath_{\rm odd}$ and $t^{(n)}_{\rm odd} v^+_{2\lambda+1}$ is an $\imath$-canonical basis element on the finite-dimensional simple $\mathbf{U}$-modules $L(2\lambda+1)$, for integers $\lambda \gg n$.

In Appendix 6, for arbitrary $\overline{\kappa} = \kappa \in \mathcal{A}$, we present closed formulae for the second $\imath$-divided powers in ${}_\mathcal{A}\mathbf{U}^\imath_{\rm ev}$ and ${}_\mathcal{A}\mathbf{U}^\imath_{\rm odd}$.

2 The $\imath$-Divided Powers $\mathrm{t}^{(n)}_{\rm ev}$ for Even Weights and Even κ

2.1 The Quantum $\mathfrak{sl}_2$

Recall the quantum group $\mathbf{U} = \mathbf{U}_q(\mathfrak{sl}_2)$ is the $\mathbb{Q}(q)$-algebra generated by F, E, K, K^{-1}, subject to the relations:

$$KK^{-1} = K^{-1}K = 1,\ EF - FE = \frac{K - K^{-1}}{q - q^{-1}},\ KE = q^2 EK,\ KF = q^{-2} FK.$$

There is an anti-involution σ of the $\mathbb{Q}$-algebra $\mathbf{U}$:

$$\sigma : \mathbf{U} \longrightarrow \mathbf{U}, \qquad E \mapsto E, \quad F \mapsto F, \quad K \mapsto K, \quad q \mapsto q^{-1}. \tag{2.1}$$

Let $\mathcal{A} = \mathbb{Z}[q, q^{-1}]$ and $\kappa \in \mathcal{A}$. Set

$$\check{E} := q^{-1} E K^{-1}, \qquad h := \frac{K^{-2} - 1}{q^2 - 1}. \tag{2.2}$$

Define, for $a \in \mathbb{Z}, n \geqslant 0$,

$$\begin{bmatrix} h; a \\ n \end{bmatrix} = \prod_{i=1}^{n} \frac{q^{4a+4i-4}K^{-2} - 1}{q^{4i} - 1}, \qquad [h; a] = \begin{bmatrix} h; a \\ 1 \end{bmatrix}. \tag{2.3}$$

Then we have, for $a \in \mathbb{Z}$, $n \in \mathbb{N}$,

$$\begin{gathered} F\check{E} - q^{-2}\check{E}F = h, \\ \begin{bmatrix} h; a \\ n \end{bmatrix} F = F \begin{bmatrix} h; a+1 \\ n \end{bmatrix}, \qquad \begin{bmatrix} h; a \\ n \end{bmatrix} \check{E} = \check{E} \begin{bmatrix} h; a-1 \\ n \end{bmatrix}. \end{gathered} \tag{2.4}$$

It follows by definition that

$$\begin{bmatrix} h; a \\ n \end{bmatrix} = \begin{bmatrix} h; 1+a \\ n \end{bmatrix} - q^{4a}K^{-2} \begin{bmatrix} h; 1+a \\ n-1 \end{bmatrix}. \tag{2.5}$$

The following formula holds for $n \geqslant 0$:

$$F\check{E}^{(n)} = q^{-2n}\check{E}^{(n)}F + \check{E}^{(n-1)} \frac{q^{3-3n}K^{-2} - q^{1-n}}{q^2 - 1}. \tag{2.6}$$

Let

$$\check{E}^{(n)} = \check{E}^n/[n]!, \quad F^{(n)} = F^n/[n]!, \quad \text{for } n \geqslant 1.$$

Then $\check{E}^{(n)} = q^{-n^2}E^{(n)}K^{-n}$. It is understood that $\check{E}^{(n)} = 0$ for $n < 0$ and $\check{E}^{(0)} = 1$.

2.2 *The Coideal Subalgebra* $\mathbf{U}^\imath$

Recall $\check{E}$ from (2.2). Let

$$t = F + \check{E} + \kappa K^{-1}. \tag{2.7}$$

Denote by $\mathbf{U}^\imath$ the $\mathbb{Q}(q)$-subalgebra of $\mathbf{U}$ generated by t. Then $\mathbf{U}^\imath$ is a coideal subalgebra of $\mathbf{U}$ and $(\mathbf{U}, \mathbf{U}^\imath)$ forms a quantum symmetric pair [6, 7]; also cf. [3].

Denote by $\dot{\mathbf{U}}$ the modified quantum group of $\mathfrak{sl}_2$ [8], which is the $\mathbb{Q}(q)$-algebra generated by $E\mathbf{1}_\lambda$, $F\mathbf{1}_\lambda$, and the idempotents $\mathbf{1}_\lambda$, for $\lambda \in \mathbb{Z}$. Let ${}_\mathcal{A}\dot{\mathbf{U}}$ (respectively, ${}_\mathcal{A}\dot{\mathbf{U}}_{\text{ev}}$, ${}_\mathcal{A}\dot{\mathbf{U}}_{\text{odd}}$) be the $\mathcal{A}$-subalgebra of $\dot{\mathbf{U}}$ generated by $E^{(n)}\mathbf{1}_\lambda$, $F^{(n)}\mathbf{1}_\lambda$, $\mathbf{1}_\lambda$, for all $n \geqslant 0$ and for $\lambda \in \mathbb{Z}$ (respectively, for λ even, for λ odd). Note ${}_\mathcal{A}\dot{\mathbf{U}} = {}_\mathcal{A}\dot{\mathbf{U}}_{\text{ev}} \oplus {}_\mathcal{A}\dot{\mathbf{U}}_{\text{odd}}$. There is a natural left action of $\mathbf{U}$ (and hence $\mathbf{U}^\imath$) on $\dot{\mathbf{U}}$ such that $K\mathbf{1}_\lambda = q^\lambda \mathbf{1}_\lambda$. For $\mu \in \mathbb{N}$, denote by $L(\mu)$ the finite-dimensional simple $\mathbf{U}$-module of highest weight μ, and denote by $L(\mu)_\mathcal{A}$ its Lusztig $\mathcal{A}$-form.

Following [3], we introduce the following $\mathcal{A}$-subalgebras of $\mathbf{U}^\imath$, for $\mathrm{p} \in \{\mathrm{ev}, \mathrm{odd}\}$:

$$\begin{aligned} {}_{\mathcal{A}}\mathbf{U}^\imath_{\mathrm{p}} &= \{x \in \mathbf{U}^\imath \mid xu \in {}_{\mathcal{A}}\dot{\mathbf{U}}_{\mathrm{p}}, \forall u \in {}_{\mathcal{A}}\dot{\mathbf{U}}_{\mathrm{p}}\} \\ &= \big\{x \in \mathbf{U}^\imath \mid xv \in L(\mu)_{\mathcal{A}}, \forall v \in L(\mu)_{\mathcal{A}}, \forall \mathrm{p} \equiv \mu \pmod 2\big\}. \end{aligned}$$

By [3], ${}_{\mathcal{A}}\mathbf{U}^\imath_{\mathrm{p}}$ is a free $\mathcal{A}$-submodule of $\mathbf{U}^\imath$ such that $\mathbf{U}^\imath = \mathbb{Q}(q) \otimes_{\mathcal{A}} {}_{\mathcal{A}}\mathbf{U}^\imath_{\mathrm{p}}$, for $\mathrm{p} \in \{\mathrm{ev}, \mathrm{odd}\}$.

2.3 *Definition of* $\mathrm{t}_{\mathrm{ev}}^{(n)}$ *for Even* κ

In this Sect. 2 we shall always take κ to be an even q-integer, i.e.,

$$\kappa = [2\ell], \qquad \text{for } \ell \in \mathbb{Z}. \tag{2.8}$$

We shall take the following as a definition of the $\imath$-divided powers $\mathrm{t}_{\mathrm{ev}}^{(n)}$.

Definition 2.3.1 Set $\mathrm{t}_{\mathrm{ev}}^{(1)} = t = F + \check{E} + \kappa K^{-1}$. The divided powers $\mathrm{t}_{\mathrm{ev}}^{(n)}$, for $n \geqslant 1$, are defined by the recursive relations:

$$\begin{aligned} t \cdot \mathrm{t}_{\mathrm{ev}}^{(2a-1)} &= [2a]\mathrm{t}_{\mathrm{ev}}^{(2a)}, \\ t \cdot \mathrm{t}_{\mathrm{ev}}^{(2a)} &= [2a+1]\mathrm{t}_{\mathrm{ev}}^{(2a+1)} + [2a]\mathrm{t}_{\mathrm{ev}}^{(2a-1)}, \qquad \text{for } a \geqslant 1. \end{aligned} \tag{2.9}$$

Equivalently, $\mathrm{t}_{\mathrm{ev}}^{(n)}$ is defined by the closed formula (1.1).

The bar involution $\psi_\imath$ on $\mathbf{U}^\imath$, which fixed t and sends $q \mapsto q^{-1}$, clearly fixes the above $\imath$-divided powers.

The following is a variant of [5, Lemma 2.2] (where $\kappa = 0$) with the same proof.

Lemma 2.1 *The anti-involution σ on $\mathbf{U}$ sends $F \mapsto F, \check{E} \mapsto \check{E}, K^{-1} \mapsto K^{-1}, q \mapsto q^{-1}$; in addition, σ sends*

$$h \mapsto -q^2 h, \quad \mathrm{t}_{\mathrm{ev}}^{(n)} \mapsto \mathrm{t}_{\mathrm{ev}}^{(n)},$$

$$\begin{bmatrix} h; a \\ n \end{bmatrix} \mapsto (-1)^n q^{2n(n+1)} \begin{bmatrix} h; 1-a-n \\ n \end{bmatrix}, \ \forall a \in \mathbb{Z}, n \in \mathbb{N}.$$

2.4 Polynomials $p_n(x)$ and $p^{(n)}(x)$

Definition 2.4.1 For $n \in \mathbb{N}$, the monic polynomial $p_n(x)$ in x of degree n is defined as

$$p_{n+1} = xp_n + q^{1-2n}[n][n-1]p_{n-1}, \qquad p_0 = 1. \tag{2.10}$$

Also set $p_n = 0$ for all $n < 0$. Note that p_n is an odd polynomial for n odd while it is an even polynomial for n even. These polynomials p_n will appear in the expansion formula for the ı-divided powers in $\mathbf{U}$.

Example 2.4.1 Here are $p_n(x)$, for the first few $n \geqslant 0$:

$$\begin{aligned} &p_0 = 1, \qquad p_1 = x, \qquad p_2 = x^2, \qquad p_3 = x^3 + (q^{-4} + q^{-2})x, \\ &p_4 = x^4 + (q^{-4} + q^{-2})(q^{-4} + q^{-2} + 2)x^2, \\ &p_5 = x^5 + (q^{-4} + q^{-2})(q^{-8} + q^{-6} + 3q^{-4} + 2q^{-2} + 3)x^3 \\ &\qquad\qquad + (q^{-4} + q^{-2})^2(q^{-8} + q^{-6} + 2q^{-4} + q^{-2} + 1)x. \end{aligned}$$

Introduce the monic polynomial of degree n:

$$g_n(x) = \begin{cases} \prod_{i=0}^{m-1}(x^2 - [2i]^2), & \text{if } n = 2m \text{ is even;} \\ x\prod_{i=1}^{m}(x^2 - [2i]^2), & \text{if } n = 2m+1 \text{ is odd.} \end{cases}$$

Define

$$p^{(n)}(x) = p_n(x)/[n]!, \qquad g^{(n)}(x) = g_n(x)/[n]!. \tag{2.11}$$

The recursion for p_n can be rewritten as

$$[n+1]p^{(n+1)} = xp^{(n)} + q^{1-2n}[n-1]p^{(n-1)}. \tag{2.12}$$

Our next goal is to show that $p^{(n)}([2\ell]) \in \mathcal{A}$ for all $\ell \in \mathbb{Z}$; cf. Proposition 2.3. This is carried out by relating $p^{(n)}$ to $g^{(n)}$ for various n, and showing that $g^{(n)}([2\ell]) \in \mathcal{A}$ for all $\ell \in \mathbb{Z}$.

Lemma 2.2 *For $\kappa = [2\ell]$ with $\ell \in \mathbb{Z}$ (see (2.8)), we have $g^{(n)}(\kappa) \in \mathcal{A}$, for all $n \geqslant 0$.* □

Proof We separate the cases for $n = 2m+1$ and $n = 2m$. Noting that

$$\kappa - [2i] = [2\ell] - [2i] = [\ell - i](q^{\ell+i} + q^{-\ell-i}), \tag{2.13}$$

we have

$$g^{(2m+1)}(\kappa) = \frac{1}{[2m+1]!} \prod_{i=-m}^{m} (\kappa - [2i]) = \begin{bmatrix} \ell + m \\ 2m+1 \end{bmatrix} \prod_{i=-m}^{m} (q^{\ell+i} + q^{-\ell-i}) \in \mathcal{A}.$$

Similarly, we have

$$\begin{aligned} g^{(2m)}(\kappa) &= \frac{1}{[2m]!} \kappa \prod_{i=1-m}^{m-1} (\kappa - [2i]) \\ &= \frac{1}{[2m]!} \prod_{i=1-m}^{m} (\kappa - [2i]) + \frac{1}{[2m-1]!} \prod_{i=1-m}^{m-1} (\kappa - [2i]) \\ &= \begin{bmatrix} \ell + m - 1 \\ 2m \end{bmatrix} \prod_{i=1-m}^{m} (q^{\ell+i} + q^{-\ell-i}) + g^{(2m-1)}(\kappa) \in \mathcal{A}. \end{aligned}$$

The lemma is proved.

Proposition 2.3 *For $m \in \mathbb{Z}_{\geqslant 1}$, we have*

$$p^{(2m)} = \sum_{a=0}^{m-1} q^{(1-2m)a} \begin{bmatrix} m-1 \\ a \end{bmatrix}_{q^2} g^{(2m-2a)},$$

$$p^{(2m-1)} = \sum_{a=0}^{m-1} q^{(3-2m)a} \begin{bmatrix} m-1 \\ a \end{bmatrix}_{q^2} g^{(2m-2a-1)}.$$

In particular, we have $p^{(n)}(\kappa) \in \mathcal{A}$, for all $\kappa = [2\ell]$ with $\ell \in \mathbb{Z}$ and all $n \geqslant 0$. □

Proof It follows from these formulae for $p^{(n)}$ and Lemma 2.2 that $p^{(n)}([2\ell]) \in \mathcal{A}$. It remains to prove these formulae. Recall

$$x \cdot p^{(n)} = [n+1]p^{(n+1)} - q^{1-2n}[n-1]p^{(n-1)}, \quad \text{for } n \geqslant 1. \tag{2.14}$$

Also recall

$$\begin{aligned} x \cdot g^{(2a-1)} &= [2a]g^{(2a)}, \\ x \cdot g^{(2a)} &= [2a+1]g^{(2a+1)} + [2a]g^{(2a-1)}, \quad \text{for } a \geqslant 1. \end{aligned} \tag{2.15}$$

We prove the formulae by induction on m, with the base cases for $p^{(1)}$ and $p^{(2)}$ being clear. We separate in two cases.

1. Let us prove the formula for $p^{(2m+1)}$, assuming the formulae for $p^{(2m-1)}$ and $p^{(2m)}$. By (2.14)–(2.15) we have

$$[2m+1]p^{(2m+1)} = xp^{(2m)} + q^{1-4m}[2m-1]p^{(2m-1)}$$

$$= \sum_{a=0}^{m-1} q^{(1-2m)a} \begin{bmatrix} m-1 \\ a \end{bmatrix}_{q^2} x \cdot g^{(2m-2a)}$$

$$+ \sum_{a=0}^{m-1} q^{(3-2m)a+1-4m} [2m-1] \begin{bmatrix} m-1 \\ a \end{bmatrix}_{q^2} g^{(2m-2a-1)}$$

$$= \sum_{a=0}^{m-1} q^{(1-2m)a} \begin{bmatrix} m-1 \\ a \end{bmatrix}_{q^2} \Big([2m-2a+1] g^{(2m-2a+1)}$$

$$+ [2m-2a] g^{(2m-2a-1)} \Big)$$

$$+ \sum_{a=0}^{m-1} q^{(3-2m)a+1-4m} [2m-1] \begin{bmatrix} m-1 \\ a \end{bmatrix}_{q^2} g^{(2m-2a-1)}.$$

By combining the like terms for $g^{(2m-2a-1)}$ above and using the identity

$$q^{(1-2m)a}[2m-2a] + q^{(3-2m)a+1-4m}[2m-1] = q^{(1-2m)(a+1)}[4m-2a-1],$$

we obtain

$$[2m+1]p^{(2m+1)} = \sum_{a=0}^{m-1} q^{(1-2m)a}[2m-2a+1] \begin{bmatrix} m-1 \\ a \end{bmatrix}_{q^2} g^{(2m-2a+1)}$$

$$+ \sum_{a=0}^{m-1} q^{(1-2m)(a+1)}[4m-2a-1] \begin{bmatrix} m-1 \\ a \end{bmatrix}_{q^2} g^{(2m-2a-1)}$$

$$= [2m+1] \sum_{a=0}^{m} q^{(1-2m)a} \begin{bmatrix} m \\ a \end{bmatrix}_{q^2} g^{(2m-2a+1)},$$

where the last equation results from shifting the second summation index from $a \to a-1$ and using the identity

$$[4m-2a+1] \begin{bmatrix} m-1 \\ a \end{bmatrix}_{q^2} + q^{(1-2m)a}[4m-2a+1] \begin{bmatrix} m-1 \\ a-1 \end{bmatrix}_{q^2} = [2m+1] \begin{bmatrix} m \\ a \end{bmatrix}_{q^2}.$$

2. We shall prove the formula for $p^{(2m+1)}$, assuming the formulae for $p^{(2m-1)}$ and $p^{(2m)}$. By (2.14)–(2.15) we have

$$[2m+2]p^{(2m+2)} = xp^{(2m+1)} + q^{-1-4m}[2m]p^{(2m)}$$

$$= \sum_{a=0}^{m} q^{(1-2m)a} \begin{bmatrix} m \\ a \end{bmatrix}_{q^2} x \cdot g^{(2m-2a+1)}$$

$$+ \sum_{a=0}^{m-1} q^{(1-2m)a-1-4m} [2m] \begin{bmatrix} m-1 \\ a \end{bmatrix}_{q^2} g^{(2m-2a)}$$

$$= \sum_{a=0}^{m} q^{(1-2m)a} [2m-2a+2] \begin{bmatrix} m \\ a \end{bmatrix}_{q^2} g^{(2m-2a+2)}$$

$$+ \sum_{a=0}^{m-1} q^{(1-2m)a-1-4m} [2m] \begin{bmatrix} m-1 \\ a \end{bmatrix}_{q^2} g^{(2m-2a)}$$

$$= [2m+2] \sum_{a=0}^{m} q^{(-1-2m)a} \begin{bmatrix} m \\ a \end{bmatrix}_{q^2} g^{(2m-2a+2)},$$

where the last equation results from shifting the second summation index from $a \to a-1$ and using the following identity

$$[2m-2a+2] \begin{bmatrix} m \\ a \end{bmatrix}_{q^2} + q^{-2-2m} [2m] \begin{bmatrix} m-1 \\ a-1 \end{bmatrix}_{q^2} = q^{-2a} [2m+2] \begin{bmatrix} m \\ a \end{bmatrix}_{q^2}.$$

The proposition is proved.

Remark 2.4 Note that $p_n(x) \in \mathbb{N}[q^{-1}][x]$ for all n. Therefore, for all non-negative even q-integer κ, we have $p_n(\kappa) \in \mathbb{N}[q, q^{-1}]$. □

2.5 *Formulae for* $\mathrm{t}_{\mathbf{ev}}^{(n)}$ *with Even* κ

Recall $p^{(n)}(x)$ from (2.10)–(2.11).

Theorem 2.5 *Assume κ is an even q-integer as in* (2.8). *Then we have, for $m \geqslant 1$,*

$$\mathrm{t}_{\mathrm{ev}}^{(2m)} = \sum_{b=0}^{2m} \sum_{a=0}^{b} \sum_{c \geqslant 0} q^{\binom{2c}{2}-b(2m-b-2c)-a(b-a)} p^{(2m-b-2c)}(\kappa) \tag{2.16}$$

$$\cdot \check{E}^{(a)} \begin{bmatrix} h; 1-m \\ c \end{bmatrix} K^{b-2m+2c} F^{(b-a)},$$

$$\mathrm{t}_{\mathrm{ev}}^{(2m-1)} = \sum_{b=0}^{2m-1} \sum_{a=0}^{b} \sum_{c \geqslant 0} q^{\binom{2c}{2}+2c-b(2m-b-2c-1)-a(b-a)} p^{(2m-b-2c-1)}(\kappa) \tag{2.17}$$

$$\cdot \check{E}^{(a)} \begin{bmatrix} h; 1-m \\ c \end{bmatrix} K^{b-2m+2c+1} F^{(b-a)}.$$

The proof of Theorem 2.5 will be given in Sect. 2.7 below. Let us list some equivalent formulae here. By applying the anti-involution σ we convert the formulae in Theorem 2.5 into the following.

Theorem 2.6 *Assume κ is an even q-integer as in* (2.8). *Then we have, for $m \geqslant 1$,*

$$\mathfrak{t}_{\mathrm{ev}}^{(2m)} = \sum_{b=0}^{2m} \sum_{a=0}^{b} \sum_{c \geqslant 0} (-1)^c q^{3c+b(2m-b-2c)+a(b-a)} p^{(2m-b-2c)}(\kappa) \cdot F^{(b-a)} K^{b-2m+2c} \begin{bmatrix} h; m-c \\ c \end{bmatrix} \check{E}^{(a)}, \tag{2.18}$$

$$\mathfrak{t}_{\mathrm{ev}}^{(2m-1)} = \sum_{b=0}^{2m-1} \sum_{a=0}^{b} \sum_{c \geqslant 0} (-1)^c q^{c+b(2m-b-2c-1)+a(b-a)} p^{(2m-b-2c-1)}(\kappa) \cdot F^{(b-a)} K^{b-2m+2c+1} \begin{bmatrix} h; m-c \\ c \end{bmatrix} \check{E}^{(a)}. \tag{2.19}$$

Proof By Lemma 2.1, the anti-involution σ fixes κ, K, $\check{E}^{(a)}$, $F^{(a)}$, and $\mathfrak{t}_{\mathrm{ev}}^{(n)}$ while mapping $q \mapsto q^{-1}$, $\begin{bmatrix} h; 1-m \\ c \end{bmatrix} \mapsto (-1)^c q^{2c(c+1)} \begin{bmatrix} h; m-c \\ c \end{bmatrix}$. The formulae (2.18)–(2.19) now follow by applying σ to the formulae (2.16)–(2.17) in Theorem 2.5.

For $c \in \mathbb{N}$, $m \in \mathbb{Z}$, we define

$$\begin{Bmatrix} m \\ c \end{Bmatrix} = \prod_{i=1}^{c} \frac{q^{4(m+i-1)} - 1}{q^{-4i} - 1} \in \mathcal{A}, \qquad \begin{Bmatrix} m \\ 0 \end{Bmatrix} = 1. \tag{2.20}$$

Note that $\begin{Bmatrix} m \\ c \end{Bmatrix}$ are q^2-binomial coefficients up to some q-powers. Let $\lambda, m \in \mathbb{Z}$ with $m \geqslant 1$. We note that

$$\begin{aligned} \begin{bmatrix} h; m-c \\ c \end{bmatrix} \mathbf{1}_{2\lambda} &= \prod_{i=1}^{c} \frac{q^{4(m-\lambda-c+i-1)} - 1}{q^{4i} - 1} \mathbf{1}_{2\lambda} \\ &= (-1)^c q^{-2c(c+1)} \begin{Bmatrix} m-\lambda-c \\ c \end{Bmatrix} \mathbf{1}_{2\lambda}. \end{aligned} \tag{2.21}$$

The following corollary is immediate from (2.21), Proposition 2.3, and Theorem 2.5.

Corollary 2.7 *We have* $\mathrm{t}_{\mathrm{ev}}^{(n)} \in {}_{\mathcal{A}}\mathbf{U}^{\imath}_{\mathrm{ev}}$, *for all* n. □

Remark 2.8 Note $p_n(0) = 0$ for $n > 0$, and $p_0(0) = 1$. The formulae in Theorems 2.5 and 2.6 in the special case for $\kappa = 0$ recover the formulae in [5, Theorem 2.5, Proposition 2.7]. □

For $n \in \mathbb{N}$, we denote

$$b^{(n)} = \sum_{a=0}^{n} q^{-a(n-a)} \check{E}^{(a)} F^{(n-a)}. \tag{2.22}$$

Example 2.5.1 The formulae for $\mathrm{t}_{\mathrm{ev}}^{(n)}$ in Theorem 2.5, for $1 \leqslant n \leqslant 3$, read as follows:

$$\mathrm{t}_{\mathrm{ev}}^{(1)} = F + \check{E} + \kappa K^{-1},$$

$$\mathrm{t}_{\mathrm{ev}}^{(2)} = \check{E}^{(2)} + q^{-1}\check{E}F + F^{(2)} + q[h;0] + \kappa(q^{-1}K^{-1}F + q^{-1}\check{E}K^{-1}) + \kappa^2 \frac{K^{-2}}{[2]},$$

$$\begin{aligned}\mathrm{t}_{\mathrm{ev}}^{(3)} &= b^{(3)} + q^3[h;-1]F + q^3\check{E}[h;-1] \\ &\quad + \big(q^{-2}\check{E}^{(2)}K^{-1} + q^{-3}\check{E}K^{-1}F + q^{-2}K^{-1}F^{(2)} + q^3[h;-1]K^{-1}\big)\kappa \\ &\quad + \frac{q^{-2}}{[2]}(\check{E}K^{-2} + K^{-2}F)\kappa^2 + \frac{\kappa^3 + (q^{-4}+q^{-2})\kappa}{[3]!}K^{-3}.\end{aligned}$$

2.6 The ı-Canonical Basis for $\dot{\mathbf{U}}^{\imath}_{\mathrm{ev}}$ for Even κ

Denote by $L(\mu)$ the simple **U**-module of highest weight $\mu \in \mathbb{N}$, with highest weight vector v_μ^+. Then $L(\mu)$ admits a canonical basis $\{F^{(a)}v_\mu^+ \mid 0 \leqslant a \leqslant \mu\}$. Following [2, 3], there exists a new bar involution $\psi_\imath$ on $L(\mu)$, which, in our current rank one setting, can be defined simply by requiring $\mathrm{t}_{\mathrm{ev}}^{(n)} v_\mu^+$ to be $\psi_\imath$-invariant for all n. As a very special case of the general results in [3, Corollary F] (also cf. [2]), we know that the $\imath$-canonical basis $\{b_a^\mu\}_{0 \leqslant a \leqslant \mu}$ of $L(\mu)$ exists and is characterized by the following 2 properties:

(iCB1) b_a^μ is $\psi_\imath$-invariant;
(iCB2) $b_a^\mu \in F^{(a)}v_\mu^+ + \sum_{0 \leqslant r < a} q^{-1}\mathbb{Z}[q^{-1}]F^{(r)}v_\mu^+$.

Theorem 2.9

(a) *Let* $n \in \mathbb{N}$. *For each integer* $\lambda \gg n$, *the element* $\mathrm{t}_{\mathrm{ev}}^{(n)} v_{2\lambda}^+$ *is an* $\imath$*-canonical basis element for* $L(2\lambda)$.
(b) *The set* $\{\mathrm{t}_{\mathrm{ev}}^{(n)} \mid n \in \mathbb{N}\}$ *forms the* $\imath$*-canonical basis for* $\mathbf{U}^\imath$ *(and an* $\mathcal{A}$*-basis in* ${}_{\mathcal{A}}\mathbf{U}^{\imath}_{\mathrm{ev}}$*).* □

Proof Let $\lambda, m \in \mathbb{N}$ with $m \geqslant 1$. We compute by Theorem 2.6(1) and (2.21) that

$$\begin{aligned}\mathrm{t}_{\mathrm{ev}}^{(2m)} v_{2\lambda}^{+} &= \sum_{b=0}^{2m} \sum_{c \geqslant 0} (-1)^c q^{3c+b(2m-b-2c)} p^{(2m-b-2c)}(\kappa) \\ &\qquad \cdot F^{(b)} K^{b-2m+2c} \begin{bmatrix} h; m-c \\ c \end{bmatrix} v_{2\lambda}^{+} \\ &= \sum_{b=0}^{2m} \sum_{c \geqslant 0} q^{(b-2\lambda)(2m-b-2c)-2c^2+c} p^{(2m-b-2c)}(\kappa) \\ &\qquad \cdot \begin{Bmatrix} m-\lambda-c \\ c \end{Bmatrix} F^{(b)} v_{2\lambda}^{+}. \end{aligned} \tag{2.23}$$

Similarly we compute by Theorem 2.6(2) that

$$\begin{aligned}\mathrm{t}_{\mathrm{ev}}^{(2m-1)} v_{2\lambda}^{+} &= \sum_{b=0}^{2m-1} \sum_{c \geqslant 0} (-1)^c q^{c+b(2m-b-2c-1)} p^{(2m-b-2c-1)}(\kappa) \\ &\qquad \cdot F^{(b)} K^{b-2m+2c+1} \begin{bmatrix} h; m-c \\ c \end{bmatrix} v_{2\lambda}^{+} \\ &= \sum_{b=0}^{2m-1} \sum_{c \geqslant 0} q^{(b-2\lambda)(2m-b-2c-1)-2c^2-c} p^{(2m-b-2c-1)}(\kappa) \\ &\qquad \cdot \begin{Bmatrix} m-\lambda-c \\ c \end{Bmatrix} F^{(b)} v_{2\lambda}^{+}. \end{aligned} \tag{2.24}$$

Note that

$$\begin{Bmatrix} m-\lambda-c \\ c \end{Bmatrix} \in \mathbb{N}[q^{-1}], \qquad \text{for } \lambda \geqslant m, c \geqslant 0. \tag{2.25}$$

Let $n \in \mathbb{N}$. Recall $\kappa = [2\ell]$, for $\ell \in \mathbb{Z}$. One computes that $\deg_q p^{(n)}(\kappa) = (2\ell-1)n - \binom{n}{2}$ and hence $q^{(b-2\lambda)(n-b-2c)-2c^2 \pm c} p^{(n-b-2c)}(\kappa) \in q^{-1}\mathbb{N}[q^{-1}]$ when $\lambda \gg n > b+2c$. It follows by (2.23), (2.24), and (2.25) that

$$\mathrm{t}_{\mathrm{ev}}^{(n)} v_{2\lambda}^{+} \in F^{(n)} v_{2\lambda}^{+} + \sum_{b<n} q^{-1}\mathbb{Z}[q^{-1}] F^{(b)} v_{2\lambda}^{+}, \qquad \text{for } \lambda \gg n. \tag{2.26}$$

Therefore, the first statement follows by the characterization properties (iCB1)–(iCB2) of the $\imath$-canonical basis for $L(2\lambda)$.

The second statement follows now from the definition of the $\imath$-canonical basis for $\mathbf{U}^\imath$ using the projective system $\{L(2\lambda)\}_{\lambda\in\mathbb{N}}$ with $\mathbf{U}^\imath$-homomorphisms $L(2\lambda+2)\to L(2\lambda)$; cf. [3, §6]; this set $\{\mathfrak{t}_{\mathrm{ev}}^{(n)} \mid n\in\mathbb{N}\}$ forms an $\mathcal{A}$-basis for ${}_{\mathcal{A}}\mathbf{U}^\imath_{\mathrm{ev}}$. The theorem is proved.

Remark 2.10 It follows by (2.23) that

$$\mathfrak{t}_{\mathrm{ev}}^{(2m)} v_{2\lambda}^+ \in F^{(2m)} v_{2\lambda}^+ + q^{2m-2\lambda-1}\kappa F^{(2m-1)} v_{2\lambda}^+ + \sum_{k\geqslant 2} \mathcal{A}\cdot F^{(2m-k)} v_{2\lambda}^+.$$

Recall $\kappa=[2\ell]$. When $\ell-1\geqslant\lambda-m\geqslant 0$, we have $q^{2m-2\lambda-1}\kappa\notin q^{-1}\mathbb{Z}[q^{-1}]$, and therefore, the nonzero element $\mathfrak{t}_{\mathrm{ev}}^{(2m)} v_{2\lambda}^+$ is not an $\imath$-canonical basis element in $L(2\lambda)$. So we cannot expect to have a stronger form of Theorem 2.9 as [5, Theorem 2.10(1)] in the case when $\kappa=0$ (i.e., $\ell=0$). □

2.7 Proof of Theorem 2.5

We prove the formulae for $\mathfrak{t}_{\mathrm{ev}}^{(n)}$ by induction on n, in two steps (1)–(2) below. The base cases when $n=1,2$ are clear.

(1) We shall prove the formula (2.16) for $\mathfrak{t}_{\mathrm{ev}}^{(2m)}$, assuming the formula (2.17) for $\mathfrak{t}_{\mathrm{ev}}^{(2m-1)}$.

Recall $[2m]\mathfrak{t}_{\mathrm{ev}}^{(2m)} = t\cdot \mathfrak{t}_{\mathrm{ev}}^{(2m-1)}$, and $t=F+\check{E}+\kappa K^{-1}$. Let us compute

$$I=\check{E}\mathfrak{t}_{\mathrm{ev}}^{(2m-1)},\qquad II=F\mathfrak{t}_{\mathrm{ev}}^{(2m-1)},\qquad III=\kappa K^{-1}\mathfrak{t}_{\mathrm{ev}}^{(2m-1)}.$$

First we have

$$\begin{aligned}
I &= \sum_{b=0}^{2m-1}\sum_{a=0}^{b}\sum_{c\geqslant 0} q^{\binom{2c}{2}+2c-b(2m-b-2c-1)-a(b-a)}[a+1]p^{(2m-b-2c-1)}(\kappa)\\
&\quad\cdot \check{E}^{(a+1)}\begin{bmatrix} h;1-m\\ c\end{bmatrix} K^{b-2m+2c+1}F^{(b-a)}\\
&= \sum_{b=0}^{2m}\sum_{a=0}^{b}\sum_{c\geqslant 0} q^{\binom{2c}{2}+2c-(b-1)(2m-b-2c)-(a-1)(b-a)}[a]p^{(2m-b-2c)}(\kappa)\\
&\quad\cdot \check{E}^{(a)}\begin{bmatrix} h;1-m\\ c\end{bmatrix} K^{b-2m+2c}F^{(b-a)},
\end{aligned}$$

where the last equation is obtained by shifting indices $a\to a-1$, $b\to b-1$ (and then adding some zero terms to make the bounds of the summations uniform throughout). Using (2.6) we have

$$
\begin{aligned}
II = & \sum_{b=0}^{2m-1} \sum_{a=0}^{b} \sum_{c\geqslant 0} q^{\binom{2c}{2}+2c-b(2m-b-2c-1)-a(b-a)} p^{(2m-b-2c-1)}(\kappa) \\
& \cdot \left(q^{-2a} \check{E}^{(a)} F + \check{E}^{(a-1)} \frac{q^{3-3a}K^{-2} - q^{1-a}}{q^2-1} \right) \begin{bmatrix} h; 1-m \\ c \end{bmatrix} K^{b-2m+2c+1} F^{(b-a)} \\
= & II^1 + II^2,
\end{aligned}
$$

where II^1 and II^2 are the two natural summands associated with the plus sign. By shifting index $b \to b-1$ and then adding some zero terms, we further have

$$
\begin{aligned}
II^1 = & \sum_{b=0}^{2m} \sum_{a=0}^{b} \sum_{c\geqslant 0} q^{\binom{2c}{2}+2c-(b-1)(2m-b-2c)-a(b-a-1)-2a+2b+4c-4m} [b-a] \\
& \cdot p^{(2m-b-2c)}(\kappa) \check{E}^{(a)} \begin{bmatrix} h; -m \\ c \end{bmatrix} K^{b-2m+2c} F^{(b-a)}.
\end{aligned}
$$

By shifting the indices $a \to a+1$, $b \to b+1$, and $c \to c-1$, we further have

$$
\begin{aligned}
II^2 = & \sum_{b=0}^{2m} \sum_{a=0}^{b} \sum_{c\geqslant 0} q^{\binom{2c-2}{2}+2c-(b+1)(2m-b-2c)-(a+1)(b-a)-2} p^{(2m-b-2c)}(\kappa) \\
& \cdot \check{E}^{(a)} \frac{q^{-3a}K^{-2} - q^{-a}}{q^2-1} \begin{bmatrix} h; 1-m \\ c-1 \end{bmatrix} K^{b-2m+2c} F^{(b-a)}.
\end{aligned}
$$

Using (2.14) we also compute

$$
\begin{aligned}
III = & \sum_{b=0}^{2m} \sum_{a=0}^{b} \sum_{c\geqslant 0} q^{\binom{2c}{2}+2c-b(2m-b-2c-1)-a(b-a)-2a} \kappa \cdot p^{(2m-b-2c-1)}(\kappa) \\
& \cdot \check{E}^{(a)} \begin{bmatrix} h; 1-m \\ c \end{bmatrix} K^{b-2m+2c} F^{(b-a)} \\
= & \sum_{b=0}^{2m} \sum_{a=0}^{b} \sum_{c\geqslant 0} q^{\binom{2c}{2}+2c-b(2m-b-2c-1)-a(b-a)-2a} [2m-b-2c] \\
& \cdot p^{(2m-b-2c)}(\kappa) \check{E}^{(a)} \begin{bmatrix} h; 1-m \\ c \end{bmatrix} K^{b-2m+2c} F^{(b-a)}
\end{aligned}
$$

$$+\sum_{b=0}^{2m}\sum_{a=0}^{b}\sum_{c\geqslant 0} q^{\binom{2c}{2}+2c-b(2m-b-2c-1)-a(b-a)-2a-4m}[2m-b-2c]$$
$$\cdot\, p^{(2m-b-2c)}(\kappa)\check{E}^{(a)}\begin{bmatrix} h;\,1-m \\ c-1 \end{bmatrix} K^{b-2m+2c-2}F^{(b-a)}.$$

(Note that we have shifted the index $c \to c-1$ in the last summand above.)

Collecting the formulae for I, II^1, II^2,

$$t\cdot \mathrm{t}_{\mathrm{ev}}^{(2m-1)} = \sum_{0\leqslant a\leqslant b\leqslant 2m}\sum_{c\geqslant 0} p^{(2m-b-2c)}(\kappa)\check{E}^{(a)}H_{a,b,c}K^{b-2m+2c}F^{(b-a)},$$

where

$$H_{a,b,c} := q^{\binom{2c}{2}+2c-(b-1)(2m-b-2c)-(a-1)(b-a)}[a]\begin{bmatrix} h;\,1-m \\ c \end{bmatrix}$$
$$+\,q^{\binom{2c}{2}+2c-(b-1)(2m-b-2c)-a(b-a-1)-2a+2b+4c-4m}[b-a]\begin{bmatrix} h;\,-m \\ c \end{bmatrix}$$
$$+\,q^{\binom{2c}{2}-2c+1-(b+1)(2m-b-2c)-(a+1)(b-a)}\frac{q^{-3a}K^{-2}-q^{-a}}{q^2-1}\begin{bmatrix} h;\,1-m \\ c-1 \end{bmatrix}$$
$$+\,q^{\binom{2c}{2}+2c-b(2m-b-2c-1)-a(b-a)-2a}[2m-b-2c]\begin{bmatrix} h;\,1-m \\ c \end{bmatrix}$$
$$-\,q^{\binom{2c}{2}+2c-b(2m-b-2c-1)-a(b-a)-2a-4m}[2m-b-2c]\begin{bmatrix} h;\,1-m \\ c-1 \end{bmatrix}K^{-2}.$$

Recall $[2m]\mathrm{t}_{\mathrm{ev}}^{(2m)} = t\cdot \mathrm{t}_{\mathrm{ev}}^{(2m-1)}$. To prove the formula (2.16), by the PBW basis theorem it suffices to prove the following identity, for all a, b, c:

$$H_{a,b,c} = q^{\binom{2c}{2}-b(2m-b-2c)-a(b-a)}[2m]\begin{bmatrix} h;\,1-m \\ c \end{bmatrix}. \tag{2.27}$$

Thanks to $q^{2m-a}[a]-[2m] = q^{-a}[a-2m]$, we can combine the RHS with the first summand of LHS (2.27). Hence, after canceling out the q-powers $q^{\binom{2c}{2}-b(2m-b-2c)-a(b-a)-a}$ on both sides, we see that (2.27) is equivalent to the following identity, for all a, b, c:

$$A+B+C+D_1+D_2=0, \tag{2.28}$$

where

$$A = [a-2m]\begin{bmatrix} h;1-m \\ c \end{bmatrix},$$

$$B = q^{4c-2m+b}[b-a]\begin{bmatrix} h;-m \\ c \end{bmatrix},$$

$$C = q^{2a-2m+1}\frac{q^{-3a}K^{-2}-q^{-a}}{q^2-1}\begin{bmatrix} h;1-m \\ c-1 \end{bmatrix},$$

$$D_1 = q^{2c+b-a}[2m-b-2c]\begin{bmatrix} h;1-m \\ c \end{bmatrix},$$

$$D_2 = -q^{2c-4m+b-a}[2m-b-2c]\begin{bmatrix} h;1-m \\ c-1 \end{bmatrix}K^{-2}.$$

Let us prove the identity (2.28). Using (2.5), we can write $B = B_1 + B_2$, where

$$B_1 = q^{4c-2m+b}[b-a]\begin{bmatrix} h;1-m \\ c \end{bmatrix}, \quad B_2 = -q^{4c-6m+b}[b-a]\begin{bmatrix} h;1-m \\ c-1 \end{bmatrix}K^{-2}.$$

Noting that

$$\frac{q^{-3a}K^{-2}-q^{-a}}{q^2-1} = q^{4m-4c-3a}\frac{(q^{4c-4m}K^{-2}-1)}{q^2-1} + q^{2m-2c-2a-1}[2m-2c-a],$$

we rewrite $C = C_1 + C_2$, where

$$C_1 = q^{-2c+2m-a}[2c]\begin{bmatrix} h;1-m \\ c \end{bmatrix}, \quad C_2 = q^{-2c}[2m-2c-a]\begin{bmatrix} h;1-m \\ c-1 \end{bmatrix}.$$

A direct computation gives us

$$A + B_1 + C_1 + D_1 = (q-q^{-1})[2c][2m-2c-a]\begin{bmatrix} h;1-m \\ c \end{bmatrix},$$

$$\begin{aligned} C_2 + (B_2 + D_2) &= C_2 - q^{2c-4m}[2m-2c-a]\begin{bmatrix} h;1-m \\ c-1 \end{bmatrix}K^{-2} \\ &= -(q-q^{-1})[2c][2m-2c-a]\begin{bmatrix} h;1-m \\ c \end{bmatrix}. \end{aligned}$$

Summing up these two equations, we have $A+B+C+D_1+D_2 = 0$, whence (2.28), completing Step (1).

(2) Assuming the formulae for $\mathrm{t}_{\mathrm{ev}}^{(n)}$ with $n \leqslant 2m$, we shall now prove the following formula for $\mathrm{t}_{\mathrm{ev}}^{(2m+1)}$ (obtained from (2.17) with m replaced by $m+1$):

$$\mathrm{t}_{\mathrm{ev}}^{(2m+1)} = \sum_{b=0}^{2m+1} \sum_{a=0}^{b} \sum_{c \geqslant 0} q^{\binom{2c}{2}+2c-b(2m-b-2c+1)-a(b-a)} p^{(2m-b-2c+1)}(\kappa) \tag{2.29}$$

$$\cdot \check{E}^{(a)} \begin{bmatrix} h; -m \\ c \end{bmatrix} K^{b-2m+2c-1} F^{(b-a)}.$$

The proof is based on the recursion $t \cdot \mathrm{t}_{\mathrm{ev}}^{(2m)} = [2m+1]\mathrm{t}_{\mathrm{ev}}^{(2m+1)} + [2m]\mathrm{t}_{\mathrm{ev}}^{(2m-1)}$. Recall $t = F + \check{E} + \kappa K^{-1}$ and $\mathrm{t}_{\mathrm{ev}}^{(2m)}$ from (2.16). We shall compute

$$\mathtt{I} = \check{E}\mathrm{t}_{\mathrm{ev}}^{(2m)}, \qquad \mathtt{II} = F\mathrm{t}_{\mathrm{ev}}^{(2m)}, \qquad \mathtt{III} = \kappa K^{-1}\mathrm{t}_{\mathrm{ev}}^{(2m)},$$

respectively. First we have

$$\begin{aligned} \mathtt{I} = & \sum_{b=0}^{2m} \sum_{a=0}^{b} \sum_{c \geqslant 0} q^{\binom{2c}{2}-b(2m-b-2c)-a(b-a)} [a+1] \\ & \cdot p^{(2m-b-2c)}(\kappa) \check{E}^{(a+1)} \begin{bmatrix} h; 1-m \\ c \end{bmatrix} K^{b-2m+2c} F^{(b-a)} \\ = & \sum_{b=0}^{2m+1} \sum_{a=0}^{b} \sum_{c \geqslant 0} q^{\binom{2c}{2}-(b-1)(2m-b-2c+1)-(a-1)(b-a)} [a] p^{(2m-b-2c+1)}(\kappa) \\ & \cdot \check{E}^{(a)} \begin{bmatrix} h; 1-m \\ c \end{bmatrix} K^{b-2m+2c-1} F^{(b-a)}, \end{aligned}$$

where the last equation is obtained by shifting indices $a \to a-1$, $b \to b-1$. We also have

$$\begin{aligned} \mathtt{II} = & \sum_{b=0}^{2m} \sum_{a=0}^{b} \sum_{c \geqslant 0} q^{\binom{2c}{2}-b(2m-b-2c)-a(b-a)} p^{(2m-b-2c)}(\kappa) \\ & \cdot \left(q^{-2a} \check{E}^{(a)} F + \check{E}^{(a-1)} \frac{q^{3-3a} K^{-2} - q^{1-a}}{q^2-1} \right) \begin{bmatrix} h; 1-m \\ c \end{bmatrix} K^{b-2m+2c} F^{(b-a)} \\ = & \mathtt{II}_1 + \mathtt{II}_2, \end{aligned}$$

where $\mathtt{II}_1$ and $\mathtt{II}_2$ are the two natural summands associated with the plus sign. By shifting the index $b \to b-1$, we have

$$\mathtt{II}_1 = \sum_{b=0}^{2m+1} \sum_{a=0}^{b} \sum_{c \geqslant 0} q^{\binom{2c}{2}-(b-1)(2m-b-2c+1)-a(b-a-1)-2a+2(b+2c-2m-1)}$$

$$\cdot [b-a] p^{(2m-b-2c+1)}(\kappa) \check{E}^{(a)} \begin{bmatrix} h; -m \\ c \end{bmatrix} K^{b-2m+2c-1} F^{(b-a)}.$$

By shifting the indices $a \to a+1$, $b \to b+1$, and $c \to c-1$, we further have

$$\mathtt{II}_2 = \sum_{b=0}^{2m+1} \sum_{a=0}^{b} \sum_{c \geqslant 0} q^{\binom{2c-2}{2}-(b+1)(2m-b-2c+1)-(a+1)(b-a)} p^{(2m-b-2c+1)}(\kappa)$$

$$\cdot \check{E}^{(a)} \frac{q^{-3a}K^{-2} - q^{-a}}{q^2-1} \begin{bmatrix} h; 1-m \\ c-1 \end{bmatrix} K^{b-2m+2c-1} F^{(b-a)}.$$

Using (2.14) we also compute

$$\mathtt{III} = \sum_{b=0}^{2m+1} \sum_{a=0}^{b} \sum_{c \geqslant 0} q^{\binom{2c}{2}-b(2m-b-2c)-a(b-a)-2a} [2m-b-2c+1]$$

$$\cdot p^{(2m-b-2c+1)}(\kappa) \check{E}^{(a)} \begin{bmatrix} h; 1-m \\ c \end{bmatrix} K^{b-2m+2c-1} F^{(b-a)}$$

$$+ \sum_{b=0}^{2m+1} \sum_{a=0}^{b} \sum_{c \geqslant 0} q^{\binom{2c}{2}-b(2m-b-2c)-a(b-a)-2a-4m} [2m-b-2c+1]$$

$$\cdot p^{(2m-b-2c+1)}(\kappa) \check{E}^{(a)} \begin{bmatrix} h; 1-m \\ c-1 \end{bmatrix} K^{b-2m+2c-3} F^{(b-a)}.$$

(Note that we have shifted the index $c \to c-1$ in the last summand above.)

Collecting the formulae for $\mathtt{I}$, $\mathtt{II}_1$, $\mathtt{II}_2$, and $\mathtt{III}$, we obtain

$$t \cdot \mathrm{t}_{\mathrm{ev}}^{(2m)} = \sum_{0 \leqslant a \leqslant b \leqslant 2m+1} \sum_{c \geqslant 0} p^{(2m-b-2c+1)}(\kappa) \check{E}^{(a)} L_{a,b,c} K^{b-2m+2c-1} F^{(b-a)},$$

where

$$L_{a,b,c} := q^{\binom{2c}{2}-(b-1)(2m-b-2c+1)-(a-1)(b-a)} [a] \begin{bmatrix} h; 1-m \\ c \end{bmatrix}$$

$$+ q^{\binom{2c}{2}-(b-1)(2m-b-2c+1)-a(b-a-1)-2a+2(b+2c-2m-1)} [b-a] \begin{bmatrix} h; -m \\ c \end{bmatrix}$$

$$+ q^{\binom{2c}{2}-4c+3-(b+1)(2m-b-2c+1)-(a+1)(b-a)} \frac{q^{-3a}K^{-2} - q^{-a}}{q^2-1} \begin{bmatrix} h; 1-m \\ c-1 \end{bmatrix}$$

$$+ q^{\binom{2c}{2}-b(2m-b-2c)-a(b-a)-2a} [2m-b-2c+1] \begin{bmatrix} h; 1-m \\ c \end{bmatrix}$$

$$+q^{\binom{2c}{2}-b(2m-b-2c)-a(b-a)-2a-4m}[2m-b-2c+1]\begin{bmatrix} h; 1-m \\ c-1 \end{bmatrix} K^{-2}.$$

On the other hand, using (2.17) (with an index shift $c \to c-1$) and (2.29) we write

$$\begin{aligned}
&[2m+1]\mathrm{t}_{\mathrm{ev}}^{(2m+1)} + [2m]\mathrm{t}_{\mathrm{ev}}^{(2m-1)} \\
&\quad = \sum_{0 \leqslant a \leqslant b \leqslant 2m+1} \sum_{c \geqslant 0} p^{(2m-b-2c+1)}(\kappa) \check{E}^{(a)} R_{a,b,c} K^{b-2m+2c-1} F^{(b-a)},
\end{aligned}$$

where

$$\begin{aligned}
R_{a,b,c} :=& q^{\binom{2c}{2}+2c-b(2m-b-2c+1)-a(b-a)}[2m+1]\begin{bmatrix} h; -m \\ c \end{bmatrix} \\
&+ q^{\binom{2c}{2}-2c+1-b(2m-b-2c+1)-a(b-a)}[2m]\begin{bmatrix} h; 1-m \\ c-1 \end{bmatrix}.
\end{aligned}$$

To prove the formula (2.29) for $\mathrm{t}_{\mathrm{ev}}^{(2m+1)}$, it suffices to show that, for all a, b, c,

$$L_{a,b,c} = R_{a,b,c}. \tag{2.30}$$

Canceling the q-powers $q^{\binom{2c}{2}+2bc-b(2m-b+1)-a(b-a)}$ on both sides, we see that the identity (2.30) is equivalent to the following identity, for all a, b, c:

$$\begin{aligned}
q^{2m-2c-a+1}[a]\begin{bmatrix} h; 1-m \\ c \end{bmatrix} &+ q^{2c-2m+b-a-1}[b-a]\begin{bmatrix} h; -m \\ c \end{bmatrix} \\
&+ q^{-2m-2c+a+2}\frac{q^{-3a}K^{-2} - q^{-a}}{q^2-1}\begin{bmatrix} h; 1-m \\ c-1 \end{bmatrix} \\
&+ q^{b-2a}[2m-b-2c+1]\begin{bmatrix} h; 1-m \\ c \end{bmatrix} \\
&+ q^{b-2a-4m}[2m-b-2c+1]\begin{bmatrix} h; 1-m \\ c-1 \end{bmatrix} K^{-2} \\
&= q^{2c}[2m+1]\begin{bmatrix} h; -m \\ c \end{bmatrix} + q^{-2c+1}[2m]\begin{bmatrix} h; 1-m \\ c-1 \end{bmatrix}.
\end{aligned} \tag{2.31}$$

By combining the second summand of LHS with the first summand of RHS as well as combining the third summand of LHS with the second summand of RHS, the identity (2.31) is reduced to the following equivalent identity, for all a, b, c:

$$W + X + Y + Z_1 + Z_2 = 0, \tag{2.32}$$

where

$$W = [a] \begin{bmatrix} h; 1-m \\ c \end{bmatrix},$$

$$X = q^{4c-2m+b-1}[-2m+b-a-1] \begin{bmatrix} h; -m \\ c \end{bmatrix},$$

$$Y = \frac{q^{-4m-a+1}K^{-2} - q^{a+1}}{q^2-1} \begin{bmatrix} h; 1-m \\ c-1 \end{bmatrix},$$

$$Z_1 = q^{2c-2m+b-a-1}[2m-b-2c+1] \begin{bmatrix} h; 1-m \\ c \end{bmatrix},$$

$$Z_2 = q^{2c-6m+b-a-1}[2m-b-2c+1] \begin{bmatrix} h; 1-m \\ c-1 \end{bmatrix} K^{-2}.$$

Let us prove the identity (2.32). Using (2.5), we can write $X = X_1 + X_2$, where

$$X_1 = q^{4c-2m+b-1}[-2m+b-a-1] \begin{bmatrix} h; 1-m \\ c \end{bmatrix},$$

$$X_2 = -q^{4c-6m+b-1}[-2m+b-a-1] \begin{bmatrix} h; 1-m \\ c-1 \end{bmatrix} K^{-2}.$$

Noting that

$$\frac{q^{-4m-a+1}K^{-2} - q^{a+1}}{q^2-1} = q^{-4c-a+1}\frac{(q^{4c-4m}K^{-2}-1)}{q^2-1} + q^{-2c}[-2c-a],$$

we rewrite $Y = Y_1 + Y_2$, where

$$Y_1 = q^{-2c-a}[2c] \begin{bmatrix} h; 1-m \\ c \end{bmatrix}, \qquad Y_2 = q^{-2c}[-2c-a] \begin{bmatrix} h; 1-m \\ c-1 \end{bmatrix}.$$

A direct computation shows that

$$W + X_1 + Y_1 + Z_1 = -(q-q^{-1})[2c+a][2c] \begin{bmatrix} h; 1-m \\ c \end{bmatrix},$$

$$\begin{aligned}(X_2 + Z_2) + Y_2 &= q^{2c-4m}[2c+a] \begin{bmatrix} h; 1-m \\ c-1 \end{bmatrix} K^{-2} + Y_2 \\ &= (q-q^{-1})[2c+a][2c] \begin{bmatrix} h; 1-m \\ c \end{bmatrix}.\end{aligned}$$

Summing up these two equations, we obtain $W + X + Y + Z_1 + Z_2 = 0$, whence (2.32), completing Step (2).

The proof of Theorem 2.5 is completed. □

3 The $\imath$-Divided Powers $\mathrm{t}_{\mathrm{odd}}^{(n)}$ for Odd Weights and Even κ

In this section we shall always take κ to be an even q-integer, i.e.,

$$\kappa = [2\ell], \qquad \text{for } \ell \in \mathbb{Z}.$$

3.1 *Definition of* $\mathrm{t}_{\mathrm{odd}}^{(n)}$ *for Even* κ

We introduce some notations. Set, for $n \geqslant 1$, $a \in \mathbb{Z}$,

$$\left[\!\!\left[\begin{matrix} h; a \\ 0 \end{matrix} \right]\!\!\right] = 1, \qquad \left[\!\!\left[\begin{matrix} h; a \\ n \end{matrix} \right]\!\!\right] = \prod_{i=1}^{n} \frac{q^{4a+4i-4}K^{-2} - q^2}{q^{4i} - 1}, \qquad [\![h; a]\!] = \left[\!\!\left[\begin{matrix} h; a \\ 1 \end{matrix} \right]\!\!\right]. \tag{3.1}$$

It follows from (3.1) that, for $n \geqslant 0$ and $a \in \mathbb{Z}$,

$$\begin{aligned} &\left[\!\!\left[\begin{matrix} h; a \\ n \end{matrix} \right]\!\!\right] F = F \left[\!\!\left[\begin{matrix} h; a+1 \\ n \end{matrix} \right]\!\!\right], \qquad \left[\!\!\left[\begin{matrix} h; a \\ n \end{matrix} \right]\!\!\right] \check{E} = \check{E} \left[\!\!\left[\begin{matrix} h; a-1 \\ n \end{matrix} \right]\!\!\right], \\ &\left[\!\!\left[\begin{matrix} h; a \\ n \end{matrix} \right]\!\!\right] = \left[\!\!\left[\begin{matrix} h; 1+a \\ n \end{matrix} \right]\!\!\right] - q^{4a} K^{-2} \left[\!\!\left[\begin{matrix} h; 1+a \\ n-1 \end{matrix} \right]\!\!\right]. \end{aligned} \tag{3.2}$$

We shall take the following as a definition of the $\imath$-divided powers $\mathrm{t}_{\mathrm{odd}}^{(n)}$.

Definition 3.1.1 Set $\mathrm{t}_{\mathrm{odd}}^{(1)} = t = F + \check{E} + \kappa K^{-1}$. The divided powers $\mathrm{t}_{\mathrm{odd}}^{(n)}$, for $n \geqslant 1$, are defined by the recursive relations:

$$\begin{aligned} t \cdot \mathrm{t}_{\mathrm{odd}}^{(2a-1)} &= [2a]\mathrm{t}_{\mathrm{odd}}^{(2a)} + [2a-1]\mathrm{t}_{\mathrm{odd}}^{(2a-2)}, \\ t \cdot \mathrm{t}_{\mathrm{odd}}^{(2a)} &= [2a+1]\mathrm{t}_{\mathrm{odd}}^{(2a+1)}, \qquad \text{for } a \geqslant 1. \end{aligned} \tag{3.3}$$

Equivalently, $\mathrm{t}_{\mathrm{odd}}^{(n)}$ is defined by the closed formula (1.2).

The following lemma is a variant of [5, Lemma 3.3] (where $\kappa = 0$) with the same proof.

Lemma 3.1 *The anti-involution σ on* $\mathbf{U}$ *fixes F, $\check{E}$, K, $t_{\mathrm{odd}}^{(n)}$, respectively. Moreover, σ sends $q \mapsto q^{-1}$, and*

$$\begin{bmatrix} h; a \\ n \end{bmatrix} \mapsto (-1)^n q^{2n(n-1)} \begin{bmatrix} h; 2-a-n \\ n \end{bmatrix}, \qquad \forall a \in \mathbb{Z},\ n \in \mathbb{N}.$$

3.2 *Formulae of* $\mathrm{t}_{\mathbf{odd}}^{(n)}$ *with Even* κ

Recall the polynomials $p^{(n)}$, for $n \geqslant 0$, from Sect. 2.4.

Theorem 3.2 *Let κ be an even q-integer. Then we have, for $m \geqslant 0$,*

$$\mathrm{t}_{\mathrm{odd}}^{(2m)} = \sum_{b=0}^{2m} \sum_{a=0}^{b} \sum_{c \geqslant 0} q^{\binom{2c}{2} - b(2m-b-2c) - a(b-a)} p^{(2m-b-2c)}(\kappa) \cdot \check{E}^{(a)} \begin{bmatrix} h; 1-m \\ c \end{bmatrix} K^{b-2m+2c} F^{(b-a)}, \tag{3.4}$$

$$\mathrm{t}_{\mathrm{odd}}^{(2m+1)} = \sum_{b=0}^{2m+1} \sum_{a=0}^{b} \sum_{c \geqslant 0} q^{\binom{2c}{2} - 2c - b(2m-b-2c+1) - a(b-a)} p^{(2m-b-2c+1)}(\kappa) \cdot \check{E}^{(a)} \begin{bmatrix} h; 1-m \\ c \end{bmatrix} K^{b-2m+2c-1} F^{(b-a)}. \tag{3.5}$$

The proof of Theorem 3.2 will be given in Sect. 3.4 below. By applying the anti-involution σ we convert the formulae in Theorem 3.2 into the following.

Theorem 3.3 *Let κ be an even q-integer. Then we have, for $m \geqslant 0$,*

$$\mathrm{t}_{\mathrm{odd}}^{(2m)} = \sum_{b=0}^{2m} \sum_{a=0}^{b} \sum_{c \geqslant 0} (-1)^c q^{-c + b(2m-b-2c) + a(b-a)} p^{(2m-b-2c)}(\kappa) \cdot F^{(b-a)} K^{b-2m+2c} \begin{bmatrix} h; 1+m-c \\ c \end{bmatrix} \check{E}^{(a)} \tag{3.6}$$

$$\mathrm{t}_{\mathrm{odd}}^{(2m+1)} = \sum_{b=0}^{2m+1} \sum_{a=0}^{b} \sum_{c \geqslant 0} (-1)^c q^{c + b(2m-b-2c+1) + a(b-a)} p^{(2m-b-2c+1)}(\kappa) \cdot F^{(b-a)} K^{b-2m+2c-1} \begin{bmatrix} h; 1+m-c \\ c \end{bmatrix} \check{E}^{(a)}. \tag{3.7}$$

Proof Recall from Lemma 3.1 that the anti-involution σ on $\mathbf{U}$ fixes $F, \check{E}, K, t_{\text{odd}}^{(n)}, \kappa$ while sending $q \mapsto q^{-1}$, $\left[\!\left[\begin{matrix} h; 1-m \\ c \end{matrix}\right]\!\right] \mapsto (-1)^c q^{2c(c-1)} \left[\!\left[\begin{matrix} h; 1+m-c \\ c \end{matrix}\right]\!\right]$. The formulae (3.6)–(3.7) now follow from (3.4)–(3.5).

Let $\lambda, m \in \mathbb{Z}$ and $c \in \mathbb{N}$. Recall $\left\{\begin{matrix} m \\ c \end{matrix}\right\}$ from (2.20). We note that

$$\left[\!\left[\begin{matrix} h; 1+m-c \\ c \end{matrix}\right]\!\right] \mathbf{1}_{2\lambda+1} = \prod_{i=1}^{c} \frac{q^{4m-4c+4i} K^{-2} - q^2}{q^{4i}-1} \mathbf{1}_{2\lambda+1}$$

$$= (-1)^c q^{-2c^2} \left\{\begin{matrix} m-\lambda-c \\ c \end{matrix}\right\} \mathbf{1}_{2\lambda+1}. \tag{3.8}$$

The following corollary is immediate from (3.8), Proposition 2.3, and Theorem 3.2.

Corollary 3.4 *We have* $\mathrm{t}_{\text{odd}}^{(n)} \in {}_{\mathcal{A}}\mathbf{U}^{\imath}_{\text{odd}}$, *for all* n. □

Example 3.2.1 The formulae of $\mathrm{t}_{\text{odd}}^{(n)}$, for $1 \leqslant n \leqslant 3$, in Theorem 3.2 read as follows.

$$\mathrm{t}_{\text{odd}}^{(1)} = F + \check{E} + \kappa K^{-1},$$

$$\mathrm{t}_{\text{odd}}^{(2)} = b^{(2)} + q[\![h; 0]\!] + \kappa(q^{-1}K^{-1}F + q^{-1}\check{E}K^{-1}) + \frac{\kappa^2}{[2]}K^{-2},$$

$$\begin{aligned}\mathrm{t}_{\text{odd}}^{(3)} &= b^{(3)} + q^{-1}[\![h; 0]\!]F + q^{-1}\check{E}[\![h; 0]\!] \\ &\quad + (q^{-2}\check{E}^{(2)}K^{-1} + q^{-3}\check{E}K^{-1}F + q^{-2}K^{-1}F^{(2)} + q^{-1}[\![h; 0]\!]K^{-1})\kappa \\ &\quad + \frac{q^{-2}}{[2]}(\check{E}K^{-2} + K^{-2}F)\kappa^2 + \frac{K^{-3}}{[3]!}\big(\kappa^3 + (q^{-4} + q^{-2})\kappa\big).\end{aligned}$$

Remark 3.5 The formula for $\mathrm{t}_{\text{odd}}^{(2m)}$ is formally obtained from $\mathrm{t}_{\text{ev}}^{(2m)}$ in Theorem 2.5, with $\begin{bmatrix} h; a \\ c \end{bmatrix}$ replaced by $\left[\!\left[\begin{matrix} h; a \\ c \end{matrix}\right]\!\right]$. The formula for $\mathrm{t}_{\text{odd}}^{(2m-1)}$ is formally obtained from $\mathrm{t}_{\text{ev}}^{(2m-1)}$ in Theorem 2.5, with $\begin{bmatrix} h; a \\ c \end{bmatrix}$ replaced by $q^{-4c} \left[\!\left[\begin{matrix} h; a+1 \\ c \end{matrix}\right]\!\right]$. □

Remark 3.6 Note $p_n(0) = 0$ for $n > 0$, and $p_0(0) = 1$. The formulae in Theorems 3.2 and 3.3 in the special case for $\kappa = 0$ recover the formulae in [5, Theorem 3.1, Proposition 3.4]. □

3.3 The $\imath$-Canonical Basis for $\dot{\mathbf{U}}_{odd}$ for Even κ

Recall from Sect. 2.6 the $\imath$-canonical basis on simple $\mathbf{U}$-modules $L(\mu)$, for $\mu \in \mathbb{N}$.

Theorem 3.7

(a) *Let $n \in \mathbb{N}$. For each integer $\lambda \gg n$, the element $\mathrm{t}_{\mathrm{odd}}^{(n)} v_{2\lambda+1}^+$ is an $\imath$-canonical basis element for $L(2\lambda+1)$.*

(b) *The set $\{\mathrm{t}_{\mathrm{odd}}^{(n)} \mid n \in \mathbb{N}\}$ forms the $\imath$-canonical basis for $\mathbf{U}^\imath$ (and an $\mathcal{A}$-basis in ${}_{\mathcal{A}}\mathbf{U}^\imath_{\mathrm{odd}}$).* □

Proof Recall $\begin{Bmatrix} m \\ c \end{Bmatrix}$ from (2.20) and $\left[\!\!\left[\begin{matrix} h; a \\ c \end{matrix}\right]\!\!\right]$ from (3.1). Let $\lambda, m \in \mathbb{N}$. It follows by a direct computation using Theorem 3.3 and (3.8) that

$$\begin{aligned}
\mathrm{t}_{\mathrm{odd}}^{(2m)} v_{2\lambda+1}^+ &= \sum_{b=0}^{2m} \sum_{c \geqslant 0} (-1)^c q^{-c+b(2m-b-2c)} p^{(2m-b-2c)}(\kappa) \\
&\quad \cdot F^{(b)} K^{b-2m+2c} \left[\!\!\left[\begin{matrix} h; 1+m-c \\ c \end{matrix}\right]\!\!\right] v_{2\lambda+1}^+ \\
&= \sum_{b=0}^{2m} \sum_{c \geqslant 0} q^{-2c^2-c+(b-2\lambda-1)(2m-b-2c)} p^{(2m-b-2c)}(\kappa) \\
&\quad \cdot \begin{Bmatrix} m-\lambda-c \\ c \end{Bmatrix} F^{(b)} v_{2\lambda+1}^+.
\end{aligned} \tag{3.9}$$

Similarly using Theorem 3.3 we have

$$\begin{aligned}
\mathrm{t}_{\mathrm{odd}}^{(2m+1)} v_{2\lambda+1}^+ &= \sum_{b=0}^{2m+1} \sum_{c \geqslant 0} (-1)^c q^{c+b(2m-b-2c+1)} p^{(2m-b-2c+1)}(\kappa) \\
&\quad \cdot F^{(b)} K^{b-2m+2c-1} \left[\!\!\left[\begin{matrix} h; 1+m-c \\ c \end{matrix}\right]\!\!\right] v_{2\lambda+1}^+ \\
&= \sum_{b=0}^{2m+1} \sum_{c \geqslant 0} q^{-2c^2+c+(b-2\lambda-1)(2m-b-2c+1)} p^{(2m-b-2c+1)}(\kappa) \\
&\quad \cdot \begin{Bmatrix} m-\lambda-c \\ c \end{Bmatrix} F^{(b)} v_{2\lambda+1}^+.
\end{aligned} \tag{3.10}$$

By a similar argument for (2.26), using (3.9)–(3.10) we obtain

$$\mathrm{t}_{\mathrm{odd}}^{(n)} v_{2\lambda+1}^{+} \in F^{(n)} v_{2\lambda+1}^{+} + \sum_{b<n} q^{-1}\mathbb{Z}[q^{-1}] F^{(b)} v_{2\lambda+1}^{+}, \qquad \text{for } \lambda \gg n.$$

The second statement follows now from the definition of the $\imath$-canonical basis for $\mathbf{U}^{\imath}$ using the projective system $\{L(2\lambda+1)\}_{\lambda\geqslant 0}$; cf. [3, §6].

3.4 Proof of Theorem 3.2

We prove the formulae for $\mathrm{t}_{\mathrm{odd}}^{(n)}$ by induction on n, in two steps (1)–(2) below. The base cases when $n = 1, 2$ are clear.

(1) We shall prove the formula (3.5) for $\mathrm{t}_{\mathrm{odd}}^{(2m+1)}$, assuming the formula (3.4) for $\mathrm{t}_{\mathrm{odd}}^{(2m)}$.

Recall $[2m+1]\mathrm{t}_{\mathrm{ev}}^{(2m+1)} = t \cdot \mathrm{t}_{\mathrm{odd}}^{(2m)}$, and $t = F + \check{E} + \kappa K^{-1}$. Let us compute

$$I = \check{E}\mathrm{t}_{\mathrm{odd}}^{(2m)}, \qquad II = F\mathrm{t}_{\mathrm{odd}}^{(2m)}, \qquad III = \kappa K^{-1}\mathrm{t}_{\mathrm{odd}}^{(2m)}.$$

First we have

$$\begin{aligned} I = &\sum_{b=0}^{2m}\sum_{a=0}^{b}\sum_{c\geqslant 0} q^{\binom{2c}{2}-b(2m-b-2c)-a(b-a)}[a+1]p^{(2m-b-2c)}(\kappa) \\ &\qquad \cdot \check{E}^{(a+1)} \begin{bmatrix} h; 1-m \\ c \end{bmatrix} K^{b-2m+2c} F^{(b-a)} \\ = &\sum_{b=0}^{2m+1}\sum_{a=0}^{b}\sum_{c\geqslant 0} q^{\binom{2c}{2}-(b-1)(2m-b-2c+1)-(a-1)(b-a)}[a]p^{(2m-b-2c+1)}(\kappa) \\ &\qquad \cdot \check{E}^{(a)} \begin{bmatrix} h; 1-m \\ c \end{bmatrix} K^{b-2m+2c-1} F^{(b-a)}, \end{aligned}$$

where the last equation is obtained by shifting indices $a \to a-1$, $b \to b-1$ (and then adding some zero terms to make the bounds of the summations uniform throughout). Using (2.6) we have

$$\begin{aligned} II = &\sum_{b=0}^{2m}\sum_{a=0}^{b}\sum_{c\geqslant 0} q^{\binom{2c}{2}-b(2m-b-2c)-a(b-a)} p^{(2m-b-2c)}(\kappa) \\ &\cdot \left(q^{-2a}\check{E}^{(a)}F + \check{E}^{(a-1)}\frac{q^{3-3a}K^{-2}-q^{1-a}}{q^2-1}\right) \begin{bmatrix} h; 1-m \\ c \end{bmatrix} K^{b-2m+2c} F^{(b-a)}, \end{aligned}$$

where II^1 and II^2 are the two natural summands associated with the plus sign. By shifting index $b \to b-1$, we further have

$$II^1 = \sum_{b=0}^{2m+1} \sum_{a=0}^{b} \sum_{c \geqslant 0} q^{\binom{2c}{2}-(b+1)(2m-b-2c+1)-a(b-a+1)} [b-a] \\ \cdot p^{(2m-b-2c+1)}(\kappa) \check{E}^{(a)} \begin{bmatrix} h; -m \\ c \end{bmatrix} K^{b-2m+2c-1} F^{(b-a)}.$$

By shifting the indices $a \to a+1$, $b \to b+1$, and $c \to c-1$, we further have

$$II^2 = \sum_{b=0}^{2m+1} \sum_{a=0}^{b} \sum_{c \geqslant 0} q^{\binom{2c-2}{2}-(b+1)(2m-b-2c+1)-(a+1)(b-a)} p^{(2m-b-2c+1)}(\kappa) \\ \cdot \check{E}^{(a)} \frac{q^{-3a}K^{-2} - q^{-a}}{q^2-1} \begin{bmatrix} h; 1-m \\ c-1 \end{bmatrix} K^{b-2m+2c-1} F^{(b-a)}.$$

Using (2.14) we also compute

$$\begin{aligned} III &= \sum_{b=0}^{2m} \sum_{a=0}^{b} \sum_{c \geqslant 0} q^{\binom{2c}{2}-b(2m-b-2c)-a(b-a)-2a} \kappa \cdot p^{(2m-b-2c)}(\kappa) \\ &\quad \cdot \check{E}^{(a)} \begin{bmatrix} h; 1-m \\ c \end{bmatrix} K^{b-2m+2c-1} F^{(b-a)} \\ &= \sum_{b=0}^{2m+1} \sum_{a=0}^{b} \sum_{c \geqslant 0} q^{\binom{2c}{2}-b(2m-b-2c)-a(b-a)-2a} [2m-b-2c+1] \\ &\quad \cdot p^{(2m-b-2c+1)}(\kappa) \check{E}^{(a)} \begin{bmatrix} h; 1-m \\ c \end{bmatrix} K^{b-2m+2c-1} F^{(b-a)} \\ &\quad - \sum_{b=0}^{2m+1} \sum_{a=0}^{b} \sum_{c \geqslant 0} q^{\binom{2c-2}{2}-(b+2)(2m-b-2c+2)-a(b-a)-2a+1} \\ &\quad \cdot [2m-b-2c+1] p^{(2m-b-2c+1)}(\kappa) \check{E}^{(a)} \begin{bmatrix} h; 1-m \\ c-1 \end{bmatrix} K^{b-2m+2c-3} F^{(b-a)}. \end{aligned}$$

(Note that we have shifted the index $c \to c-1$ in the last summand above.)

Collecting the formulae for I, II^1, II^2, and III gives us

$$t \cdot \mathrm{t}_{\mathrm{odd}}^{(2m)} = \sum_{0 \leqslant a \leqslant b \leqslant 2m+1} \sum_{c \geqslant 0} p^{(2m-b-2c+1)}(\kappa) \check{E}^{(a)} \mathcal{H}_{a,b,c} K^{b-2m+2c-1} F^{(b-a)},$$

where

$$\mathcal{H}_{a,b,c} := q^{\binom{2c}{2}-(b-1)(2m-b-2c+1)-(a-1)(b-a)}[a] \left[\!\!\left[\begin{matrix} h; 1-m \\ c \end{matrix}\right]\!\!\right]$$
$$+ q^{\binom{2c}{2}-(b+1)(2m-b-2c+1)-a(b-a+1)}[b-a] \left[\!\!\left[\begin{matrix} h; -m \\ c \end{matrix}\right]\!\!\right]$$
$$+ q^{\binom{2c-2}{2}-(b+1)(2m-b-2c+1)-(a+1)(b-a)} \frac{q^{-3a}K^{-2} - q^{-a}}{q^2-1} \left[\!\!\left[\begin{matrix} h; 1-m \\ c-1 \end{matrix}\right]\!\!\right]$$
$$+ q^{\binom{2c}{2}-b(2m-b-2c)-a(b-a)-2a}[2m-b-2c+1] \left[\!\!\left[\begin{matrix} h; 1-m \\ c \end{matrix}\right]\!\!\right]$$
$$- q^{\binom{2c-2}{2}-(b+2)(2m-b-2c+2)-a(b-a)-2a+1}[2m-b-2c+1] \left[\!\!\left[\begin{matrix} h; 1-m \\ c-1 \end{matrix}\right]\!\!\right] K^{-2}.$$

Recall $[2m+1]\mathrm{t}_{\mathrm{odd}}^{(2m+1)} = t \cdot \mathrm{t}_{\mathrm{odd}}^{(2m)}$. To prove the formula (3.5) for $\mathrm{t}_{\mathrm{odd}}^{(2m+1)}$, by the PBW basis theorem and the inductive assumption it suffices to prove the following identity, for all a, b, c:

$$\mathcal{H}_{a,b,c} = q^{\binom{2c}{2}-2c-b(2m-b-2c+1)-a(b-a)}[2m+1] \left[\!\!\left[\begin{matrix} h; 1-m \\ c \end{matrix}\right]\!\!\right]. \tag{3.11}$$

Thanks to $q^{2m-a+1}[a] - [2m+1] = q^{-a}[a-2m-1]$, we can combine the RHS with the first summand of LHS (3.11). Hence, after canceling out $q^{\binom{2c}{2}-2c-b(2m-b-2c+1)-a(b-a)-a}$ on both sides, we see that (3.11) is equivalent to the following identity, for all a, b, c:

$$\mathcal{A} + \mathcal{B} + \mathcal{C} + \mathcal{D}_1 + \mathcal{D}_2 = 0, \tag{3.12}$$

where

$$\mathcal{A} = [a-2m-1] \left[\!\!\left[\begin{matrix} h; 1-m \\ c \end{matrix}\right]\!\!\right],$$
$$\mathcal{B} = q^{4c-2m+b-1}[b-a] \left[\!\!\left[\begin{matrix} h; -m \\ c \end{matrix}\right]\!\!\right],$$
$$\mathcal{C} = q^{2a-2m+2} \frac{q^{-3a}K^{-2} - q^{-a}}{q^2-1} \left[\!\!\left[\begin{matrix} h; 1-m \\ c-1 \end{matrix}\right]\!\!\right],$$
$$\mathcal{D}_1 = q^{2c+b-a}[2m-b-2c+1] \left[\!\!\left[\begin{matrix} h; 1-m \\ c \end{matrix}\right]\!\!\right],$$

$$\mathcal{D}_2 = -q^{2c-4m+b-a}[2m-b-2c+1]\begin{bmatrix} h; 1-m \\ c-1 \end{bmatrix} K^{-2}.$$

Let us prove the identity (3.12). Using (3.2), we can write $\mathcal{B} = \mathcal{B}_1 + \mathcal{B}_2$, where

$$\mathcal{B}_1 = q^{4c-2m+b-1}[b-a]\begin{bmatrix} h; -m \\ c \end{bmatrix}, \quad \mathcal{B}_2 = -q^{4c-6m+b-1}[b-a]\begin{bmatrix} h; -m \\ c \end{bmatrix}.$$

Noting that

$$\frac{q^{-3a}K^{-2} - q^{-a}}{q^2-1} = q^{4m-4c-3a}\frac{(q^{4c-4m}K^{-2} - q^2)}{q^2-1} + q^{2m-2c-2a}[2m-2c-a+1],$$

we rewrite $\mathcal{C} = \mathcal{C}_1 + \mathcal{C}_2$, where

$$\mathcal{C}_1 = q^{2m-2c-a+1}[2c]\begin{bmatrix} h; 1-m \\ c \end{bmatrix}, \quad \mathcal{C}_2 = q^{2-2c}[2m-2c-a+1]\begin{bmatrix} h; 1-m \\ c-1 \end{bmatrix}.$$

A direct computation gives us

$$\mathcal{A} + \mathcal{B}_1 + \mathcal{C}_1 + \mathcal{D}_1 = (q-q^{-1})[2c][2m-2c-a+1]\begin{bmatrix} h; 1-m \\ c \end{bmatrix},$$

$$\begin{aligned} \mathcal{C}_2 + (\mathcal{B}_2 + \mathcal{D}_2) &= \mathcal{C}_2 - q^{2c-4m}[2m-2c-a+1]\begin{bmatrix} h; 1-m \\ c-1 \end{bmatrix} K^{-2} \\ &= -(q-q^{-1})[2c][2m-2c-a+1]\begin{bmatrix} h; 1-m \\ c \end{bmatrix}. \end{aligned}$$

Summing up these two equations, we have $\mathcal{A}+\mathcal{B}+\mathcal{C}+\mathcal{D}_1+\mathcal{D}_2 = 0$, whence (3.12), completing Step (1).

(2) Assuming the formulae for $\mathrm{t}_{\mathrm{odd}}^{(n)}$ with $n \leqslant 2m+1$, we shall now prove the following formula (3.13) for $\mathrm{t}_{\mathrm{odd}}^{(2m+2)}$ (obtained with m replaced by $m+1$ in (3.4)):

$$\mathrm{t}_{\mathrm{odd}}^{(2m+2)} = \sum_{b=0}^{2m+2}\sum_{a=0}^{b}\sum_{c\geqslant 0} q^{\binom{2c}{2}-b(2m-b-2c+2)-a(b-a)} p^{(2m-b-2c+2)}(\kappa) \cdot \check{E}^{(a)}\begin{bmatrix} h; -m \\ c \end{bmatrix} K^{b-2m+2c-2} F^{(b-a)}. \tag{3.13}$$

The proof is based on the recursion $t\cdot \mathrm{t}_{\mathrm{odd}}^{(2m+1)} = [2m+2]\mathrm{t}_{\mathrm{odd}}^{(2m+2)} + [2m+1]\mathrm{t}_{\mathrm{odd}}^{(2m)}$. Recall $t = F + \check{E} + \kappa K^{-1}$ and $\mathrm{t}_{\mathrm{odd}}^{(2m+1)}$ from (3.5). We shall compute

$$\mathtt{I} = \check{E}\mathrm{t}_{\mathrm{odd}}^{(2m+1)}, \qquad \mathtt{II} = F\mathrm{t}_{\mathrm{odd}}^{(2m+1)}, \qquad \mathtt{III} = \kappa K^{-1}\mathrm{t}_{\mathrm{odd}}^{(2m+1)},$$

respectively. First by shifting indices $a \to a-1$ and $b \to b-1$, we have

$$\mathtt{I} = \sum_{b=0}^{2m+1} \sum_{a=0}^{b} \sum_{c\geqslant 0} q^{\binom{2c}{2}-2c-b(2m-b-2c+1)-a(b-a)} [a+1] p^{(2m-b-2c+1)}(\kappa)$$
$$\cdot \check{E}^{(a+1)} \begin{bmatrix} h; 1-m \\ c \end{bmatrix} K^{b-2m+2c-1} F^{(b-a)}$$
$$= \sum_{b=0}^{2m+2} \sum_{a=0}^{b} \sum_{c\geqslant 0} q^{\binom{2c}{2}-2c-(b-1)(2m-b-2c+2)-(a-1)(b-a)}$$
$$\cdot [a] p^{(2m-b-2c+2)}(\kappa) \check{E}^{(a)} \begin{bmatrix} h; 1-m \\ c \end{bmatrix} K^{b-2m+2c-2} F^{(b-a)}.$$

Using (2.6) we also have

$$\mathtt{II} = \sum_{b=0}^{2m+1} \sum_{a=0}^{b} \sum_{c\geqslant 0} q^{\binom{2c}{2}-2c-b(2m-b-2c+1)-a(b-a)} p^{(2m-b-2c+1)}(\kappa)$$
$$\cdot \left(q^{-2a} \check{E}^{(a)} F + \check{E}^{(a-1)} \frac{q^{3-3a}K^{-2} - q^{1-a}}{q^2-1} \right) \begin{bmatrix} h; 1-m \\ c \end{bmatrix} K^{b-2m+2c-1} F^{(b-a)}$$
$$= \mathtt{II}_1 + \mathtt{II}_2,$$

where $\mathtt{II}_1$ and $\mathtt{II}_2$ are the two natural summands associated with the plus sign. By shifting the index $b \to b-1$, we have

$$\mathtt{II}_1 = \sum_{b=0}^{2m+2} \sum_{a=0}^{b} \sum_{c\geqslant 0} q^{\binom{2c}{2}-2c-(b+1)(2m-b-2c+2)-a(b-a-1)-2a} [b-a]$$
$$\cdot p^{(2m-b-2c+2)}(\kappa) \check{E}^{(a)} \begin{bmatrix} h; -m \\ c \end{bmatrix} K^{b-2m+2c-2} F^{(b-a)}.$$

By shifting the indices $a \to a+1$, $b \to b+1$, and $c \to c-1$, we further have

$$\mathtt{II}_2 = \sum_{b=0}^{2m+2} \sum_{a=0}^{b} \sum_{c\geqslant 0} q^{\binom{2c-2}{2}-2c-(b+1)(2m-b-2c+2)-(a+1)(b-a)+2}$$

$$\cdot\, p^{(2m-b-2c+2)}(\kappa)\check{E}^{(a)}\frac{q^{-3a}K^{-2}-q^{-a}}{q^2-1}\begin{bmatrix} h;1-m \\ c-1\end{bmatrix} K^{b-2m+2c-2}F^{(b-a)}.$$

Using (4.5) we also compute

$$\begin{aligned}
\mathtt{III} &= \sum_{b=0}^{2m+1}\sum_{a=0}^{b}\sum_{c\geqslant 0} q^{\binom{2c}{2}-2c-b(2m-b-2c+1)-a(b-a)-2a}\kappa\cdot p^{(2m-b-2c+1)}(\kappa)\\
&\quad\cdot \check{E}^{(a)}\begin{bmatrix} h;1-m \\ c\end{bmatrix} K^{b-2m+2c-2}F^{(b-a)}\\
&= \sum_{b=0}^{2m+2}\sum_{a=0}^{b}\sum_{c\geqslant 0} q^{\binom{2c}{2}-2c-b(2m-b-2c+1)-a(b-a)-2a}[2m-b-2c+2]\\
&\quad\cdot p^{(2m-b-2c+2)}(\kappa)\check{E}^{(a)}\begin{bmatrix} h;1-m \\ c\end{bmatrix} K^{b-2m+2c-2}F^{(b-a)}\\
&\quad-\sum_{b=0}^{2m+2}\sum_{a=0}^{b}\sum_{c\geqslant 0} q^{\binom{2c-2}{2}-2c-(b+2)(2m-b-2c+3)-a(b-a)-2a+3}\\
&\quad\cdot[2m-b-2c+2]p^{(2m-b-2c+2)}(\kappa)\check{E}^{(a)}\begin{bmatrix} h;1-m \\ c-1\end{bmatrix} K^{b-2m+2c-4}F^{(b-a)}.
\end{aligned}$$

(Note that we have shifted the index $c \to c-1$ in the last summand above.)

Collecting the formulae for $\mathtt{I}$, $\mathtt{II}_1$, $\mathtt{II}_2$, and $\mathtt{III}$, we obtain

$$\begin{aligned}
&t\cdot \mathrm{t}_{\mathrm{odd}}^{(2m+1)}\\
&\qquad= \sum_{0\leqslant a\leqslant b\leqslant 2m+2}\sum_{c\geqslant 0} p^{(2m-b-2c+2)}(\kappa)\check{E}^{(a)}\mathcal{L}_{a,b,c}K^{b-2m+2c-2}F^{(b-a)},
\end{aligned}$$

where

$$\begin{aligned}
\mathcal{L}_{a,b,c} :=\ & q^{\binom{2c}{2}-2c-(b-1)(2m-b-2c+2)-(a-1)(b-a)}[a]\begin{bmatrix} h;1-m \\ c\end{bmatrix}\\
&+q^{\binom{2c}{2}-2c-(b+1)(2m-b-2c+2)-a(b-a-1)-2a}[b-a]\begin{bmatrix} h;-m \\ c\end{bmatrix}\\
&+q^{\binom{2c-2}{2}-2c-(b+1)(2m-b-2c+2)-(a+1)(b-a)+2}\frac{q^{-3a}K^{-2}-q^{-a}}{q^2-1}\begin{bmatrix} h;1-m \\ c-1\end{bmatrix}\\
&+q^{\binom{2c}{2}-2c-b(2m-b-2c+1)-a(b-a)-2a}[2m-b-2c+2]\begin{bmatrix} h;1-m \\ c\end{bmatrix}
\end{aligned}$$

$$- q^{\binom{2c-2}{2}-2c-(b+2)(2m-b-2c+3)-a(b-a)-2a+3}[2m-b-2c+2]\times \begin{bmatrix} h; 1-m \\ c-1 \end{bmatrix} K^{-2}.$$

On the other hand, using (3.4) (with an index shift $c \to c-1$) and (3.13) we write

$$[2m+2]\mathrm{t}_{\mathrm{odd}}^{(2m+2)} + [2m+1]\mathrm{t}_{\mathrm{odd}}^{(2m)} = \sum_{0\leqslant a\leqslant b\leqslant 2m+2} \sum_{c\geqslant 0} p^{(2m-b-2c+2)}(\kappa)\check{E}^{(a)}\mathcal{R}_{a,b,c}K^{b-2m+2c-2}F^{(b-a)},$$

where

$$\mathcal{R}_{a,b,c} := q^{\binom{2c}{2}-b(2m-b-2c+2)-a(b-a)}[2m+2] \begin{bmatrix} h; -m \\ c \end{bmatrix} + q^{\binom{2c-2}{2}-b(2m-b-2c+2)-a(b-a)}[2m+1] \begin{bmatrix} h; 1-m \\ c-1 \end{bmatrix}.$$

To prove the formula (3.13) for $\mathrm{t}_{\mathrm{odd}}^{(2m+2)}$, it suffices to show that, for all a, b, c,

$$\mathcal{L}_{a,b,c} = \mathcal{R}_{a,b,c}. \tag{3.14}$$

Canceling the q-powers $q^{\binom{2c}{2}-2c-(b-1)(2m-b-2c+2)-(a-1)(b-a)}$ on both sides, we see that the identity (3.14) is equivalent to the following identity, for all a, b, c:

$$\begin{aligned}
[a] \begin{bmatrix} h; 1-m \\ c \end{bmatrix} &+ q^{4c-4m+b-4}[b-a] \begin{bmatrix} h; -m \\ c \end{bmatrix} \\
&+ q^{2a-4m+1}\frac{q^{-3a}K^{-2}-q^{-a}}{q^2-1} \begin{bmatrix} h; 1-m \\ c-1 \end{bmatrix} \\
&+ q^{2c-2m+b-a-2}[2m-b-2c+2] \begin{bmatrix} h; 1-m \\ c \end{bmatrix} \\
&- q^{2c-6m+b-a-2}[2m-b-2c+2] \begin{bmatrix} h; 1-m \\ c-1 \end{bmatrix} K^{-2} \\
&= q^{4c-2m+a-2}[2m+2] \begin{bmatrix} h; -m \\ c \end{bmatrix} + q^{a-2m+1}[2m+1] \begin{bmatrix} h; 1-m \\ c-1 \end{bmatrix}.
\end{aligned} \tag{3.15}$$

By combining the second summand of LHS with the first summand of RHS as well as combining the third summand of LHS with the second summand of RHS, the identity (3.15) is reduced to the following equivalent identity, for all a, b, c:

$$\mathcal{W} + \mathcal{X} + \mathcal{Y} + \mathcal{Z}_1 + \mathcal{Z}_2 = 0, \tag{3.16}$$

where

$$\mathcal{W} = [a] \left[\!\!\left[\begin{matrix} h; 1-m \\ c \end{matrix} \right]\!\!\right],$$

$$\mathcal{X} = q^{4c-2m+b-2}[b-a-2m-2] \left[\!\!\left[\begin{matrix} h; -m \\ c \end{matrix} \right]\!\!\right],$$

$$\mathcal{Y} = \frac{q^{-a-4m+1}K^{-2} - q^{a+3}}{q^2-1} \left[\!\!\left[\begin{matrix} h; 1-m \\ c-1 \end{matrix} \right]\!\!\right],$$

$$\mathcal{Z}_1 = q^{2c-2m+b-a-2}[2m-b-2c+2] \left[\!\!\left[\begin{matrix} h; 1-m \\ c \end{matrix} \right]\!\!\right],$$

$$\mathcal{Z}_2 = -q^{2c-6m+b-a-2}[2m-b-2c+2] \left[\!\!\left[\begin{matrix} h; 1-m \\ c-1 \end{matrix} \right]\!\!\right] K^{-2}.$$

Let us finally prove the identity (3.16). Using (3.2), we can write $\mathcal{X} = \mathcal{X}_1 + \mathcal{X}_2$, where

$$\mathcal{X}_1 = q^{4c-2m+b-2}[b-a-2m-2] \left[\!\!\left[\begin{matrix} h; 1-m \\ c \end{matrix} \right]\!\!\right],$$

$$\mathcal{X}_2 = -q^{4c-6m+b-2}[b-a-2m-2] \left[\!\!\left[\begin{matrix} h; 1-m \\ c-1 \end{matrix} \right]\!\!\right] K^{-2}.$$

Noting that

$$\frac{q^{-a-4m+1}K^{-2} - q^{a+3}}{q^2-1} = q^{-4c-a+1} \frac{(q^{4c-4m}K^{-2} - q^2)}{q^2-1} + q^{2-2c}[-2c-a],$$

we rewrite $\mathcal{Y} = \mathcal{Y}_1 + \mathcal{Y}_2$, where

$$\mathcal{Y}_1 = q^{-2c-a}[2c] \left[\!\!\left[\begin{matrix} h; 1-m \\ c \end{matrix} \right]\!\!\right], \qquad \mathcal{Y}_2 = q^{2-2c}[-2c-a] \left[\!\!\left[\begin{matrix} h; 1-m \\ c-1 \end{matrix} \right]\!\!\right].$$

A direct computation shows that

$$\mathcal{W} + \mathcal{X}_1 + \mathcal{Y}_1 + \mathcal{Z}_1 = -(q - q^{-1})[2c + a][2c] \begin{bmatrix} h; 1-m \\ c \end{bmatrix},$$

$$(\mathcal{X}_2 + \mathcal{Z}_2) + \mathcal{Y}_2 = q^{2c-4m}[2c + a] \begin{bmatrix} h; 1-m \\ c-1 \end{bmatrix} K^{-2} + \mathcal{Y}_2$$
$$= (q - q^{-1})[2c + a][2c] \begin{bmatrix} h; 1-m \\ c \end{bmatrix}.$$

Summing up these two equations, we obtain $\mathcal{W} + \mathcal{X} + \mathcal{Y} + \mathcal{Z}_1 + \mathcal{Z}_2 = 0$, whence (3.16), completing Step (2).

The proof of Theorem 3.2 is completed. □

4 The $\imath$-Divided Powers $t_{\mathrm{ev}}^{(n)}$ for Even Weights and Odd κ

In this Sect. 4 we shall always take κ to be an odd q-integer, i.e.,

$$\kappa = [2\ell - 1], \qquad \text{for } \ell \in \mathbb{Z}. \tag{4.1}$$

4.1 Definition of $t_{\mathrm{ev}}^{(n)}$ for Odd κ

We shall take the following as a definition of the $\imath$-divided powers $t_{\mathrm{ev}}^{(n)}$ with odd κ, which is formally identical to the formulae for $\mathrm{t}_{\mathrm{odd}}^{(n)}$ with even κ.

Definition 4.1.1 Set $t_{\mathrm{ev}}^{(1)} = t = F + \check{E} + \kappa K^{-1}$. The divided powers $t_{\mathrm{ev}}^{(n)}$, for $n \geqslant 1$, are defined by the recursive relations:

$$\begin{aligned} t \cdot t_{\mathrm{ev}}^{(2a-1)} &= [2a] t_{\mathrm{ev}}^{(2a)} + [2a-1] t_{\mathrm{ev}}^{(2a-2)}, \\ t \cdot t_{\mathrm{ev}}^{(2a)} &= [2a+1] t_{\mathrm{ev}}^{(2a+1)}, \qquad \text{for } a \geqslant 1. \end{aligned} \tag{4.2}$$

Equivalently, we have the following closed formula for $t_{\mathrm{ev}}^{(n)}$, for $a \in \mathbb{N}$:

$$t_{\mathrm{ev}}^{(n)} = \begin{cases} \frac{1}{[2a]!}(t - [-2a+1])(t - [-2a+3]) \cdots (t - [2a-3])(t - [2a-1]), \\ \hfill \text{if } n = 2a, \\ \\ \frac{t}{[2a+1]!}(t - [-2a+1])(t - [-2a+3]) \cdots (t - [2a-3])(t - [2a-1]), \\ \hfill \text{if } n = 2a+1. \end{cases}$$

Note the above formulae are formally the same as (1.2) for $\mathfrak{t}_{\text{odd}}^{(n)} \in {}_{\mathcal{A}}\mathbf{U}_{\text{odd}}^{\imath}$ for κ an even q-integer.

4.2 Polynomials $\mathfrak{p}_n(x)$ and $\mathfrak{p}^{(n)}(x)$

We define a sequence of polynomials $\mathfrak{p}_n(x)$ in a variable x, for $n \in \mathbb{N}$, by letting

$$\mathfrak{p}_{n+1} = x\mathfrak{p}_n + q^{2-2n}[n][n-2]\mathfrak{p}_{n-1}, \qquad \mathfrak{p}_0 = 1. \tag{4.3}$$

Note $\mathfrak{p}_n$ is a monic polynomial in x of degree n. Also set $\mathfrak{p}_n = 0$ for $n < 0$. These polynomials $\mathfrak{p}_n$ will appear in the expansion formula for the $\imath$-divided powers in $\mathbf{U}$.

Example 4.2.1 Here are the polynomials $\mathfrak{p}_n(x)$, for $1 \leqslant n \leqslant 5$:

$$\mathfrak{p}_1(x) = x, \qquad \mathfrak{p}_2(x) = x^2 - 1, \qquad \mathfrak{p}_3(x) = x^3 - x,$$
$$\mathfrak{p}_4(x) = x^4 + (q^{-4}[3] - 1)x^2 - q^{-4}[3],$$
$$\mathfrak{p}_5(x) = x(x^2 - 1)(x^2 + q^{-5}[3]! + q^{-6}[5]).$$

A simple induction using the recursive formula (4.3) shows that $\mathfrak{p}_n/(x-1) \in \mathbb{N}[q^{-1}][x]$, for $n \geqslant 2$. In particular, for κ any positive odd q-integer, we always have $\mathfrak{p}_n(\kappa) \in \mathbb{N}[q, q^{-1}]$.

Introduce the monic polynomials $\mathfrak{g}_n(x)$ of degree n:

$$\mathfrak{g}_n(x) = \begin{cases} \prod_{i=1}^m (x^2 - [2i-1]^2), & \text{if } n = 2m \text{ is even;} \\ x \prod_{i=1}^m (x^2 - [2i-1]^2), & \text{if } n = 2m+1 \text{ is odd.} \end{cases} \tag{4.4}$$

Define

$$\mathfrak{p}^{(n)}(x) = \mathfrak{p}_n(x)/[n]!, \qquad \mathfrak{g}^{(n)}(x) = \mathfrak{g}_n(x)/[n]!.$$

Then Eq. (4.3) implies that

$$x \cdot \mathfrak{p}^{(n)} = [n+1]\mathfrak{p}^{(n+1)} - q^{2-2n}[n-2]\mathfrak{p}^{(n-1)} \quad (n \geqslant 1). \tag{4.5}$$

Our next goal is to prove that $\mathfrak{p}^{(n)}(\kappa)$ are integral, i.e., $\mathfrak{p}^{(n)}([2\ell - 1]) \in \mathcal{A}$, for all $n, \ell \in \mathbb{Z}$; see Proposition 4.2. This is achieved by relating $\mathfrak{p}^{(n)}$ to $\mathfrak{g}^{(n)}$ for varied n.

Lemma 4.1 *For $\kappa = [2\ell - 1]$ as in* (4.1), *we have $\mathfrak{g}^{(n)}(\kappa) \in \mathcal{A}$, for all $n \in \mathbb{N}$.* □

Proof We separate the cases for $n = 2m+1$ and $n = 2m$. Noting that

$$\kappa - [2i-1] = [2\ell - 1] - [2i-1] = [\ell - i](q^{\ell+i-2} + q^{-\ell-i+2}), \tag{4.6}$$

we have

$$\begin{aligned}\mathfrak{g}^{(2m)}(\kappa) &= \frac{1}{[2m]!}\prod_{i=1-m}^{m}(\kappa-[2i-1])\\ &= \begin{bmatrix}\ell+m-1\\ 2m\end{bmatrix}\prod_{i=1-m}^{m}(q^{\ell+i-2}+q^{-\ell-i+2})\in\mathcal{A}.\end{aligned}$$

Similarly, using (4.6) we have

$$\begin{aligned}\mathfrak{g}^{(2m+1)}(\kappa) &= \frac{1}{[2m+1]!}\kappa\prod_{i=1-m}^{m}(\kappa-[2i-1])\\ &= \frac{1}{[2m+1]!}\prod_{i=-m}^{m}(\kappa-[2i-1]) - \frac{1}{[2m]!}\prod_{i=1-m}^{m}(\kappa-[2i-1])\\ &= \begin{bmatrix}\ell+m\\ 2m+1\end{bmatrix}\prod_{i=-m}^{m}(q^{\ell+i-2}+q^{-\ell-i+2})\\ &\quad + \begin{bmatrix}\ell+m-1\\ 2m\end{bmatrix}\prod_{i=1-m}^{m}(q^{\ell+i-2}+q^{-\ell-i+2})\in\mathcal{A}.\end{aligned}$$

The lemma is proved.

Proposition 4.2 *For $m\in\mathbb{N}$, we have*

$$\begin{aligned}\mathfrak{p}^{(2m)} &= \sum_{a=0}^{m-1} q^{(3-2m)a}\begin{bmatrix}m-1\\ a\end{bmatrix}_{q^2}\mathfrak{g}^{(2m-2a)},\\ \mathfrak{p}^{(2m+1)} &= \sum_{a=0}^{m-1} q^{(1-2m)a}\begin{bmatrix}m-1\\ a\end{bmatrix}_{q^2}\mathfrak{g}^{(2m-2a+1)}.\end{aligned}$$

In particular, we have $\mathfrak{p}^{(n)}(\kappa)\in\mathcal{A}$, for all $n\in\mathbb{N}$ and all $\kappa=[2\ell-1]$ with $\ell\in\mathbb{Z}$.

□

Proof It follows from these formulae for $\mathfrak{p}^{(n)}$ and Lemma 4.1 that $\mathfrak{p}^{(n)}([2\ell-1])\in\mathcal{A}$.

Let us prove these formulae. Note that

$$\begin{aligned}x\cdot\mathfrak{g}^{(2a-1)} &= [2a]\mathfrak{g}^{(2a)}+[2a-1]\mathfrak{g}^{(2a-2)},\\ x\cdot\mathfrak{g}^{(2a)} &= [2a+1]\mathfrak{g}^{(2a+1)},\qquad \text{for } a\geqslant 1.\end{aligned}\tag{4.7}$$

We prove by induction on m, with the base cases for $\mathfrak{p}^{(1)}$ and $\mathfrak{p}^{(2)}$ being clear. We separate in two cases.

(1) Let us prove the formula for $\mathfrak{p}^{(2m+1)}$, assuming the formulae for $\mathfrak{p}^{(2m-1)}$ and $\mathfrak{p}^{(2m)}$. By (4.5) and (4.7) we have

$$\begin{aligned}
[2m+1]\mathfrak{p}^{(2m+1)} &= x\mathfrak{p}^{(2m)} + q^{2-4m}[2m-2]\mathfrak{p}^{(2m-1)} \\
&= \sum_{a=0}^{m-1} q^{(3-2m)a} \begin{bmatrix} m-1 \\ a \end{bmatrix}_{q^2} x \cdot \mathfrak{g}^{(2m-2a)} \\
&\quad + \sum_{a=0}^{m-2} q^{(3-2m)a+2-4m}[2m-2] \begin{bmatrix} m-2 \\ a \end{bmatrix}_{q^2} \mathfrak{g}^{(2m-2a-1)} \\
&= \sum_{a=0}^{m-1} q^{(3-2m)a}[2m-2a+1] \begin{bmatrix} m-1 \\ a \end{bmatrix}_{q^2} \mathfrak{g}^{(2m-2a+1)} \\
&\quad + \sum_{a=0}^{m-2} q^{(3-2m)a+2-4m}[2m-2] \begin{bmatrix} m-2 \\ a \end{bmatrix}_{q^2} \mathfrak{g}^{(2m-2a-1)} \\
&= \sum_{a=0}^{m-1} q^{(3-2m)a}[2m-2a+1] \begin{bmatrix} m-1 \\ a \end{bmatrix}_{q^2} \mathfrak{g}^{(2m-2a+1)} \\
&\quad + \sum_{a=1}^{m-1} q^{(3-2m)(a-1)+2-4m}[2m-2] \begin{bmatrix} m-2 \\ a-1 \end{bmatrix}_{q^2} \mathfrak{g}^{(2m-2a+1)} \\
&= [2m+1] \sum_{a=0}^{m-1} q^{(1-2m)a} \begin{bmatrix} m-1 \\ a \end{bmatrix}_{q^2} \mathfrak{g}^{(2m-2a+1)},
\end{aligned}$$

where the last equation results from shifting the second summation by $a \to a-1$ and using the identity

$$q^{2a}[2m-2a+1] \begin{bmatrix} m-1 \\ a \end{bmatrix}_{q^2} + q^{2a-2m-1}[2m-2] \begin{bmatrix} m-2 \\ a-1 \end{bmatrix}_{q^2} = [2m+1] \begin{bmatrix} m-1 \\ a \end{bmatrix}_{q^2}.$$

(2) We shall prove the formula for $\mathfrak{p}^{(2m+2)}$, assuming the formulae for $\mathfrak{p}^{(2m+1)}$ and $\mathfrak{p}^{(2m)}$. By (4.5) and (4.7) we have

$$\begin{aligned}
&[2m+2]\mathfrak{p}^{(2m+2)} \\
&= x\mathfrak{p}^{(2m+1)} + q^{-4m}[2m-1]\mathfrak{p}^{(2m)} \\
&= \sum_{a=0}^{m-1} q^{(1-2m)a} \begin{bmatrix} m-1 \\ a \end{bmatrix}_{q^2} x \cdot \mathfrak{g}^{(2m-2a+1)}
\end{aligned}$$

$$+\sum_{a=0}^{m-1} q^{(3-2m)a-4m}[2m-1]\begin{bmatrix} m-1 \\ a \end{bmatrix}_{q^2} \mathfrak{g}^{(2m-2a)}$$

$$= \sum_{a=0}^{m-1} q^{(1-2m)a} \begin{bmatrix} m-1 \\ a \end{bmatrix}_{q^2} \Big([2m-2a+2]\mathfrak{g}^{(2m-2a+2)} + [2m-2a+1]\mathfrak{g}^{(2m-2a)}\Big)$$

$$+\sum_{a=0}^{m-1} q^{(3-2m)a-4m}[2m-1]\begin{bmatrix} m-1 \\ a \end{bmatrix}_{q^2} \mathfrak{g}^{(2m-2a)}.$$

By combining the like terms for $\mathfrak{g}^{(2m-2a)}$ and using the identity

$$q^{(1-2m)a}[2m-2a+1] + q^{(3-2m)a-4m}[2m-1] = q^{(1-2m)(a+1)}[4m-2a],$$

we have

$$[2m+2]\mathfrak{p}^{(2m+2)} = \sum_{a=0}^{m-1} q^{(1-2m)a}[2m-2a+2]\begin{bmatrix} m-1 \\ a \end{bmatrix}_{q^2} \mathfrak{g}^{(2m-2a+2)}$$

$$+\sum_{a=0}^{m-1} q^{(1-2m)(a+1)}[4m-2a]\begin{bmatrix} m-1 \\ a \end{bmatrix}_{q^2} \mathfrak{g}^{(2m-2a)}$$

$$= [2m+2]\sum_{a=0}^{m} q^{(1-2m)a} \begin{bmatrix} m \\ a \end{bmatrix}_{q^2} \mathfrak{g}^{(2m-2a+2)},$$

where the last equation results from shifting the second summation index from $a \to a-1$ and using the identity

$$[2m-2a+2]\begin{bmatrix} m-1 \\ a \end{bmatrix}_{q^2} + [4m-2a+2]\begin{bmatrix} m-1 \\ a-1 \end{bmatrix}_{q^2} = [2m+2]\begin{bmatrix} m \\ a \end{bmatrix}_{q^2}.$$

The proposition is proved.

4.3 Formulae for $t_{\mathbf{ev}}^{(n)}$ with Odd κ

Theorem 4.3 *Let κ be an odd q-integer. Then, for $m \geqslant 0$, we have*

$$t_{\rm ev}^{(2m)} = \sum_{b=0}^{2m}\sum_{a=0}^{b}\sum_{c\geqslant 0} q^{\binom{2c}{2}+2c-b(2m-b-2c)-a(b-a)}\mathfrak{p}^{(2m-b-2c)}(\kappa) \cdot \check{E}^{(a)}\begin{bmatrix} h;1-m \\ c \end{bmatrix} K^{b-2m+2c}F^{(b-a)}, \tag{4.8}$$

$$t_{\rm ev}^{(2m+1)} = \sum_{b=0}^{2m+1}\sum_{a=0}^{b}\sum_{c\geqslant 0} q^{\binom{2c}{2}-b(2m-b-2c+1)-a(b-a)}\mathfrak{p}^{(2m-b-2c+1)}(\kappa) \cdot \check{E}^{(a)}\begin{bmatrix} h;1-m \\ c \end{bmatrix} K^{b-2m+2c-1}F^{(b-a)}. \tag{4.9}$$

The proof of Theorem 4.3 will be given in Sect. 4.5 below. By applying the anti-involution σ we convert the formulae in Theorem 4.3 as follows.

Theorem 4.4 *Let κ be an odd q-integer. Then we have, for $m \geqslant 1$,*

$$t_{\rm ev}^{(2m)} = \sum_{b=0}^{2m}\sum_{a=0}^{b}\sum_{c\geqslant 0} (-1)^c q^{c+b(2m-b-2c)+a(b-a)}\mathfrak{p}^{(2m-b-2c)}(\kappa) \cdot F^{(b-a)}K^{b-2m+2c}\begin{bmatrix} h;m-c \\ c \end{bmatrix}\check{E}^{(a)} \tag{4.10}$$

$$t_{\rm ev}^{(2m+1)} = \sum_{b=0}^{2m+1}\sum_{a=0}^{b}\sum_{c\geqslant 0} (-1)^c q^{3c+b(2m-b-2c+1)+a(b-a)}\mathfrak{p}^{(2m-b-2c+1)}(\kappa) \cdot F^{(b-a)}K^{b-2m+2c-1}\begin{bmatrix} h;m-c \\ c \end{bmatrix}\check{E}^{(a)}. \tag{4.11}$$

Proof By Lemma 2.1, the anti-involution σ fixes $\kappa, K, \check{E}^{(a)}, F^{(a)}$, and $\mathrm{t}_{\rm ev}^{(n)}$ and sends $\begin{bmatrix} h;1-m \\ c \end{bmatrix} \mapsto (-1)^c q^{2c(c+1)}\begin{bmatrix} h;m-c \\ c \end{bmatrix}$, $q \mapsto q^{-1}$. The formulae (4.10)–(4.11) now follow from (4.8)–(4.9) in Theorem 4.3.

The following corollary is immediate from (2.21), Proposition 4.2, and Theorem 4.3.

Corollary 4.5 *We have $t_{\rm ev}^{(n)} \in {}_{\mathcal{A}}\mathbf{U}^{\imath}_{\rm ev}$, for all n.* □

Remark 4.6 Note $\mathfrak{p}_n(1) = 0$ for $n \geqslant 2$, and $\mathfrak{p}_0(1) = \mathfrak{p}_1(1) = 1$. The formulae in Theorems 4.3 and 4.4 in the special case for $\kappa = 1$ recover the formulae in [5, Theorem 4.1, Proposition 4.3]. □

Example 4.3.1 The formulae of $t_{\rm ev}^{(n)}$, for $1 \leqslant n \leqslant 3$, in Theorem 4.3 read as follows.

$$t_{\rm ev}^{(1)} = F + \check{E} + K^{-1},$$

$$t_{\rm ev}^{(2)} = b^{(2)} + q^3[h;0] + \kappa(q^{-1}K^{-1}F + q^{-1}\check{E}K^{-1}) + \frac{\kappa^2-1}{[2]}K^{-2},$$

$$t_{\rm ev}^{(3)} = b^{(3)} + q\check{E}[h;0] + q[h;0]F$$
$$+\kappa(q^{-2}\check{E}^{(2)}K^{-1} + q^{-3}\check{E}K^{-1}F + q^{-2}K^{-1}F^{(2)} + q[h;0]K^{-1})$$
$$+q^{-2}\frac{\kappa^2-1}{[2]}(\check{E}K^{-2} + K^{-2}F) + \frac{\kappa^3-\kappa}{[3]!}K^{-3}.$$

4.4 The ı-Canonical Basis for $\dot{\mathbf{U}}_{\rm ev}$ with Odd κ

Recall from Sect. 2.6 the $\imath$-canonical basis on simple $\mathbf{U}$-modules $L(\mu)$, for $\mu \in \mathbb{N}$.

Theorem 4.7

(a) *Let $n \in \mathbb{N}$. For each integer $\lambda \gg n$, the element $t_{\rm ev}^{(n)} v_{2\lambda}^+$ is an $\imath$-canonical basis element for $L(2\lambda)$.*
(b) *The set $\{t_{\rm ev}^{(n)} \mid n \in \mathbb{N}\}$ forms the $\imath$-canonical basis for $\mathbf{U}^\imath$ (and an $\mathcal{A}$-basis in ${}_{\mathcal{A}}\mathbf{U}^\imath_{\rm ev}$).* □

Proof Let $\lambda, m \in \mathbb{N}$. Recall $\begin{Bmatrix} m \\ c \end{Bmatrix}$ from (2.20) and $\begin{bmatrix} h; a \\ c \end{bmatrix}$ from (2.3). It follows by a direct computation using Theorem 4.4 and (2.21) that

$$t_{\rm ev}^{(2m)} v_{2\lambda}^+ = \sum_{b=0}^{2m}\sum_{c\geqslant 0}(-1)^c q^{c+b(2m-b-2c)}\mathfrak{p}^{(2m-b-2c)}(\kappa) \tag{4.12}$$
$$\cdot F^{(b)}K^{b-2m+2c}\begin{bmatrix} h; m-c \\ c \end{bmatrix} v_{2\lambda}^+$$

$$= \sum_{b=0}^{2m}\sum_{c\geqslant 0} q^{-2c^2-c+(b-2\lambda)(2m-b-2c)}\mathfrak{p}^{(2m-b-2c)}(\kappa) \tag{4.13}$$

$$\cdot \begin{Bmatrix} m-\lambda-c \\ c \end{Bmatrix} F^{(b)} v_{2\lambda}^+. \tag{4.14}$$

Similarly using Theorem 4.4 we have

$$t_{\rm ev}^{(2m+1)} v_{2\lambda}^+ = \sum_{b=0}^{2m+1}\sum_{c\geqslant 0}(-1)^c q^{3c+b(2m-b-2c+1)}\mathfrak{p}^{(2m-b-2c+1)}(\kappa)$$

$$\cdot F^{(b)} K^{b-2m+2c-1} \begin{bmatrix} h; m-c \\ c \end{bmatrix} v_{2\lambda}^{+}$$

$$= \sum_{b=0}^{2m+1} \sum_{c \geqslant 0} q^{-2c^2+c+(b-2\lambda)(2m-b-2c+1)} \mathfrak{p}^{(2m-b-2c+1)}(\kappa)$$

$$\cdot \begin{Bmatrix} m-\lambda-c \\ c \end{Bmatrix} F^{(b)} v_{2\lambda}^{+}. \tag{4.15}$$

By a similar argument for (2.26), using (4.14)–(4.15) we obtain

$$t_{\rm ev}^{(n)} v_{2\lambda}^{+} \in F^{(n)} v_{2\lambda}^{+} + \sum_{b<n} q^{-1}\mathbb{Z}[q^{-1}] F^{(b)} v_{2\lambda}^{+}, \qquad \text{for } \lambda \gg n.$$

The second statement follows now from the definition of the $\imath$-canonical basis for $\mathbf{U}^\imath$ using the projective system $\{L(2\lambda)\}_{\lambda \geqslant 0}$; cf. [3, §6].

4.5 *Proof of Theorem 4.3*

We prove the formulae for $t_{\rm ev}^{(n)}$ by induction on n, in two steps (1)–(2) below. The base cases when $n=1,2$ are clear.

(1) We shall prove the formula (4.9) for $t_{\rm ev}^{(2m+1)}$, assuming the formula (4.8) for $t_{\rm ev}^{(2m)}$.

Recall $[2m+1] t_{\rm ev}^{(2m+1)} = t \cdot t_{\rm ev}^{(2m)}$, and $t = F + \check{E} + \kappa K^{-1}$. Let us compute

$$I = \check{E} t_{\rm ev}^{(2m)}, \qquad II = F t_{\rm ev}^{(2m)}, \qquad III = \kappa K^{-1} t_{\rm ev}^{(2m)}.$$

First we have

$$I = \sum_{b=0}^{2m} \sum_{a=0}^{b} \sum_{c \geqslant 0} q^{\binom{2c}{2}+2c-b(2m-b-2c)-a(b-a)} [a+1] \mathfrak{p}^{(2m-b-2c)}(\kappa)$$

$$\cdot \check{E}^{(a+1)} \begin{bmatrix} h; 1-m \\ c \end{bmatrix} K^{b-2m+2c} F^{(b-a)},$$

$$= \sum_{b=0}^{2m+1} \sum_{a=0}^{b} \sum_{c \geqslant 0} q^{\binom{2c}{2}+2c-(b-1)(2m-b-2c+1)-(a-1)(b-a)} [a] \mathfrak{p}^{(2m-b-2c+1)}(\kappa)$$

$$\cdot \check{E}^{(a)} \begin{bmatrix} h; 1-m \\ c \end{bmatrix} K^{b-2m+2c-1} F^{(b-a)},$$

where the last equation is obtained by shifting indices $a \to a-1$, $b \to b-1$. Using (2.6) we have

$$
\begin{aligned}
II &= \sum_{b=0}^{2m}\sum_{a=0}^{b}\sum_{c\geqslant 0} q^{\binom{2c}{2}+2c-b(2m-b-2c)-a(b-a)}\mathfrak{p}^{(2m-b-2c)}(\kappa) \\
&\quad \cdot \left(q^{-2a}\check{E}^{(a)}F + \check{E}^{(a-1)}\frac{q^{3-3a}K^{-2}-q^{1-a}}{q^2-1}\right)\begin{bmatrix} h;1-m \\ c \end{bmatrix} K^{b-2m+2c}F^{(b-a)} \\
&= II^1 + II^2,
\end{aligned}
$$

where II^1 and II^2 are the two natural summands associated with the plus sign. By shifting index $b \to b-1$, we further have

$$
\begin{aligned}
II^1 = \sum_{b=0}^{2m+1}\sum_{a=0}^{b}\sum_{c\geqslant 0} & q^{\binom{2c}{2}+2c-(b+1)(2m-b-2c+1)-a(b-a+1)}[b-a] \\
& \cdot \mathfrak{p}^{(2m-b-2c+1)}(\kappa)\check{E}^{(a)}\begin{bmatrix} h;-m \\ c \end{bmatrix} K^{b-2m+2c-1}F^{(b-a)}.
\end{aligned}
$$

By shifting the indices $a \to a+1$, $b \to b+1$ and $c \to c-1$, we further have

$$
\begin{aligned}
II^2 = \sum_{b=0}^{2m+1}\sum_{a=0}^{b}\sum_{c\geqslant 0} & q^{\binom{2c-2}{2}+2c-(b+1)(2m-b-2c+1)-(a+1)(b-a)-2} \\
& \cdot \mathfrak{p}^{(2m-b-2c+1)}(\kappa)\check{E}^{(a)}\frac{q^{-3a}K^{-2}-q^{-a}}{q^2-1}\begin{bmatrix} h;1-m \\ c-1 \end{bmatrix} K^{b-2m+2c-1}F^{(b-a)}.
\end{aligned}
$$

Using (4.5) we also compute

$$
\begin{aligned}
III &= \sum_{b=0}^{2m}\sum_{a=0}^{b}\sum_{c\geqslant 0} q^{\binom{2c}{2}+2c-b(2m-b-2c)-a(b-a)-2a}\kappa \cdot \mathfrak{p}^{(2m-b-2c)}(\kappa) \\
&\qquad \cdot \check{E}^{(a)}\begin{bmatrix} h;1-m \\ c \end{bmatrix} K^{b-2m+2c-1}F^{(b-a)} \\
&= \sum_{b=0}^{2m+1}\sum_{a=0}^{b}\sum_{c\geqslant 0} q^{\binom{2c}{2}+2c-b(2m-b-2c)-a(b-a)-2a}[2m-b-2c+1] \\
&\qquad \cdot \mathfrak{p}^{(2m-b-2c+1)}(\kappa)\check{E}^{(a)}\begin{bmatrix} h;1-m \\ c \end{bmatrix} K^{b-2m+2c-1}F^{(b-a)}
\end{aligned}
$$

$$- \sum_{b=0}^{2m+1} \sum_{a=0}^{b} \sum_{c \geqslant 0} q^{\binom{2c-2}{2}+2c-(b+2)(2m-b-2c+2)-a(b-a)-2a} [2m-b-2c]$$
$$\cdot \mathfrak{p}^{(2m-b-2c+1)}(\kappa) \check{E}^{(a)} \begin{bmatrix} h; 1-m \\ c-1 \end{bmatrix} K^{b-2m+2c-3} F^{(b-a)}.$$

(Note that we have shifted the index $c \to c-1$ in the last summand above.)
Collecting the formulae for I, II^1, II^2, and III gives us

$$t \cdot t_{\text{ev}}^{(2m)} = \sum_{\substack{0 \leqslant a \leqslant b \leqslant 2m+1 \\ c \geqslant 0}} \mathfrak{p}^{(2m-b-2c+1)}(\kappa) \check{E}^{(a)} \mathcal{H}_{a,b,c} K^{b-2m+2c-1} F^{(b-a)},$$

where

$$\mathcal{H}_{a,b,c} := q^{\binom{2c}{2}+2c-(b-1)(2m-b-2c+1)-(a-1)(b-a)} [a] \begin{bmatrix} h; 1-m \\ c \end{bmatrix}$$
$$+ q^{\binom{2c}{2}+2c-(b+1)(2m-b-2c+1)-a(b-a+1)} [b-a] \begin{bmatrix} h; -m \\ c \end{bmatrix}$$
$$+ q^{\binom{2c-2}{2}+2c-(b+1)(2m-b-2c+1)-(a+1)(b-a)-2} \frac{q^{-3a} K^{-2} - q^{-a}}{q^2-1} \begin{bmatrix} h; 1-m \\ c-1 \end{bmatrix}$$
$$+ q^{\binom{2c}{2}+2c-b(2m-b-2c)-a(b-a)-2a} [2m-b-2c+1] \begin{bmatrix} h; 1-m \\ c \end{bmatrix}$$
$$+ q^{\binom{2c-2}{2}+2c-(b+2)(2m-b-2c+2)-a(b-a)-2a} [2m-b-2c] \begin{bmatrix} h; 1-m \\ c-1 \end{bmatrix} K^{-2}.$$

Recall $[2m+1] \mathfrak{t}_{\text{odd}}^{(2m+1)} = t \cdot t_{\text{ev}}^{(2m)}$. To prove the formula (4.9) for $\mathfrak{t}_{\text{odd}}^{(2m+1)}$, by the PBW basis theorem and the inductive assumption it suffices to prove the following identity, for all a, b, c:

$$\mathcal{H}_{a,b,c} = q^{\binom{2c}{2}-b(2m-b-2c+1)-a(b-a)} [2m+1] \begin{bmatrix} h; 1-m \\ c \end{bmatrix}. \tag{4.16}$$

Thanks to $q^{2m-a+1}[a] - [2m+1] = q^{-a}[a-2m-1]$, we can combine the RHS(4.16) with the first summand of LHS(4.16). Hence, after canceling $q^{\binom{2c}{2}-b(2m-b-2c+1)-a(b-a)-a}$ on both sides, we see that (4.16) is equivalent to the following identity, for all a, b, c:

$$\mathcal{A} + \mathcal{B} + \mathcal{C} + \mathcal{D}_1 + \mathcal{D}_2 = 0, \tag{4.17}$$

where

$$\mathcal{A} = [a - 2m - 1]\begin{bmatrix} h; 1-m \\ c \end{bmatrix},$$

$$\mathcal{B} = q^{4c-2m+b-1}[b-a]\begin{bmatrix} h; -m \\ c \end{bmatrix},$$

$$\mathcal{C} = q^{2a-2m}\frac{q^{-3a}K^{-2} - q^{-a}}{q^2-1}\begin{bmatrix} h; 1-m \\ c-1 \end{bmatrix},$$

$$\mathcal{D}_1 = q^{2c+b-a}[2m-b-2c+1]\begin{bmatrix} h; 1-m \\ c \end{bmatrix},$$

$$\mathcal{D}_2 = -q^{2c-4m+b-a-1}[2m-b-2c]\begin{bmatrix} h; 1-m \\ c-1 \end{bmatrix}K^{-2}.$$

Let us prove the identity (4.17). Using (2.5), we can write $\mathcal{B} = \mathcal{B}_1 + \mathcal{B}_2$, where

$$\mathcal{B}_1 = q^{4c-2m+b-1}[b-a]\begin{bmatrix} h; -m \\ c \end{bmatrix}, \quad \mathcal{B}_2 = -q^{4c-6m+b-1}[b-a]\begin{bmatrix} h; -m \\ c \end{bmatrix}.$$

Noting that

$$q^{2a-2m}\frac{q^{-3a}K^{-2} - q^{-a}}{q^2-1} = q^{2m-4c-a}\frac{(q^{4c-4m}K^{-2} - 1)}{q^2-1} + q^{-2c-1}[2m-2c-a],$$

we rewrite $\mathcal{C} = \mathcal{C}_1 + \mathcal{C}_2$, where

$$\mathcal{C}_1 = q^{2m-2c-a-1}[2c]\begin{bmatrix} h; 1-m \\ c \end{bmatrix}, \quad \mathcal{C}_2 = q^{-2c-1}[2m-2c-a]\begin{bmatrix} h; 1-m \\ c-1 \end{bmatrix}.$$

A direct computation gives us

$$\mathcal{A} + \mathcal{B}_1 + \mathcal{C}_1 + \mathcal{D}_1 = (1-q^{-2})[2c][2m-2c-a]\begin{bmatrix} h; 1-m \\ c \end{bmatrix},$$

$$\begin{aligned}\mathcal{C}_2 + (\mathcal{B}_2 + \mathcal{D}_2) &= \mathcal{C}_2 - q^{2c-4m-1}[2m-2c-a]\begin{bmatrix} h; 1-m \\ c-1 \end{bmatrix}K^{-2} \\ &= -(1-q^{-2})[2c][2m-2c-a]\begin{bmatrix} h; 1-m \\ c \end{bmatrix}.\end{aligned}$$

Summing up these two equations, we have $\mathcal{A}+\mathcal{B}+\mathcal{C}+\mathcal{D}_1+\mathcal{D}_2 = 0$, whence (4.17), completing Step (1).

(2) Assuming the formulae for $t_{\mathrm{ev}}^{(n)}$ with $n \leqslant 2m+1$, we shall now prove the following formula (4.18) for $t_{\mathrm{ev}}^{(2m+2)}$ (obtained with m replaced by $m+1$ in (4.8)):

$$t_{\mathrm{ev}}^{(2m+2)} = \sum_{b=0}^{2m+2} \sum_{a=0}^{b} \sum_{c\geqslant 0} q^{\binom{2c}{2}+2c-b(2m-b-2c+2)-a(b-a)} \mathfrak{p}^{(2m-b-2c+2)}(\kappa) \cdot \check{E}^{(a)} \begin{bmatrix} h; -m \\ c \end{bmatrix} K^{b-2m+2c-2} F^{(b-a)}. \tag{4.18}$$

The proof is based on the recursion $t \cdot t_{\mathrm{ev}}^{(2m+1)} = [2m+2] t_{\mathrm{ev}}^{(2m+2)} + [2m+1] t_{\mathrm{ev}}^{(2m)}$. Recall $t = F + \check{E} + \kappa K^{-1}$ and $t_{\mathrm{ev}}^{(2m+1)}$ from (4.9). We shall compute

$$\mathtt{I} = \check{E} t_{\mathrm{ev}}^{(2m+1)}, \qquad \mathtt{II} = F t_{\mathrm{ev}}^{(2m+1)}, \qquad \mathtt{III} = \kappa K^{-1} t_{\mathrm{ev}}^{(2m+1)},$$

respectively. First by shifting indices $a \to a-1$ and $b \to b-1$, we have

$$\begin{aligned}
\mathtt{I} &= \sum_{b=0}^{2m+1} \sum_{a=0}^{b} \sum_{c\geqslant 0} q^{\binom{2c}{2}-b(2m-b-2c+1)-a(b-a)} [a+1] \mathfrak{p}^{(2m-b-2c+1)}(\kappa) \\
&\qquad \cdot \check{E}^{(a+1)} \begin{bmatrix} h; 1-m \\ c \end{bmatrix} K^{b-2m+2c-1} F^{(b-a)} \\
&= \sum_{b=0}^{2m+2} \sum_{a=0}^{b} \sum_{c\geqslant 0} q^{\binom{2c}{2}-(b-1)(2m-b-2c+2)-(a-1)(b-a)} [a] \mathfrak{p}^{(2m-b-2c+2)}(\kappa) \\
&\qquad \cdot \check{E}^{(a)} \begin{bmatrix} h; 1-m \\ c \end{bmatrix} K^{b-2m+2c-2} F^{(b-a)}.
\end{aligned}$$

Using (2.6) we also have

$$\begin{aligned}
\mathtt{II} &= \sum_{b=0}^{2m+1} \sum_{a=0}^{b} \sum_{c\geqslant 0} q^{\binom{2c}{2}-b(2m-b-2c+1)-a(b-a)} \mathfrak{p}^{(2m-b-2c+1)}(\kappa) \\
&\quad \cdot \Big(q^{-2a} \check{E}^{(a)} F + \check{E}^{(a-1)} \frac{q^{3-3a} K^{-2} - q^{1-a}}{q^2-1} \Big) \begin{bmatrix} h; 1-m \\ c \end{bmatrix} K^{b-2m+2c-1} F^{(b-a)} \\
&= \mathtt{II}_1 + \mathtt{II}_2,
\end{aligned}$$

where $\mathtt{II}_1$ and $\mathtt{II}_2$ are the two natural summands associated with the plus sign. By shifting the index $b \to b-1$, we further have

$$\mathtt{II}_1 = \sum_{b=0}^{2m+2} \sum_{a=0}^{b} \sum_{c \geqslant 0} q^{\binom{2c}{2} - (b+1)(2m-b-2c+2) - a(b-a-1) - 2a} [b-a]$$

$$\cdot \mathfrak{p}^{(2m-b-2c+2)}(\kappa) \check{E}^{(a)} \begin{bmatrix} h; -m \\ c \end{bmatrix} K^{b-2m+2c-2} F^{(b-a)}.$$

By shifting the indices $a \to a+1$, $b \to b+1$, and $c \to c-1$, we further have

$$\mathtt{II}_2 = \sum_{b=0}^{2m+2} \sum_{a=0}^{b} \sum_{c \geqslant 0} q^{\binom{2c-2}{2} - (b+1)(2m-b-2c+2) - (a+1)(b-a)} \mathfrak{p}^{(2m-b-2c+2)}(\kappa)$$

$$\cdot \check{E}^{(a)} \frac{q^{-3a} K^{-2} - q^{-a}}{q^2 - 1} \begin{bmatrix} h; 1-m \\ c-1 \end{bmatrix} K^{b-2m+2c-2} F^{(b-a)}.$$

Using (4.5) we also compute

$$\mathtt{III} = \sum_{b=0}^{2m+1} \sum_{a=0}^{b} \sum_{c \geqslant 0} q^{\binom{2c}{2} - b(2m-b-2c+1) - a(b-a) - 2a} \kappa \cdot \mathfrak{p}^{(2m-b-2c+1)}(\kappa)$$

$$\cdot \check{E}^{(a)} \begin{bmatrix} h; 1-m \\ c \end{bmatrix} K^{b-2m+2c-2} F^{(b-a)}$$

$$= \sum_{b=0}^{2m+2} \sum_{a=0}^{b} \sum_{c \geqslant 0} q^{\binom{2c}{2} - b(2m-b-2c+1) - a(b-a) - 2a} [2m-b-2c+2]$$

$$\cdot \mathfrak{p}^{(2m-b-2c+2)}(\kappa) \check{E}^{(a)} \begin{bmatrix} h; 1-m \\ c \end{bmatrix} K^{b-2m+2c-2} F^{(b-a)}$$

$$- \sum_{b=0}^{2m+2} \sum_{a=0}^{b} \sum_{c \geqslant 0} q^{\binom{2c-2}{2} - (b+2)(2m-b-2c+3) - a(b-a) - 2a + 2} [2m-b-2c+1]$$

$$\cdot \mathfrak{p}^{(2m-b-2c+2)}(\kappa) \check{E}^{(a)} \begin{bmatrix} h; 1-m \\ c-1 \end{bmatrix} K^{b-2m+2c-4} F^{(b-a)}.$$

(Note that we have shifted the index $c \to c-1$ in the last summand above.)

Collecting the formulae for $\mathtt{I}$, $\mathtt{II}_1$, $\mathtt{II}_2$, and $\mathtt{III}$, we obtain

$$t \cdot t_{\mathrm{ev}}^{(2m+1)} = \sum_{\substack{0 \leqslant a \leqslant b \leqslant 2m+2 \\ c \geqslant 0}} \mathfrak{p}^{(2m-b-2c+2)}(\kappa) \check{E}^{(a)} \mathcal{L}_{a,b,c} K^{b-2m+2c-2} F^{(b-a)},$$

where

$$\mathcal{L}_{a,b,c} := q^{\binom{2c}{2}-(b-1)(2m-b-2c+2)-(a-1)(b-a)}[a]\begin{bmatrix} h; 1-m \\ c \end{bmatrix}$$
$$+ q^{\binom{2c}{2}-(b+1)(2m-b-2c+2)-a(b-a-1)-2a}[b-a]\begin{bmatrix} h; -m \\ c \end{bmatrix}$$
$$+ q^{\binom{2c-2}{2}-(b+1)(2m-b-2c+2)-(a+1)(b-a)}\frac{q^{-3a}K^{-2}-q^{-a}}{q^2-1}\begin{bmatrix} h; 1-m \\ c-1 \end{bmatrix}$$
$$+ q^{\binom{2c}{2}-b(2m-b-2c+1)-a(b-a)-2a}[2m-b-2c+2]\begin{bmatrix} h; 1-m \\ c \end{bmatrix}$$
$$- q^{\binom{2c-2}{2}-(b+2)(2m-b-2c+3)-a(b-a)-2a+2}[2m-b-2c+1]\begin{bmatrix} h; 1-m \\ c-1 \end{bmatrix}K^{-2}.$$

On the other hand, using (4.8) (with an index shift $c \to c-1$) and (4.18) we write

$$[2m+2]t_{\mathrm{ev}}^{(2m+2)} + [2m+1]t_{\mathrm{ev}}^{(2m)}$$
$$= \sum_{0\leqslant a\leqslant b\leqslant 2m+2}\sum_{c\geqslant 0}\mathfrak{p}^{(2m-b-2c+2)}(\kappa)\check{E}^{(a)}\mathcal{R}_{a,b,c}K^{b-2m+2c-2}F^{(b-a)},$$

where

$$\mathcal{R}_{a,b,c} := q^{\binom{2c}{2}+2c-b(2m-b-2c+2)-a(b-a)}[2m+2]\begin{bmatrix} h; -m \\ c \end{bmatrix}$$
$$+ q^{\binom{2c-2}{2}+2c-b(2m-b-2c+2)-a(b-a)-2}[2m+1]\begin{bmatrix} h; 1-m \\ c-1 \end{bmatrix}.$$

To prove the formula (3.13) for $t_{\mathrm{ev}}^{(2m+2)}$, it suffices to show that, for all a, b, c,

$$\mathcal{L}_{a,b,c} = \mathcal{R}_{a,b,c}. \tag{4.19}$$

Canceling the q-powers $q^{\binom{2c}{2}-(b-1)(2m-b-2c+2)-(a-1)(b-a)}$ on both sides, we see that the identity (4.19) is equivalent to the following identity, for all a, b, c:

$$[a]\begin{bmatrix} h; 1-m \\ c \end{bmatrix} + q^{4c-4m+b-4}[b-a]\begin{bmatrix} h; -m \\ c \end{bmatrix}$$
$$+ q^{2a-4m-1}\frac{q^{-3a}K^{-2}-q^{-a}}{q^2-1}\begin{bmatrix} h; 1-m \\ c-1 \end{bmatrix}$$

$$+ q^{2c-2m+b-a-2}[2m-b-2c+2]\begin{bmatrix} h; 1-m \\ c \end{bmatrix}$$

$$- q^{2c-6m+b-a-3}[2m-b-2c+1]\begin{bmatrix} h; 1-m \\ c-1 \end{bmatrix} K^{-2}$$

$$= q^{4c-2m+a-2}[2m+2]\begin{bmatrix} h; -m \\ c \end{bmatrix} + q^{a-2m-1}[2m+1]\begin{bmatrix} h; 1-m \\ c-1 \end{bmatrix}. \tag{4.20}$$

By combining the second summand of LHS with the first summand of RHS as well as combining the third summand of LHS with the second summand of RHS, the identity (4.20) is reduced to the following equivalent identity, for all a, b, c:

$$\mathcal{W} + \mathcal{X} + \mathcal{Y} + \mathcal{Z}_1 + \mathcal{Z}_2 = 0, \tag{4.21}$$

where

$$\mathcal{W} = [a]\begin{bmatrix} h; 1-m \\ c \end{bmatrix},$$

$$\mathcal{X} = q^{4c-2m+b-2}[b-a-2m-2]\begin{bmatrix} h; -m \\ c \end{bmatrix},$$

$$\mathcal{Y} = \frac{q^{-a-4m-1}K^{-2} - q^{a+1}}{q^2-1}\begin{bmatrix} h; 1-m \\ c-1 \end{bmatrix},$$

$$\mathcal{Z}_1 = q^{2c-2m+b-a-2}[2m-b-2c+2]\begin{bmatrix} h; 1-m \\ c \end{bmatrix},$$

$$\mathcal{Z}_2 = -q^{2c-6m+b-a-3}[2m-b-2c+1]\begin{bmatrix} h; 1-m \\ c-1 \end{bmatrix} K^{-2}.$$

Let us finally prove the identity (4.21). Using (2.5), we can write $\mathcal{X} = \mathcal{X}_1 + \mathcal{X}_2$, where

$$\mathcal{X}_1 = q^{4c-2m+b-2}[b-a-2m-2]\begin{bmatrix} h; 1-m \\ c \end{bmatrix},$$

$$\mathcal{X}_2 = -q^{4c-6m+b-2}[b-a-2m-2]\begin{bmatrix} h; 1-m \\ c-1 \end{bmatrix} K^{-2}.$$

Noting that

$$\frac{q^{-a-4m-1}K^{-2} - q^{a+1}}{q^2-1} = q^{-4c-a-1}\frac{(q^{4c-4m}K^{-2}-1)}{q^2-1} + q^{-2c-1}[-2c-a-1],$$

we rewrite $\mathcal{Y} = \mathcal{Y}_1 + \mathcal{Y}_2$, where

$$\mathcal{Y}_1 = q^{-2c-a-2}[2c]\begin{bmatrix} h;1-m \\ c \end{bmatrix}, \qquad \mathcal{Y}_2 = -q^{-2c-1}[2c+a+1]\begin{bmatrix} h;1-m \\ c-1 \end{bmatrix}.$$

A direct computation shows that

$$\mathcal{W} + \mathcal{X}_1 + \mathcal{Y}_1 + \mathcal{Z}_1 = -(1-q^{-2})[2c+a+1][2c]\begin{bmatrix} h;1-m \\ c \end{bmatrix},$$

$$\begin{aligned}(\mathcal{X}_2 + \mathcal{Z}_2) + \mathcal{Y}_2 &= q^{2c-4m-1}[2c+a+1]\begin{bmatrix} h;1-m \\ c-1 \end{bmatrix} K^{-2} + \mathcal{Y}_2 \\ &= (1-q^{-2})[2c+a+1][2c]\begin{bmatrix} h;1-m \\ c \end{bmatrix}.\end{aligned}$$

Summing up these two equations, we obtain $\mathcal{W} + \mathcal{X} + \mathcal{Y} + \mathcal{Z}_1 + \mathcal{Z}_2 = 0$, whence (4.21), completing Step (2).

The proof of Theorem 4.3 is completed. □

5 The $\imath$-Divided Powers $t_{\mathrm{odd}}^{(n)}$ for Odd Weights and Odd κ

In this Sect. 5 we always take κ to be an odd q-integer, i.e.,

$$\kappa = [2\ell - 1], \qquad \text{for } \ell \in \mathbb{Z}.$$

5.1 *Definition of $t_{\mathrm{odd}}^{(n)}$ for Odd κ*

Definition 5.1.1 Set $t_{\mathrm{odd}}^{(1)} = t = F + \check{E} + \kappa K^{-1}$. The divided powers $t_{\mathrm{odd}}^{(n)}$, for $n \geqslant 1$, are defined by the recursive relations:

$$\begin{aligned} t \cdot t_{\mathrm{odd}}^{(2a-1)} &= [2a] t_{\mathrm{odd}}^{(2a)}, \\ t \cdot t_{\mathrm{odd}}^{(2a)} &= [2a+1] t_{\mathrm{odd}}^{(2a+1)} + [2a] t_{\mathrm{odd}}^{(2a-1)}, \qquad \text{for } a \geqslant 1. \end{aligned} \tag{5.1}$$

Equivalently, we have the following closed formula for $t_{\mathrm{odd}}^{(n)}$:

$$t_{\rm odd}^{(n)} = \begin{cases} \frac{t}{[2a]!}(t-[-2a+2])(t-[-2a+4])\cdots(t-[2a-4])(t-[2a-2]), & \text{if } n=2a, \\ \frac{1}{[2a+1]!}(t-[-2a])(t-[-2a+2])\cdots(t-[2a-2])(t-[2a]), & \text{if } n=2a+1. \end{cases} \tag{5.2}$$

Note the formulae for $t_{\rm odd}^{(n)}$ with odd κ are formally identical to the formulae for $\mathrm{t}_{\rm ev}^{(n)}$ with even κ.

5.2 Formulae for $t_{\rm odd}^{(n)}$ with Odd κ

Recall the polynomials $\mathfrak{p}^{(n)}$, for $n \geqslant 0$, from Sect. 4.2.

Theorem 5.1 *Assume κ is an odd q-integer. Then we have, for $m \geqslant 1$,*

$$t_{\rm odd}^{(2m)} = \sum_{b=0}^{2m}\sum_{a=0}^{b}\sum_{c\geqslant 0} q^{\binom{2c}{2}-2c-b(2m-b-2c)-a(b-a)}\mathfrak{p}^{(2m-b-2c)}(\kappa) \cdot \check{E}^{(a)} \left[\!\!\left[\begin{matrix} h;2-m \\ c\end{matrix}\right]\!\!\right] K^{b-2m+2c}F^{(b-a)}, \tag{5.3}$$

$$t_{\rm odd}^{(2m-1)} = \sum_{b=0}^{2m-1}\sum_{a=0}^{b}\sum_{c\geqslant 0} q^{\binom{2c}{2}-b(2m-b-2c-1)-a(b-a)}\mathfrak{p}^{(2m-b-2c-1)}(\kappa) \cdot \check{E}^{(a)} \left[\!\!\left[\begin{matrix} h;2-m \\ c\end{matrix}\right]\!\!\right] K^{b-2m+2c+1}F^{(b-a)}. \tag{5.4}$$

The proof of Theorem 5.1 will be given in Sect. 5.4 below. By applying the anti-involution σ we convert the formulae in Theorem 5.1 as follows.

Theorem 5.2 *Assume κ is an odd q-integer. Then we have, for $m \geqslant 1$,*

$$t_{\rm odd}^{(2m)} = \sum_{b=0}^{2m}\sum_{a=0}^{b}\sum_{c\geqslant 0} (-1)^c q^{c+b(2m-b-2c)+a(b-a)}\mathfrak{p}^{(2m-b-2c)}(\kappa) \cdot F^{(b-a)}K^{b-2m+2c}\left[\!\!\left[\begin{matrix} h;m-c \\ c\end{matrix}\right]\!\!\right] \check{E}^{(a)}, \tag{5.5}$$

$$t_{\rm odd}^{(2m-1)} = \sum_{b=0}^{2m-1}\sum_{a=0}^{b}\sum_{c\geqslant 0} (-1)^c q^{-c+b(2m-b-2c-1)+a(b-a)}\mathfrak{p}^{(2m-b-2c-1)}(\kappa) \tag{5.6}$$

$$\cdot F^{(b-a)} K^{b-2m+2c+1} \begin{bmatrix} h; m-c \\ c \end{bmatrix} \check{E}^{(a)}.$$

Proof Recall from Lemma 3.1 that the anti-involution σ on $\mathbf{U}$ fixes $F, \check{E}, K, t_{\text{odd}}^{(n)}, \kappa$ while sending $q \mapsto q^{-1}$, $\begin{bmatrix} h; 2-m \\ c \end{bmatrix} \mapsto (-1)^c q^{2c(c-1)} \begin{bmatrix} h; m-c \\ c \end{bmatrix}$. The formulae (5.5)–(5.6) now follow by applying σ to the formulae (5.3)–(5.4) in Theorem 5.1.

The following corollary is immediate from (3.8), Proposition 4.2, and Theorem 5.1.

Corollary 5.3 *We have* $t_{\text{odd}}^{(n)} \in {}_{\mathcal{A}}\mathbf{U}^{\imath}_{\text{odd}}$*, for all* n. □

Remark 5.4 Note $\mathfrak{p}_n(1) = 0$ for $n \geqslant 2$, and $\mathfrak{p}_0(1) = \mathfrak{p}_1(1) = 1$. The formulae in Theorems 5.1 and 5.2 in the special case for $\kappa = 1$ recover the formulae in [5, Theorem 5.1, Proposition 5.3]. □

Example 5.2.1 The formulae of $t_{\text{odd}}^{(n)}$, for $1 \leqslant n \leqslant 3$, in Theorem 5.1 read as follows.

$$\begin{aligned}
t_{\text{odd}}^{(1)} &= F + \check{E} + \kappa K^{-1}, \\
t_{\text{odd}}^{(2)} &= \check{E}^{(2)} + q^{-1}\check{E}F + F^{(2)} + q^{-1}[[h;1]] + \kappa(q^{-1}K^{-1}F + q^{-1}\check{E}K^{-1}) \\
&\quad + \frac{\kappa^2-1}{[2]}K^{-2}, \\
t_{\text{odd}}^{(3)} &= b^{(3)} + q[[h;0]]F + q\check{E}[[h;0]] \\
&\quad + \big(q^{-2}\check{E}^{(2)}K^{-1} + q^{-3}\check{E}K^{-1}F + q^{-2}K^{-1}F^{(2)} + q[[h;0]]K^{-1}\big)\kappa \\
&\quad + \frac{(\kappa^2-1)}{[2]}(q^{-2}\check{E}K^{-2} + q^{-2}K^{-2}F) + \frac{\kappa^3-\kappa}{[3]!}K^{-3}.
\end{aligned}$$

5.3 The $\imath$-Canonical Basis for $\dot{\mathbf{U}}_{odd}$ with Odd κ

Recall from Sect. 2.6 the $\imath$-canonical basis on simple $\mathbf{U}$-modules $L(\mu)$, for $\mu \in \mathbb{N}$.

Theorem 5.5

(a) *Let* $n \in \mathbb{N}$*. For each integer* $\lambda \gg n$*, the element* $t_{\text{odd}}^{(n)} v_{2\lambda+1}^{+}$ *is an* $\imath$*-canonical basis element for* $L(2\lambda+1)$*.*
(b) *The set* $\{t_{\text{odd}}^{(n)} \mid n \in \mathbb{N}\}$ *forms the* $\imath$*-canonical basis for* $\mathbf{U}^{\imath}$ *(and an* $\mathcal{A}$*-basis in* ${}_{\mathcal{A}}\mathbf{U}^{\imath}_{\text{odd}}$*).* □

Proof Let $\lambda, m \in \mathbb{N}$. Recall $\begin{Bmatrix} m \\ c \end{Bmatrix}$ from (2.20). It follows by a direct computation using Theorem 5.2 and (3.8) that

$$\begin{aligned} & t_{\text{odd}}^{(2m)} v_{2\lambda+1}^{+} \\ & = \sum_{b=0}^{2m} \sum_{c \geqslant 0} (-1)^c q^{c+b(2m-b-2c)} \mathfrak{p}^{(2m-b-2c)}(\kappa) F^{(b)} K^{b-2m+2c} \left[\!\left[\begin{matrix} h; m-c \\ c \end{matrix} \right]\!\right] v_{2\lambda+1}^{+} \\ & = \sum_{b=0}^{2m} \sum_{c \geqslant 0} q^{-2c^2+c+(b-2\lambda)(2m-b-2c)} \mathfrak{p}^{(2m-b-2c)}(\kappa) \begin{Bmatrix} m-\lambda-c \\ c \end{Bmatrix} F^{(b)} v_{2\lambda+1}^{+}. \end{aligned} \tag{5.7}$$

Similarly using Theorem 5.2 we have

$$\begin{aligned} & t_{\text{odd}}^{(2m-1)} v_{2\lambda+1}^{+} \\ & = \sum_{b=0}^{2m-1} \sum_{c \geqslant 0} (-1)^c q^{-c+b(2m-b-2c-1)} \mathfrak{p}^{(2m-b-2c-1)}(\kappa) \\ & \quad \cdot F^{(b)} K^{b-2m+2c+1} \left[\!\left[\begin{matrix} h; m-c \\ c \end{matrix} \right]\!\right] v_{2\lambda+1}^{+} \\ & = \sum_{b=0}^{2m-1} \sum_{c \geqslant 0} q^{-2c^2-c+(b-2\lambda)(2m-b-2c+1)} \mathfrak{p}^{(2m-b-2c+1)}(\kappa) \\ & \quad \cdot \begin{Bmatrix} m-\lambda-c \\ c \end{Bmatrix} F^{(b)} v_{2\lambda+1}^{+}. \end{aligned} \tag{5.8}$$

By a similar argument for (2.26), using (5.7)–(5.8) we obtain

$$t_{\text{odd}}^{(n)} v_{2\lambda+1}^{+} \in F^{(n)} v_{2\lambda+1}^{+} + \sum_{b<n} q^{-1} \mathbb{Z}[q^{-1}] F^{(b)} v_{2\lambda+1}^{+}, \qquad \text{for } \lambda \gg n.$$

The second statement follows now from the definition of the $\imath$-canonical basis for $\mathbf{U}^\imath$ using the projective system $\{L(2\lambda+1)\}_{\lambda \geqslant 0}$; cf. [3, §6].

5.4 Proof of Theorem 5.1

We prove the formulae for $t_{\text{odd}}^{(n)}$ by induction on n, in two separate cases (1)–(2) below. The base cases when $n = 1, 2$ are clear.

(1) We shall prove the formula (5.3) for $t_{\rm odd}^{(2m)}$, assuming the formula (5.4) for $t_{\rm odd}^{(2m-1)}$.

Recall $[2m]t_{\rm odd}^{(2m)} = t \cdot t_{\rm odd}^{(2m-1)}$, and $t = F + \check{E} + \kappa K^{-1}$. Let us compute

$$I = \check{E} t_{\rm odd}^{(2m-1)}, \qquad II = F t_{\rm odd}^{(2m-1)}, \qquad III = \kappa K^{-1} t_{\rm odd}^{(2m-1)}.$$

We compute

$$\begin{aligned} I = & \sum_{b=0}^{2m-1} \sum_{a=0}^{b} \sum_{c\geqslant 0} q^{\binom{2c}{2} - b(2m-b-2c-1) - a(b-a)} [a+1] \mathfrak{p}^{(2m-b-2c-1)}(\kappa) \\ & \cdot \check{E}^{(a+1)} \begin{bmatrix} h; 2-m \\ c \end{bmatrix} K^{b-2m+2c+1} F^{(b-a)} \\ = & \sum_{b=0}^{2m} \sum_{a=0}^{b} \sum_{c\geqslant 0} q^{\binom{2c}{2} - (b-1)(2m-b-2c) - (a-1)(b-a)} [a] \mathfrak{p}^{(2m-b-2c)}(\kappa) \\ & \cdot \check{E}^{(a)} \begin{bmatrix} h; 2-m \\ c \end{bmatrix} K^{b-2m+2c} F^{(b-a)}, \end{aligned}$$

where the last equation is obtained by shifting indices $a \to a-1$, $b \to b-1$. Using (2.6) we have

$$\begin{aligned} II = & \sum_{b=0}^{2m-1} \sum_{a=0}^{b} \sum_{c\geqslant 0} q^{\binom{2c}{2} - b(2m-b-2c-1) - a(b-a)} \mathfrak{p}^{(2m-b-2c-1)}(\kappa) \\ & \cdot \left(q^{-2a} \check{E}^{(a)} F + \check{E}^{(a-1)} \frac{q^{3-3a} K^{-2} - q^{1-a}}{q^2 - 1} \right) \begin{bmatrix} h; 2-m \\ c \end{bmatrix} K^{b-2m+2c+1} F^{(b-a)} \\ = & II^1 + II^2, \end{aligned}$$

where II^1 and II^2 are the two natural summands associated with the plus sign. By shifting the index $b \to b-1$ and then adding some zero terms, we obtain

$$\begin{aligned} II^1 = & \sum_{b=0}^{2m} \sum_{a=0}^{b} \sum_{c\geqslant 0} q^{\binom{2c}{2} - (b+1)(2m-b-2c) - a(b-a+1)} [b-a] \mathfrak{p}^{(2m-b-2c)}(\kappa) \\ & \cdot \check{E}^{(a)} \begin{bmatrix} h; 1-m \\ c \end{bmatrix} K^{b-2m+2c} F^{(b-a)}. \end{aligned}$$

By shifting the indices $a \to a+1$, $b \to b+1$ and then adding some zero terms, we also obtain

$$II^2 = \sum_{b=0}^{2m}\sum_{a=0}^{b}\sum_{c\geqslant 0} q^{\binom{2c-2}{2}-(b+1)(2m-b-2c)-(a+1)(b-a)}\mathfrak{p}^{(2m-b-2c)}(\kappa)$$

$$\cdot \check{E}^{(a)}\frac{q^{-3a}K^{-2}-q^{-a}}{q^2-1}\begin{bmatrix} h;2-m \\ c-1 \end{bmatrix} K^{b-2m+2c}F^{(b-a)}.$$

By the identity (4.5) we have

$$III = \sum_{b=0}^{2m-1}\sum_{a=0}^{b}\sum_{c\geqslant 0} q^{\binom{2c}{2}-b(2m-b-2c-1)-a(b-a)-2a}\kappa\cdot\mathfrak{p}^{(2m-b-2c-1)}(\kappa)$$

$$\cdot \check{E}^{(a)}\begin{bmatrix} h;2-m \\ c \end{bmatrix} K^{b-2m+2c}F^{(b-a)}$$

$$= \sum_{b=0}^{2m}\sum_{a=0}^{b}\sum_{c\geqslant 0} q^{\binom{2c}{2}-b(2m-b-2c-1)-a(b-a)-2a}[2m-b-2c]$$

$$\cdot\mathfrak{p}^{(2m-b-2c)}\check{E}^{(a)}\begin{bmatrix} h;2-m \\ c \end{bmatrix} K^{b-2m+2c}F^{(b-a)}$$

$$-\sum_{b=0}^{2m}\sum_{a=0}^{b}\sum_{c\geqslant 0} q^{\binom{2c-2}{2}-(b+2)(2m-b-2c+1)-a(b-a)-2a+2}[2m-b-2c-1]$$

$$\cdot\mathfrak{p}^{(2m-b-2c)}\check{E}^{(a)}\begin{bmatrix} h;2-m \\ c-1 \end{bmatrix} K^{b-2m+2c-2}F^{(b-a)}.$$

(Note that we have shifted the index $c\to c-1$ in the last summand above.)

Collecting the formulae for I, II^1, II^2, III gives us

$$t\cdot t_{\text{odd}}^{(2m-1)} = \sum_{0\leqslant a\leqslant b\leqslant 2m}\sum_{c\geqslant 0}\mathfrak{p}^{(2m-b-2c)}(\kappa)\check{E}^{(a)}\mathfrak{H}_{a,b,c}K^{b-2m+2c}F^{(b-a)},$$

where

$$\mathfrak{H}_{a,b,c} := q^{\binom{2c}{2}-(b-1)(2m-b-2c)-(a-1)(b-a)}[a]\begin{bmatrix} h;2-m \\ c \end{bmatrix}$$

$$+q^{\binom{2c}{2}-(b+1)(2m-b-2c)-a(b-a+1)}[b-a]\begin{bmatrix} h;1-m \\ c \end{bmatrix}$$

$$+q^{\binom{2c-2}{2}-(b+1)(2m-b-2c)-(a+1)(b-a)}\frac{q^{-3a}K^{-2}-q^{-a}}{q^2-1}\begin{bmatrix} h;2-m \\ c-1 \end{bmatrix}$$

$$+q^{\binom{2c}{2}-b(2m-b-2c-1)-a(b-a)-2a}[2m-b-2c]\begin{bmatrix} h;2-m \\ c \end{bmatrix}$$

$$-q^{\binom{2c-2}{2}-(b+2)(2m-b-2c+1)-a(b-a)-2a+2}[2m-b-2c-1]\begin{bmatrix} h;2-m \\ c-1 \end{bmatrix}K^{-2}.$$

Recall $[2m]t_{\rm odd}^{(2m)}=t\cdot t_{\rm odd}^{(2m-1)}$. To prove the formula (5.3) for $t_{\rm odd}^{(2m)}$, by the PBW basis theorem and the inductive assumption it suffices to prove the following identity, for all a, b, c:

$$\mathfrak{H}_{a,b,c}=q^{\binom{2c}{2}-2c-b(2m-b-2c)-a(b-a)}[2m]\begin{bmatrix} h;2-m \\ c \end{bmatrix}. \tag{5.9}$$

Thanks to $q^{2m-a}[a]-[2m]=q^{-a}[a-2m]$, we can combine the RHS(5.9) with the first summand of the LHS(5.9). Hence, after canceling out $q^{\binom{2c}{2}-2c-b(2m-b-2c)-a(b-a)-a}$ on both sides, we see that (5.9) is equivalent to the following identity, for all a, b, c:

$$\mathfrak{A}+\mathfrak{B}+\mathfrak{C}+\mathfrak{D}_1+\mathfrak{D}_2=0, \tag{5.10}$$

where

$$\mathfrak{A}=[a-2m]\begin{bmatrix} h;2-m \\ c \end{bmatrix},$$

$$\mathfrak{B}=q^{b+4c-2m}[b-a]\begin{bmatrix} h;1-m \\ c \end{bmatrix},$$

$$\mathfrak{C}=q^{2a-2m+3}\frac{q^{-3a}K^{-2}-q^{-a}}{q^2-1}\begin{bmatrix} h;2-m \\ c-1 \end{bmatrix},$$

$$\mathfrak{D}_1=q^{2c+b-a}[2m-b-2c]\begin{bmatrix} h;2-m \\ c \end{bmatrix},$$

$$\mathfrak{D}_2=-q^{2c-4m+b-a+3}[2m-b-2c-1]\begin{bmatrix} h;2-m \\ c-1 \end{bmatrix}K^{-2}.$$

Let us prove the identity (5.10). Using (3.2), we can write $\mathfrak{B}=\mathfrak{B}_1+\mathfrak{B}_2$, where

$$\mathfrak{B}_1=q^{b+4c-2m}[b-a]\begin{bmatrix} h;2-m \\ c \end{bmatrix},$$

$$\mathfrak{B}_2=-q^{b+4c-6m+4}[b-a]\begin{bmatrix} h;2-m \\ c-1 \end{bmatrix}K^{-2}.$$

Noting that

$$\frac{q^{-3a}K^{-2}-q^{-a}}{q^2-1} = q^{4m-4c-3a-4}\frac{(q^{4c-4m+4}K^{-2}-q^2)}{q^2-1} + q^{2m-2c-2a-2}[2m-2c-a-1],$$

we rewrite $\mathfrak{C} = \mathfrak{C}_1 + \mathfrak{C}_2$, where

$$\mathfrak{C}_1 = q^{2m-a-2c-2}[2c]\begin{bmatrix} h;2-m \\ c \end{bmatrix},$$

$$\mathfrak{C}_2 = q^{1-2c}[2m-2c-a-1]\begin{bmatrix} h;2-m \\ c-1 \end{bmatrix}.$$

A direct computation gives us

$$\begin{aligned}
&\mathfrak{A}+\mathfrak{B}_1+\mathfrak{C}_1+\mathfrak{D}_1 \\
&\quad = (1-q^{-2})[2c][2m-2c-a-1]p^{(2m-b-2c)}(\kappa)\begin{bmatrix} h;2-m \\ c \end{bmatrix}, \\
&\mathfrak{C}_2+(\mathfrak{B}_2+\mathfrak{D}_2) \\
&\quad = \mathfrak{C}_2 - q^{2c-4m+3}[2m-2c-a-1]p^{(2m-b-2c)}(\kappa)\begin{bmatrix} h;2-m \\ c-1 \end{bmatrix}K^{-2} \\
&\quad = -(1-q^{-2})[2c][2m-2c-a-1]p^{(2m-b-2c)}(\kappa)\begin{bmatrix} h;2-m \\ c \end{bmatrix}.
\end{aligned}$$

Summing up these two equations, we have $\mathfrak{A}+\mathfrak{B}+\mathfrak{C}+\mathfrak{D}_1+\mathfrak{D}_2 = 0$, whence (5.10), completing Step (1).

(2) Assuming the formulae for $t_{\text{odd}}^{(n)}$ with $n \leqslant 2m$, we shall now prove the following formula for $t_{\text{odd}}^{(2m+1)}$ (obtained from (5.4) with m replaced by $m+1$):

$$t_{\text{odd}}^{(2m+1)} = \sum_{b=0}^{2m+1}\sum_{a=0}^{b}\sum_{c\geqslant 0} q^{\binom{2c}{2}-b(2m-b-2c+1)-a(b-a)}p^{(2m-b-2c+1)}(\kappa) \cdot \check{E}^{(a)}\begin{bmatrix} h;1-m \\ c \end{bmatrix}K^{b-2m+2c-1}F^{(b-a)}. \tag{5.11}$$

We first need to compute $t \cdot t_{\text{odd}}^{(2m)}$, where we recall $t = F + \check{E} + \kappa K^{-1}$ and $t_{\text{odd}}^{(2m)}$ from (5.3). We shall compute

$$\mathtt{I} = \check{E} t_{\mathrm{odd}}^{(2m)}, \qquad \mathtt{II} = F t_{\mathrm{odd}}^{(2m)}, \qquad \mathtt{III} = \kappa K^{-1} t_{\mathrm{odd}}^{(2m)},$$

respectively. First we have

$$\begin{aligned}\mathtt{I} = & \sum_{b=0}^{2m} \sum_{a=0}^{b} \sum_{c \geqslant 0} q^{\binom{2c}{2} - 2c - b(2m-b-2c) - a(b-a)} [a+1] \mathfrak{p}^{(2m-b-2c)}(\kappa) \\ & \cdot \check{E}^{(a+1)} \begin{bmatrix} h; 2-m \\ c \end{bmatrix} K^{b-2m+2c} F^{(b-a)} \\ = & \sum_{b=0}^{2m+1} \sum_{a=0}^{b} \sum_{c \geqslant 0} q^{\binom{2c}{2} - 2c - (b-1)(2m-b-2c+1) - (a-1)(b-a)} [a] \\ & \mathfrak{p}^{(2m-b-2c+1)}(\kappa) \cdot \check{E}^{(a)} \begin{bmatrix} h; 2-m \\ c \end{bmatrix} K^{b-2m+2c-1} F^{(b-a)},\end{aligned}$$

where the last equation is obtained by shifting the indices $a \to a-1, b \to b-1$. We also have

$$\begin{aligned}\mathtt{II} = & \sum_{b=0}^{2m} \sum_{a=0}^{b} \sum_{c \geqslant 0} q^{\binom{2c}{2} - 2c - b(2m-b-2c) - a(b-a)} \mathfrak{p}^{(2m-b-2c)}(\kappa) \\ & \cdot \left(q^{-2a} \check{E}^{(a)} F + \check{E}^{(a-1)} \frac{q^{3-3a} K^{-2} - q^{1-a}}{q^2-1} \right) \begin{bmatrix} h; 2-m \\ c \end{bmatrix} K^{b-2m+2c} F^{(b-a)} \\ = & \mathtt{II}_1 + \mathtt{II}_2,\end{aligned}$$

where $\mathtt{II}_1$ and $\mathtt{II}_2$ are the two natural summands associated with the plus sign. By shifting the index $b \to b-1$, we further have

$$\begin{aligned}\mathtt{II}_1 = & \sum_{b=0}^{2m+1} \sum_{a=0}^{b} \sum_{c \geqslant 0} q^{\binom{2c}{2} - 2c - (b+1)(2m-b-2c+1) - a(b-a+1)} \\ & [b-a] \mathfrak{p}^{(2m-b-2c+1)}(\kappa) \cdot \check{E}^{(a)} \begin{bmatrix} h; 1-m \\ c \end{bmatrix} K^{b-2m+2c-1} F^{(b-a)}.\end{aligned}$$

By shifting the indices $a \to a+1$, $b \to b+1$, and $c \to c-1$, we further have

$$\begin{aligned}\mathtt{II}_2 = & \sum_{b=0}^{2m+1} \sum_{a=0}^{b} \sum_{c \geqslant 0} q^{\binom{2c-2}{2} - 2c - (b+1)(2m-b-2c+1) - (a+1)(b-a) + 2} \\ & \cdot \mathfrak{p}^{(2m-b-2c+1)}(\kappa) \cdot \check{E}^{(a)} \frac{q^{-3a} K^{-2} - q^{-a}}{q^2-1} \begin{bmatrix} h; 2-m \\ c-1 \end{bmatrix} K^{b-2m+2c-1} F^{(b-a)}.\end{aligned}$$

Using the identity (4.5) we also compute

$$\mathtt{III} = \sum_{b=0}^{2m}\sum_{a=0}^{b}\sum_{c\geqslant 0} q^{\binom{2c}{2}-2c-b(2m-b-2c)-a(b-a)-2a} \kappa \cdot \mathfrak{p}^{(2m-b-2c)}(\kappa)$$
$$\cdot \check{E}^{(a)} \begin{bmatrix} h;2-m \\ c \end{bmatrix} K^{b-2m+2c-1} F^{(b-a)}$$
$$= \sum_{b=0}^{2m+1}\sum_{a=0}^{b}\sum_{c\geqslant 0} q^{\binom{2c}{2}-2c-b(2m-b-2c)-a(b-a)-2a} [2m-b-2c+1]$$
$$\cdot \mathfrak{p}^{(2m-b-2c+1)}(\kappa) \check{E}^{(a)} \begin{bmatrix} h;2-m \\ c \end{bmatrix} K^{b-2m+2c-1} F^{(b-a)}$$
$$- \sum_{b=0}^{2m+1}\sum_{a=0}^{b}\sum_{c\geqslant 0} q^{\binom{2c-2}{2}-2c-(b+2)(2m-b-2c+2)-a(b-a)-2a+4}$$
$$[2m-b-2c]\mathfrak{p}^{(2m-b-2c+1)}(\kappa)\check{E}^{(a)} \begin{bmatrix} h;2-m \\ c-1 \end{bmatrix} K^{b-2m+2c-3} F^{(b-a)}.$$

(Note that we have shifted the index $c \to c-1$ in the last summand above.)

Collecting the formulae for $\mathtt{I}, \mathtt{II}_1, \mathtt{II}_2, \mathtt{III}$, we obtain an expression of the form

$$t \cdot t_{\rm odd}^{(2m)} = \sum_{0\leqslant a\leqslant b\leqslant 2m+1} \sum_{c\geqslant 0} \mathfrak{p}^{(2m-b-2c+1)}(\kappa)\check{E}^{(a)} \mathfrak{L}_{a,b,c} K^{b-2m+2c-1} F^{(b-a)},$$

where

$$\mathfrak{L}_{a,b,c} := q^{\binom{2c}{2}-2c-(b-1)(2m-b-2c+1)-(a-1)(b-a)} [a] \begin{bmatrix} h;2-m \\ c \end{bmatrix}$$
$$+ q^{\binom{2c}{2}-2c-(b+1)(2m-b-2c+1)-a(b-a+1)} [b-a] \begin{bmatrix} h;1-m \\ c \end{bmatrix}$$
$$+ q^{\binom{2c-2}{2}-2c-(b+1)(2m-b-2c+1)-(a+1)(b-a)+2} \frac{q^{-3a}K^{-2}-q^{-a}}{q^2-1} \begin{bmatrix} h;2-m \\ c-1 \end{bmatrix}$$
$$+ q^{\binom{2c}{2}-2c-b(2m-b-2c)-a(b-a)-2a} [2m-b-2c+1] \begin{bmatrix} h;2-m \\ c \end{bmatrix}$$
$$- q^{\binom{2c-2}{2}-2c-(b+2)(2m-b-2c+2)-a(b-a)-2a+4} [2m-b-2c] \begin{bmatrix} h;2-m \\ c-1 \end{bmatrix} K^{-2}.$$

On the other hand, using (5.4) (with an index shift $c \to c-1$) and (5.11) we write

$$[2m+1]t_{\rm odd}^{(2m+1)} + [2m]t_{\rm odd}^{(2m-1)}$$
$$= \sum_{0\leqslant a\leqslant b\leqslant 2m+1} \sum_{c\geqslant 0} \mathfrak{p}^{(2m-b-2c+1)}(\kappa)\check{E}^{(a)}\mathfrak{R}_{a,b,c}K^{b-2m+2c-1}F^{(b-a)},$$

where

$$\mathfrak{R}_{a,b,c} := q^{\binom{2c}{2}-b(2m-b-2c+1)-a(b-a)}[2m+1]\begin{bmatrix} h;1-m \\ c \end{bmatrix}$$
$$+ q^{\binom{2c-2}{2}-b(2m-b-2c+1)-a(b-a)}[2m]\begin{bmatrix} h;2-m \\ c-1 \end{bmatrix}.$$

To prove the formula (5.11) for $t_{\rm odd}^{(2m+1)}$, it suffices to show that, for all a, b, c,

$$\mathfrak{L}_{a,b,c} = \mathfrak{R}_{a,b,c}. \tag{5.12}$$

Canceling the q-powers $q^{\binom{2c}{2}-2c-(b-1)(2m-b-2c+1)-(a-1)(b-a)}$ on both sides, we see that the identity (5.12) is equivalent to the following identity, for all a, b, c:

$$\begin{aligned}
&[a]\begin{bmatrix} h;2-m \\ c \end{bmatrix} + q^{4c-4m+b-2}[b-a]\begin{bmatrix} h;1-m \\ c \end{bmatrix} \\
&+ q^{2a-4m+3}\frac{q^{-3a}K^{-2}-q^{-a}}{q^2-1}\begin{bmatrix} h;2-m \\ c-1 \end{bmatrix} \\
&+ q^{2c-2m+b-a-1}[2m-b-2c+1]\begin{bmatrix} h;2-m \\ c \end{bmatrix} \\
&- q^{2c-6m+b-a+2}[2m-b-2c]\begin{bmatrix} h;2-m \\ c-1 \end{bmatrix}K^{-2} \\
&= q^{4c-2m+a-1}[2m+1]\begin{bmatrix} h;1-m \\ c \end{bmatrix} + q^{a-2m+2}[2m]\begin{bmatrix} h;2-m \\ c-1 \end{bmatrix}.
\end{aligned} \tag{5.13}$$

By combining the second summand of the LHS(5.13) with the first summand of the RHS(5.13) as well as combining the third summand of the LHS with the second summand of the RHS, the identity (5.13) is reduced to the following equivalent identity, for all a, b, c:

$$\mathfrak{W} + \mathfrak{X} + \mathfrak{Y} + \mathfrak{Z}_1 + \mathfrak{Z}_2 = 0, \tag{5.14}$$

where

$$\mathfrak{W} = [a] \left[\!\left[\begin{matrix} h; 2-m \\ c \end{matrix} \right]\!\right],$$

$$\mathfrak{X} = q^{4c-2m+b-1}[b-a-2m-1] \left[\!\left[\begin{matrix} h; 1-m \\ c \end{matrix} \right]\!\right],$$

$$\mathfrak{Y} = \frac{q^{3-a-4m}K^{-2} - q^{3+a}}{q^2-1} \left[\!\left[\begin{matrix} h; 2-m \\ c-1 \end{matrix} \right]\!\right],$$

$$\mathfrak{Z}_1 = q^{2c-2m+b-a-1}[2m-b-2c+1] \left[\!\left[\begin{matrix} h; 2-m \\ c \end{matrix} \right]\!\right],$$

$$\mathfrak{Z}_2 = -q^{2c-6m+b-a+2}[2m-b-2c] \left[\!\left[\begin{matrix} h; 2-m \\ c-1 \end{matrix} \right]\!\right] K^{-2}.$$

Let us finally prove the identity (5.14). Using (3.2), we can write $\mathfrak{X} = \mathfrak{X}_1 + \mathfrak{X}_2$, where

$$\mathfrak{X}_1 = q^{4c-2m+b-1}[b-a-2m-1] \left[\!\left[\begin{matrix} h; 2-m \\ c \end{matrix} \right]\!\right],$$

$$\mathfrak{X}_2 = -q^{4c-6m+b+3}[b-a-2m-1] \left[\!\left[\begin{matrix} h; 2-m \\ c-1 \end{matrix} \right]\!\right] K^{-2}.$$

Noting that

$$\frac{q^{3-a-4m}K^{-2} - q^{3+a}}{q^2-1} = q^{-4c-a-1}\frac{(q^{4c-4m+4}K^{-2} - q^2)}{q^2-1} + q^{1-2c}[-2c-a-1],$$

we rewrite $\mathfrak{Y} = \mathfrak{Y}_1 + \mathfrak{Y}_2$, where

$$\mathfrak{Y}_1 = q^{-2c-a-2}[2c] \left[\!\left[\begin{matrix} h; 2-m \\ c \end{matrix} \right]\!\right], \qquad \mathfrak{Y}_2 = -q^{1-2c}[2c+a+1] \left[\!\left[\begin{matrix} h; 2-m \\ c-1 \end{matrix} \right]\!\right].$$

A direct computation shows that

$$\mathfrak{W} + \mathfrak{X}_1 + \mathfrak{Y}_1 + \mathfrak{Z}_1 = -(1-q^{-2})[2c+a+1][2c] \left[\!\left[\begin{matrix} h; 2-m \\ c \end{matrix} \right]\!\right],$$

$$\begin{aligned}(\mathfrak{X}_2 + \mathfrak{Z}_2) + \mathfrak{Y}_2 &= q^{-2m-a+1}[2c+a] \left[\!\left[\begin{matrix} h; 2-m \\ c-1 \end{matrix} \right]\!\right] K^{-2} + \mathfrak{Y}_2 \\ &= (1-q^{-2})[2c+a+1][2c] \left[\!\left[\begin{matrix} h; 2-m \\ c \end{matrix} \right]\!\right].\end{aligned}$$

Summing up these two equations, we obtain $\mathfrak{W}+\mathfrak{X}+\mathfrak{Y}+\mathfrak{Z}_1+\mathfrak{Z}_2=0$, whence (5.14), completing Step (2).

The proof of Theorem 5.1 is completed. □

6 On the $\imath$-Divided Powers for Generic κ

In this appendix we provide closed formulae for the second $\imath$-divided power for $\mathbf{U}^\imath$ with an arbitrary parameter $\overline{\kappa}=\kappa\in\mathcal{A}$.

For $\overline{\kappa}=\kappa\in\mathcal{A}$, we can write

$$\kappa=\sum_{i\in\mathbb{Z}}c_i q^i,\qquad \text{where } c_i=c_{-i},\ \forall i.$$

Define

$$\diamondsuit=\diamondsuit(\kappa)=\sum_{i\in\mathbb{Z}}(-1)^i c_{2i}. \tag{6.1}$$

Lemma 6.1 *Let $\overline{\kappa}=\kappa\in\mathcal{A}$. We have $\frac{1}{[2]}(\kappa-\diamondsuit)\in\mathcal{A}$.* □

Proof Note that $q^{2i+1}+q^{-2i-1}\in[2]\mathcal{A}$ and $q^{2i}+q^{-2i}-2(-1)^i\in[2]\mathcal{A}$ for all $i\in\mathbb{Z}$. The lemma follows.

Recall from (2.7) that $t=F+\check{E}+\kappa K^{-1}$. We shall denote the second $\imath$-divided power in ${}_\mathcal{A}\mathbf{U}^\imath_{\rm ev}$ (and respectively, ${}_\mathcal{A}\mathbf{U}^\imath_{\rm odd}$) by $t^{(2)}_{\rm ev}$ (and respectively, $t^{(2)}_{\rm odd}$), which is by definition the $\imath$-canonical basis element with a leading term $F^{(2)}$ (see [3]). The following formula was obtained with help from Huanchen Bao.

Proposition 6.2 *Let $\overline{\kappa}=\kappa\in\mathcal{A}$. For $\lambda\in\mathbb{Z}$, we have*

$$t^{(2)}_{\rm ev}=\frac{t^2-\diamondsuit^2}{[2]},\qquad t^{(2)}_{\rm odd}=\frac{t^2-1+\diamondsuit^2}{[2]}.$$

Proof A direct computation shows that, for $\mu\in\mathbb{Z}$,

$$t^2/[2]\cdot v^+_\mu=F^{(2)}v^+_\mu+q^{1-\mu}\kappa Fv^+_\mu+\frac{\frac{1-q^{-2\mu}}{1-q^{-2}}+q^{-2\mu}\kappa^2}{q+q^{-1}}v^+_\mu. \tag{6.2}$$

Hence we obtain by using (6.2) that

$$\begin{aligned}(t^2-\diamondsuit^2)/[2]\cdot v^+_{2\lambda}&=F^{(2)}v^+_{2\lambda}+q^{1-2\lambda}\kappa Fv^+_{2\lambda}+\frac{q^{1-2\lambda}[2\lambda]+q^{-4\lambda}\kappa^2-\diamondsuit^2}{q+q^{-1}}v^+_{2\lambda}\\&=F^{(2)}v^+_{2\lambda}+q^{1-2\lambda}\kappa Fv^+_{2\lambda}\end{aligned}$$

$$+\frac{q^{1-2\lambda}[2\lambda]+q^{-4\lambda}(\kappa^2-\diamondsuit^2)+(q^{-4\lambda}-1)\diamondsuit^2}{q+q^{-1}}v^+_{2\lambda}.$$

Note that (each of the 3 summands of) the last fraction above lies in $\mathcal{A}$ thanks to Lemma 6.1. Hence $(t^2-\diamondsuit^2)/[2]$ is integral, i.e., it lies in ${}_\mathcal{A}\mathbf{U}^\imath_{\mathrm{ev}}$. Moreover, the lower terms above all have coefficients in $q^{-1}\mathbb{Z}[q^{-1}]$ for $\lambda \gg 0$. Hence it follows by definition [3] that $(t^2-\diamondsuit^2)/[2]$ is an $\imath$-canonical basis element in ${}_\mathcal{A}\mathbf{U}^\imath_{\mathrm{ev}}$.

On the other hand, by using (6.2) again we have

$$\begin{aligned}(t^2-1+\diamondsuit^2)/[2]\cdot v^+_{2\lambda+1} &= F^{(2)}v^+_{2\lambda+1}+q^{1-2\lambda}\kappa F v^+_{2\lambda+1}\\ &\quad+\frac{q^{-2\lambda-1}[2\lambda]+q^{-4\lambda-2}\kappa^2+\diamondsuit^2}{q+q^{-1}}v^+_{2\lambda+1}\\ &= F^{(2)}v^+_{2\lambda+1}+q^{1-2\lambda}\kappa F v^+_{2\lambda+1}\\ &\quad+\frac{q^{-2\lambda-1}[2\lambda]+q^{-4\lambda-2}(\kappa^2-\diamondsuit^2)+(q^{-4\lambda-2}+1)\diamondsuit^2}{q+q^{-1}}v^+_{2\lambda+1}.\end{aligned}$$

Note that (each of the 3 summands of) the fraction above lies in $\mathcal{A}$ thanks to Lemma 6.1. Hence $(t^2-1+\diamondsuit^2)/[2]\mathbf{1}_{2\lambda}$ is integral, i.e., it lies in ${}_\mathcal{A}\mathbf{U}^\imath_{\mathrm{ev}}$. Moreover, the lower terms above all have coefficients in $q^{-1}\mathbb{Z}[q^{-1}]$ for $\lambda \gg 0$. Hence it follows by definition [3] that $(t^2-1+\diamondsuit^2)/[2]$ is an $\imath$-canonical basis element in ${}_\mathcal{A}\mathbf{U}^\imath_{\mathrm{odd}}$.

Recall $[h;0]=\frac{K^{-2}-1}{q^4-1}$, and $[\![h;0]\!]=\frac{K^{-2}-q^2}{q^4-1}$. We leave the verification of the following formulae using Proposition 6.2 to the reader.

Proposition 6.3 *Let $\overline{\kappa}=\kappa\in\mathcal{A}$. The following formulae for $t^{(2)}_{\mathrm{ev}}$ and $t^{(2)}_{\mathrm{odd}}$ in $\mathbf{U}$ hold:*

$$\begin{aligned}t^{(2)}_{\mathrm{ev}} &= \check{E}^{(2)}+q^{-1}\check{E}F+F^{(2)}+q^{-1}\kappa K^{-1}F+q^{-1}\kappa\check{E}K^{-1}\\ &\quad+\big(q+(q^3-q)\kappa^2\big)\frac{K^{-2}-1}{q^4-1}+\frac{\kappa^2-\diamondsuit^2}{[2]},\\ t^{(2)}_{\mathrm{odd}} &= \check{E}^{(2)}+q^{-1}\check{E}F+F^{(2)}+q^{-1}\kappa K^{-1}F+q^{-1}\kappa\check{E}K^{-1}\\ &\quad+\big(q+(q^3-q)\kappa^2\big)\frac{K^{-2}-q^2}{q^4-1}+\frac{q^2(\kappa^2-\diamondsuit^2)}{[2]}+q\diamondsuit^2.\end{aligned}$$

Example 6.0.1 Assume κ is a q-integer, i.e., $\kappa=[n]$ for some $n\in\mathbb{Z}$. Then $\diamondsuit$ in (6.1) becomes

$$\diamondsuit=\begin{cases}0, & \text{for } n \text{ even},\\ 1, & \text{for } n\equiv 1 \pmod 4,\\ -1, & \text{for } n\equiv -1 \pmod 4.\end{cases}$$

In this case, the formulae in Propositions 6.2 and 6.3 reduce to those obtained in the previous sections.

It is an open question to find closed formula for general $\imath$-divided powers of t with an arbitrary bar invariant parameter $\kappa \in \mathcal{A}$.

It is also an open question to find closed formulae for the $\imath$-canonical bases for the simple $\mathbf{U}$-modules $L(\lambda)$, for $\lambda \in \mathbb{N}$, even for κ being an arbitrary q-integer as in Sects. 2–5. The explicit formulae for such $\imath$-canonical bases for $L(\lambda)$ were known in [5] for $\kappa = 0$ or 1; in this case, they coincide with the nonzero images of the $\imath$-divided powers acting on v_λ^+.

Acknowledgments WW thanks Huanchen Bao for his insightful collaboration. The formula in the Appendix for the second $\imath$-divided power with arbitrary parameter κ (which was obtained with help from Huanchen) was crucial to this project, and to a large extent this paper grows by exploring for what values for the parameter κ reasonable formulae for higher divided powers can be obtained. The research of WW and the undergraduate research of CB are partially supported by a grant DMS-1702254 from National Science Foundation. Mathematica was used intensively in this work.

References

1. M. Balagovic and S. Kolb, *Universal K-matrix for quantum symmetric pairs*, J. Reine Angew. Math. **747** (2019), 299–353, DOI 10.1515/crelle-2016-0012, arXiv:1507.06276v2.
2. H. Bao and W. Wang, *A new approach to Kazhdan-Lusztig theory of type B via quantum symmetric pairs*, Astérisque **402** (2018), vii+134 pp, arXiv:1310.0103v3.
3. H. Bao and W. Wang, *Canonical bases arising from quantum symmetric pairs*, Invent. Math. **213** (2018), 1099–1177, arXiv:1610.09271v2.
4. H. Bao, W. Wang and H. Watanabe, *Multiparameter quantum Schur duality of type B*, Proc. AMS **146** (2018), 3203–3216, arXiv:1609.01766.
5. C. Berman and W. Wang, *Formulae of $\imath$-divided powers in* $\mathbf{U}_q(\mathfrak{sl}_2)$, J. Pure Appl. Algebra **222** (2018), 2667–2702, arXiv:1703.00602.
6. T. Koornwinder, *Askey-Wilson polynomials as zonal spherical functions on the $SU(2)$ quantum group*, SIAM J. Math. Anal. **24** (1993), 795–813.
7. G. Letzter, *Symmetric pairs for quantized enveloping algebras*, J. Algebra **220** (1999), 729–767.
8. G. Lusztig, *Introduction to quantum groups*, Modern Birkhäuser Classics, Reprint of the 1993 Edition, Birkhäuser, Boston, 2010.

Longest Weyl Group Elements in Action

Yiqiang Li and Yan Ling

Abstract Inspired from graded quiver varieties, we attach a Weyl-like group $W^{\langle n \rangle}$ to a Cartan matrix and a nonnegative integer n. As the integer varies, we get a projective system. In type A_ℓ, the group $W^{\langle n \rangle}$ is related to reduced Burau representations, and in general to complex reflection groups. Behavior of the longest Weyl group elements in $W^{\langle n \rangle}$ is studied in detail. The results will be used in the analysis by Li (in preparation) of graded/cyclic quiver varieties in the spirit of Li (Represent Theory 23:1–56, 2019).

1 Introduction

The Weyl groups and their longest elements have played, and are still playing, important roles in representation theory [4, 5]. In a recent paper [6], the reflection functor, say S_{w_0}, attached to the longest Weyl group element w_0 serves as a crucial ingredient in defining two classes of fixed-point loci of Nakajima quiver varieties. In a sense, these two new classes of quiver varieties can be thought of as the "Cartan decomposition" of Nakajima varieties, since they correspond to the Cartan decomposition of the associated complex semisimple Lie algebras. It is natural to expect that the graded version of Nakajima varieties [11] admits a similar decomposition. To this end, one needs to understand the interaction between the reflection functor S_{w_0} and the action of the finite abelian group involved in a Nakajima variety. The analysis of the numerical data, similar to [6, (23)], so that the fixed-point locus in the graded Nakajima varieties setting makes sense, is nontrivial and is studied in this paper. The results will be used in a forthcoming paper [7]. They form the main results in the paper and are given in Propositions 3.1, 3.2, and 4.4 (see also Corollary 4.6).

Y. Li (✉)
The State University of New York at Buffalo, Buffalo, NY, USA
e-mail: yiqiang@buffalo.edu

Y. Ling
Buffalo, NY, USA

J. Greenstein et al. (eds.), *Interactions of Quantum Affine Algebras with Cluster Algebras, Current Algebras and Categorification*, Progress in Mathematics 337,
https://doi.org/10.1007/978-3-030-63849-8_8

Further, these results can and shall be put into a broader framework. As such, we introduce a Weyl-like group, say $W^{\langle n\rangle}$, attached to a generalized Cartan matrix C and a cyclic group $\mathbb{Z}/n\mathbb{Z}$, which has its root in graded Nakajima varieties. Its definition resembles the original definition of a Weyl group and when $n = 1$ or 2 they are Weyl groups. The main results in this paper are then deduced from a detailed analysis of the behaviors of (the images of) the longest Weyl group elements in $W^{\langle n\rangle}$. Besides these, the group $W^{\langle n\rangle}$ itself is of independent interest. It can be regarded as an intermediate group sitting in between the Weyl group and its associated braid group: it receives a surjective morphism from the braid group and surjects onto the Weyl group. In particular, when C is of type A_ℓ and $n = 0$, the group $W^{\langle n\rangle}$ comes from the reduced Burau representation [3]. When $4 \nmid n$, the group $W^{\langle n\rangle}$ are complex reflection groups. Fixing the Cartan matrix but letting n vary, we obtain a projective system. All these suggest that the group $W^{\langle n\rangle}$ deserves a closer look, such as its order and its precise relationship with graded Nakajima varieties.

Proofs of Propositions 3.1, 3.2, and 4.4 in exceptional cases, especially E_7 and E_8, involve computations exceeding human power. Instead we resort to computer for help. The computer programs, using R-codes, are available in the first author's website. It is interesting to see if one can provide a conceptual proof.

2 The Group $W^{\langle n\rangle}$

In this section, we associate with a Cartan matrix and a nonnegative integer a Weyl-like group $W^{\langle n\rangle}$. We show that these groups form a projective system as n varies. Moreover, we relate the group $W^{\langle n\rangle}$ to the Burau representations and to complex reflection groups.

2.1 Definition and a Projective System

Let I be a finite index set. Let $C = (c_{ij})_{i,j\in I}$ be a generalized Cartan matrix, i.e.,

- $c_{ii} = 2$ for all $i \in I$,
- $c_{ij} \in \mathbb{Z}_{\leqslant 0}$ if $i \neq j$,
- $c_{ij} = 0$ if and only if $c_{ji} = 0$,
- there exists $(d_i)_{i\in I} \in \mathbb{Z}^I$ such that $d_i c_{ij} = d_j c_{ji}$ for all $i, j \in I$.

Let $s_i : \mathbb{Z}[I] \to \mathbb{Z}[I]$ be the simple reflection defined by $s_i(\zeta) = \zeta'$ with $\zeta'_j = \zeta_j - c_{ji}\zeta_i$ for all $j \in I$. Let W be the Weyl group generated by the simple reflections s_i for all $i \in I$.

Now set $\mathbb{Z}_n = \mathbb{Z}/n\mathbb{Z}$ for any nonnegative integer n. Arising from the study of graded quiver varieties [11], there is an "n-deformed" Cartan matrix $C^{\langle n\rangle} =$

$\left(c^{\langle n\rangle}_{(i,a),(j,b)}\right)_{(i,a),(j,b)\in I\times\mathbb{Z}_n}$ where

$$c^{\langle n\rangle}_{(i,a),(j,b)} = \begin{cases} 1 & \text{if } i=j,\, a-b\in\{0,2\},\\ c_{ij} & \text{if } i\neq j,\, a-b=1,\\ 0 & \text{otherwise.}\end{cases} \tag{2.1}$$

Let $x \in \mathbb{Z}[I \times \mathbb{Z}_n]$. We denote by $x_i(a)$ the (i, a) coefficient of x. For each $i \in I$, we define the "n-deformed" simple reflection to be the $\mathbb{Z}$-linear function

$$s_i^{\langle n\rangle} : \mathbb{Z}[I \times \mathbb{Z}_n] \to \mathbb{Z}[I \times \mathbb{Z}_n] \tag{2.2}$$

by $s_i^{\langle n\rangle}(x) = y$ where

$$y_j(b) = x_j(b) - \sum_{a\in\mathbb{Z}_n} c^{\langle n\rangle}_{(j,b),(i,a)} x_i(a) \overset{(2.1)}{=} \begin{cases} -x_i(b-2) & \text{if } j=i,\\ x_j(b)-c_{ji}x_i(b-1) & \text{if } j\neq i.\end{cases} \tag{2.3}$$

We refer to Sects. 2.4–(2.20) for a more conventional look of $s_i^{\langle n\rangle}$. The function $s_i^{\langle n\rangle}$ is bijective whose inverse $(s_i^{\langle n\rangle})^{-1}$ is given by $(s_i^{\langle n\rangle})^{-1}(x) = y$ where

$$y_j(b) = \begin{cases} -x_i(b+2) & \text{if } j=i,\\ x_j(b)-c_{ji}x_i(b+1) & \text{if } j\neq i.\end{cases}$$

Let $W^{\langle n\rangle}$ be the subgroup of $\mathrm{Aut}(\mathbb{Z}[I \times \mathbb{Z}_n])$ generated by $s_i^{\langle n\rangle}$ for all $i \in I$. Clearly, we have

$$W^{\langle 1\rangle} = W. \tag{2.4}$$

If $n|m$, then the natural quotient $\mathbb{Z}_m \to \mathbb{Z}_n$ induces an embedding of $\mathbb{Z}$-modules

$$\mathbb{Z}[I \times \mathbb{Z}_n] \hookrightarrow \mathbb{Z}[I \times \mathbb{Z}_m], \quad x \mapsto y \tag{2.5}$$

such that $y_i(a) = x_i(a), \forall i \in I, a \in \mathbb{Z}_m$. In particular $y_i(a) = y_i(a + n)$.

Remark 2.1 The embedding (2.5) is ill-defined when $m = 0$ and $n \neq 0$. In this case, it can be rectified as the following embedding, which we shall use whenever it shows up.

$$\mathbb{Z}[I \times \mathbb{Z}_n] \hookrightarrow \mathbb{Z}^{I\times\mathbb{Z}}, \quad x \mapsto y. \tag{2.6}$$

It can be checked from definition that the restriction of $s_i^{\langle m\rangle}$ to $\mathbb{Z}[I \times \mathbb{Z}_n]$ via the above embedding is $s_i^{\langle n\rangle}$:

$$s_i^{\langle m\rangle}|_{\mathbb{Z}[I\times\mathbb{Z}_n]} = s_i^{\langle n\rangle}.$$

Thus, there is a surjective group homomorphism via restriction

$$\pi_{m,n} : W^{\langle m\rangle} \to W^{\langle n\rangle}, \quad w \mapsto w|_{\mathbb{Z}[I\times\mathbb{Z}_n]}, \quad \forall n|m. \tag{2.7}$$

sending $s_i^{\langle m\rangle}$ to $s_i^{\langle n\rangle}$. Define a partial order on $\mathbb{Z}_{\geqslant 0}$ by declaring $n \preceq m$ if $n|m$. Clearly, $(W^{\langle n\rangle}, \pi_{m,n})$ forms a projective system of groups with respect to the partial order $\preceq$. Let $\varprojlim_{n\in\mathbb{Z}_{\geqslant 0}}(W^{\langle n\rangle}, \pi_{m,n})$ be its projective limit. Since any positive integer n divides 0, we have that the projective limit of the projective system is exactly $W^{\langle 0\rangle}$:

$$W^{\langle 0\rangle} = \varprojlim_{n\in\mathbb{Z}_{\geqslant 0}} (W^{\langle n\rangle}, \pi_{m,n}).$$

We propose the following question, for which it is unclear if there is a positive answer.

Questio 2.2 Is it true that $W^{\langle 0\rangle} = \varprojlim_{n\in\mathbb{Z}_{>0}}(W^{\langle n\rangle}, \pi_{m,n})$? □

2.3 Relations in $W^{\langle n\rangle}$

In this section we determine a set of relations in the group $W^{\langle n\rangle}$.

Proposition 2.2 *The braid relations hold in* $W^{\langle n\rangle}$, *that is*

$$\underbrace{s_i^{\langle n\rangle} s_j^{\langle n\rangle} s_i^{\langle n\rangle} \cdots}_{m_{ij} \text{ terms}} = \underbrace{s_j^{\langle n\rangle} s_i^{\langle n\rangle} s_j^{\langle n\rangle} \cdots}_{m_{ji} \text{ terms}} \quad \forall i \neq j. \tag{2.8}$$

where m_{ij} *is a Coxeter matrix element defined by*

$$m_{ij} = \begin{cases} 2 & \text{if } c_{ij}c_{ji} = 0, \\ 3 & \text{if } c_{ij}c_{ji} = 1, \\ 4 & \text{if } c_{ij}c_{ji} = 2, \\ 6 & \text{if } c_{ij}c_{ji} = 3. \end{cases}$$

Proof Let $e_i(a)$ for $i \in I$ and $a \in \mathbb{Z}_n$ be the standard basis of $\mathbb{Z}[I \times \mathbb{Z}_n]$. We only need to show that the evaluation of both sides of (2.8) at $e_i(a)$ matches for all $i \in I$ and $a \in \mathbb{Z}_n$. Let

$$z : \mathbb{Z}[I \times \mathbb{Z}_n] \to \mathbb{Z}[I \times \mathbb{Z}_n], e_i(a) \mapsto e_i(a+1) \tag{2.9}$$

be the shift function. It is clear that z commutes with $s_i^{\langle n \rangle}$. Hence we only need to show that (2.8) holds when evaluated at $e_k(0)$ for all $k \in I$. Note that we have

$$s_i^{\langle n \rangle}(e_i(a)) = -e_i(a+2) + \sum_{j \neq i} -c_{ji} e_j(a+1), \quad s_i^{\langle n \rangle} e_k(a) = e_k(a), \quad \text{if } k \neq i. \tag{2.10}$$

For $k \neq i, j$, both sides of the braid relation (2.8) remain $e_k(0)$ when evaluated at $e_k(0)$ by (2.10). So it suffices to show that (2.8) holds when evaluated at $e_i(0)$ and $e_j(0)$, which we shall check in the following.

Case 1 $m_{ij} = 2$. This case is obviously true.

Case 2 $m_{ij} = 3$, it is $(c_{ij}, c_{ji}) = (-1, -1)$. Hence, by using (2.10), we have

$$\begin{aligned} s_j^{\langle n \rangle} s_i^{\langle n \rangle}(e_i(0)) &= s_j^{\langle n \rangle}(-e_i(2) + e_j(1) + \sum_{k \neq i,j} -c_{k,i} e_k(1)) \\ &= -e_j(3) + \sum_{k \neq i,j} -c_{kj} e_k(2) + \sum_{k \neq i,j} -c_{k,i} e_k(1). \end{aligned}$$

Thus, in light of (2.10), we have

$$s_i^{\langle n \rangle} s_j^{\langle n \rangle} s_i^{\langle n \rangle}(e_i(0)) = s_j^{\langle n \rangle} s_i^{\langle n \rangle}(e_i(0)) = s_j^{\langle n \rangle} s_i^{\langle n \rangle} s_j^{\langle n \rangle}(e_i(0)).$$

Now by switching the role of i and j in the above equation, it yields

$$s_j^{\langle n \rangle} s_i^{\langle n \rangle} s_j^{\langle n \rangle}(e_j(0)) = s_i^{\langle n \rangle} s_j^{\langle n \rangle} s_i^{\langle n \rangle}(e_j(0)).$$

Therefore, (2.8) holds when evaluated at $e_i(0)$ and $e_j(0)$, finishing the proof of (2.8) for $m_{ij} = 3$.

Case 3 $m_{ij} = 4$. There are two subcases to consider: the type B_n where $(c_{ij}, c_{ji}) = (-2, -1)$ or the type C_n where $(c_{ij}, c_{ji}) = (-1, -2)$. We start with the case $(c_{ij}, c_{ji}) = (-2, -1)$. As in case $m_{ij} = 3$, it is enough to check (2.8) when evaluated at $e_i(0)$ and $e_j(0)$. We have

$$s_i^{\langle n \rangle} s_j^{\langle n \rangle} s_i^{\langle n \rangle}(e_i(0)) = s_i^{\langle n \rangle} s_j^{\langle n \rangle}(-e_i(0) + e_j(1) + \sum_{k \neq i,j} -c_{ki} e_k(1))$$

$$= s_i(e_i(2) - e_j(3) + \sum_{k\neq i,j} -c_{kj}e_k(2) + \sum_{k\neq i,j} -c_{ki}e_k(1))$$

$$= -e_i(4) + \sum_{k\neq i,j} -c_{ki}(e_k(3) + e_k(1)) + \sum_{k\neq i,j} -c_{kj}e_k(2).$$

Thus, by the above equality and using (2.10) twice, we have

$$s_i^{\langle n\rangle}s_j^{\langle n\rangle}s_i^{\langle n\rangle}s_j^{\langle n\rangle}(e_i(0)) = s_i^{\langle n\rangle}s_j^{\langle n\rangle}s_i^{\langle n\rangle}(e_i(0)) = s_j^{\langle n\rangle}s_i^{\langle n\rangle}s_j^{\langle n\rangle}s_i^{\langle n\rangle}(e_i(0)). \tag{2.11}$$

Since $c_{ij} \neq c_{ji}$, we cannot use the technique we applied in case $m_{ij} = 3$. This time we use the following computation:

$$s_j^{\langle n\rangle}s_i^{\langle n\rangle}s_j^{\langle n\rangle}(e_j(0)) = s_j^{\langle n\rangle}s_i^{\langle n\rangle}(-e_j(2) + 2e_i(1) + \sum_{k\neq i,j} -c_{kj}e_k(1))$$

$$= s_j^{\langle n\rangle}(e_j(2) - 2e_i(3) + 2\sum_{k\neq i,j} -c_{ki}e_k(2) + \sum_{k\neq i,j} -c_{kj}e_k(1)$$

$$= -e_j(4) + \sum_{k\neq i,j} -c_{kj}(e_k(3) + e_k(1)) + \sum_{k\neq i,j} -c_{ki}e_k(2).$$

Thanks to (2.10), this leads to

$$s_i^{\langle n\rangle}s_j^{\langle n\rangle}s_i^{\langle n\rangle}s_j^{\langle n\rangle}(e_j(0)) = s_j^{\langle n\rangle}s_i^{\langle n\rangle}s_j^{\langle n\rangle}(e_j(0)) = s_j^{\langle n\rangle}s_i^{\langle n\rangle}s_j^{\langle n\rangle}s_i^{\langle n\rangle}(e_j(0)). \tag{2.12}$$

By switching i and j in the above analysis, we see that (2.11) and (2.12) still holds when $(c_{ij}, c_{ji}) = (-1, -2)$. Therefore when evaluated at $e_i(0)$ and $e_j(0)$, the braid relation (2.8) holds for $m_{ij} = 4$, and hence holds generally.

Case 4 $m_{ij} = 6$. We set $(c_{ij}, c_{ji}) = (-3, -1)$. We have the following computations:

$$s_i^{\langle n\rangle}s_j^{\langle n\rangle}s_i^{\langle n\rangle}s_j^{\langle n\rangle}s_i^{\langle n\rangle}e_i(0) = s_i^{\langle n\rangle}s_j^{\langle n\rangle}s_i^{\langle n\rangle}s_j^{\langle n\rangle}(-e_i(2) + e_j(1)$$

$$+ \sum_{k\neq i,j} -c_{ki}e_k(1))$$

$$= s_i^{\langle n\rangle}s_j^{\langle n\rangle}s_i^{\langle n\rangle}(2e_i(2) - e_j(3) + 3\sum_{k\neq i,j} -c_{kj}e_k(2)$$

$$+ \sum_{k\neq i,j} -c_{ki}e_k(1))$$

$$= s_i^{\langle n\rangle}s_j^{\langle n\rangle}(-2e_i(4) + e_j(3) + 3\sum_{k\neq i,j} -c_{kj}e_k(2)$$

$$+ \sum_{k\neq i,j} -c_{ki}(e_k(1) + 2e_k(3)))$$

$$
\begin{aligned}
&= s_i^{\langle n\rangle}(e_i(4) - e_j(5) + 3\sum_{k\neq i,j} -c_{kj}(e_k(2) + 2e_k(4)) \\
&\qquad + \sum_{k\neq i,j} -c_{ki}(e_k(1) + 2e_k(3))) \\
&= -e_i(6) + 3\sum_{k\neq i,j} -c_{kj}(e_k(2) + 2e_k(4)) \\
&\qquad + \sum_{k\neq i,j} -c_{ki}(e_k(1) + 2e_k(3) + e_k(4)).
\end{aligned} \tag{2.13}
$$

By applying (2.10), we have

$$
\begin{aligned}
(s_i^{\langle n\rangle}s_j^{\langle n\rangle})^3(e_i(0)) &= (s_i^{\langle n\rangle}s_j^{\langle n\rangle})^2 s_i^{\langle n\rangle}(e_i(0)) \\
&= s_j^{\langle n\rangle}(s_i^{\langle n\rangle}s_j^{\langle n\rangle})^2 s_i^{\langle n\rangle}(e_i(0)) = (s_j^{\langle n\rangle}s_i^{\langle n\rangle})^3(e_i(0)).
\end{aligned} \tag{2.14}
$$

Similarly, we have

$$
\begin{aligned}
s_j^{\langle n\rangle}s_i^{\langle n\rangle}s_j^{\langle n\rangle}s_i^{\langle n\rangle}s_j^{\langle n\rangle}e_j(0) &= s_j^{\langle n\rangle}s_i^{\langle n\rangle}s_j^{\langle n\rangle}s_i^{\langle n\rangle}(-e_j(2) + 3e_i(1) \\
&\qquad + 3\sum_{k\neq i,j} -c_{kj}e_k(1)) \\
&= s_j^{\langle n\rangle}s_i^{\langle n\rangle}s_j^{\langle n\rangle}(-3e_i(3) + 2e_j(2) + 3\sum_{k\neq i,j} -c_{ki}e_k(2) + 3\sum_{k\neq i,j} -c_{kj}e_k(1)) \\
&= s_j^{\langle n\rangle}s_i^{\langle n\rangle}(-2e_j(4) + 3e_i(3) + 3\sum_{k\neq i,j} -c_{ki}e_k(2) \\
&\qquad + 3\sum_{k\neq i,j} -c_{kj}(e_k(1) + 2e_k(3))) \\
&= s_j^{\langle n\rangle}(-3e_i(5) + e_j(4) + 3\sum_{k\neq i,j} -c_{ki}(e_k(2) + e_k(4)) \\
&\qquad + 3\sum_{k\neq i,j} -c_{kj}(e_k(1) + 2e_k(3))) \\
&= -e_j(6) + 3\sum_{k\neq i,j} -c_{ki}(e_k(2) + e_k(4)) \\
&\qquad + 3\sum_{k\neq i,j} -c_{kj}(e_k(1) + 2e_k(3) + e_k(5)).
\end{aligned} \tag{2.15}
$$

Thus, in light of (2.10), one has

$$(s_i^{\langle n\rangle} s_j^{\langle n\rangle})^3 e_j(0) = s_i^{\langle n\rangle} s_j^{\langle n\rangle} (s_i^{\langle n\rangle} s_j^{\langle n\rangle})^2 e_j(0) \overset{(2.15)}{=} s_j^{\langle n\rangle} (s_i^{\langle n\rangle} s_j^{\langle n\rangle})^2 e_j(0)$$
$$= (s_j^{\langle n\rangle} s_i^{\langle n\rangle})^3 e_j(0). \tag{2.16}$$

The equalities (2.14) and (2.16) together imply that (2.8) holds when $m_{ij} = 6$. The proposition is thus proved.

We determine the order of the generators $s_i^{\langle n\rangle}$ as follows.

Proposition 2.3 *Suppose that i is isolated, i.e., there exists no vertex adjacent to i. Then*

$$\left|s_i^{\langle n\rangle}\right| = \begin{cases} 2n & \text{if } n \text{ is odd,} \\ n & \text{if } n \text{ is even.} \end{cases} \tag{2.17}$$

Suppose that i is not isolated. Then the order of $s_i^{\langle n\rangle}$ is given by

$$\left|s_i^{\langle n\rangle}\right| = \begin{cases} 2n & \text{if } n \text{ is odd,} \\ n & \text{if } n = 2n',\ n' \text{ is odd,} \\ \infty & \text{if } 4|n. \end{cases} \tag{2.18}$$

Proof By a simple induction, we have $(s_i^{\langle n\rangle})^k(x) = y$ with

$$y_j(b) = \begin{cases} (-1)^k x_i(b-2k) & \text{if } j = i, \\ x_j(b) - c_{ji} \sum_{l=1}^k (-1)^{l-1} x_i(b-(2l-1)) & \text{if } j \neq i. \end{cases} \tag{2.19}$$

So if i is isolated, the order of $s_i^{\langle n\rangle}$ is as in the assumption. Now we consider the case where i is not isolated and n is odd. If we take $k = 2n$ in the above formula, we obtain $y_i(b) = x_i(b)$. So it remains only to check if $y_j(b) = x_j(b)$ for $j \neq i$ such that $c_{ji} \neq 0$. Note that

$$\sum_{l=1}^{2n} (-1)^{l-1} x_i(b-(2l-1))$$
$$= \sum_{l=1}^{n} (-1)^{l-1} x_i(b-(2l-1)) + \sum_{l=n+1}^{2n} (-1)^{l-1} x_i(b-(2l-1))$$
$$= \sum_{l=1}^{n} (-1)^{l-1} x_i(b-(2l-1)) + \sum_{l'=1}^{n} (-1)^{l'+n-1} x_i(b-(2l'+2n-1)),$$

$$l = n + l'$$

$$= \sum_{l=1}^{n} (-1)^{l-1} x_i (b - (2l - 1)) - \sum_{l'=1}^{n} (-1)^{l'-1} x_i (b - (2l' - 1)) = 0.$$

This implies that $(s_i^{\langle n \rangle})^{2n} = 1$. Note that the terms in $\sum_{l=1}^{n} (-1)^{l-1} x_i (b - (2l - 1))$ are all distinct, and hence $2n$ is the order of $s_i^{\langle n \rangle}$ when n is odd.

Now we assume that i is not isolated and $n = 2n'$ with n' being odd. We have

$$\sum_{l=1}^{n} (-1)^{l-1} x_i (b - (2l - 1))$$

$$= \sum_{l=1}^{n'} (-1)^{l-1} x_i (b - (2l - 1)) + \sum_{l=n'+1}^{n} (-1)^{l-1} x_i (b - (2l - 1))$$

$$= \sum_{l=1}^{n'} (-1)^{l-1} x_i (b - (2l - 1)) + \sum_{l'=1}^{n'} (-1)^{l'+n'-1} x_i (b - (2l' + 2n' - 1)),$$

$$l = l' + n'$$

$$= \sum_{l=1}^{n'} (-1)^{l-1} x_i (b - (2l - 1)) - \sum_{l'=1}^{n'} (-1)^{l'-1} x_i (b - (2l' - 1)) = 0.$$

This implies that $(s_i^{\langle n \rangle})^n = 1$. Moreover, the terms in $\sum_{l=1}^{n'} (-1)^{l-1} x_i (b - (2l - 1))$ are distinct in general and cannot be cancelled out. Hence n is the order of $s_i^{\langle n \rangle}$ in this case.

Finally, we assume that $4|n$. By (2.7), we only need to show that the order of $s_i^{\langle n \rangle}$ is ∞ for $n = 4$. We have

$$\sum_{l=1}^{k} (-1)^{l-1} x_i (b - (2l - 1))$$

$$= \sum_{l \text{ odd}} (-1)^{l-1} x_i (b - (2l - 1)) + \sum_{l \text{ even}} (-1)^{l-1} x_i (b - (2l - 1))$$

$$= \left(\lfloor k/2 \rfloor + \frac{1 - (-1)^k}{2} \right) x_j (b - 1) - \lfloor k/2 \rfloor x_j (b + 1).$$

This implies that $(s_i^{\langle n\rangle})^k \neq (s_i^{\langle n\rangle})^{k'}$ if $k \neq k'$. So the order of $s_i^{\langle n\rangle}$ is infinity. The proposition is thus proved.

Corollary 2.4 *We have* $W^{\langle 2\rangle} \cong W^{\langle 1\rangle} = W$. □

Proof Due to (2.17)–(2.18), we have $|s_i^{\langle 2\rangle}| = 2$. Hence there is a surjective algebra homomorphism from $W^{\langle 1\rangle} = W \to W^{\langle 2\rangle}$. Clearly, this is the inverse of $\pi_{2,1}$ in (2.7), which proves the Corollary.

2.4 The Dual Picture

To facilitate the discussion in this section, we rewrite the standard basis element $e_i(a)$ as $\varpi_i(a)$. We set

$$\alpha_i^\vee(a) = \sum_{(j,b)\in I\times\mathbb{Z}_n} c_{(j,b),(i,a)}^{\langle n\rangle}\varpi_j(b), \forall (i,a) \in I \times \mathbb{Z}_n.$$

Let $\alpha_i(a)$ for all $(i,a) \in I \times \mathbb{Z}_n$ be unrelated variables. Let $X_{\langle n\rangle} = \mathbb{Z}[\alpha_i(a)]_{(i,a)\in I\times\mathbb{Z}_n}$ be the free abelian group generated by $\alpha_i(a)$ and set $X_{\langle n\rangle}^\vee = \mathbb{Z}[\varpi_i(a)]_{(i,a)\in I\times\mathbb{Z}_n}$. Define a perfect pairing $\langle -,-\rangle_{\langle n\rangle} : X_{\langle n\rangle} \times X_{\langle n\rangle}^\vee \to \mathbb{Z}$ by

$$\langle \alpha_i(a), \varpi_j(b)\rangle_{\langle n\rangle} = \delta_{ij}\delta_{ab}, \quad \forall (i,a),(j,b) \in I \times \mathbb{Z}_n.$$

Then the n-shifted simple reflection $s_i^{\langle n\rangle}$ in (2.2) can be rewritten as

$$s_i^{\langle n\rangle}(\lambda) = \lambda - \sum_{a\in\mathbb{Z}_n} \langle \alpha_i(a), \lambda\rangle_{\langle n\rangle}\alpha_i^\vee(a), \quad \forall \lambda \in X_{\langle n\rangle}^\vee. \tag{2.20}$$

Notice here that by setting $n = 1$, we go back to the standard definition of the Weyl group W. Let $r_i^{\langle n\rangle} : X_{\langle n\rangle} \to X_{\langle n\rangle}$ be the function defined by

$$r_i^{\langle n\rangle}(\mu) = \mu - \sum_{a\in\mathbb{Z}_n} \langle \mu, \alpha_i^\vee(a)\rangle_{\langle n\rangle}\alpha_i(a), \quad \forall \mu \in X_{\langle n\rangle}. \tag{2.21}$$

One can check that $r_i^{\langle n\rangle}$ is bijective (see the following proof of Proposition 2.5). Let $W^{*\langle n\rangle}$ be the group generated by $r_i^{\langle n\rangle}$ for all $i \in I$. When $n = 0$, the groups $W^{\langle n\rangle}$ and $W^{*\langle n\rangle}$ are two different incarnations of Weyl groups. For a general n, we have

Proposition 2.5 *There is an isomorphism* $W^{\langle n\rangle} \cong W^{*\langle n\rangle}$, $s_i^{\langle n\rangle} \mapsto (r_i^{\langle n\rangle})^{-1}$, $\forall i \in I$. □

Proof We observe that

$$\langle r_i^{\langle n\rangle}(\mu), \lambda\rangle_{\langle n\rangle} = \langle \mu, s_i^{\langle n\rangle}(\lambda)\rangle_{\langle n\rangle}, \forall \mu \in X_{\langle n\rangle}, \lambda \in X_{\langle n\rangle}^{\vee}. \tag{2.22}$$

Write $\mu = \sum_{(i,a)\in I\times\mathbb{Z}_n} \mu_i(a)\alpha_i(a)$ and $\lambda = \sum_{(i,a)\in I\times\mathbb{Z}_n} \lambda_i(a)\varpi_i(a)$. A direct computation yields

$$\langle r_i^{\langle n\rangle}(\mu), \lambda\rangle_{\langle n\rangle} = \sum_{j,b} \mu_j(b)\lambda_j(b) - \sum_{a\in\mathbb{Z}_n}\sum_{j,b} \mu_j(b)c_{(j,b),(i,a)}^{\langle n\rangle}\lambda_i(a)$$
$$= \langle \mu, s_i^{\langle n\rangle}(\lambda)\rangle_{\langle n\rangle}.$$

If we write $r_i^{\langle n\rangle}, s_i^{\langle n\rangle}$ as matrices with respect to the standard basis, then they are transpose to each other in light of (2.22). Taking transpose defines an anti-isomorphism from $W^{\langle n\rangle}$ to $W^{*\langle n\rangle}$ sending $s_i^{\langle n\rangle}$ to $r_i^{\langle n\rangle}$, while taking inverse defines another anti-automorphism on $W^{*\langle n\rangle}$ itself. By composing the two anti-homomorphisms, the Proposition is proved.

2.5 *Reduced Burau Representation*

In this section, we consider the case when $\mathbb{Z}_n = \mathbb{Z}$ for $n = 0$ and the Cartan matrix is of type A_ℓ for $\ell \geqslant 1$. We shall use the indexing of a Dynkin diagram as in (3.1). We note that there is an isomorphism of free $\mathbb{Z}[t, t^{-1}]$-modules of rank ℓ:

$$X_{\langle 0\rangle} \cong \mathbb{Z}[t, t^{-1}][e_i]_{1\leqslant i\leqslant \ell} \cong \mathbb{Z}[t, t^{-1}]^{\ell}, \quad \alpha_i(a) \mapsto t^a e_i,$$

where e_i is an indeterminate. Note also that $s_i^{\langle 0\rangle}$ commutes with the shift function z in (2.9). So $s_i^{\langle 0\rangle}$ is $\mathbb{Z}[t, t^{-1}]$-linear. Via the above isomorphism, the shifted simple reflection $r_i^{\langle 0\rangle}$ in (2.21) for $n = 0$ is naturally an element in $\mathrm{GL}_\ell(\mathbb{Z}[t^{\pm 1}])$, which admits a matrix presentation as follows:

$$r_1^{\langle 0\rangle}(t) = \left[\begin{array}{cc|c} -t^2 & t & 0 \\ 0 & 1 & 0 \\ \hline 0 & 0 & I_{\ell-2} \end{array}\right],$$

$$r_i^{\langle 0\rangle}(t) = \left[\begin{array}{c|ccc|c} I_{i-2} & 0 & 0 & 0 & 0 \\ \hline 0 & 1 & 0 & 0 & 0 \\ 0 & t & -t^2 & t & 0 \\ 0 & 0 & 0 & 1 & 0 \\ \hline 0 & 0 & 0 & 0 & I_{\ell-i-1} \end{array}\right], \quad 2 \leqslant i \leqslant \ell-1, \tag{2.23}$$

$$r_\ell^{\langle 0\rangle}(t) = \left[\begin{array}{c|cc} I_{\ell-2} & 0 & 0 \\ \hline 0 & 1 & 0 \\ 0 & t & -t^2 \end{array}\right].$$

Recall that the reduced Burau representation [3, 10] of a braid group on $X_{\langle 0\rangle}$ has the generators σ_i presented as the following matrices:

$$\sigma_1(t) = \left[\begin{array}{cc|c} -t & 1 & 0 \\ 0 & 1 & 0 \\ \hline 0 & 0 & I_{\ell-2} \end{array}\right],$$

$$\sigma_i(t) = \left[\begin{array}{c|ccc|c} I_{i-2} & 0 & 0 & 0 & 0 \\ \hline 0 & 1 & 0 & 0 & 0 \\ 0 & t & -t & 1 & 0 \\ 0 & 0 & 0 & 1 & 0 \\ \hline 0 & 0 & 0 & 0 & I_{\ell-i-1} \end{array}\right], \quad 2 \leqslant i \leqslant \ell-1, \tag{2.24}$$

$$\sigma_\ell(t) = \left[\begin{array}{c|cc} I_{\ell-2} & 0 & 0 \\ \hline 0 & 1 & 0 \\ 0 & t & -t \end{array}\right].$$

It turns out that the matrices in (2.23) and (2.24) are similar. Indeed, let D be the diagonal matrix of size $\ell \times \ell$ whose i-diagonal entry is t^i. Then we have

$$\sigma_i(t^2) = D \cdot r_i^{\langle 0\rangle}(t) \cdot D^{-1}, \quad \forall 1 \leqslant i \leqslant \ell. \tag{2.25}$$

This implies the following proposition.

Proposition 2.6 *The braid group representation $X_{\langle 0\rangle}$ defined by $r_i^{\langle 0\rangle}(t)$ for $1 \leqslant i \leqslant \ell$ is isomorphic to the Burau representation defined by $\sigma_i(t^2)$ for all $1 \leqslant i \leqslant \ell$.* □

Remark 2.7 It is well known (e.g., [10]) that a reduced Burau representation is not faithful in general. So neither is $W^{\langle 0\rangle}$ an Artin group in general. In particular, the relations (2.8), (2.17), and (2.18) do not provide a presentation of $W^{\langle n\rangle}$ for a general

n. It seems interesting to investigate under what assumption on n these relations provide a presentation of $W^{\langle n\rangle}$. □

2.6 $W^{\langle n\rangle}$ as Complex Reflection Groups

In this section, we assume that $n > 0$. Let us fix an n-th primitive root of unity ζ_n. Then there is a ring homomorphism

$$\mathbb{Z}[\mathbb{Z}_n] \to \mathbb{Z}[\zeta_n], x = (x(a))_{0\leqslant a\leqslant n-1} \mapsto \sum_{a=0}^{n-1} x(a)\zeta_n^a.$$

This map is clearly surjective, but not injective. The shift function z commutes with the multiplication by ζ. Further we have a $\mathbb{Z}$-linear map $X_{\langle n\rangle} \to \mathbb{Z}[\zeta_n]^\ell$. So each $s_i^{\langle n\rangle}$ defines an element $\bar{s}_i^{\langle n\rangle} \in \mathrm{GL}_\ell(\mathbb{Z}[\zeta_n])$. This assignment is compatible with function composition. Hence there is a group homomorphism

$$\mathrm{ev} : W^{\langle n\rangle} \to \mathrm{GL}_\ell(\mathbb{Z}[\zeta_n]).$$

Let $\bar{W}^{\langle n\rangle}$ be the image of the homomorphism ev. In particular $\bar{W}^{\langle n\rangle}$ is generated by $\bar{s}_i^{\langle n\rangle}$ for all i, whose order is finite when $4 \nmid n$ thanks to Proposition 2.3. This proves the following Proposition:

Proposition 2.8 *If $4 \nmid n$, then $\bar{s}_i^{\langle n\rangle}$ is a pseudo-reflection, i.e., it is of finite order and* $\ker(1 - \bar{s}_i^{\langle n\rangle})$ *is a hyperplane in the complexified vector space.* □

By definition, we have $\ker(1 - s_i^{\langle n\rangle}) = \mathrm{span}_{\mathbb{C}}\{e_j(0) : j \neq i\}$. Hence we have the following Proposition by passage to quotient.

Proposition 2.9 *If $4 \nmid n$, the group $\bar{W}^{\langle n\rangle}$ is a complex reflection group.* □

The following two natural questions seem quite interesting.

Questio 2.7 Is $W^{\langle n\rangle}$ finite when n is not divisible by 4? □

One can define a group, denoted by $\sqrt{W^{\langle n\rangle}}$, by conjugating the generators by a diagonal matrix like (2.25) and substituting t^2 by a second parameter. The group seems to be more natural than $W^{\langle n\rangle}$.

3 The Identity $w_0^{\langle n\rangle} = -z^{\mathrm{c}}\theta$

In the section, we assume that the generalized Cartan matrix C is a genuine Cartan matrix. Let w_0 be the longest element in W. It is well-known that $w_0 = -\theta$ where

θ is a certain diagram involution on $\mathbb{Z}[I]$. In this section, we establish a similar identity in $W^{\langle n \rangle}$.

3.1 The Statement

Let w_0 be the longest element of W of length N. Let $R(w_0)$ be the set of sequences $\mathbf{i} = (i_1, i_2, \cdots, i_N) \in I^N$ such that $s_{i_1} s_{i_2} \cdots s_{i_N} = w_0$. Let $\Gamma(w_0)$ be a graph with vertex set $R(w_0)$ and two vertices $\mathbf{i}$ and $\mathbf{j}$ are joined by an edge if $\mathbf{j}$ can be obtained from $\mathbf{i}$ by one of the following rules.

- replace a consecutive pair (i, j) in $\mathbf{i}$ by (j, i) if $c_{ij} = 0$.
- replace a consecutive triple (i, j, i) in $\mathbf{i}$ by (j, i, j) if $c_{ij} c_{ji} = 1$.
- replace a consecutive quadruple (i, j, i, j) in $\mathbf{i}$ by (j, i, j, i) if $c_{ij} c_{ji} = 2$.
- replace a consecutive sextuple (i, j, i, j, i, j) in $\mathbf{i}$ by (j, i, j, i, j, i) if $c_{ij} c_{ji} = 3$.

It is known that $\Gamma(w_0)$ is connected, due to Matsumoto [9] (see also at [2, Theorem 3.3.1] or [8, 2.1.2]).

For any sequence $\mathbf{i} = (i_j)_{1 \leqslant j \leqslant N} \in R(w_0)$, we set

$$w_0^{\langle n \rangle} = s_{i_1}^{\langle n \rangle} s_{i_2}^{\langle n \rangle} \cdots s_{i_N}^{\langle n \rangle}.$$

In light of (2.8) and the fact that $\Gamma(w_0)$ is connected, $w_0^{\langle n \rangle}$ is independent of the choice of $\mathbf{i} \in R(w_0)$.

Recall the shift function z from (2.9). Let $\theta : I \to I$ be the diagram involution defined by $w_0(i) = -\theta(i)$ for all $i \in I$. Clearly, θ extends to an involution on $\mathbb{Z}[I \times \mathbb{Z}_n]$, still denoted by θ, such that $\theta(e_i(a)) = e_{\theta(i)}(a)$ for all $i \in I$.

Proposition 3.1 *We have $w_0^{\langle n \rangle} = -z^{\mathbf{c}}\theta$ where $\mathbf{c}$ is the Coxeter number of W. In particular, $w_0^{\langle n \rangle}$ is a central element in $W^{\langle n \rangle}$ if $\theta = 1$.* □

The proof of this Proposition is given in Sects. 3.2 for type A_ℓ, 3.3 for type B_ℓ, 3.4 for type C_ℓ, 3.5 for type D_ℓ, 3.6 for type E_6, 3.7 for type E_7, E_8, F_4, and 3.8 for type G_2, using a case-by-case analysis. As a byproduct, these also lead to the following result.

Proposition 3.2 *There is a reduced expression $w_0 = s_{i_1} \cdots s_{i_N} \in W$ such that the i_{j-1} entry of $s_{i_j} \cdots s_{i_N} (\sum_{i=1}^{\ell} e_i(0))$ is positive for all $2 \leqslant j \leqslant N+1$.* □

Proof The reduced expression of w_0 is the one we used in the proof of Proposition 3.1. For type A_ℓ, this is due to (3.2)–(3.3). For type B_ℓ, this is due to Lemma 3.4 and (3.5). For type C_ℓ, this is due to Lemma 3.5 and an analogue of (3.5). For type D_ℓ, this is due to Lemma 3.6 and (3.9). For type F_4 and G_2, it can be checked directly that the reduced expression listed in [1, Table 1] satisfies the condition. For exceptional cases E_6, E_7, and E_8, it is checked by using computer, and the R-code can be found at the first author's website.

Remark 3.3 Proposition 3.2 is expected to hold for any reduced expression of w_0. □

3.2 *Proof of Proposition 3.1: Type A_ℓ*

We use the following index of the type A_ℓ Dynkin diagram:

$$\underset{1}{\circ} \text{———} \underset{2}{\circ} \text{———} \cdots \text{———} \underset{\ell-1}{\circ} \text{———} \underset{\ell}{\circ} \tag{3.1}$$

We observe that for all $\mathbf{v} \in \mathbb{Z}[I \times \mathbb{Z}_n]$,

$$(s_i^{\langle n\rangle} \circ \cdots \circ s_\ell^{\langle n\rangle} \mathbf{v})_k(a) = \begin{cases} \mathbf{v}_i(a) & \text{if } k \leqslant i-2, \\ \sum_{j=i-1}^{\ell} \mathbf{v}_j(a-j+i-1) & \text{if } k = i-1, \\ -\sum_{j=i}^{\ell} \mathbf{v}_j(a-j+i-2) & \text{if } k = i, \\ \mathbf{v}_{k-1}(a-1) & \text{if } k \geqslant i+1. \end{cases} \tag{3.2}$$

Inductively using (3.2), one has

$$\begin{aligned} &[(s_i^{\langle n\rangle} \cdots s_\ell^{\langle n\rangle})(s_{i-1}^{\langle n\rangle} \cdots s_\ell^{\langle n\rangle}) \cdots (s_1^{\langle n\rangle} \cdots s_\ell^{\langle n\rangle})\mathbf{v}]_k(a) \\ &= \begin{cases} -\mathbf{v}_{\ell-k+1}(a-\mathbf{c}) & \text{if } k \leqslant i-1, \\ -\sum_{j=i}^{\ell} \mathbf{v}_{j+1-i}(a-j-1) & \text{if } k = i, \\ \mathbf{v}_{k-i}(a-i) & \text{if } k \geqslant i+1. \end{cases} \end{aligned} \tag{3.3}$$

Observe that $w_0 = s_\ell \circ (s_{\ell-1}s_\ell) \circ \cdots \circ (s_1 \cdots s_\ell)$. We have Proposition 3.1 of type A_ℓ by setting $i = \ell$ in (3.3).

3.3 *Proof of Proposition 3.1: Type B_ℓ*

We shall use the following indexing for type B_ℓ Dynkin diagram:

$$\underset{1}{\circ} \text{———} \underset{2}{\circ} \text{———} \cdots \text{———} \underset{\ell-1}{\circ} \Longrightarrow \underset{\ell}{\circ}$$

We start with a lemma. Set $\mathbf{c}(i) = 2\ell - 2(i-1)$ for all $1 \leqslant i \leqslant \ell$.

Lemma 3.4 *For all $\mathbf{v} \in \mathbb{Z}[I \times \mathbb{Z}_n]$ and $1 \leqslant i \leqslant n$, we have*

$$s_i^{\langle n\rangle} s_{i+1}^{\langle n\rangle} \cdots s_\ell^{\langle n\rangle} \cdots s_{i+1}^{\langle n\rangle} s_i^{\langle n\rangle}(\mathbf{v}) = \tilde{\mathbf{v}},$$

where

$$\tilde{\mathbf{v}}_i(a) = -\mathbf{v}_i(a - \mathbf{c}(i)) - \sum_{j=i+1}^{\ell} \mathbf{v}_j(a - j + i - 2) - \sum_{j=i+1}^{\ell} \mathbf{v}_j(a - \mathbf{c}(i) + j - i),$$

$$\tilde{\mathbf{v}}_k(a) = \begin{cases} \mathbf{v}_k(a) & \text{if } k < i - 1, \\ \mathbf{v}_{i-1}(a) + \mathbf{v}_i(a-1) - \tilde{\mathbf{v}}_i(a+1) & \text{if } k = i - 1, \\ \mathbf{v}_k(a-2) & \text{if } k > i. \end{cases}$$

Proof We prove this lemma by induction on i. If $i = \ell$, it follows from the definition.

Assume that the formula holds for all $k > i$. Let $s_i^{\langle n \rangle}\mathbf{v} = \mathbf{v}'$ and apply induction hypothesis to $s_{i+1}^{\langle n \rangle} \cdots s_\ell^{\langle n \rangle} \cdots s_{i+1}^{\langle n \rangle} s_i^{\langle n \rangle}(\mathbf{v}) = s_{i+1}^{\langle n \rangle} \cdots s_\ell^{\langle n \rangle} \cdots s_{i+1}^{\langle n \rangle}(\mathbf{v}') = \mathbf{v}''$. We then have

$$\mathbf{v}_k''(a) = \mathbf{v}_k'(a) = \mathbf{v}_k(a), \forall k \leqslant i - 2; \quad \mathbf{v}_k''(a) = \mathbf{v}_k'(a-2) = \mathbf{v}_k(a-2), \forall k \geqslant i + 2.$$

We also have

$$\begin{aligned} \mathbf{v}_{i-1}''(a) &= \mathbf{v}_{i-1}'(a) = \mathbf{v}_{i-1}(a) + \mathbf{v}_i(a-1), \\ \mathbf{v}_i''(a) &= \mathbf{v}_i'(a) + \mathbf{v}_{i+1}'(a-1) - \tilde{\mathbf{v}}_{i+1}'(a+1) \\ &= \mathbf{v}_{i+1}(a-1) - \tilde{\mathbf{v}}_{i+1}'(a+1), \\ \mathbf{v}_{i+1}''(a) &= \tilde{\mathbf{v}}_{i+1}'(a). \end{aligned}$$

Applying $\mathbf{s}_i^{\langle n \rangle}$ to $\mathbf{v}''$, the values $\mathbf{v}_k''(a)$ will only change when $k = i-1, i, i+1$. Observe that $\mathbf{v}_k''(a) = \tilde{\mathbf{v}}_k(a)$ if $k \neq i-1, i, i+1$. If we can show that

$$-\mathbf{v}_i''(a-2) = \tilde{\mathbf{v}}_i(a), \tag{3.4}$$

then the lemma follows. Indeed, at the $(i-1, a)$-th entry of $s_i^{\langle n \rangle}\mathbf{v}''$, we get

$$\mathbf{v}_{i-1}''(a) + \mathbf{v}_i''(a-1) \overset{(3.4)}{=} \mathbf{v}_{i-1}(a) + \mathbf{v}_i(a-1) - \tilde{\mathbf{v}}_i(a+1),$$

and at the $(i+1, a)$-th entry of $s_i^{\langle n \rangle}\mathbf{v}''$, we get

$$\tilde{\mathbf{v}}_{i+1}'(a) + \mathbf{v}_{i+1}(a-2) - \tilde{\mathbf{v}}_{i+1}'(a) = \mathbf{v}_{i+1}(a-2).$$

The equality (3.4) can be checked as follows:

$$\begin{aligned}\mathbf{v}_i''(a-2) &= \mathbf{v}_{i+1}(a-3) - \tilde{\mathbf{v}}_{i+1}'(a-1)\\ &= \mathbf{v}_{i+1}(a-3) + \mathbf{v}_{i+1}'(a-1-\mathbf{c}(i+1))\\ &\quad + \sum_{j=i+2}^{\ell} \mathbf{v}_j(a-1-j+(i+1)-2)\\ &\quad + \sum_{j=i+2}^{\ell} \mathbf{v}_j(a-1-\mathbf{c}(i+1)+j-i-1)\\ &= \mathbf{v}_{i+1}(a-3) + \mathbf{v}_{i+1}(a-\mathbf{c}(i)+1) + \mathbf{v}_i(a-\mathbf{c}(i))\\ &\quad + \sum_{j=i+2}^{\ell} \mathbf{v}_j(a-j+i-2) + \sum_{j=i+2}^{\ell} \mathbf{v}_j(a-\mathbf{c}(i)+j-i)\\ &= \tilde{\mathbf{v}}_i(a).\end{aligned}$$

This finishes the induction and the lemma follows.

By applying Lemma 3.4 repetitively, we obtain the following formula:

$$\begin{gathered}(s_i^{\langle n\rangle} s_{i+1}^{\langle n\rangle} \cdots s_\ell^{\langle n\rangle} \cdots s_{i+1}^{\langle n\rangle} s_i^{\langle n\rangle})(s_{i+1}^{\langle n\rangle} \cdots s_\ell^{\langle n\rangle} \cdots s_{i+1}^{\langle n\rangle}) \cdots (s_1^{\langle n\rangle} \cdots \\ s_\ell^{\langle n\rangle} \cdots s_1^{\langle n\rangle})\mathbf{v} = \mathbf{v}',\\ \mathbf{v}_k'(a) = \begin{cases} -\mathbf{v}_k(a-\mathbf{c}) & \text{if } k < i,\\ \tilde{\mathbf{v}}_i(a-2(i-1)) & \text{if } k = i,\\ \mathbf{v}_k(a-2i) & \text{if } k > i.\end{cases}\end{gathered} \tag{3.5}$$

The formula (3.5) can be obtained by an induction on i in light of Lemma 3.4 and the equality:

$$\tilde{\mathbf{v}}_i(a-2i+2)+\mathbf{v}_{i+1}(a-2i-1)-\tilde{\mathbf{v}}_{i+1}(a-2i+1)=-\mathbf{v}_i(a-\mathbf{c}), \quad \forall 1 \leqslant i \leqslant \ell. \tag{3.6}$$

The verification of the latter equality is straightforward and omitted.

From [1, Table 1], we know that w_0 admits a reduced expression of the form

$$w_0 = (s_\ell)(s_{\ell-1}s_\ell s_{\ell-1}) \cdots (s_i s_{i+1} \cdots s_\ell \cdots s_{i+1} s_i) \cdots (s_1 s_2 \cdots s_\ell \cdots s_2 s_1). \tag{3.7}$$

Taking $i = \ell$ in (3.5) leads to Proposition 3.1 in type B_ℓ.

3.4 Proof of Proposition 3.1: Type C_ℓ

In this section, we deal with type C_ℓ. We fix an indexing of type C_ℓ Dynkin diagram.

$$\underset{1}{\circ} \text{---} \underset{2}{\circ} \text{---} \cdots \text{---} \underset{\ell-1}{\circ} \Longleftarrow \underset{\ell}{\circ}$$

The proof follows the same line of arguments in type B_ℓ. We begin with a lemma similar to Lemma 3.4. Recall that $\mathbf{c}(i) = 2\ell - 2(i-1)$.

Lemma 3.5 *For all $\mathbf{v} \in \mathbb{Z}[I \times \mathbb{Z}_n]$ and $1 \leqslant i < \ell$, we have $s_i^{\langle n \rangle} s_{i+1}^{\langle n \rangle} \cdots s_\ell^{\langle n \rangle} \cdots s_{i+1}^{\langle n \rangle} s_i^{\langle n \rangle} \mathbf{v} \equiv \tilde{\mathbf{v}}$ where*

$$\tilde{\mathbf{v}}_i(a) = -\sum_{j=i+1}^{\ell-1} \mathbf{v}_j(a-j+i-2) - \sum_{j=i}^{\ell} \mathbf{v}_j(a-\mathbf{c}(i)+j-i),$$

$$\tilde{\mathbf{v}}_k(a) = \begin{cases} \mathbf{v}_k(a) & \text{if } k < i-1, \\ \mathbf{v}_{i-1}(a) + \mathbf{v}_i(a-1) - \check{\mathbf{v}}_i(a+1) & \text{if } k = i-1, \\ \mathbf{v}_k(a-2) & \text{if } k > i. \end{cases}$$

Note that $i \neq \ell$ in the assumption and $j \neq \ell$ in the first sum of $\tilde{\mathbf{v}}_i(a)$ in the above lemma.

Proof The proof is similar to that of Lemma 3.4. First one observes that the formula holds when $i = \ell - 1$. Assume that it holds for all j such that $i < j < \ell$. Following the argument as in the proof of Lemma 3.4, it is reduced to show that

$$-\mathbf{v}_{i+1}(a-3) - \mathbf{v}_i(a-\mathbf{c}(i)) + \tilde{\mathbf{v}}_{i+1}(a) = \tilde{\mathbf{v}}_i(a), \tag{3.8}$$

which can be verified directly. (Here $\tilde{\mathbf{v}}_{i+1}(a)$ is obtained by substituting i in $\tilde{\mathbf{v}}_i(a)$ by $i+1$ in the Lemma.)

Observe that (3.6) still holds in here, and so by induction we obtain the counterpart of (3.5). Taking $i = \ell$ in (3.5) yields the proof of Proposition 3.1 in type C_ℓ since w_0 still has a reduced expression (3.7).

3.5 Proof of Proposition 3.1: Type D_ℓ

We fix an indexing of type D_ℓ Dynkin diagram.

We set $\mathbf{c}'(i) = 2\ell - 2i = \mathbf{c}(i) + 2$. The following is a type D_ℓ counterpart of Lemmas 3.4 and 3.5. Set $\mathbf{r}_i = s_i^{\langle n\rangle} s_{i+1}^{\langle n\rangle} \cdots s_{\ell-2}^{\langle n\rangle} s_\ell^{\langle n\rangle} s_{\ell-1}^{\langle n\rangle} s_{\ell-2}^{\langle n\rangle} \cdots s_{i+1}^{\langle n\rangle} s_i^{\langle n\rangle}$ for $1 \leqslant i \leqslant \ell - 2$. We also set $\mathbf{r}_{\ell-1} = s_{\ell-1}^{\langle n\rangle}$ and $\mathbf{r}_\ell = s_\ell^{\langle n\rangle}$.

Lemma 3.6 *For all* $\mathbf{v} \in \mathbb{Z}[I \times \mathbb{Z}_n]$ *and* $1 \leqslant i \leqslant \ell - 2$, *we have* $\mathbf{r}_i(\mathbf{v}) = \tilde{\mathbf{v}}$, *where*

$$\tilde{\mathbf{v}}_i(a) = -\sum_{j=i+1}^{\ell-2} \mathbf{v}_j(a - j + i - 2) - \sum_{j=i}^{\ell-2} \mathbf{v}_j(a - \mathbf{c}'(i) + j - i) - \sum_{j=\ell-1}^{\ell} \mathbf{v}_j(a - \mathbf{c}'(i) + \ell - i - 1),$$

$$\tilde{\mathbf{v}}_k(a) = \begin{cases} \mathbf{v}_k(a) & \text{if } k < i - 1, \\ \mathbf{v}_{i-1}(a) + \mathbf{v}_i(a-1) - \tilde{\mathbf{v}}_i(a+1) & \text{if } k = i - 1, \\ \mathbf{v}_k(a-2) & \text{if } i < k \leqslant \ell - 2, \\ \mathbf{v}_\ell(a-2) & \text{if } k = \ell - 1, \\ \mathbf{v}_{\ell-1}(a-2) & \text{if } k = \ell. \end{cases}$$

The lemma can be proved in a similar way as that of Lemma 3.4 once we check the case $i = \ell - 2$ and have the identity obtained from (3.8) by replacing $\mathbf{c}(i)$ with $\mathbf{c}'(i)$.

Note that (3.6) still holds here. By using Lemma 3.6 and (3.6), we obtain via induction the following formula. For all $i = \ell - 2$,

$$\mathbf{r}_i \mathbf{r}_{i-1} \cdots \mathbf{r}_1(\mathbf{v}) = \mathbf{v}',$$
$$\mathbf{v}'_k(a) = \begin{cases} -\mathbf{v}_k(a - \mathbf{c}) & \text{if } k < i, \\ \tilde{\mathbf{v}}_i(a - 2(i-1)) & \text{if } k = i, \\ \mathbf{v}_{k'}(a - 2i) & \text{if } k > i, \end{cases} \tag{3.9}$$

where $k' = k$ if i is even and if i is odd, it is defined by $k' = k$ if $k \leqslant \ell - 2$ and

$$k' = \begin{cases} \ell & \text{if } k = \ell - 1, \\ \ell - 1 & \text{if } k = \ell. \end{cases} \tag{3.10}$$

Note that when $i = \ell - 2$, the $(\ell-2, a)$-th entry of $\mathbf{r}_i\mathbf{r}_{i-1}\cdots\mathbf{r}_1(\mathbf{v})$ is $-\mathbf{v}_{\ell-2}(a-\mathbf{c})-\mathbf{v}_{\ell-1}(a-\mathbf{c}+1)-\mathbf{v}_\ell(a-\mathbf{c}+1)$. The entry at $(\ell-1, a)$ and (ℓ, a) of $\mathbf{r}_i\mathbf{r}_{i-1}\cdots\mathbf{r}_1(\mathbf{v})$ is either $\mathbf{v}_{\ell-1}(a-\mathbf{c}+2)$ or $\mathbf{v}_\ell(a-\mathbf{c}+2)\}$. From [1], we see that $w_0^{\langle n\rangle} = \mathbf{r}_\ell\mathbf{r}_{\ell-1}\cdots\mathbf{r}_1$ and hence by taking $i = \ell - 2$ in (3.9) and apply $\mathbf{r}_\ell\mathbf{r}_{\ell-1}$, we have Proposition 3.1 in type D_ℓ as desired.

3.6 *Proof of Proposition 3.1: Type E_6*

We shall use the following indexing of type E_6 diagram:

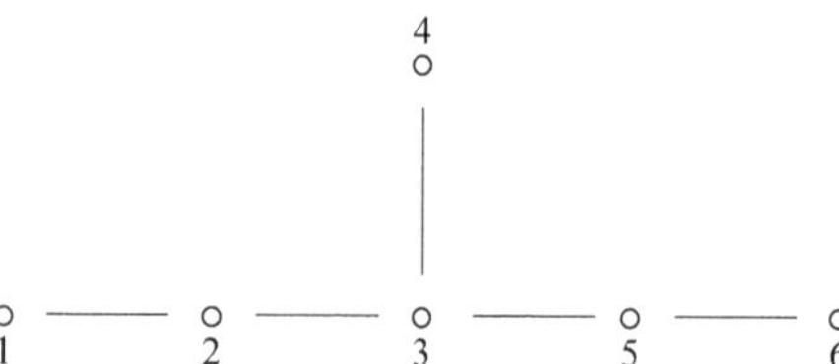

Recall from [1] that

$$w_0 = \mathbf{r}_5\mathbf{r}_4\cdots\mathbf{r}_1\mathbf{r}_6,$$

with $\mathbf{r}_i$ for $i = 1, \cdots, 5$ the same as in type D_5 and $\mathbf{r}_6$ is the product of $s_i^{\langle n\rangle}$ in the following order:

$$(6, 5, 3, 4, 2, 1, 3, 2, 5, 3, 4, 6, 5, 3, 2, 1).$$

By a direct computation, we have

$$(\mathbf{r}_6\mathbf{v})_i(a) = \begin{cases} \mathbf{v}_6(a-4) & \text{if } i = 1, \\ \mathbf{v}_5(a-4) & \text{if } i = 2, \\ \mathbf{v}_3(a-4) & \text{if } i = 3, \\ \mathbf{v}_2(a-4) & \text{if } i = 4, \\ \mathbf{v}_4(a-4) & \text{if } i = 5, \\ \beta & \text{if } i = 6, \end{cases} \tag{3.11}$$

where

$$\begin{aligned}\beta &= \mathbf{v}_4(a-5) + \mathbf{v}_3(a-6) + \mathbf{v}_2(a-6) + \mathbf{v}_5(a-6) \\ &\quad + \mathbf{v}_6(a-7) + \mathbf{v}_3(a-7) + \mathbf{v}_5(a-8) + \mathbf{v}_4(a-8) + \mathbf{v}_3(a-9) + \mathbf{v}_2(a-10) \\ &\quad + \mathbf{v}_1(a-11).\end{aligned}$$

So we have

$$(w_0^{\langle n\rangle}\mathbf{v})_i(a) = (w_0^{\langle n\rangle}(D_5)\mathbf{r}_6\mathbf{v})_i(a) = \begin{cases} -\mathbf{v}_6(a-4-\mathbf{c}') & \text{if } i=1, \\ -\mathbf{v}_5(a-4-\mathbf{c}') & \text{if } i=2, \\ -\mathbf{v}_3(a-4-\mathbf{c}') & \text{if } i=3, \\ -\mathbf{v}_4(a-4-\mathbf{c}') & \text{if } i=4, \\ -\mathbf{v}_2(a-4-\mathbf{c}') & \text{if } i=5, \\ \beta' & \text{if } i=6, \end{cases} \tag{3.12}$$

for some β' and $\mathbf{c}'$ is the Coxeter number for D_5. By the diagram involution of E_6, we see that β' must be $-\mathbf{v}_1(a-4-\mathbf{c}')$. The proof of Proposition 3.1 for type E_6 follows once we know that $\mathbf{c} = \mathbf{c}' + 4$.

3.7 *Proof of Proposition 3.1: Type E_7, E_8, and F_4*

For the exceptional cases E_7, E_8, and F_4, we use computer software R to verify Proposition 3.1. The R-codes can be found at the first author's website. Since the "n-deformed" simple reflection $s_i^{\langle n\rangle}$ commutes with the shift function z, we only need to check the formula when evaluated at $e_i(0)$ for various i.

Note that we use the reduced expression of w_0 in [1, Table 1]. But since we are using computer it might be easier to write the code by using the expression $w_0 = \chi^{h/2}$ where χ is a Coxeter element written as a product of two commuting involutions and h is a Coxeter number of type E_7, E_8, or F_4. We refer to J. E. Humphreys' notes [5] for further details.

Note that the indexing of type E_7, E_8, and F_4 Dynkin diagrams used in the programming are as follows:

- E_7:

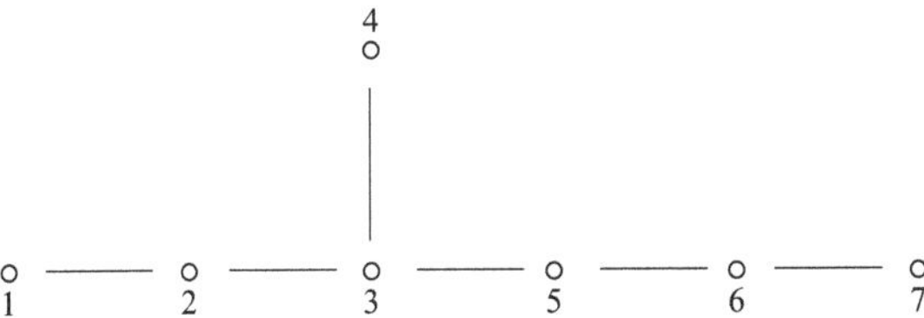

- E_8:

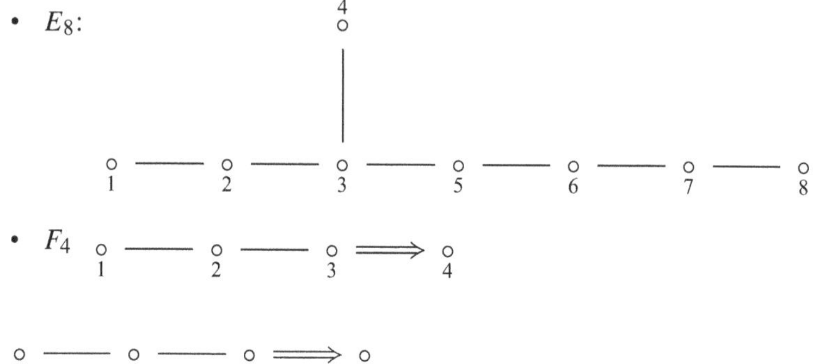

- F_4 $\underset{1}{\circ} \text{———} \underset{2}{\circ} \text{———} \underset{3}{\circ} \Longrightarrow \underset{4}{\circ}$

3.8 Proof of Proposition 3.1: Type G_2

We use the following index of type G_2 Dynkin diagram $\underset{1}{\circ} \Rrightarrow \underset{2}{\circ}$. This is due to (2.13) and (2.15) since $w_0^{\langle n\rangle} = (s_1^{\langle n\rangle} s_2^{\langle n\rangle})^3$. The proof of Proposition 3.1 is complete.

4 An Explicit Description of $w_0 *_{\mathbf{w}}^{\langle n\rangle} \mathbf{v}$

In this section, we study a non-linear $W^{\langle n\rangle}$-action which is an algebraization of the one used in graded quiver varieties.

4.1 A Characterization

For each $i \in I$, we define a non-linear function

$$s_i *_{\mathbf{w}}^{\langle n\rangle} : \mathbb{Z}[I \times \mathbb{Z}_n] \to \mathbb{Z}[I \times \mathbb{Z}_n]$$

by the following rule: for all $\mathbf{v} \in \mathbb{Z}[I \times \mathbb{Z}_n]$,

$$(s_i *_{\mathbf{w}}^{\langle n\rangle} \mathbf{v})_j(b) = \begin{cases} \mathbf{v}_j(b) & \text{if } j \neq i, \\ (z\mathbf{w} - C^{\langle n\rangle}\mathbf{v} + \mathbf{v})_i(a) & \text{if } (j, b) = (i, a). \end{cases}$$

Note that we have

$$(z\mathbf{w} - C^{\langle n\rangle}\mathbf{v} + \mathbf{v})_i(a) = \mathbf{w}_i(a-1) + \sum_{h:h'=i,h''=j} \mathbf{v}_j(a-1) - \mathbf{v}_i(a-2).$$

The following lemma is an analogue of a well-known identity used in quiver varieties.

Lemma 4.1 *For any* $\mathbf{v} \in \mathbb{Z}[I \times \mathbb{Z}_n]$, *we have* $s_i^{\langle n \rangle}(C^{\langle n \rangle}\mathbf{v} - z\mathbf{w}) = C^{\langle n \rangle} s_i *_{\mathbf{w}}^{\langle n \rangle} \mathbf{v} - z\mathbf{w}$ □

Proof The proof is indeed formal. We have the following calculation. For all $(j, a) \in I \times \mathbb{Z}_n$,

$$\begin{aligned}
(C^{\langle n \rangle} s_i *_{\mathbf{w}}^{\langle n \rangle} \mathbf{v} - z\mathbf{w})_j(a) &= \sum_{(k,b)\in I\times\mathbb{Z}_n} C^{\langle n \rangle}_{(j,a),(k,b)} (s_i *_{\mathbf{w}}^{\langle n \rangle} \mathbf{v})_k(b) - \mathbf{w}_j(a-1) \\
&= \sum_{(k,b):k\neq i} C^{\langle n \rangle}_{(j,a),(k,b)} \mathbf{v}_k(b) \\
&\quad + \sum_{b\in\mathbb{Z}_n} C^{\langle n \rangle}_{(j,a),(i,b)} \Big(\mathbf{w}_i(b-1) - \sum_{c\in\mathbb{Z}_n} C^{\langle n \rangle}_{(i,b),(k,c)} \mathbf{v}_k(c) + \mathbf{v}_i(b)\Big) - \mathbf{w}_j(a-1) \\
&= \sum_{(k,b)} C^{\langle n \rangle}_{(j,a),(k,b)} \mathbf{v}_k(b) - \mathbf{w}_j(a-1) \\
&\quad + \sum_{b\in\mathbb{Z}_n} C^{\langle n \rangle}_{(j,a),(i,b)} \Big(\sum_{c\in\mathbb{Z}_n} C^{\langle n \rangle}_{(i,b),(k,c)} \mathbf{v}_k(c) - \mathbf{w}_i(b-1)\Big) \\
&= (C^{\langle n \rangle}\mathbf{v} - z\mathbf{w})_j(a) + \sum_{b\in\mathbb{Z}_n} C^{\langle n \rangle}_{(j,a),(i,b)} (C^{\langle n \rangle}\mathbf{v} - z\mathbf{w})_i(b) \\
&= (s_i^{\langle n \rangle}(C^{\langle n \rangle}\mathbf{v} - z\mathbf{w}))_j(a).
\end{aligned} \tag{4.1}$$

The lemma follows at once.

It is convenient to rewrite the identity in Lemma 4.1 as

$$C^{\langle n \rangle} s_i *_{\mathbf{w}}^{\langle n \rangle} \mathbf{v} = s_i^{\langle n \rangle}(C^{\langle n \rangle}\mathbf{v} - z\mathbf{w}) + z\mathbf{w}. \tag{4.2}$$

We denote by $w_0 *_{\mathbf{w}}^{\langle n \rangle}$ the composition of the $s_i *_{\mathbf{w}}^{\langle n \rangle}$'s for a fixed reduced expression. Certainly, one must show that the definition is independent of the choice of reduced expression. We shall prove this after the following proposition.

Proposition 4.2 *When C is a Cartan matrix, the matrix $C^{\langle 0 \rangle}$ is injective.* □

Proof It is enough to show that $C^{\langle 0 \rangle}$ can be row-reduced to an upper triangular matrix. This can be done by writing $C^{\langle 0 \rangle}$ as a block matrix with the i-th diagonal blocks being the sum of the standard basis matrices $E_{(i,a),(i,a)}$ and $E_{(i,a),(i,a-2)}$ for all $a \in \mathbb{Z}$. The off-diagonals are identity matrices, up to a shift. One can use the latter to make the diagonal blocks to be identity matrix, that is to get rid of $E_{(i,a),(i,a-2)}$ and then get rid of the nonzero entries below in the lower triangular part. This procedure transforms matrix $C^{\langle 0 \rangle}$ to the desire upper triangular form.

Lemma 4.3 *The definition of* $w_0 *_{\mathbf{w}}^{\langle n\rangle}$ *is independent of the choice of the reduced expression of* w_0. □

Proof Recall there is an embedding $\mathbb{Z}[I \times \mathbb{Z}_n] \to \mathbb{Z}^{I\times\mathbb{Z}}$ in (2.5) and (2.6), so it makes sense to say that $\mathbf{w} \in \mathbb{Z}[I \times \mathbb{Z}_n]$ if $\mathbf{w} \in \mathbb{Z}^{I\times\mathbb{Z}}$. It can be checked that

$$s_i *_{\mathbf{w}}^{\langle 0\rangle} |_{\mathbb{Z}[I\times\mathbb{Z}_n]} = s_i *_{\mathbf{w}}^{\langle n\rangle}, \quad \text{if } \mathbf{w} \in \mathbb{Z}[I \times \mathbb{Z}_n]. \tag{4.3}$$

Thus it suffices to show that $w *_0^{\langle 0\rangle}$ is independent of the choice of expression of w_0. In light of Lemma 4.1 and Proposition 3.1, the vector $C^{\langle 0\rangle} w_0 *^{\langle 0\rangle} \mathbf{v}$ is equal to $w_0^{\langle 0\rangle}(C^{\langle 0\rangle}\mathbf{v} - z\mathbf{w}) + z\mathbf{w}$, which is independent of the choice of the reduced expression of w_0. From Proposition 4.2, we have that $C^{\langle 0\rangle}$ is injective, which implies the desired result.

By (4.2) and Proposition 3.1, there is

$$C^{\langle n\rangle} w_0 *_{\mathbf{w}}^{\langle n\rangle} \mathbf{v} = -C^{\langle n\rangle} z^{\mathbf{c}}\theta\mathbf{v} + z^{\mathbf{c}+1}\theta\mathbf{w} + z\mathbf{w}. \tag{4.4}$$

$$w_0 *_{\mathbf{w}}^{\langle n\rangle} \mathbf{v} = -z^{\mathbf{c}}\theta\mathbf{v} + (C^{\langle n\rangle})^{-1}(z^{\mathbf{c}+1}\theta\mathbf{w} + z\mathbf{w}), \quad \text{if } C^{\langle n\rangle} \text{ is invertible.} \tag{4.5}$$

Recall that $\mathbf{c}$ is the Coxeter number of Cartan matrix. With Proposition 4.2 and (4.3), the equality (4.5) can be strengthened as follows.

Proposition 4.4 *We have*

$$w_0 *_{\mathbf{w}}^{\langle n\rangle} \mathbf{v} = -z^{\mathbf{c}}\theta\mathbf{v} + B^{\langle n\rangle}\mathbf{w}, \tag{4.6}$$

for some matrix $B^{\langle n\rangle}$, *independent of* $\mathbf{v}$, *such that* $C^{\langle n\rangle} B^{\langle n\rangle}\mathbf{w} = z^{\mathbf{c}+1}\theta\mathbf{w} + z\mathbf{w}$. □

Proof Consider the two parameter function

$$f(\mathbf{w}, \mathbf{v}) = C^{\langle n\rangle}(w_0 *_{\mathbf{w}}^{\langle n\rangle} \mathbf{v} + z^{\mathbf{c}}\theta\mathbf{v})$$

By (4.4), $f(\mathbf{w}, \mathbf{v})$ is equal to $z^{\mathbf{c}+1}\theta\mathbf{w} + z\mathbf{w}$, hence independent of $\mathbf{v}$. If $C^{\langle n\rangle}$ is injective, then the sum $w_0 *_{\mathbf{w}}^{\langle n\rangle} \mathbf{v} + z^{\mathbf{c}}\theta\mathbf{v}$ is also independent of $\mathbf{v}$, and by the definition of $w_0 *_{\mathbf{w}}^{\langle n\rangle} \mathbf{v}$, it must be a linear sum of $\mathbf{w}$. In other words, there is a square matrix $B^{\langle n\rangle}$ of size $(I \times \mathbb{Z}_n) \times (I \times \mathbb{Z}_n)$, independent of $\mathbf{v}$, such that $w_0 *_{\mathbf{w}}^{\langle n\rangle} \mathbf{v} + z^{\mathbf{c}}\theta\mathbf{v} = B^{\langle n\rangle}\mathbf{w}$, which is exactly (4.6). If $C^{\langle n\rangle}$ is not invertible, we still have $B^{\langle 0\rangle}$ since $C^{\langle 0\rangle}$ is injective by Proposition 4.2. Let $B^{\langle n\rangle} = B^{\langle 0\rangle}|_{\text{ mod } n}$. Then the equality (4.6) still holds in view of (4.3) via restriction. This finishes the proof.

Remark 4.5 In the following Sects. 4.2, 4.3, and 4.4, we shall give an explicit description of $B^{\langle n\rangle}$ (or rather of $B^{\langle n\rangle}\mathbf{w}$) of type A_ℓ, D_ℓ, E_6, E_7, and E_8. Note that coefficients in $B^{\langle n\rangle}$ are all nonnegative integers. Formulas for other non-simply-laced type can be obtained similarly. □

The following numerical criterion, which will be used in the analysis of graded/cyclic quiver varieties, is clearly due to (4.6).

Corollary 4.6 *The condition* $\mathbf{v}' = w_0 *^{\langle n \rangle}_{\mathbf{w}} \mathbf{v}$ *with* $\mathbf{v}, \mathbf{v}', \mathbf{w} \in \mathbb{Z}[I \times \mathbb{Z}_n]$ *if and only if*

$$\mathbf{v}' + z^{\mathbf{c}}\theta\mathbf{v} = B^{\langle n \rangle}\mathbf{w}. \tag{4.7}$$

When $n = 1$, $C^{\langle 1 \rangle} = C$ and $z = 1$. In this case, the equality 4.7 reads

$$\mathbf{v}' + \theta\mathbf{v} = C^{-1}(\mathbf{w} + \theta\mathbf{w}).$$

Remark 4.7 In type A_1, the condition (4.7) reads: $\mathbf{v}'_i(a) + \mathbf{v}_i(a-2) = \mathbf{w}_i(a-1)$ for all a. □

In what follows, we shall provide a precise description of $w_0 *^{\langle n \rangle}_{\mathbf{w}} \mathbf{v}$, in other words, to provide an explicit description of the vector $B^{\langle n \rangle}\mathbf{w}$ in (4.6).

4.2 Formula in Type A_ℓ

In this section, we provide an explicit formula for $w_0 *^{\langle n \rangle}_{\mathbf{w}} \mathbf{v}$ in type A_ℓ.

Lemma 4.8 *For* $1 \leqslant i \leqslant j \leqslant \ell$, *we have*

$$\begin{aligned}
&(s_i *^{\langle n \rangle}_{\mathbf{w}} s_{i+1} *^{\langle n \rangle}_{\mathbf{w}} \cdots s_j *^{\langle n \rangle}_{\mathbf{w}} \mathbf{v})_i(a)\\
&= \sum_{k=i}^{j} \mathbf{w}_k(a-k+i-1) + \mathbf{v}_{j+1}(a-j+i-1) - \mathbf{v}_j(a-j+i-2) + \mathbf{v}_{i-1}(a-1).
\end{aligned} \tag{4.8}$$

Proof We prove by induction on the number $(j-i)$. When $j-i=0$, the formula follows from the definition. By applying induction hypothesis,

$$\begin{aligned}
&(s_i *^{\langle n \rangle}_{\mathbf{w}} s_{i+1} *^{\langle n \rangle}_{\mathbf{w}} \cdots s_j *^{\langle n \rangle}_{\mathbf{w}} \mathbf{v})_i(a)\\
&= \mathbf{w}_i(a-1) + \mathbf{v}_{i-1}(a-1) - \mathbf{v}_i(a-2) + (s_{i+1} *^{\langle n \rangle}_{\mathbf{w}} \cdots s_j *^{\langle n \rangle}_{\mathbf{w}} \mathbf{v})_{i+1}(a-1)\\
&= \mathbf{w}_i(a-1) + \mathbf{v}_{i-1}(a-1) - \mathbf{v}_i(a-2) + \sum_{k=i+1}^{j} \mathbf{w}_k(a-k+i-1)\\
&\quad + \mathbf{v}_i(a-2) + \mathbf{v}_{j+1}(a-j+i-1) - \mathbf{v}_j(a) - j + i - 2)\\
&= \sum_{k=i}^{j} \mathbf{w}_k(a-k+i-1) + \mathbf{v}_{j+1}(a-j+i-1) - \mathbf{v}_j(a-j+i-2) + \mathbf{v}_{i-1}(a-1).
\end{aligned}$$

The lemma follows.

For convenience, we set

$$\mathbf{v}^{[i]} = \mathbf{r}_i *^{\langle n\rangle} \mathbf{r}_{i-1} *^{\langle n\rangle} \cdots \mathbf{r}_1 *^{\langle n\rangle}_{\mathbf{w}} \mathbf{v}, \quad 1 \leqslant i \leqslant \ell.$$

The following lemma will be used in the induction step in deriving the formula. Recall that in type A_ℓ, we have $\theta(i) = \ell + 1 - i$.

Lemma 4.9 *For* $1 \leqslant j \leqslant i \leqslant \ell$, *we have*

$$\begin{aligned}
&\mathbf{v}_{i-j}^{[i-j]}(a-j) - \mathbf{v}_{\theta(j)}^{[i-j]}(a-j-\theta(i)) \\
&= \sum_{k=1}^{\ell-i+1} \mathbf{w}_{i-j+k-1}(a-j-k) + \mathbf{v}_{i-(j+1)}^{[i-(j+1)]}(a-(j+1)) \\
&\quad - \mathbf{v}_{\theta(j+1)}^{[i-(j+1)]}(a-(j+1)-\theta(i)).
\end{aligned}$$

Proof This is obtained by using Lemma 4.8.

Recall the Coxeter number $\mathbf{c} = \ell + 1$ in type A_ℓ. The formula for $w_0 *^{\langle n\rangle}_{\mathbf{w}} \mathbf{v}$ is as follows.

Proposition 4.10 *For any* $1 \leqslant i \leqslant \ell$, *the number* $(w_0 *^{\langle n\rangle}_{\mathbf{w}} \mathbf{v})_i(a)$ *is given by*

$$(w_0 *^{\langle n\rangle}_{\mathbf{w}} \mathbf{v})_i(a) = \sum_{j=1}^{i} \sum_{k=1}^{\ell-i+1} \mathbf{w}_{j+k-1}(a-i+j-k) - \mathbf{v}_{\theta(i)}(a-\mathbf{c}). \tag{4.9}$$

Proof Recall $w_0 = \mathbf{r}_\ell \mathbf{r}_{\ell-1} \cdots \mathbf{r}_1$ where $\mathbf{r}_i = s_i \cdots s_\ell$. When $i = 1$, $(w_0 *^{\langle n\rangle}_{\mathbf{w}} \mathbf{v})_1(a) = (\mathbf{v}_1^{[1]})_1(a)$ and the proposition follows from Lemma 4.8. In general, we have $(w_0 *^{\langle n\rangle}_{\mathbf{w}} \mathbf{v})_i(a) = (\mathbf{v}^{[i]})_i(a)$. Hence, by using Lemmas 4.8 and 4.9, we have

$$\begin{aligned}
(\mathbf{v}^{[i]})_i(a) &= \sum_{k=1}^{\ell-i+1} \mathbf{w}_{i+k-1}(a-k) + \mathbf{v}_{i-1}^{[i-1]}(a-1) - \mathbf{v}_{\ell}^{[i-1]}((a-1)-\theta(i-1)) \\
&= \sum_{k=1}^{\ell-i+1} \mathbf{w}_{i+k-1}(a-k) + \mathbf{w}_{(i-1)+k-1}((a-1)-k) \\
&\quad + \mathbf{v}_{i-2}^{[i-2]}(a-2) - \mathbf{v}_{\theta(2)}^{[n-2]}((a-2)-\theta(i-2)) \\
&= \cdots = \sum_{j=1}^{i} \sum_{k=1}^{\ell-i+1} \mathbf{w}_{j+k-1}(a-i+j-k) - \mathbf{v}_{\theta}(a-\mathbf{c}).
\end{aligned}$$

Proposition is thus proved.

4.3 Formula in Type D_ℓ

This section is devoted to deducing the formula of $w_0 *_{\mathbf{w}}^{\langle n\rangle} \mathbf{v}$ in type D_ℓ. Recall, in type D_ℓ, $\mathbf{r}_i = s_i s_{i+1} \cdots s_{\ell-2} s_\ell s_{\ell-1} s_{n-2} \cdots s_{i+1} s_i$ for $1 \leqslant i \leqslant \ell - 2$. Then $\mathbf{r}_i *_{\mathbf{w}}^{\langle n\rangle} \mathbf{v}$ is the composition of $s_j *_{\mathbf{w}}^{\langle n\rangle}$ in the order in $\mathbf{r}_i$. Recall also $\mathbf{c} = 2\ell - 2$.

Lemma 4.11 *For $1 \leqslant i \leqslant \ell - 2$, we have*

$$(\mathbf{r}_i *_{\mathbf{w}}^{\langle n\rangle} \mathbf{v})_i(a) = \mathbf{w}_i^{[i]}(a) + \mathbf{v}_{i-1}(a-1) + \mathbf{v}_{i-1}(b-1) - \mathbf{v}_i(b-2),$$

where

$$\mathbf{w}_i^{[i]}(a) = \sum_{j=i}^{\ell-2} \mathbf{w}_j(a-j+i-1) + \mathbf{w}_j(a-\mathbf{c}+i+j-1) + \sum_{j=\ell-1}^{\ell} \mathbf{w}_j(a-\ell+i), \tag{4.10}$$

$$b \equiv b(i,a) = a - \mathbf{c} + 2i. \tag{4.11}$$

Here we set $\mathbf{v}_0(a) \equiv 0$ for all a. □

Proof Clearly, $\mathbf{r}_i *_{\mathbf{w}}^{\langle n\rangle} \mathbf{v} = s_i *_{\mathbf{w}}^{\langle n\rangle} \mathbf{r}_{i+1} *_{\mathbf{w}}^{\langle n\rangle} s_i *_{\mathbf{w}}^{\langle n\rangle} (\mathbf{v})$. By definition and induction on i decreasingly, we have the following calculation:

$$\begin{aligned}
\mathbf{r}_i *_{\mathbf{w}}^{\langle n\rangle} \mathbf{v} &= \mathbf{w}_i(a-1) + \mathbf{v}_{i-1}(a-1) + (\mathbf{r}_{i+1} *_{\mathbf{w}}^{\langle n\rangle} s_i *_{\mathbf{w}}^{\langle n\rangle} \mathbf{v})_{i+1}(a-1) \\
&\quad - (s_i *_{\mathbf{w}}^{\langle n\rangle} \mathbf{v})_i(a-2) \\
&= \mathbf{w}_i(a-1) + \mathbf{w}_{i+1}^{[i+1]}(a-1) + (s_i *_{\mathbf{w}}^{\langle n\rangle} \mathbf{v})_i(b(i+1,a-1)-1) \\
&\quad - \mathbf{v}_{i+1}(b(i+1,a-1)-2) + \mathbf{v}_{i-1}(a-1) \\
&= \mathbf{w}_i(a-1) + \mathbf{w}_{i+1}^{[i+1]}(a-1) + \mathbf{w}_i(b-1) + \mathbf{v}_{i-1}(b-1) \\
&\quad - \mathbf{v}_i(b-2) + \mathbf{v}_{i-1}(a-1) \\
&= \mathbf{w}_i^{[i]}(a) + \mathbf{v}_{i-1}(a-1) + \mathbf{v}_{i-1}(b-1) - \mathbf{v}_i(b-2).
\end{aligned}$$

It is then reduced to show the initial step of the induction for $i = \ell - 2$. In this case, it follows easily from the calculation as follows:

$$\begin{aligned}(\mathbf{r}_{\ell-2} *_{\mathbf{w}}^{\langle n\rangle} \mathbf{v})_{\ell-2}(a) &= \mathbf{w}_{\ell-2}(a-1) + \mathbf{v}_{\ell-3}(a-1) + (s_{\ell-1} *_{\mathbf{w}}^{\langle n\rangle} \mathbf{v})_{\ell-1}(a-1) \\ &\quad + (s_{\ell} *_{\mathbf{w}}^{\langle n\rangle} \mathbf{v})_{\ell}(a-1) - (s_{\ell-2} *_{\mathbf{w}}^{\langle n\rangle} \mathbf{v})_{\ell-2}(a-2) \\ &= \mathbf{w}_{\ell-2}(a-1) + \mathbf{w}_{\ell-1}(a-2) + \mathbf{w}_{\ell}(a-2) + (s_{\ell-2} *_{\mathbf{w}}^{\langle n\rangle} \mathbf{v})_{\ell-2}(a-2) \\ &\quad - \mathbf{v}_{\ell-1}(a-3) - \mathbf{v}_{\ell}(a-3) + \mathbf{v}_{\ell-3}(a-1) \\ &= \mathbf{w}_{\ell-2}^{[\ell-2]}(a) + \mathbf{v}_{\ell-3}(a-1) + \mathbf{v}_{\ell-3}(a-3) - \mathbf{v}_{\ell-2}(a-4).\end{aligned} \tag{4.12}$$

This matches with the formula once we see that $b(\ell-2, a) = a-2$. The lemma follows.

For convenience, we set

$$\mathbf{v}^{[i]} = \mathbf{r}_i *_{\mathbf{w}}^{\langle n\rangle} \mathbf{r}_{i-1} *_{\mathbf{w}}^{\langle n\rangle} \cdots \mathbf{r}_1 *_{\mathbf{w}}^{\langle n\rangle} (\mathbf{v}), \quad \forall 0 \leqslant i \leqslant \ell-2. \tag{4.13}$$

Lemma 4.12 *For $1 \leqslant j \leqslant i \leqslant \ell-2$, we have*

$$\begin{aligned}\mathbf{v}_{i-j}^{[i-j]}(b-j) - \mathbf{v}_i^{[i-j]}(b-2j) = U_i^{[j]}(b-j) \\ + \mathbf{v}_{i-j-1}^{[i-j-1]}(b-j-1) - \mathbf{v}_i^{[i-j-1]}(b-2j-2) \\ - \mathbf{v}_{i-j-1}^{[i-j-1]}(b-3j-1) + \mathbf{v}_{i-j}^{[i-j-1]}(b-3j-2),\end{aligned} \tag{4.14}$$

where

$$\mathbf{u}_i^{[j]}(b-j) = \sum_{k=1}^{i} \mathbf{w}_{i-j+k-1}(b-j-k) - \mathbf{w}_{i-j+k-1}(b-3j+k-1). \tag{4.15}$$

Proof By definition, we deduce that the left-hand side of (4.14) is equal to

$$\begin{aligned}\sum_{k=1}^{i} \mathbf{w}_{i-j+k-1}(b-j-k) + \mathbf{v}_{i-j-1}^{[i-j-1]}(b-j-1) \\ - (s_{i-1} *_{\mathbf{w}}^{\langle n\rangle} \cdots s_{i-j} *_{\mathbf{w}}^{\langle n\rangle} \mathbf{v}^{[i-j-1]})_{i-1}(b-2j-1) \\ - \mathbf{v}_i^{[i-j]}(b-2j).\end{aligned}$$

Applying definition on the term $(s_{i-1} *_{\mathbf{w}}^{\langle n\rangle} \cdots s_{i-j} *_{\mathbf{w}}^{\langle n\rangle} \mathbf{v}^{[i-j-1]})_{i-1}(b-2j-1)$ in the above expression leads to the lemma.

By combining Lemmas 4.11 with 4.12, we obtain

Lemma 4.13 *For $1 \leqslant j \leqslant i \leqslant \ell - 2$ with $b = a - \mathbf{c} + 2i$, there is*

$$\mathbf{v}_{i-j}^{[i-j]}(a-j) + \mathbf{v}_{i-j}^{[i-j]}(b-j) - \mathbf{v}_{i}^{[i-j]}(b-2j) = \mathbf{w}_{i-j}^{[i-j]}(a-j) + \mathbf{u}_{i}^{[j]}(b-j)$$
$$+\mathbf{v}_{i-j-1}^{[i-j-1]}(a-j-1) + \mathbf{v}_{i-j-1}^{[i-j-1]}(b-j-1) - \mathbf{v}_{i}^{[i-j-1]}(b-2j-2). \tag{4.16}$$

Proof We apply Lemma 4.11 to $\mathbf{v}_{i-j}^{[i-j]}(a-j)$ to get

$$\begin{aligned}\mathbf{v}_{i-j}^{[i-j]}(a-j) &= (\mathbf{r}_{i-j} *_{\mathbf{w}}^{\langle n\rangle} \mathbf{v}^{[i-j-1]})_{i-j}(a-j)\\ &= \mathbf{w}_{i-j}^{[i-j]}(a-j) + \mathbf{v}_{i-j-1}^{[i-j-1]}(a-j-1)\\ &\quad + \mathbf{v}_{i-j-1}^{[i-j-1]}(b(i-j,a-j)-1) - \mathbf{v}_{i-j}^{[i-j-1]}(b(i-j,a-j)-2).\end{aligned}$$

Clearly, we get $b(i-j, a-j) = b - 3j$. With this, a sum of the above equality with that in Lemma 4.12 gives rise to (4.16).

We finally have the explicit description of $w_0 *_{\mathbf{w}}^{\langle n\rangle} \mathbf{v}$ for type D_ℓ and $1 \leqslant i \leqslant \ell - 2$.

Proposition 4.14 *For $1 \leqslant i \leqslant \ell - 2$, the number $(w_0 *_{\mathbf{w}}^{\langle n\rangle} \mathbf{v})_i(a)$ can be described as follows:*

$$(w_0 *_{\mathbf{w}}^{\langle n\rangle} \mathbf{v})_i(a) = \sum_{j=0}^{i-1} \mathbf{w}_{i-j}^{[i-j]}(a-j) + \sum_{j=1}^{i-1} \mathbf{u}_{i}^{[j]}(b-j) - \mathbf{v}_i(a-\mathbf{c}). \tag{4.17}$$

Proof We first observe that $(w_0 *_{\mathbf{w}}^{\langle n\rangle} \mathbf{v})_i(a) = \mathbf{v}_i^{[i]}(a)$. By using Lemma 4.11, we get

$$\mathbf{v}_i^{[i]}(a) = \mathbf{v}_i^{[i]}(a) + \mathbf{v}_{i-1}^{[i-1]}(a-1) + \mathbf{v}_{i-1}^{[i-1]}(b-1) - \mathbf{v}_i^{[i-1]}(b-2).$$

Now apply Lemma 4.13 to obtain the proposition.

To reach the case when $i = \ell - 1$ and ℓ the following lemma is needed.

Lemma 4.15 *For $x \in \{\ell - 1, \ell\}$ and $j \geqslant 1$, we have*

$$\begin{aligned}&\mathbf{v}_{(\ell-1)-j}^{[(\ell-1)-j]}(a-j) - \mathbf{v}_{x}^{[(\ell-1)-j]}(a-2j)\\ &\quad = \mathbf{u}_x^{[j]} + \mathbf{v}_{(\ell-1)-j-1}^{[(\ell-1)-j-1]}(a-j-1) - \mathbf{v}_{\theta(x)}^{[(\ell-1)-j-1]}(a-2j-2),\end{aligned}$$

where $\mathbf{u}_x^{[j]} = \sum_{k=1}^{j} \mathbf{w}_{(\ell-1)-j+k-1}(a-j-k) + \mathbf{w}_{\theta(x)}(a-2j-1)$. □

Proof A direct computation yields that the left-hand side of the equality in Lemma 4.15 is

$$\sum_{k=1}^{j-1} \mathbf{w}_{(\ell-1)-j+k-1}(a-j-k) + \mathbf{v}_{(\ell-1)-j-1}^{[(\ell-1)-j-1]}(a-j-1) - \mathbf{v}_x^{[(\ell-1)-j]}(a-2j)$$
$$+ (s_{\ell-2} * s_x * s_{\theta(x)} * s_{\ell-2} * s_{\ell-3} * \cdots s_{(\ell-1)-j}\mathbf{v}^{[(\ell-1)-j-1]})_{\ell-2}(a-2j+1)$$
$$- (s_{\ell-3} * \cdots s_{(\ell-1)-j}\mathbf{v}^{[(\ell-1)-j-1]})_{\ell-3}(a-2j). \tag{4.18}$$

The last three terms in (4.18) can further be simplified to

$$\mathbf{w}_{\ell-2}(a-2j) + \mathbf{w}_{\theta(x)}(a-2j-1) - \mathbf{v}_{\theta(x)}^{[(\ell-1)-j-1]}(a-2j-2). \tag{4.19}$$

The lemma follows by (4.18) and (4.19).

Now we can state the formula for $i = \ell-1, \ell$, finishing the description of $w_0 *_{\mathbf{w}}^{\langle n\rangle} \mathbf{v}$ in D_ℓ.

Proposition 4.16 *For $x = \ell-1, \ell$, we have*

$$(w_0 *_{\mathbf{w}}^{\langle n\rangle} \mathbf{v})_x(a) = \mathbf{w}_x(a-1) + \sum_{j=1}^{\ell-1} \mathbf{u}_{\theta^{j-1}(x)}^{[j]} - \mathbf{v}_{\theta(x)}(a-\mathbf{c}). \tag{4.20}$$

Proof By definition and using Lemma 4.15, we have

$$\begin{aligned}
(w_0 *_{\mathbf{w}}^{\langle n\rangle} \mathbf{v})_x(a) &= (s_x *_{\mathbf{w}}^{\langle n\rangle} \mathbf{v}^{[\ell-2]})_x(a) \\
&= \mathbf{w}_x(a-1) + \mathbf{v}_{\ell-2}^{[\ell-2]}(a-1) - \mathbf{v}_x^{[\ell-2]}(a-2) \\
&= \mathbf{w}_x(a-1) + \mathbf{u}_x^{[1]} + \mathbf{v}_{(\ell-1)-2}^{[(n-1)-2]}(a-2) - \mathbf{v}_{\theta(x)}^{[(\ell-1)-2]}(a-4) \\
&= \cdots \\
&= \mathbf{w}_x(a-1) + \sum_{j=1}^{\ell-1} \mathbf{u}_{\theta^{j-1}(x)}^{[j]} - \mathbf{v}_{\theta(x)}(a-\mathbf{c}).
\end{aligned}$$

The proposition is thus followed.

4.4 An Algorithm to Compute $B^{\langle n\rangle}\mathbf{w}$

Recall that $w_0 *_{\mathbf{w}}^{\langle n\rangle} \mathbf{v} + z^{\mathbf{c}}\theta\mathbf{v} = B^{\langle n\rangle}\mathbf{w}$. We are interested in determining explicitly $w_0 *_{\mathbf{w}}^{\langle n\rangle} \mathbf{v}$, and hence $B^{\langle n\rangle}\mathbf{w}$, in type E_6, E_7, and E_8. The computation is too involved to be done by hand, so instead we resort to computer for this duty. In this section, we derive a formula so that computer can be used to do the remaining job. If we write $\mathbf{w} = \sum_{(j,b)\in I\times\mathbb{Z}_n} \mathbf{w}_j(b)e_j(b)$, then the number $(B^{\langle n\rangle}\mathbf{w})_i(a)$ can be expressed as follows:

$$
\begin{aligned}
(B^{\langle n\rangle}\mathbf{w})_i(a) &= (\sum_{j,b} w_j(b)B^{\langle n\rangle}e_j(b))_i(a) \\
&= \sum_{1\leqslant j\leqslant \ell}\sum_{b\in\mathbb{Z}_n} \mathbf{w}_j(b)(B^{\langle n\rangle}e_j(b))_i(a) \\
&= \sum_{1\leqslant j\leqslant \ell}\sum_{b\in\mathbb{Z}_n} \mathbf{w}_j(b)(w_0 *_{e_j(b)}^{\langle n\rangle} 0)_i(a) \\
&= \sum_{1\leqslant j\leqslant \ell}\sum_{b\in\mathbb{Z}_n} \mathbf{w}_j(b)(z^b w_0 *_{e_j(0)}^{\langle n\rangle} 0)_i(a) \\
&= \sum_{1\leqslant j\leqslant \ell}\sum_{b\in\mathbb{Z}_n} \mathbf{w}_j(b)(w_0 *_{e_j(0)}^{\langle n\rangle} 0)_i(a-b) \\
&= \sum_{1\leqslant j\leqslant \ell}\sum_{b\in\mathbb{Z}_n} \mathbf{w}_j(a-b)(w_0 *_{e_j(0)}^{\langle n\rangle} 0)_i(b).
\end{aligned}
\tag{4.21}
$$

So to determine $(B^{\langle n\rangle}\mathbf{w})_i(a)$, it is reduced to determine the vector

$$
B_j^{\langle n\rangle} := w_0 *_{e_j(0)}^{\langle n\rangle} 0, \quad \forall 1 \leqslant j \leqslant \ell. \tag{4.22}
$$

Conveniently, $B_j^{\langle n\rangle}$ can be presented as a matrix of size $\mathbb{Z}_n \times I$, whose (b,i)-th entry is $(w_0 *_{e_j(0)}^{\langle n\rangle} 0)_i(b)$. From this presentation, the coefficient $(B^{\langle n\rangle}\mathbf{w})_i(a)$ as in (4.21) is a proper linear combination of the i-th column vectors of $B_j^{\langle n\rangle}$ with $\mathbf{w}_j(b)$'s. It can be seen that the (b,i)-th entry is zero unless $1 \leqslant b \leqslant \mathbf{c} - 1$ when $n = 0$. The matrix $B_j^{\langle n\rangle}$ for $1 \leqslant j \leqslant \ell$ is computable by using computer for all exceptional types. The results can be found in the first author's website.

Acknowledgments We thank Jim Humphreys for helpful discussions. We also thank the anonymous referee for very helpful suggestions in improving the paper. Y. Li is partially supported by the NSF grant DMS 1801915.

References

1. G. Benkart, S.-J. Kang, S.-J. Oh and E. Park, *Construction of irreducible representations over Khovanov-Lauda-Rouquier algebras of finite classical type*, arxiv:1108.1048.
2. A. Björner and F. Brenti, *Combinatorics of Coxeter groups*, Graduate Texts in Mathematics **231**, Springer, New York, 2005.
3. W. Burau, *Über Zopfgruppen und gleichsinnig verdrillte Verkettungen*, Abh. Math. Sem. Univ. Hamburg. **11** (1935), 179–186.
4. J.E. Humphreys, *Reflection Groups and Coxeter Groups*, Cambridge Univ. Press, 1990.
5. J.E. Humphreys, *Longest element of a finite Coxeter group*, available at the following website. http://people.math.umass.edu/ jeh/pub/longest.pdf.
6. Y. Li, *Quiver varieties and symmetric pairs*, Representation Theory, **23** (2019), 1–56.
7. Y. Li, in preparation.
8. G. Lusztig, *Introduction to quantum groups*, Progress in Math. **110**, Birkhäuser, 1993.
9. H. Matsumoto, *Générateurs et relations des groupes de Weyl généralisées.* C. R. Acad. Sci. Paris. **258** (1964), 3419–3422.
10. J. Moody, *The faithfulness question for the Burau representation*, Proceedings of the American Mathematical Society **119** (1993), no. 2, 671–679.
11. H. Nakajima, *Quiver varieties and finite dimensional representations of quantum affine algebras*, JAMS **14** (2000) no. 1, 145–238.

Dual Kashiwara Functions for the $B(\infty)$ Crystal in the Bipartite Case

Anthony Joseph

Abstract Let $\mathfrak{g}$ be a simple algebra. Let r be the number of its positive roots. The Kashiwara $B(\infty)$ crystal is an important combinatorial object that describes a crystal basis for the simple finite dimensional $\mathfrak{g}$ modules. After [A. Berenstein and A. Zelevinsky, Tensor product multiplicities, canonical bases and totally positive varieties. Invent. Math. **143** (2001), no. 1, 77–128], it is a polyhedral subset of the integer points of an r-dimensional affine space, and moreover, the linear functions describing this polyhedral subset are given by "trails" in the fundamental modules of lowest weight of the Langlands dual. In general, trails depend on a reduced decomposition of the longest element of the Weyl group and are very difficult to compute. In the present work, it is shown for $\mathfrak{g}$ classical, when the reduced decomposition is given by a power of a suitably chosen Coxeter element, that the set of trails is in natural bijection with the crystal of a suitable highest weight fundamental module.

The proofs are based on a specially developed theory of S-sets and do not require any of the results in Berenstein–Zelevinsky (loc cit).

1 Introduction

This paper is part of a project to understand the combinatorics behind the Kashiwara $B(\infty)$ crystal and on which several papers have been published so far [6–12]. In this, S. Zelikson has played a key, if not essential, role. I have attempted on each occasion for him to agree to joint authorship, and I regret he did not accept even this time. In any case following a lead given by Nakashima–Zelevinsky [17, Sects. 5,6], for which type A and type A affine were found to simplify for a certain periodic reduced decomposition, Zelikson conjectured that in the "bipartite case" the Berenstein–Zelevinsky trails should themselves form the structure of a crystal (of a fundamental

A. Joseph (✉)
Department of Mathematics, The Weizmann Institute of Science, Rehovot, Israel
e-mail: anthony.joseph@weizmann.ac.il

J. Greenstein et al. (eds.), *Interactions of Quantum Affine Algebras with Cluster Algebras, Current Algebras and Categorification*, Progress in Mathematics 337,
https://doi.org/10.1007/978-3-030-63849-8_9

module). He backed this up with many computer computations not least in type F_4 for which the correctness of his suggestion remains a startling "unexplained" fact. On the other hand, outside the bipartite case the corresponding assertion is completely false and even fails for $\mathfrak{sl}(4)$.

Here, we establish the Zelikson conjecture when $\mathfrak{g}$ is classical. Moreover, we describe rather explicitly all the trails.

In general, one would like a "global" combinatorial object to describe all trails, as in the present example of a crystal.

Some of the present analysis might be deemed to be somewhat complicated, though it is entirely elementary. The reader might find it useful to read [11] in which the relatively easy minuscule case is treated. A key new combinatorial object used in the present theory is the notion of an S-graph on which a study was made particularly in [12] with several illustrations. These have a totally unexpected relationship with the Chevalley–Serre relations in Demazure modules [9], and this should extend [9, Sect.7] the result of Berenstein–Zelevinsky [1] from the finite to the infinite case. Unfortunately, the proof does not so far go through because of the possible existence of "false trails". In the case treated here, false trails are rather common yet despite this we are able to proceed at least for $\mathfrak{g}$ classical, though only by some prodigious efforts.

1.1 Root Data and Kac–Moody Lie Algebras

For each positive integer n, set $[1, n] := \{1, 2, \dots, n\}$. For each rational number q, let $[q]$ denote the largest integer $\leqslant q$.

Let $\mathfrak{g}$ be a Kac–Moody Lie algebra of rank ℓ. Set $I = [1, \ell]$. Fix a Cartan subalgebra $\mathfrak{h}$ of $\mathfrak{g}$, and choose a set $\pi = \{\alpha_i\}_{i \in I}$ (resp. $\pi^\vee = \{\alpha_i^\vee\}_{i \in I}$) of simple roots (resp. coroots). Let $\mathfrak{b}$ (resp. $\mathfrak{b}^-$) denote the Borel subalgebra of $\mathfrak{g}$ associated with these choices, that is to say containing $\mathfrak{h}$ and whose roots lie in $\mathbb{N}\pi$ (resp. $-\mathbb{N}\pi$). We can assume without loss of generality that the Dynkin diagram of π is *connected*. This avoids some minor technicalities.

Fix a Chevalley basis for $\mathfrak{g}$. For each $s \in I$, let e_s (resp. f_s) be the element of this basis of weight α_s (resp. $-\alpha_s$) and set $h_s = [e_s, f_s]$. We shall sometimes drop the s subscript. We recall that (e, h, f) satisfy the relations $[h, e] = 2e, [h, f] = -2f, [e, f] = h$ and, hence, their linear span is an $\mathfrak{sl}(2)$ subalgebra. It is called an s-triple.

For all $i \in I$, let ϖ_i (resp. $\varpi_i^\vee$) denote the corresponding fundamental weight (resp. coweight). Let s_i denote the simple reflection defined by $\alpha_i : i \in I$. By definition, they form a set of generators of the Weyl group W associated to $\mathfrak{g}$. It is a finite group if and only if $\mathfrak{g}$ is semisimple.

1.2 The Crystal $B(\infty)$

From the above data, one may construct, [6, 14, 15], the Kashiwara crystal $B(\infty)$ as a purely combinatorial object. It depends on a choice of reduced decomposition as a set, but not as a crystal. Again the latter is associated to the dual Verma module of highest weight 0 for $\mathfrak{g}$, whilst set dependence is related to the Bott–Samelson resolution that also depends on a choice of reduced decomposition.

Thus, $B(\infty)$ is an intricate combinatorial object, which demands understanding. After Gleizer and Postnikov [3], it is polyhedral in type A. After Berenstein and Zelevinsky [1], this also holds for finite type. In the latter case, the linear inequalities that define $B(\infty)$ as a set are given by $\mathbf{i}$-trails (which we shall simply call trails) in the lowest weight fundamental module $V(-\varpi_i^\vee)$ for the Langlands dual of $\mathfrak{g}$. Trails also depend on a choice of reduced decomposition. They are neither combinatorially defined nor easy (not to say impossible) to compute. Again the theory of Berenstein–Zelevinsky does not extend to infinite type.

Our approach to describing $B(\infty)$ was founded on some commutation relations coming from a duality on $B(\infty)$ developed by Kashiwara [15] and Joseph [6, 2.5]. We used these relations to inductively construct sets of dual Kashiwara functions [10], using a specially developed theory of S-graphs [7, 12].

In the above, each set Z_t of dual Kashiwara functions is associated to a simple root $\alpha_t : t \in I$.

1.3 Adjoining Faces to Trails

In Sect. 3.8, we describe a process of adjoining "faces" to trails. This is analogous to, though more, in general, complicated than, the action of the Kashiwara operations on a crystal.

Here, following a suggestion of Zelikson, we consider the simplifying condition that the reduced decomposition, which is used to construct $B(\infty)$ as a set, is periodic, that is to say given by a power of a Coxeter element. Eventually, we must restrict to a very special Coxeter element, which we refer to as being "bipartite", a concept we explain in Sects. 2.2 and 4.6.

The first steps needed for the present theory are made here and applied to the case when the reduced decomposition defining the trails is bipartite.

In order that Z_t describes the dual Kashiwara parameter $\varepsilon_t^\star$, it must be a t-semi-invariant set in the sense of [10, Prop. 5.1]. This will play no role here except that it tells us that Z_t should be constructed by adjoining faces corresponding to S-sets at each induction step.

We check here (and this is not so easy) that for $\mathfrak{g}$ classical, this constructs all trails corresponding to a bipartite reduced decomposition of unique longest element of the Weyl group.

We further show that, almost miraculously, the set of trails with adjunction of faces is just a suitable Demazure subcrystal associated to the corresponding fundamental highest weight module of the Langlands dual with an action of the Kashiwara operators. This is not quite an isomorphism because adjoining faces is less demanding than crystal structure (see Remark 2 of Proposition in Sect. 6.4).

1.4 Comparison with Crystal Operators

For the above purpose, we prove a comparison lemma (Lemma in Sect. 4.7) for the data needed to describe the appropriate S-graphs, with the data needed to describe the action of the crystal operators. *Notably, it is special to the periodic case.* Yet this is far from enough to settle the issue. Indeed as we note in Sects. 5.4, 6.11 and Fig. 3 that even in this simple case, adjoining S-graphs and crystal elements follow slightly different rules.

For type A, the result is of less interest because we already have a precise description of the dual Kashiwara functions either from [1] or in a more elementary fashion by Joseph [11]. Ignoring type A, we can assume that the Coxeter number c is even. Then the Coxeter element σ must be chosen so that $\sigma^{c/2}$ is the longest element w_0 of the Weyl group and so defines a reduced decomposition of w_0.

Owing to a failure of equality in the last line of Eq. (14), we must make a further restriction on the choice of Coxeter element, namely that it be bipartite (see Sect. 2.2).

We analyse in Sects. 6.5–6.8 the “false trails” that can arise under these very special circumstances. In this manner, the aim outlined in the last paragraph of Sect. 1.3 is achieved when the coefficient of α_t in the highest root is $\leqslant 2$ and gives an elegant description of Z_t. This is sufficient for $\mathfrak{g}$ of classical type (Theorem in Sect. 6.3). The analogous result holds trivially in type G_2 and has been verified by Zelikson for the two remaining cases in type F_4 using a computer generated description of the crystal associated to the corresponding fundamental module. In this case, the result is even more remarkable and mysterious.

2 Coxeter Elements

2.1 Coxeter Elements in Finitely Generated Groups

Recall that a Coxeter element σ of W is a product of all the simple reflections taken once and in any order. It is a remarkable fact due to Speyer [18] that if the Dynkin diagram is connected and W is infinite, then any power of σ is a reduced word.

The above result fails if W is finite, yet as we note below we do have the next best thing.

Following Bourbaki [2, Chap. IV], associate to any finitely generated group Γ a graph (or Dynkin diagram D) whose set $V(D)$ of vertices labels the given generators $g_v \in \Gamma$ such that v, v' are joined by an edge if and only if $g_v, g_{v'}$ do not commute. Fix a total order on $V(D)$. Call a product of the $g_v : v \in V(D)$ defined by that order, a Coxeter element of Γ.

Assume that D has no cycles and is connected, for example, if Γ is the Weyl group of a simple Lie algebra, assumed of rank > 1 to avoid trivialities. Then, one easily checks (cf [2, Chap. IV, Prop. 2] that $V(D)$ admits a, unique up to permutation, decomposition into disjoint subsets V_1, V_2 such that for $i = 1, 2$, the $g_v : v \in V_i$ commute. Obviously, the product σ_i of the $g_v : v \in V_i$ is independent of order and $\sigma := \sigma_1\sigma_2$ is a Coxeter element of Γ.

Again under the above hypothesis on D, the Coxeter elements of Γ are all conjugate [2, Chap. V, Sect. 6, Lemma 1]. Briefly, the argument is as follows. Since the graph D has no cycles and is finite, it admits [2, Chap. IV, Prop. 2(i)] a vertex v with only one neighbour, say v'. Fix a Coxeter element τ. Up to conjugation, we can assume that τ ends in g_v. Now the product g_ξ of the generators in σ, lying strictly between $g_{v'}$ and g_v, commutes with g_v, so taking g_ξ through g_v and the resulting product of elements lying after $g_{v'}g_v$ to the left we conclude that τ, and so any Coxeter element of Γ, is conjugate to an element that ends in $g_{v'}g_v$. Finally, consider the Dynkin diagram of the group Γ' with generators $(g_{v''} : v'' \in V(D) \setminus \{v', v\}) \cup \{g_{v'}g_v\}$. It is obtained by deleting v from D, so is of cardinality one less and has no cycles. Then the assertion results by induction on $|\Gamma|$.

2.2 Bipartite Coxeter Elements

Now assume that W is the Weyl group of a simple Lie algebra. Then as noted in [2, Chap. V, Sect. 4, Prop. 8], the Dynkin diagram of W is finite with no cycles. Thus, all the Coxeter elements of W are conjugate. In particular, the order of a Coxeter element is independent of choice. It is called the Coxeter number c. A Coxeter element acts on the set Δ of roots, and each orbit has cardinality c. Moreover, the number of orbits is exactly $|\pi|$. These last two classical facts may be conveniently proved by considering the Coxeter group $< \sigma_1, \sigma_2 >$ generated by the pair σ_1, σ_2, defined two paragraphs above. Moreover, as noted above, the unique longest element of this Coxeter subgroup of W is exactly the unique longest element w_0 of W, [10, Lemma 2.8].

We call $\sigma := \sigma_1\sigma_2$ a bipartite Coxeter element. When the Coxeter number is even, say $c = 2m$, the assertions of the previous paragraph imply that $\sigma^m = w_0$, a result that can also be found in [2].

Finally, the Coxeter number only fails to be even in type A_{2k}. Since trails are completely understood in type A, we may just assume that c is even. Thus, we may write w_0 as a power of σ, and the resulting expression is reduced. When $w_0 = -1$, so in particular central in W, then since all the Coxeter elements are conjugate, the previous result holds for any choice of Coxeter element. However, it fails in general. For example in type A_3 with the Bourbaki labelling, $s_1s_2s_3s_1s_2s_3$ is not a reduced decomposition, and so in particular not equal to w_0.

Yet even then for (12) to hold, we also need σ to be bipartite.

3 The Kashiwara Functions

3.1 Preliminaries

A reduced sequence J is a sequence $(\dots, i_j, i_{j-1}, \dots, i_1)$ of elements of I such that for all $k \in \mathbb{N}^+$ the product $w_k := s_{i_k}s_{i_{k-1}} \cdots s_{i_1}$ is reduced. If the Weyl group is finite, then $\mathfrak{g}$ has $r < \infty$ positive roots and $j \mapsto i_j$ is bijection of $[1, r]$ onto J. Otherwise, it is a bijection of $\mathbb{N}^+$ onto J. We shall often identify these sets in bijection. In the finite (resp. infinite) case, we set $\hat{J} = [1, r + 1]$ (resp. $\hat{J} = J$).

For a given Kac–Moody algebra, we generally assume J fixed, eventually making special choices of J.

Fix $t \in I$. A Berenstein–Zelevinsky trail associated to t is a sequence of *non-zero* vectors in the lowest weight fundamental module $v_{\gamma_j^K} \in V(-\varpi_t^\vee) : j \in J$ for the Langlands dual of $\mathfrak{g}$ satisfying the auxiliary conditions (T), (B) defined below. Our personal preference is to consider trails as lying in the lowest weight fundamental module $v_{\gamma_j^K} \in V(-\varpi_t)$ for $\mathfrak{g}$. This means that we have to interchange roots and coroots in the construction of the dual Kashiwara functions in terms of trails in $V(-\varpi_t)$. As we shall see, the construction of the resulting functions uses sums of successive differences of Kashiwara functions (in which roots and coroots are interchanged).

3.2 Kashiwara Functions

Consider the set B_J of all elements of the form $b = (\dots, m_n, m_{n-1}, \dots, m_1) : m_j \in \mathbb{N}$ with all but finitely many $m_j : j \in J$ equal to zero. Sometimes we consider the $\{m_j\}_{j \in \hat{J}}$ as coordinate functions on B_J taking values in $\mathbb{N}$. If W is finite, we always set $m_{r+1} = 0$.

It is convenient to represent the subscript $j \in J$ by the pair $(s, k) \in I \times \mathbb{N}^+$, where for a fixed $s \in I$, we let $k \in \mathbb{N}^+$ denote the number of times $i_j = s$ counting from the right. In this, we set $m_s^k = m_j$.

Observe that the natural linear order on $J \subseteq \mathbb{N}^+$ induces a linear order on the image of the map $j \mapsto (s, k)$. Then identifying J with its image, we may write $j > (s, k)$ to mean $j > j'$, whenever (s, k) is the image of j'.

In this notation, the Kashiwara function described in [6, 2.3.2, 2.4.1] can be written as

$$r_s^k(b) = m_s^k + \sum_{j \in J | j > (s,k)} \alpha_s^\vee(\alpha_{i_j}) m_j, \forall s \in I, k \in \mathbb{N}^+, \tag{1}$$

where $b \in B_J$ is as above. It is a linear function on B_J.

We call r_s^k the kth Kashiwara function of type $s \in I$.

The sum in (1) is finite since all but finitely many m_j are equal to zero.

When we consider m_j as the jth coordinate function on B_J, the expression m_j in (4) should be written as $m_j(b)$. Alternatively, we may replace $r_s^k(b)$ by r_s^k in the left hand side of (1). Observe that r_s^k is a linear function on B_J, but not locally finite, that is to say not a finite sum of the coordinate functions $m_j : j \in J$.

3.3 A Crystal Structure on B_J

The Kashiwara functions were used by Kashiwara to give B_J a crystal structure. This goes as follows.

Given $b = (\ldots, m_n, m_{n-1}, \ldots, m_1) \in B_J$, set $\operatorname{wt} b = -\sum_{j \in J} m_j \alpha_{i_j}$. It is a finite sum. Set $\varepsilon_s(b) = \max_{k \in \mathbb{N}^+} r_s^k(b)$ and $\varphi_s(b) = \varepsilon_s(b) + \alpha_s^\vee(\operatorname{wt} b)$. We call these as the Kashiwara parameters associated to $s \in I$. If J is infinite, then $\varepsilon_s(b) \geqslant 0$ for all $b \in B_J$. Otherwise, this may fail and a small technical adjustment is made. In this case, suppose $s \in I$ appears $k_s < \infty$ times in J and set $r_s^{k_s+1}(b) = 0$, for all $s \in I, b \in B_J$. In other words in our previous notation, we view $r+1$ as (s, k_s+1), for all $s \in I$. This is justified by the way Kashiwara uses his embedding theorem [6, Thm. 2.5.7]. Indeed, this required him to take b_∞ as the $(r+1)^{th}$ entry of B_J. Omitting this seemingly innocuous factor can make a difference already in type A_2—see [8, 2.5.26].

To describe how the Kashiwara operators $\tilde{e}_s, \tilde{f}_s : s \in I$ act on an element of $b \in B_J$, let $\ell_s(b)$ (resp. $r_s(b)$) be the largest (resp. smallest) value of k such that $r_s^k(b) = \varepsilon_s(b)$. Then, $\tilde{e}_s b$ (resp. $\tilde{f}_s b$) is obtained from b by decreasing $m_s^{\ell_s(b)}$ (resp. increasing $m_s^{r_s(b)}$) by one, and deeming an element of B_J to be zero if it has a negative entry. In this, we say that $\tilde{e}_s$ (resp. $\tilde{f}_s$) enters b at the $\ell_s(b)^{th}$ (resp. $r_s(b)^{th}$) place.

Observe that $\tilde{f}_s$ acts injectively, but $\tilde{e}_s$ does not. Using the fact that the Cartan matrix has 2 on the diagonal, one may check that $\tilde{e}_s \tilde{f}_s$ is the identity on B_J and $\tilde{f}_s \tilde{e}_s b = b$ if $\tilde{e}_s b \neq 0$.

3.4 Dual Kashiwara Operators

Let $\tilde{\mathcal{E}}$ (resp. $\tilde{\mathcal{F}}$) denote the monoid generated by $\{\tilde{e}_s\}_{s \in I}$ (resp. $\{\tilde{f}_s\}_{s \in I}$). Let b_∞ be the element of B_J in which all entries are zero. Set $B_J(\infty) = \tilde{\mathcal{F}} b_\infty$. Remarkably, $\tilde{\mathcal{E}} B_J(\infty) = B_J(\infty)$, that is to say $B_J(\infty)$ is a crystal. More precisely, it is a strict subcrystal of B_J, that is to say the embedding of $B_J(\infty) \hookrightarrow B_J$ is strict in the language of Kashiwara [15, Sect. 1]. (In terms of the crystal graphs, $B_J(\infty)$ is a component of B_J and indeed the unique connected component containing b_∞.) Moreover, *as a crystal*, $B_J(\infty)$ is independent of the choice of J and this common crystal is denoted as $B(\infty)$. It is upper normal, that is to say $\varepsilon_s(b) = \max_{k\in\mathbb{N}}\{\tilde{e}_s^k b \neq 0\}$, for all $b \in B(\infty)$. Finally, it admits an involution $\star$, with the crucial property that the transposed operators $\tilde{e}_i^\star, \tilde{f}_i^\star$ commute with the $\tilde{e}_j, \tilde{f}_j$, for all distinct pairs $i, j \in I$. All these results are due to Kashiwara [15] using the quantum group, so requiring the Cartan matrix to be symmetrizable, and were extended using the Littelmann path model to the general case in [6]. Note that in [6] we wrote $\tilde{e}_s, \tilde{f}_s$ simply as e_s, f_s.

The dual Kashiwara parameters and dual Kashiwara operators are obtained by transport of structure through $\star$. In particular, we must have $\varepsilon_t^\star(b) = \max_{k\in\mathbb{N}}\{(\tilde{e}_s^\star)^k b \neq 0\}$, for all $b \in B(\infty)$. By Joseph [6, 3.2.3], if we use the parametrization of $B_J(\infty)$ given by Sect. 3.3, then $B_J(\infty)$ is determined rather concretely as a polyhedral subset of B_J, given for each $t \in I$, a set Z_t of *linear* functions on $B_J(\infty)$ for which

$$\varepsilon_t^\star(b) = \max_{z\in Z_t} z(b), \forall b \in B_J(\infty).$$

We call the elements $z_t^k : k \in \mathbb{N}^+$ of Z_t, the dual Kashiwara functions associated to $t \in I$. We cannot simply take them to be the Kashiwara functions associated to $t \in I$ transposed under $\star$ because we do not know that $\star$ is a linear map. Worse still $\star$ is only defined $B_J(\infty)$ that does not obviously inherit the additive structure of B_J, so the question of linearity does not even make sense. Again $\star$ is, surprisingly, utterly complicated, and we have no idea of what such a procedure would give.

For the highlighted relation above to hold, it is enough by the commutation relations noted in Sect. 1.2 that the set Z_t be t-semi-invariant in the sense of [10, Prop. 5.1]. This is where S-sets come into play and why they should determine the trails ultimately by an inductive construction. (The difficulties in realizing this construction are discussed in [10, 5.3].)

3.5 Berenstein–Zelevinsky Trails

Fix $t \in I$. A Berenstein–Zelevinsky trail K associated to t is a sequence of *non-zero* vectors $v_{\gamma_j^K} \in V(-\varpi_t) : j \in \hat{J}$ of weight γ_j^K satisfying the following rules:

(T). **The Trail Condition**. For all $j \in J$, there exists $n_j \in \mathbb{N}$ such that $e_{i_j}^{n_j} v_{\gamma_j^K} = v_{\gamma_{j+1}^K}$.

Thus a trail is given by an appropriate sequence of monomials $e_j^K : j \in J$ in the simple root vectors $e_i : i \in I$. We denote this sequence briefly by $\mathbf{e}^K$. Note that if $j = (s, k)$, then $i_j = s$ and $\mathbf{e}^K$ has a factor of $e_s^{n_j}$ at position (s, k), by definition.

(B). **The Boundary Conditions.**

(i). $\gamma_1^K = -s_t\varpi_t$.

(ii). For every trail K, there exists $\varphi(K) \in \mathbb{N}^+$ such that $\gamma_{j+1}^K = -w_j\varpi_t$, for all $j \in J | j \geqslant \varphi(K)$.

The set of all Berenstein–Zelevinsky trails associated to $t \in I$ is denoted by $\mathcal{K}_t^{BZ}$, or simply $\mathcal{K}_t$. We shall say that a trail K trivializes at $w_j : j \in J$, or simply at $j \in J$, if $j \geqslant \varphi(K)$. From then on, the trail is just the appropriate sequence of extremal vectors and so is uniquely determined by the form it takes to the vector $v_{-w_j\varpi_t}$. Thus, we may regard $\mathbf{e}^K$ as a finite sequence since it is determined for $j \geqslant \varphi(K)$. However, one should note that $\mathbf{e}^K$ is not quite determined by $e_{\varphi(K)}^K$.

Observe in particular that $e_{\varphi(K)}^K v_{-\varpi_t} = v_{-w_{\varphi(K)}\varpi_t}$, up to a non-zero scalar. The latter is the generator of the Demazure submodule $F_{w_{\varphi(K)}}(-\varpi_t) := U(\mathfrak{b}^-)v_{-w_{\varphi(K)}\varpi_t}$ of $V(-\varpi_t)$.

Let $\mathcal{K}_t^j$ denote the subset of $\mathcal{K}_t$ of trails that trivialize at $j \in J$.

From this definition, it follows that the subsets $\mathcal{K}_t^j$ of $\mathcal{K}_t$ are increasing in j.

When W is finite, we can take $\varphi(K) = \ell(w_0)$, for all $K \in \mathcal{K}_t$ in the above. Notice that this means that $\gamma_{\ell(w_0)+1}^K = -w_0\varpi_t$. However, a trail may trivialize before w_0 is reached.

The second boundary condition means that the trail has to move fast and efficiently.[1] For example in $V(-\varpi)$ with ϖ a *regular* dominant weight, one could *only* have $\gamma_{j+1}^K = -w_j\varpi_t$, for all $j \in J$. This imposed efficiency and the required non-vanishing of the vectors make trails particularly difficult to determine.

3.6 *The Initial Driving Trail*

The initial "driving trail" associated to t is defined as follows.

Set $u = (t, 1)$, so then $i_u = t$. Set

$$\gamma_j^{K_t^1} := \begin{cases} -s_t\varpi_t, & \text{if } j \leqslant u+1, \\ -s_{i_{j-1}} \dots s_{i_{u+1}} s_t\varpi_t, & \text{if } j > u+1. \end{cases} \tag{2}$$

The driving trail K_t^1 consists just of extremal vectors; but it only trivializes at $(t, 1)$ because $\gamma_{j+1}^{K_t^1} = -w_j\varpi_t$, if and only if $j \geqslant u = (t, 1)$. We also shall refer to K_t^1 as the (open) face F_t^1 associated to t.

Lemma

(i) *Take $j \in J$. If $j < (t, 1)$, then $\mathcal{K}_t^j$ is empty.*

---

[1] As recommended by the music hall song, "Don't dilly-dally on the way".

(ii) $\mathcal{K}_t^{(t,1)} = \{F_t^1\}$. □

Proof Take $K \in \mathcal{K}_t^j$. If $j < (t, 1)$, then $w_j\varpi_t = \varpi_t$, so then $\gamma_{j+1}^K = -\varpi_t$. On the other hand by (T) and $B(i)$, we have $\gamma_{j+1}^K \in \mathbb{N}\pi + \alpha_t - \varpi_t$. This contradiction gives (i). Again if $j = (t, 1)$, then $w_j\varpi_t = s_t\varpi_t$, so then $\gamma_{j+1}^K = -s_t\varpi_t$. Then by (T) and $B(i)$, we obtain $\gamma_{j+1}^K = -s_t\varpi_t$, for all $j \leqslant (t, 1)$. Thus, $\gamma_j^K = \gamma_j^{K_t^1}$, for $j > (t, 1)$, by the assumption that K trivializes at $(t, 1)$.

3.7 *Trail Functions*

Following [1], we associate to a trail K, a linear function z^K on B_J as follows. Set $\delta_j^K = \frac{1}{2}(\gamma_j^K + \gamma_{j+1}^K)$ and

$$z^K := \sum_{j \in J} \alpha_{i_j}^\vee(\delta_j^K)m_j. \tag{3}$$

One checks that $z_t^1 := z^{F_t^1} = m_t^1 + \sum_{j<(t,1)} \alpha_{i_j}^\vee(\alpha_t)m_j$. *After interchanging roots and coroots*, it coincides with the "initial driving function", introduced in [10, 4.7] following quite different reasoning.

Remark 1 If K trivializes at k, so in particular $\gamma_{k+2}^K = s_{i_{k+1}}\gamma_{k+1}^K$, then $m_j = 0$, for all $j > k$.

Note the trail function z^K determines K. We shall not always distinguish between the trail K and the trail function z^K.

Remark 2 For counting purposes, it is convenient to introduce the trivial trail K_∞ defined by $z^{K_\infty} = 0$.

3.8 *Faces and the Parametrization of Trails*

3.8.1 Faces and Face Functions

We define the (closed) face $F_s^{k+1} : s \in I, k \in \mathbb{N}^+$ through the face function $z^{F_s^{k+1}}$ defined to be the difference of successive Kashiwara functions $r_s^k - r_s^{k+1}$ *in which roots and coroots are interchanged*. Precisely, one has

$$z^{F_s^{k+1}} := r_s^k - r_s^{k+1} = m_s^{k+1} + \sum_{(s,k)<j<(s,k+1)} \alpha_{i_j}^\vee(\alpha_s)m_j + m_s^k. \tag{4}$$

In particular, it is a locally finite linear function on B_J. It has support in $[(s,k),(s,k+1)]$.

Adjoining the face F_s^{k+1} to a trail $K \in \mathcal{K}_t^{BZ}$ means taking

$$z^{F_s^{k+1}+K} = z^{F_s^{k+1}} + z^K.$$

Comparing this with (3) and (4), one may check that it exactly corresponds to shifting a factor of e_s in $\mathbf{e}^K$ from position $(s,k+1)$ to position (s,k) through the intermediate simple root vectors to obtain $\mathbf{e}^{K+F_s^{k+1}}$. This coincidence holds because we have interchanged roots and coroots in writing down (4).

In the above, we must show that the $v_{\gamma_j^{K+F_s^{k+1}}} : j \in J$ do not vanish. This is a delicate issue *and the most important one in the theory of trails*. When this holds, we say that the face F_s^{k+1} can be adjoined to the trail K to obtain the trail $K+F_s^{k+1}$.

Obviously, $F_t^1 + K_\infty = K_t^1$.

We remark that the pictorial meaning of faces and their adjunction (open or closed) can be appreciated from [10, Figures 1, 2].

3.8.2 Parametrization of Trails

Fix $s \in I$, and set $J^s := \{j \in J | i_j = s\}$. Given $j \in J^s$, we can write $j = (s,n)$ for some $n \in \mathbb{N}^+$. Of course, n depends on $j \in J$, but we shall stick to the same symbol throughout. Then, the sequence $\mathbf{e}^K$, for $K \in \mathcal{K}_t^{BZ}$ trivializing at some $w_j : j \in J^s$, defines an n-tuple $\mathbf{e}$ whose first term v_{-a_1} is a lowest weight vector for the $\mathfrak{sl}(2)$ subalgebra defined by s and whose $k^{th} : k > 1$ term is the product $e_{-a_k} := \prod_{(s,k)<j<(s,k+1)} e_{i_j}^{n_j}$. Taking $w = w_j$, this means that we are realizing the extremal vector $v_{-w\varpi_t}$ in the form

$$v_{-w\varpi_t} := \overline{v}_{\mathbf{k}} = e^{k_n} e_{-a_n} e^{k_{n-1}} e_{-a_{n-1}} \cdots e^{k_1} v_{-a_1}. \tag{5}$$

Here, the subscript $-a_k$ denotes the h_s weight where we note that $a_k \in \mathbb{N}$ and $\mathbf{k}$ is the n-tuple $(k_n, k_{n-1}, \dots, k_1)$ of non-negative integers. It is convenient to denote the n-tuple $(a_n, a_{n-1}, \dots, a_1)$ of non-negative integers by $\mathbf{a}$.

We remark that (5) is just [9, Eq. (16)].

As in [9, Eq. (6)], partial sums are denoted by a bracketed superscript. For example, $a^{(j)} := \sum_{i=1}^{j} a_i$.

Let $\mathbf{E}_s^j$ be the set of all finite tuples so obtained. For all $\mathbf{e} \in \mathbf{E}_s^j$, let $T_s(\mathbf{e})$ be the set of $\mathbf{e}^K$ with $K \in \mathcal{K}_t^{BZ}$ in which $\mathbf{e}$ is fixed and only the powers of e_s may differ at positions $(s,k) \leqslant j$. We may also designate an element of $T_s(\mathbf{e})$ by the trail K it defines or by the n-tuple $\mathbf{k}$ defined by K.

We may also designate $T_s(\mathbf{e})$ by $T_s(\mathbf{a})$, as is sometimes done in [9].

Through the n-tuples $\mathbf{k}$, $\mathbf{a}$, we may define

$$v_{\mathbf{k}} := e^{k_n} v_{-a_n} \otimes e^{k_{n-1}} v_{-a_{n-1}} \otimes \cdots \otimes e^{k_1} v_{-a_1}, \tag{6}$$

viewed as in the n-fold tensor product $V(-a_n) \otimes V(-a_{n-1}) \otimes \cdots \otimes V(-a_1)$ given an action of $\mathfrak{sl}(2)$, by the Leibnitz rule. It is clear that there exists an $\mathfrak{sl}(2)$ module homomorphism taking $v_{\mathbf{k}}$ to $\overline{v}_{\mathbf{k}}$. One knows that the former is non-zero if and only if $k_i \leqslant a_i$, for all $i = 1, 2, \ldots, n$. By contrast, it is *much more* difficult to say when the latter is non-zero. Indeed, this is the main problem in the theory of trails.

Set $\mathbf{E}_s = \cup_{j \in J^s} \mathbf{E}_s^j$. The sets $T_s(\mathbf{e}) : \mathbf{e} \in \mathbf{E}_s$ are disjoint, and their union is $\mathcal{K}_t^{BZ}$, by definition of the latter.

The basic task is to describe each $T_s(\mathbf{e})$. The main idea is to select a "minimal" trail in $T_s(\mathbf{e})$ and to adjoin to it the faces defined by the S-graph it defines. In this, we show, following [9, Lemma 5.3.4(i)], that our choice of minimal trail trivializes at w_{j-1}, and so in principle, one obtains an inductive procedure to construct $\mathcal{K}_t^{BZ}$.

Let us recall that an element $\overline{v}_{\mathbf{k}} : \mathbf{k} \in T_s(\mathbf{e})$ is a *non-zero* vector in $V(-\varpi_t)$. Let $M_s(\mathbf{e})$ denote the subspace generated by $\{\overline{v}_{\mathbf{k}}\}_{\mathbf{k} \in T_s(\mathbf{e})}$. It is clear that $M_s(\mathbf{e})$ is a module for the $\mathfrak{sl}(2)$ subalgebra spanned by the s-triple (e, h, f). By Joseph [9, 4.2], it is a simple module. The proof uses crucially the boundary condition $(B)(ii)$, specifically that the trails in $T_s(\mathbf{e})$ trivialize at w_j and an elementary property of Demazure modules. It means that $\overline{v}_{\mathbf{k}} : \mathbf{k} \in T_s(\mathbf{e})$ is determined up to a non-zero scalar by its weight and so by the sum $\sum_{i=1}^{n} k_i$, a fact that is a priori surprising. This was used in [9, 5.2, Remark 2] with consequences for [9, 5.3.1,5.3.2]. Similarly, [9, 5.2, Remark 2] is used in the proof of Lemma in Sect. 4.1.4.

In the present paper, we observe (Lemma in Sect. 4.7) that the rules for adjoining faces become very similar to the rules for crystal construction in the case when J is a periodic reduced expression, that is to say J is given by a power of a suitable Coxeter element. This gives a rather explicit model for $\mathcal{K}_t^{BZ}$, namely the highest weight crystal $B(\varpi_t)$ associated to the highest weight module $V(\varpi_t)$ of highest weight ϖ_t. In this, not all the crystal elements are to be used, except for one value of t depending on J, as there is a cut-off at the final function ([11, 2.6] and Sect. 4.5).

3.9 *The Positivity Condition*

It is the condition

$$\gamma_j^K \in \gamma_j^{K_t^1} + \mathbb{N}\pi, \forall j \in J.$$

Given a trail $K \in \mathcal{K}_t^{BZ}$ with $K + F_s^{k+1} \in \mathcal{K}_t^{BZ}$, set $u = (s, k+1)$, $v = (s, k)$. Then by Joseph [9, Eq. (3)], one has

$$\gamma_j^{F_s^k} := \begin{cases} \gamma_j^K + \alpha_s, & \text{if } v < j \leqslant u, \\ \gamma_j^K, & \text{otherwise.} \end{cases}$$

Thus, the positivity condition is automatically satisfied if every trail is obtained by adjoining faces to the driving trail. Thus, it might seem advisable to impose it from the start. However, this is not necessary in the minuscule case [11, Remark 3.10] and indeed superfluous if an inductive process is used to obtain trails by adjoining trails to the driving trails. Thus, it will *not* be imposed.

4 Adjoining Faces to Trails

Fix $t \in I$ throughout.

4.1 *Preliminaries*

4.1.1 S-Graphs

Set $\mathbf{F}_t := \{z^K\}_{K \in \mathcal{K}_t^{BZ}}$. It lies in the set

$$X_t = z_t^1 + \sum_{(s,k)\in I\times\mathbb{N}^+} \mathbb{N}(r_s^k - r_s^{k+1}). \tag{7}$$

Given a subset F of $\mathbf{F}_t$, we may consider its $\mathbb{Z}$ convex hull, that is to say the subset of non-negative rational linear combinations of elements of F lying in X_t, in the sense of (7). Conversely given a convex subset F of X_t, we may consider its subset $E(F)$ of extremal elements.

As noted in Sect. 1.2, S-graphs were introduced in [7] to resolve the invariance property required of the dual Kashiwara functions. They are defined by an n-tuple $\mathbf{c} = (c_n, c_{n-1}, \ldots, c_1)$ of non-negative integers, called a coefficient set. The vertices of an S-graph form an S-set $Z(\mathbf{c})$ of functions. When S of type $s \in I$, these are sums of successive differences of the Kashiwara functions $r_s^k - r_s^{k+1}$, the sums being dependent on $\mathbf{c}$ but not on s.

Let $K(\mathbf{c})$ (resp. $K_{\mathbb{Z}}(\mathbf{c})$) denote the convex (resp. $\mathbb{Z}$ convex) hull of $Z(\mathbf{c})$. The latter is described directly and explicitly in [8, 1.5], where it is also shown that $Z(\mathbf{c})$ is just the set of extremal elements of $K(\mathbf{c})$. (This results in an easier more practical way to describe $Z(\mathbf{c})$. It will not be used here because the S-sets that occur are rather simple.)

4.1.2 The Inductive Construction and ℓ-Minimal Trails

Let j be a positive integer. A linear function z on B_J is said to have support in $[1, j]$ if the coefficients of $m_{j'} : j' \in J$ vanish for $j' > j$, and we write $\operatorname{Supp} z \subseteq [1, j]$. In particular, $\operatorname{Supp} z_t^1 \subseteq [1, (t, 1)]$.

Recall Lemma in Sect. 3.6 and fix $j \geqslant (t, 1)$. Set $\mathbf{F}_t^j := \{z^K\}_{K \in \mathcal{K}_t^j}$.

By Remark 1 of 3.7, the elements of $\mathbf{F}_t^j$ have support in $[1, j]$.

We now propose an inductive construction, $(*)$ below, of $\mathbf{F}_t^j$ and show that the elements of $\mathbf{F}_t$ with support in $[1, j]$ lie in $\mathbf{F}_t^j$, that is to say we show that the converse of the previous assertion holds.

Given $j \in J$, we may assume $j > (t, 1)$ by Lemma in Sect. 3.6 and we set $i_j = s$. By definition, we may write $j = (s, n)$ for some $n \in \mathbb{N}^+$.

Recall Sect. 3.8. Take $\mathbf{e} \in \mathbf{E}_s^j$ and give $T_s(\mathbf{e})$ the linear order obtained from the lexicographic ordering defined by powers of e_s [9, 4.5], that is to say on the $n_{(s,k)}$: $k = 1, 2, \ldots,$ defined in (T).

Let $K_{\ell\,\min}$, or simply L, be the unique ℓ-minimal trail in $T_s(\mathbf{e})$ given by this total order. Equivalently, L defined through (5) by an n-tuple $\mathbf{l}$ of exponents of e_s with the property that if K is a trail in $T_s(\mathbf{e}) \setminus \{K_{\ell\,\min}\}$ defined by an n-tuple $\mathbf{k}$, then there exists $i' \in N$ such that $k_i = \ell_i$ for all $i < i'$ and $k_{i'} > \ell_{i'}$.

One easily shows, [9, Eq. (26)], that the coefficients $c_k^{K_{\ell\,\min}}$, or simply c_k, of the $-m_s^k : k \in \mathbb{N}^+$ in $z^{K_{\ell\,\min}}$ are given (in the notation of Sect. 3.5) by

$$c_k = a^{(k)} - (\ell^{(k)} + \ell^{(k-1)}), \forall k \in \mathbb{N}^+. \tag{8}$$

In addition, an exercise just using the Chevalley–Serre relations shows that they satisfy [9, 5.3.3,5.3.4] the conditions $c_k \geqslant 0$ with $c_k = 0$ for all $k \geqslant n$.

Thus, $z := z^{K_{\ell\,\min}}$ provides a coefficient set in the sense of Sect. 4.1.1. We define z to be the driving function (of type s) for $T_s(\mathbf{e})$.

Recall the notation of Sect. 3.8.2. One may remark that if $\mathbf{e} \in \mathbf{E}_s^j$, then the trails $K \in T_s(\mathbf{e})$ trivialize at w_j.

Let $T_s^-(\mathbf{e})$ denote the subset of $T_s(\mathbf{e})$ of trails that trivialize at w_{j-1}. In particular, the assertion $c_n = 0$ implies that the minimal trail trivializes at w_{j-1} by Joseph [9, Lemma 5.3.4(i)]. Then, if the n-tuple $\mathbf{l}$ defining $L = K_{\ell\,\min}$ satisfies (8), for $k = n$ we obtain $a^{(n)} = \ell_n + 2\ell^{(n-1)}$. Yet by (5), we have $\alpha_s^\vee(-w_j\varpi_t) = 2\ell^{(n)} - a^{(n)} = \ell_n$ and so

$$v_{-w_j\varpi_t} = e_s^{\ell_n} v_{-w_{j-1}\varpi_t}. \tag{9}$$

In particular, $L \in T_s^-(\mathbf{e})$. (In this, we remark that $T_s^-(\mathbf{e})$ need not be reduced to $K_{\ell\,\min}$, [9, Sect. 7].)

Our inductive construction is

$(*)$ to obtain $T_s(\mathbf{e})$ adjoin sums of faces of the form $F = \sum_{k \leqslant n} d_k F_s^k : d_k \in \mathbb{N}$ to the ℓ-minimal trail L in $T_s(\mathbf{e})$.

Observe that $K_{\ell\,\min} + F$ has support in $[1, j]$ (and not in $[1, j-1]$) (resp. trivializes at w_j (and not at w_{j-1})) if and only if $d_n \neq 0$.

This proves the required converse.

We make this construction more precise by requiring that the sum of faces we adjoin is determined by the S-set $Z(\mathbf{c})$ of type $s = i_j$ and where each such sum consists of an element of $\mathbb{Z}$ convex hull $K_{\mathbb{Z}}(\mathbf{c})$ of $Z(\mathbf{c})$. Of course what we also have to prove is that this gives *all* of $T_s(\mathbf{e})$. This is carried out in the present special case in Sects. 6.6–6.9. Even that is not so easy partly because of the presence of "false trails".

It is already a total leap of faith to believe that all the trails in $T_s(\mathbf{e})$ are obtained by just *adding* faces to $K_{\ell \min}$. This is shown in Sects. 6.7–6.9 for the present rather simple case and is already quite difficult. We regard this as an important test case.

Notice that this construction means that if $j \in J : j > (t, 1)$ is maximal such that the coefficient of m_j in z^K is non-zero, then this coefficient d is positive and d copies of $F_s^i : j = (s, i)$ possibly together with faces $F_s^{i'} : i' < i$ (as determined by the S-set in question) may be removed from K, to obtain a previously constructed trail. We use this in 6.9.3 where we show that fortunately lower order terms are not needed in the particular cases studied.

Notice from the point of view of trails, we can equally well designate $Z(\mathbf{c})$ as $Z(\mathbf{e})$ (and $K_{\mathbb{Z}}(\mathbf{c})$, by $K_{\mathbb{Z}}(\mathbf{e})$), since $\mathbf{e} \in \mathbf{E}_s^j$ determines $z^{K_{\ell \min}}$ and its coefficient set $\mathbf{c}$, as was done in [9].

4.1.3 False Trails

Definition A false trail in $T_s(\mathbf{e})$ is a trail that does not obtain from the ℓ-minimal trail L by adjoining faces prescribed by the S-set defined by $\mathbf{l}$.

The main result of [9] is to show that if there are no false trails in the subset $T_s^-(\mathbf{e})$ of $T_s(\mathbf{e})$ of trails that trivialize at w_{j-1}, then there are no false trails in $T_s(\mathbf{e})$ and moreover, $T_s(\mathbf{e})$ is obtained by adjoining the sums of faces in $K_{\mathbb{Z}}(\mathbf{c})$ to $K_{\ell \min}$. This uses $\mathfrak{sl}(2)$ theory (specifically [9, Lemma 3.2]) and a detailed analysis of S-sets. In the case that the $c_j : j \in [1, n-1]$ are increasing, $T_s^-(\mathbf{e})$ is reduced to $K_{\ell \min}$ and the proof is rather easy [9, Sect. 6]. In general, it is a difficult result.

Because of the possible existence of false trails, we need to push through at least part of the analysis of [9] without the assumption that false trails are absent. This is carried out below.

4.1.4 $\mathbb{Z}$-Skins and Trails

The method used in [9, 5.3] to prove the positivity of the c_j also establishes the following result. Both are mainly an exercise in the use of the Chevalley–Serre relations.

Take a trail $K \in T_s(\mathbf{e})$ defined by an n-tuple $\mathbf{k}$ through (5). Take $j \in [1, n-1]$ such that $k_i = \ell_i$, for all $i < j$. Then, $\overline{v}_{\mathbf{k}}$ defined by (5) may be written in the form $\overline{v}_{\mathbf{k}} = e^{k_n} e_{-a_n} \cdots e^{k_j} v_{-d_j}$, where $v_{-d_j} := e_{-a_j} e^{\ell_{j-1}} e_{-a_{j-1}} \cdots e^{\ell_1} v_{-a_1}, j \in$

$[1, n-1]$, which one checks has weight $-d_j$, where $d_j = c_j + \ell_j$, with respect to h of the s-triple (e, h, f).

Let $\mathbf{c} = (c_n, c_{n-1}, \ldots, c_1)$ be the coefficient set defined by L, recalling Sect. 4.1.2 that the c_i are non-negative integers with $c_n = 0$. We set $c_0 = 0$. Recall that the pointed chain [9, 6.3] of the S-set defined by $\mathbf{c}$ is defined through the set of n-tuples $\{\mathbf{k}(j)\}_{j=1}^n$, where

$$\mathbf{k}(j)_i = \begin{cases} \ell_i + c_i - c_{i-1}, & \text{if } i \in [j+1, n], \\ \ell_i + c_i, & \text{if } i = j, \\ \ell_i, & \text{if } i \in [1, j-1]. \end{cases}$$

Here, we may remark that in terms of the coefficients $c'_i := k^{(i)} - \ell^{(i)}$, as first defined in [9, Eq. (30)], we have for $\mathbf{k} = \mathbf{k}(j)$, that

$$c'_i = \begin{cases} c_i, & \text{if } i \in [j, n], \\ 0, & \text{if } i \in [1, j-1]. \end{cases}$$

Definition The $\mathbb{Z}$-skin of the pointed chain is given by the set of n-tuples given by

$$c'_i = \begin{cases} c_i, & \text{if } i \in [j+1, n], \\ \in [0, c_i], & \text{if } i = j, \\ 0, & \text{if } i \in [1, j-1], \end{cases}$$

where j runs from 1 to n.

The $\mathbb{Z}$-skin of the pointed chain has a natural linear order, $\mathbf{c}' \geqslant \mathbf{c}''$ given by $c'_i = c''_i, n \geqslant i > j, c'_j > c''_j$. It contains the pointed chain and is contained in the $\mathbb{Z}$ convex hull of the pointed chain. Again the number of elements in the $\mathbb{Z}$-skin is $\sum_{i=1}^n c_i$, an expression that we shall see again in (12).

Lemma *Take $j \in [1, n-1]$. Suppose in the n-tuple $\mathbf{k}$, one has $k_i = \ell_i$, for all $i < j$.*

(i). *If $k_j > d_j$. Then, $\overline{v}_{\mathbf{k}} = 0$.*
(ii). *Conversely suppose that an n-tuple $\mathbf{k}$ with $k_i = \ell_i$ for all $i \leqslant j$ defines a trail. Then if $c_{j+1} = 0$, one may move up to c_j copies of e to the right through e_{-a_j} in $v_{\mathbf{k}}$ and still obtain a trail.*
(iii). *The elements of the $\mathbb{Z}$-skin are trails.* □

Proof

(i). The proof is by induction following closely the analysis of [9, 5.3.2]. One has $d_1 = a_1$, whilst $e^{k_1} v_{-a_1} = 0$ if $k_1 > a_1$, since v_{-a_1} is a lowest weight vector. Thus, the assertion holds for $j = 1$.

Let $\overline{v}_{\mathbf{k}}$ and v_{-d_j} be as above.

Recall that $\mathbf{l}$ defines the ℓ-minimal trail $K_{\ell\,\min}$, so in particular ℓ_{j-1} cannot be decreased. Noting that $a_j - \ell_{j-1} \geqslant 0$ by Joseph [9, Lemma 5.3.1], then by Joseph [9, Remark 2 of 5.2], there exists an integer $q_j \geqslant k_j + \ell_{j-1} - a_{j-1}$ so that we may take q_j copies of e to the right through e_{-a_j} and in so doing only change $\overline{v}_{\mathbf{k}}$ by a non-zero scalar. Up to this scalar $\overline{v}_{\mathbf{k}} = e^{k_n} \cdots e^{k_j - q_j} e_{-a_j} e^{q_j + \ell_{j-1}} v_{-d_{j-1}}$.

Thus by the induction hypothesis, non-vanishing of this last expression implies that $q_j \leqslant c_{j-1}$. Consequently, $k_j \leqslant a_j - \ell_{j-1} + c_{j-1} = c_j + \ell_j = d_j$, as required.

(ii). One has $a_{j+1} - \ell_{j+1} - \ell_j = c_{j+1} - c_j = -c_j$. By minimality, ℓ_j cannot be decreased, whilst by (i) it cannot be increased beyond $\ell_j + c_j$.

Then by Joseph [9, Lemma 5.2], we can move any $q_j \in [0, c_j]$ copies of e through $e_{-a_{j+1}}$, only changing $\overline{v}_{\mathbf{k}}$ by a non-zero scalar. Hence (ii).

(iii). The proof of (iii) follows closely that of (ii) and proceeds by decreasing induction. Since $c_n = 0$, it holds for $j = n - 1$. Now assume it holds for some $n - 1 \geqslant j > 1$, and let $\mathbf{k}$ be some $\mathbf{k}(j)$. Then, $k_j = c_j + \ell_j$. By minimality, $k_{j-1} = \ell_{j-1}$ cannot be decreased, and by (i), it cannot be increased beyond $c_{j-1} + \ell_{j-1}$. On the other hand, $k_j + \ell_{j-1} - a_{j-1} = c_j + \ell_j + \ell_{j-1} - a_{j-1} = c_{j-1}$. Then by Joseph [9, Lemma 5.2], we can take $e^{q_j} : q_j \in [0, c_{j-1}]$ to the right through e_{-a_j}, only changing $\overline{v}_{\mathbf{k}}$ by a non-zero scalar. Hence, the assertion.

Remark 1 Notice that (i) falls short of showing that $e^{d_j+1} v_{-d_j} = 0$. The latter would in turn imply by $\mathfrak{sl}(2)$ theory that v_{-d_j} is a lowest weight vector. Given such a result (for all $j \in [1, n-1]$ and noting that it holds for $j = n$ by Joseph [9, Lemma 5.3.4(i)]), we could conclude that $K_{\ell \min}$ is a minimal trail in the sense of [9, Sect. 4]. Yet although we cannot say that $f v_{-d_j} = 0$, we have the next best thing, namely $e^{k_n} \cdots e^{k_j} f v_{-d_j} = 0$. Indeed, the action of f on v_{-d_j} is to reduce one factor of e by 1. Then, the above vanishing results from $\mathbf{l}$ being an ℓ-minimal trail.

We shall see in Sect. 6 that (i) limits the possible false trails in $T_s(\mathbf{e})$.

Remark 2 Notice that by (i) we can obtain a stronger result, should some $c_j : j \in [1, n-1]$ vanish. Indeed, we may decompose $\mathbf{c}$ into connected components such that $c_j \neq 0$ in each such component, say $\mathbf{c}_{j,j'} := (c_j, c_{j-1}, \dots, c_{j'})$. Then by definition, $c_{j+1} = 0$. Starting from $j + 1$ instead of n, the exact same argument given in (iii) shows that each element of the $\mathbb{Z}$-skin of the S-set defined by $\mathbf{c}_{j,j'}$ is a trail. We call this set the extended $\mathbb{Z}$-skin of the S-set.

Remark 3 If we believe that all trails are obtained by adjoining faces to the initial driving trail, then there is an alternative definition of a minimal trail L in $T_s(\mathbf{e})$, namely that we cannot remove a face or sum of faces of type s from L. If we take $\mathbf{l}$ to be the n-tuple that defines L, it follows easily from [9, Lemma 5.3.4(i)] that this is equivalent to saying for any $i \in [1, n]$ that ℓ_i cannot be decreased in $\mathbf{l}$, which of course is also true of the unique ℓ-minimal trail. However for the latter, we have the stronger property that this is also true even after increasing some $\ell_j : j > i$.

We were unable to establish analogues of [9, Lemma 5.3.2] and the above lemma for minimality in this weaker sense.

The trails in $T_s(\mathbf{e})$ are of the form $L + \sum_{i=1}^{n-1} c_i' F_s^{i+1}$ with c_i' integer. Part of the definition (see Sect. 6.4) of there being no false trails (at the previous induction step) is that $c_i' \geqslant 0$, for all $i < n - 1$. In [9, Lemma 7.4], we used $\mathfrak{sl}(2)$ action to show that then this also holds when $i = n - 1$.

The condition $c_j' \geqslant 0$, for all $j \in [1, n - 1]$, is equivalent to there being no distinction between the above two notions of minimality.

4.2 Assumptions

From now on, we assume that W is the Weyl group of a simple Lie algebra $\mathfrak{g}$ whose Coxeter number c is even. Recall the notation of 1.1.

Fix σ a Coxeter element $s_{i_\ell} s_{i_{\ell-1}} \cdots s_{i_1}$, where by definition $\{i_\ell, i_{\ell-1}, \ldots, i_1\} = \pi$, and assume that every power of σ up to $c/2$, which is an integer by hypothesis, is a reduced decomposition. This holds for σ defined in 2.1. We shall say that J is periodic if it is given by a reduced decomposition of w_0 obtained from powers of a Coxeter element.

Apart from Sect. 4.8, we shall assume in the remainder of Sect. 4 that J is periodic and relabel π so that $i_j = j : j = 1, 2, \ldots, \ell$. This relabelling is **not** used in the tables and figures.

4.3 Some Computations

Fix a positive integer $j > (t, 1)$ and assume $\mathcal{K}_t^{j-1}$ has been constructed as a sum with non-negative coefficients of closed faces $F_u^v : u \in I, v > 1$ adjoined to the open face F_t^1. Now, F_u^v has support in $[1, (u, v)]$, whilst the elements of $\mathcal{K}_t^{j-1}$ have support in $[1, j - 1]$ by Sect. 4.1.2.

Now set $s = i_j$, then by Sect. 4.1.2, the driving function given by the unique ℓ-minimal element of $T_s(\mathbf{e})$ lies in $T_s^-(\mathbf{e})$ and, hence, in $\mathcal{K}_t^{j-1}$. Let F^z denote the corresponding driving trail (defined by $z^{F^z} = z$).

Recall (Sect. 4.2) how a reduced sequence is given by the power of a suitable Coxeter element.

Let n be the positive integer defined by $\ell \geqslant j - (n - 1)\ell > 0$, explicitly that $n - 1 = [(j - 1)/\ell]$. Observe that $s = i_j = j - (n - 1)\ell$.

Again a closed face F_u^v has support in $[1, j - 1]$ if and only if either $v = n, 0 < u < j - 1 - (n - 1)\ell = s$ or $1 < v < n, u \in I$. Consequently, the closed faces occurring in F^z must satisfy these conditions.

The sums of closed faces F_u^v to be adjoined at the next induction step must have support in $[1, j]$ and so must satisfy either that $v = n, u \leqslant j - (n - 1)\ell$ or $1 < v < n, u \in I$. In this, we require $u = s$. Moreover, unless the coefficient of F_s^n is non-zero, the resulting terms already lie in $\mathcal{K}_t^{j-1}$.

For the precise sums to be adjoined, we must compute the coefficients of $-m_s^i$ in z as these determine the coefficient set $\mathbf{c}$.

Now by (4), the face $F_{u'}^{v'}$, for $u' \in I \setminus \{u\}, v' \leqslant n$, makes the contribution $\alpha_u^\vee(\alpha_{u'})$ to m_u^v given that

$$(u', v'-1) < (u, v) < (u', v').$$

In the present periodic case, one has $(u, v) = u + (v-1)\ell$, so this last equation simplifies to

$$u' + (v'-2)\ell < u + (v-1)\ell < u' + (v'-1)\ell. \tag{10}$$

Yet $0 \leqslant |u - u'| < \ell$. Thus if $u < u'$, one has $u' - \ell < u$, so the contribution $\alpha_u^\vee(\alpha_{u'})$ to m_u^v is made exactly when $v = v'$, whilst if $u > u'$, one has $u - \ell < u'$, so the contribution $\alpha_u^\vee(\alpha_{u'})$ to m_u^v is made exactly when $v = v' - 1$.

Again by (4), the face $F_u^{v'}$ makes the contribution of 1 to m_u^v given that either $v = v'$ or $v = v' - 1$.

We conclude that the overall contribution d_u^v to the coefficient of $m_u^v : v \leqslant n$ coming from the driving trail $F^z := \sum_{u' \in I, v' \leqslant n} a_{u'}^{v'} F_{u'}^{v'}$ (where we note that $a_{u'}^n = 0$ if $\ell \geqslant u' \geqslant s$) of type u corresponding to z is just

$$d_u^v = a_u^v + a_u^{v+1} + \sum_{u'=1}^{u-1} \alpha_u^\vee(\alpha_{u'}) a_{u'}^{v+1} + \sum_{u'=u+1}^{\ell} \alpha_u^\vee(\alpha_{u'}) a_{u'}^v, \tag{11}$$

using the convention that $a_{u'}^n = 0$, if $\ell \geqslant u' \geqslant s$.

4.4 The Crystal $B(\varpi_t)$ as a Subcrystal of $B(\infty)$

We may consider the crystal $B(\varpi_t)$ of the highest weight module $V(\varpi_t)$ as a subcrystal of $B(\infty)$. This means that the crystal graph of $B(\varpi_t)$ is a subgraph of the crystal graph of $B(\infty)$. However, the embedding $B(\varpi_t) \hookrightarrow B(\infty)$ is not a strict in the sense of Kashiwara [15, Sect. 1]. Indeed, the latter would mean that the crystal graph of $B(\varpi_t)$ is the connected component of the crystal graph of $B(\infty)$ containing b_∞. In order to obtain this latter property, we may proceed as in [6, 2.5.9] by introducing the crystal $S(\varpi_t)$ having just one element s_{ϖ_t} of weight ϖ_t satisfying $\varphi_i(s_{\varpi_t}) = 0$, for all $i \in I$. Then, indeed [6, Lemma 2.5.9], $B(\varpi_t)$ is a strict subcrystal of $B(\infty) \otimes S(\varpi_t)$. In this, there is a very slight difference in computing the action of the Kashiwara operators $\tilde{f}_i : i \in I$. Explicitly for all $b \in B(\infty)$, one has $\tilde{f}_i(b \otimes s_{\varpi_t}) = b \otimes \tilde{f}_i s_{\varpi_t} = 0$, if and only if $\varphi_i(b) \leqslant -\alpha_i^\vee(\varpi_t)$, which is what is required for a strict embedding (see proof of [6, Lemma 2.5.9]). For example, one has $\varphi_i(b_\infty) = 0$, for all $i \in I$ and so $\tilde{f}_i(b \otimes s_{\varpi_t}) \neq 0$, if and only if $i = t$.

Notice that in this presentation of $B(\varpi_t)$, we are taking $B(\infty)$ to be $B_J(\infty)$. In order to indicate this, we shall write the former as $B_J(\varpi_t)$, though to avoid heavy notation this J subscript is sometimes dropped.

4.5 The Map θ

Let $\mathbf{F}$ denote the additive semigroup of formal sums of faces $F_{u'}^{v'} : u' \in I, v' \in \mathbb{N}^+$ with non-negative integer coefficients. Let $1_{u'}^{v'}$ be the element of B_J having 1 in the (u', v') place and zero elsewhere. Let Φ denote the additive bijection from $\mathbf{F} \to B_J$ defined by $\Phi(F_{u'}^{v'}) = 1_{u'}^{v'}$. Given $b \in B_J$, let $\operatorname{Supp} b$ denote the set of places of its non-zero entries. By (4), $F_{u'}^{v'}$ has support in $[(u', v'-1), (u', v')]$, whilst the support of its image is just $\{(u', v')\} \subseteq [(u', v'-1), (u', v')]$. Thus if $\operatorname{Supp}(\mathbf{F}) \subseteq [1, k]$, then $\operatorname{Supp}(\Phi(\mathbf{F})) \subseteq [1, k]$.

Again let $\mathbf{L}$ denote the $\mathbb{Z}$ module of locally finite additive functions on B_J with values in $\mathbb{Z}$. Then by (4) and the definition of $z^{F_t^1}$, it follows that $F \mapsto z^F$ is an additive bijection of $\mathbf{F}$ onto $\mathbf{L}$.

It is convenient to view the inverse image of b_∞ as the trivial trail K_∞.

Recall (Sect. 4.1.1), the subset $\mathbf{F}_t$ of $\mathbf{F}$. It may be identified with $\mathcal{K}_t$. Set $\hat{\mathcal{K}}_t = \mathcal{K}_t \cup \{K_\infty\}$ and, correspondingly, $\hat{\mathbf{F}}_t = \mathbf{F}_t \cup \{0\}$.

An interesting open question is to describe $\Phi(\hat{\mathbf{F}}_t)$ as a subset of B_J. The result depends on the choice of J and the form of the final function z_t^f, which by [11, Lemma 2.6] equals $m_{\theta(t)}$, where $\theta(t)$ is the unique value of $j \in J$ such that $w_j\alpha_t = -\alpha_{i_j}$. The trivial case is when $\theta(t) = 1$ (which arises when $i_1 = t$) and the most complicated case is when $\theta(t) = r$. For any choice of J, the latter always arises for some (unique) $t \in I$. This is clear for example from the paragraph following [11, Eq. (11)].

Let $F_w(\varpi_t)$ denote the Demazure submodule $U(\mathfrak{b})v_{w\varpi_t}$.

In all cases we have examined, one always has $|\Phi(\hat{\mathbf{F}}_t)| \geqslant \dim F_{w_{\theta(t)}}(\varpi_t)$ and this inequality is generally strict.

One may compute θ when the Coxeter element $\sigma = \sigma_1\sigma_2$ is bipartite; that is, it takes the form given in the first paragraph of Sect. 2.2. Let the roots be labelled so that $\sigma_1 = s_\ell s_{\ell-1} \cdots s_{m+1}, \sigma_2 = s_m s_{m-1} \cdots s_1$. Assume that the Coxeter number c is even. Then, $\sigma^{c/2} = w_0$. One has $w_0(\alpha_i) = -\alpha_{\kappa(i)}$, which defines κ as a Dynkin diagram involution. Because $c = 2m$ is even, $\mathfrak{g}$ is not of type A_{2k}, and then one easily checks that $\kappa(i) \leqslant m$ if and only if $i \leqslant m$.

Lemma *For all $i \in I$, one has the following:*

(i) $\theta(i) = i$, *if* $i \leqslant m$.
(ii) $\theta(i) = \ell(w_0) - (\ell - \kappa(i))$, *if* $i > m$. □.

Proof (i) is obvious. If $i > m$, then $j := \kappa(i) > m$.

Yet $w_{\ell(w_0)-(\ell-j)} = s_j s_{j-1} \cdots s_{m+1}\sigma_2\sigma^{c/2-1} = s_{j+1}s_{j+2} \cdots s_\ell w_0$.

On the other hand, $s_{j+1}s_{j+2}\cdots s_\ell w_0\alpha_i = -s_{j+1}s_{j+2}\cdots s_\ell \alpha_{\kappa(i)} = -\alpha_{\kappa(i)}$.

Remark Thus, $\theta(i) = r$ exactly when $\kappa(i) = \ell$.

4.6 Corollary of Lemma in Sect. 4.5

We shall need the following consequence of the above lemma. In this, we let Supp w denote the set of simple reflections that occur in some (and hence any) reduced decomposition of w.

Corollary *Suppose* $\ell \geqslant u > t$.

(i) *If* $t \leqslant m$, *one has* $\alpha_u^\vee(\alpha_t) = 0$ *for all* $s_u \in$ Supp $w_{\theta(t)}$.
(ii) *If* $t > m$, *one has* $\alpha_u^\vee(\alpha_t) = 0$ *for all* $u > t$. □

Proof (i) (resp. (ii)) follows from (i) (resp. (ii)) of Lemma in Sect. 4.5.

Remark and Definition The result of this corollary is needed in Proposition in Sect. 4.9. It seems unlikely that it holds for an arbitrary Coxeter element.

We say J is periodic (resp. bipartite) if w_0 is given by the power of a Coxeter (resp. bipartite Coxeter) element.

4.7 A Comparison Lemma

We may consider the "string" of crystal elements $b_i := \tilde{f}_u^i(b \otimes s_{\varpi_t}) \in B_J(\varpi_t)$, with $i = 0, 1, \dots, m$, where m is the largest integer for which this expression does not vanish. Since $B_J(\varpi_t)$ is lower normal, one has $m = \varphi_u(b \otimes \varpi_t)$ and this equals $\varphi_u(b) + \delta_{u,t}$, where δ is the Kronecker delta.

Our present goal is to show that $\{b_i\}_{i=1}^m$ is the image under Φ of the $\mathbb{Z}$-skin of the S-set defined by $\mathbf{c}^z$, and specifically that

$$m = \sum_{i=1}^{n} c_i^z, \tag{12}$$

when J is bipartite.

Towards this aim, we first prove the following comparison lemma.

Take $(u, v) \leqslant k$ and set $\tilde{d}_u^v = (r_u^v - r_u^{v+1})(b)$. Here, we shall be calculating the latter using (1) that is to say using the Kashiwara functions in which roots and coroots *are not interchanged*.

Lemma *Assume that J is periodic. Fix a driving trail $F^z \in \mathcal{K}_t^{j-1}$ and assume that $\Phi(F^z) \otimes s_{\varpi_t} \in B_J(\varpi_t)$. Then,*

$$d_u^v = \tilde{d}_u^v, \quad \forall (u, v) \leqslant j - 1. \tag{13}$$

Proof By the linearity of the Kashiwara functions, it is enough to add the contributions coming from each face. The contribution from $\Phi(F_{u'}^{v'}) = 1_{u'}^{v'}$ to r_u^v, that is to say $r_u^v(1_{u'}^{v'})$, is given by (1). When $u' = u$, its value is 2 (resp. 1, 0) when $v' > v$ (resp. $v' = v, v' < v$). Thus, $(r_u^v - r_u^{v+1})(\sum_{v' \leqslant n} a_u^{v'} 1_u^{v'})$ is just $a_u^v + a_u^{v+1}$, using the same convention as in 4.3. This is the sum of the first two terms of (11).

The contribution from $1_{u'}^{v'}$ to r_u^v when $u' \neq u$ is $\alpha_u^\vee(\alpha_{u'})$ (resp. 0) when $(u', v') > (u, v)$ (resp. $(u', v') < (u, v)$). For our particular reduced decomposition, one has $(u, v) = u + (v - 1)\ell$, and so $(u', v') > (u, v)$ if and only if $u' < u, \quad v' > v$ or $u' > u, \quad v' \geqslant v$. Thus, the contribution $\tilde{d}_u^v$ from $\sum_{u' \in I \setminus \{u\} v' \leqslant n} a_{u'}^{v'} 1_{u'}^{v'}$, to $r_u^v - r_u^{v+1}$ is just $\sum_{u' < u} \alpha_u^\vee(\alpha_{u'}) a_{u'}^{v+1} + \sum_{u' > u} \alpha_u^\vee(\alpha_{u'}) a_{u'}^v$, using the same convention as in 4.3. These are the same sums as in (11).

Combined with the first part and recalling the definition of b, this proves the lemma.

Remark The conclusion of the lemma may be expressed as follows. The contribution of the face function $z^{F_{u'}^{v'+1}} = r_{u'}^{v'} - r_{u'}^{v'+1}$ to the coefficient of m_u^v is the same as the contribution of its image in B_J, namely $\Phi(F_{u'}^{v'+1}) = 1_{u'}^{v'+1}$ to the coefficient of $r_u^v - r_u^{v+1}$. The former determines the coefficient set of a driving sum F^z and hence the S-set it defines, whilst the latter determines in which place of $\Phi(F^z) \in B_J$, an element $\tilde{f}_s : s \in I$ enters.

This coincidence is special to periodic decompositions.

4.8 The Importance of Periodicity: An Example

The above lemma is of itself not quite sufficient to obtain (12).

In this section, we just note that the lemma fails if the decomposition is not periodic as might be expected from Remark in Sect. 4.7. This is shown in the following example.

Take $\mathfrak{g}$ of type A_3 and adopt the Bourbaki labelling. Consider the reduced decomposition of the longest element given in shorthand notation by 123212, in which i denotes s_i. It is not a periodic decomposition. The initial driving face associated to 3 is just F_3^1. It is a driving face of type 2. The contribution of F_3^1 makes to the coefficient d_2^j of m_2^j is $(0, -1, -1)$, for $j = 3, 2, 1$, respectively. The S-set it defines is $\{F_3^1, F_2^3 + F_3^1, F_2^2 + F_2^3 + F_3^1\}$. Then, Z_3 consists of this set together with $F_1^2 + F_2^2 + F_2^3 + F_3^1$. Their images in $B_J(\infty)$ under Φ are, respectively, $(0, 0, 1, 0, 0, 0)$, $(0, 1, 1, 0, 0, 0)$, $(0, 1, 1, 1, 0, 0)$, $(1, 1, 1, 1, 0, 0)$.

On the other hand, the contribution of $\Phi(F_3^1)$, which is 1 in the (3, 1) place, to r_2^j equals $(0, 0, -1, -1)$ for $j = 4, 3, 2, 1$, respectively, and so the contribution $\tilde{d}_2^j$ to $r_2^j - r_2^{j+1}$ is $(0, -1, 0)$, for $j = 3, 2, 1$, respectively.

In particular, $d_2^1 \neq \tilde{d}_2^1$.

Then, the crystal rules for $B(\infty)$ give the elements $\tilde{f}_3 b_\infty = (0, 0, 1, 0, 0, 0)$, $\tilde{f}_2\tilde{f}_3 b_\infty = (0, 1, 1, 0, 0, 0)$, which so far correspond to the above sums of faces; but then $\tilde{f}_2^2\tilde{f}_3 b_\infty = (0, 1, 1, 0, 0, 1)$, which corresponds to the face sum $F_2^1 + F_2^3 + F_3^1$, which differs from the third element of the S-set in the paragraph above. Worse still this element has zero image in $B(\varpi_3)$ because $\varphi_2(\tilde{f}_2^2\tilde{f}_3 b_\infty) = 0 = \alpha_2^\vee(\varpi_3)$. Again $\dim V(\varpi_3) = 4$, whilst the number of dual Kashiwara functions (including the zero function) is 5. Notice also that $\tilde{f}_1\tilde{f}_2\tilde{f}_3 b_\infty = (1, 1, 1, 0, 0, 0)$, which is not the image of an element of $\mathbf{F}_t$.

Further examples indicate that Lemma in Sect. 4.7 practically always fails if J is not periodic.

4.9 *The Image of* $\mathbb{Z}$*-Skin of an S-Set in the Periodic Case*

Recall that a driving function z with support in $[1, k]$ determines a coefficient set $\mathbf{c}^z$, or simply $\mathbf{c}$, and that we adjoin the functions in the S-set $Z(\mathbf{c})$.

Here and in the next section, we establish (12).

In general, an S-set $Z(\mathbf{c})$ has the structure of a hypercube that is in general more complicated than that of an $\mathfrak{sl}(2)$-string. Rather the latter should correspond to the $\mathbb{Z}$-skin of the pointed chain of $Z(\mathbf{c})$.

Set $c^j = \sum_{i=j}^n c_i$.

Proposition *Assume J periodic. Then, the $\mathbb{Z}$-skin of the S-set $Z(\mathbf{c})$ maps under Φ to $\{\tilde{f}_u^i b = b_i\}_{i=1}^{c^1}$. Moreover, if $\alpha_u^\vee(\alpha_t) = 0$ given $u > t$, then (12) holds.*

Proof Fix a driving sum F of type u. Let $\mathbf{c}^z = \{c_i : i = 1, 2, \dots, n\}$ be the coefficient set defined by the corresponding driving function $z = z^F$ (of type u). Recall that $c_n = 0$, by Joseph [9, Lemma 5.3.4(ii)]. Augment this set by taking $c_j = 0 : j > n$. By definition, the pointed chain consists of the set of formal sums $\{F + \sum_{i=j}^n c_i F_u^{i+1}\}_{j=1}^{n+1}$.

On the other hand by definition, the coefficient of m_u^i is $-c_i$, for all $i \in \mathbb{N}^+$. Set $b = \Phi(F_u)$. By Lemma in Sect. 4.7, one has $r_u^i(b) - r_u^{i+1}(b) = -c_i$, for all $i \in \mathbb{N}^+$, so the r_u^i are increasing i. On the other hand, $r_u^{r+1} = 0$. Thus, $r_u^i(b) \leqslant 0$, for all $i \in [1, r+1]$, and then by definition, $\varepsilon_u(b) = 0$. On the other hand, the only face in F of the form $F_s^1 : s \in I$ is F_t^1. Thus, $\Phi(F_t^1)$ has entry 1 in the $(t, 1)^{th}$ place and zeros elsewhere. Since the reduced decomposition is periodic, we also have $(s, k+1) > \ell$, for all $k \in \mathbb{N}^+$. Thus for the first ℓ places of $b = \Phi(F_u)$ counting from the right are all zero except $(t, 1)$, which has entry 1. We conclude that

$$r_u^1(b) := \begin{cases} -\alpha_u^\vee(\mathrm{wt}\, b), & \text{if } u < t, \\ 1 - \alpha_u^\vee(\mathrm{wt}\, b + \alpha_t) = -1 - \alpha_u^\vee(\mathrm{wt}\, b) & \text{if } u = t, \\ -\alpha_u^\vee(\mathrm{wt}\, b + \alpha_t) \geqslant -\alpha_u^\vee(\mathrm{wt}\, b) & \text{if } u > t. \end{cases} \tag{14}$$

Consequently, $\varphi_u(b) + \alpha_u^\vee(\varpi_t) = \alpha_u^\vee(\mathrm{wt}\, b + \varpi_t) \geqslant -r_u^1(b) = \sum_{i=1}^n c_i = c^1$.
We conclude that the $f_u^i(b \otimes s_{\varpi_t}) \neq 0$, for all $i = 1, 2, \ldots, c^1$. More precisely, $f_u^i(b \otimes s_{\varpi_t}) \neq 0$ if and only if $0 \leqslant i \leqslant \sum_{i=1}^n c_i + d_{u,t}$, where

$$d_{u,t} = \begin{cases} -\alpha_u^\vee(\alpha_t), & \text{if } u > t, \\ 0 & \text{if } u \leqslant t. \end{cases} \tag{15}$$

On the other hand, since $r_u^i(b) - r_u^{i+1}(b) = -c_i \leqslant 0$, for all $i \in \mathbb{N}^+$, it follows from the crystal rules [6, 2.3.2] that the first c_n powers of $\tilde{f}_u$ enter b at the $(n+1)^{th}$ place, then the next c_{n-1} powers of $\tilde{f}_u$ enter b at the nth place and so on. Thus, the set $\{\tilde{f}_u^{c^j}(b \otimes s_{\varpi_t})\}_{j=n+1}^1$ contains the image of the pointed chain. More precisely, the set of crystal elements $\{b_j = \tilde{f}_u^j(b \otimes s_{\varpi_t})\}_{j=0}^{c^1}$ is the image of the $\mathbb{Z}$-skin of $Z(\mathbf{c})$, as required. Finally under the hypothesis of the last part, $c^1 = m$, hence (12).

4.10 Corollary of Proposition in Sect. 4.9

Retain the notation and hypotheses of Sect. 4.7 in particular that $b := \Phi(F^z)$ satisfies $b \otimes s_{\varpi_t} \in B_J(\varpi_t)$.

Corollary *If J is bipartite, then* (12) *holds.* □

Proof This follows from Proposition in Sect. 4.9 and Corollary in Sect. 4.6, noting that we need never go beyond the final function.

Remark Suppose that the Coxeter element is chosen as in Sect. 2.2. Then, if $t \leqslant m$, one has $z_t^f = m_t^1$ and Z_t is reduced to this one element.

4.11 Concluding Remarks

In Remark 2 of Sect. 4.1.4, we noted that if the coefficient set $\mathbf{c}$ has more than one connected component of non-zero elements, then there are additional trails coming from the remaining part of the extended $\mathbb{Z}$-skins (as defined in Remark 2 of Sect. 4.1.4). A priori it is *not* obvious that these lie in $\Phi^{-1}(B_J(\varpi_t))$. For this, we will need to apply the induction hypothesis (described in Sect. 6.4 below), the point being that these trails and hence crystal elements appear at a previous induction step.

It leads to a fundamental difficulty (see Sect. 6.12) that would appear to exclude a simple proof of our main theorem.

5 Demazure Crystals

In this section, we assume J fixed, though not necessarily of bipartite type or even periodic, and omit the J subscript.

We need to describe the subset of B_J obtained as $\Phi(\mathbf{F}_t^j)$. The natural choice is the Demazure "subcrystal" of $B_J(\varpi_t)$ defined below.

5.1 Demazure Modules

Let λ be a dominant weight. As is well-known, there is a simple $U(\mathfrak{g})$ module $V(\lambda)$ with highest weight λ. Its multi-set of weights is stable under W. Consequently, for all $w \in W$, there is a unique up to scalars vector $v_{w\lambda} \in V(\lambda)$ of weight $w \in W$. The $U(\mathfrak{b})$ module $F_w(\lambda) := U(\mathfrak{b})v_{w\lambda}$ is called the Demazure module associated to the pair (w, λ).

Take any reduced decomposition $s_{i_j} \cdots s_{i_1}$ of w and set

$$\mathcal{F}_w = \bigcup_{n_j \in \mathbb{N}, \forall j} f_{i_j}^{n_j} \cdots f_{i_1}^{n_1}, \tag{16}$$

a notation justified by the fact that the right hand side is independent of choice of reduced decomposition.

This last assertion was proven by Bernstein–Gelfand–Gelfand who showed that

$$\mathcal{F}_w v_\lambda = F_w(\lambda), \tag{17}$$

for all dominant weights λ—see [5, Cor. 4.4.6].

One may remark that we used (17) in [9, 4.2] to establish a matching property for adjoining faces to trails ([9, Remark 4.2]).

5.2 The Crystal Analogue

The above results have a crystal analogue due to Kashiwara in the symmetrizable case and extended in [4] to the general Kac–Moody case using the Littelmann path model. Here, we shall use [4] as a convenient reference.

For all $w \in W$, define $\tilde{\mathfrak{F}}_w$ by replacing f_i by $\tilde{f}_i$, in (16), *viewed as maps of* $B(\infty)$.

The notation is justified by the fact [4, 2.10] that the resulting set is independent of the choice of reduced decomposition *viewed as maps of* $B(\infty)$ *but not for an arbitrary crystal*. Set $B_w(\lambda) := \tilde{\mathfrak{F}}_w b_\lambda$, which is what we call a Demazure subcrystal of $B(\lambda) := \tilde{\mathfrak{F}} b_\lambda$, though it is *only nearly* a subcrystal as made precise by the lemma below. Here, we view b_λ as $b_\infty \otimes s_\lambda$.

Fix J as in Sect. 3.1. Then, $w_j : j \in J$, as defined in Sect. 3.1, also comes with a reduced decomposition.

The following is an immediate consequence of the insertion rules for the Kashiwara operators $\tilde{e}, \tilde{f}_i : i \in I$.

Lemma *With either* λ *a dominant weight or* ∞, *one has the following:*

(i) $B_{w_j}(\lambda) = \{b \in B(\lambda) | \operatorname{Supp} b \in [1, j]\}$.
(ii) $\tilde{e}_i B_{w_j}(\lambda) \subseteq B_{w_j}(\lambda)$, *for all* $i \in I$.
(iii) $\tilde{f}_i B_{w_j}(\lambda) \subseteq B_{w_j}(\lambda)$, *if and only if* $s_i w_j < w_j$ *(in the Bruhat order).* □

5.3 The Demazure Property

A significant fact for our purposes is the "Demazure property" also due to Kashiwara (in the symmetrizable case).

Corollary *Suppose* $b \in B_w(\lambda)$ *satisfies* $\tilde{f}_i b \in B_w(\lambda)$, *then* $\tilde{f}_i^n b \in B_w(\lambda)$, *for all* $n \in \mathbb{N}$. □

Proof By Lemma in Sect. 5.2 (iii) and independence of reduced decomposition, we can assume $w = w_j$ and $i = i_{j+1}$. If $\tilde{f}_i$ enters at the $(j+1)$th place (that is increasing the value of m_i^{j+1} from 0 to 1), then $\tilde{f}_i b \notin B_w(\lambda)$. Hence, it must enter at an earlier place (that is at the kth place for $k < j$), then by the crystal rules in Sect. 3.3, further factors of $\tilde{f}_i$ must enter at earlier places. Hence, the assertion.

Remark This was first proven by Kashiwara (see, for example, [5, 6.3.3]) by a more complicated argument using the almost commutativity of the dual operators. For these, we found in this work another use.

5.4 A Question Coming from Trails

Assume that J is bipartite. Take $w = w_j$ and $i = s := i_{j+1}$ in Sect. 5.3. New trails are obtained from the S-set of a driving function z of type s.

Set $b := \Phi(z)$, which lies in B_J. Here by an induction argument (which will become explicit in Sect. 6.4), we can assume that $b \in B_{w_j}(\varpi_t)$.

The property that the coefficients of $m_s^k : k \in \mathbb{N}^+$ are all non-positive translates to $b := \Phi(z) \in B_{w_j}(\varpi_t)$ having the property that the $r_s^k(b)$ decrease as k decreases. As a consequence, $\tilde{e}_s b = 0$. Unfortunately, the converse is false since $\tilde{e}_s b = 0$ only requires that $r_s^k(b) \leqslant 0$, for all $k \in \mathbb{N}^+$. On the other hand, we only need to consider the new elements (that is to say not lying in $B_{w_j}(\varpi_t)$) obtained by the action of $\tilde{f}_s$ on b. Compatibility with what we expect from trails then leads to the question.

(**Q**). For all $s, t \in I$, $w \in W$, $b \in B_w(\varpi_t)$, one has $\tilde{f}_s b \notin B_w(\varpi_t)$ only if the $r_s^k(b)$ decrease as k decreases.

Remark Notice that by Corollary in Sect. 5.3, the hypothesis implies that $\tilde{e}_s b = 0$.

We prove in Sect. 6.10 that **Q** holds for classical type given a periodic decomposition of w_0 of Coxeter type. Yet computer computations of Zelikson show that it fails in type F_4. This indicates a distinction between the action of the Kashiwara operators and the adjunction of faces following S-sets. A further distinction is that an S-set is generally larger than the $\mathbb{Z}$ convex hull of its pointed chain. These distinctions do not necessarily mean that the conclusion of our main theorem should fail in exceptional type—see Sect. 4.11. Indeed in type F_4, its conclusion still holds, via computer calculations of Zelikson.

6 Main Theorem

6.1 An Upper Bound on Coefficients

Let J be a reduced sequence but not necessarily periodic. Let z^K be a dual Kashiwara function given by a trail $K \in \mathcal{K}_t^{BZ}$. Let $\mathcal{M}_t$ be the largest value that the coefficient of α_t can take in a real root for $\mathfrak{g}$.

Lemma *The coefficient d_s^k of m_s^k in $z^K : K \in \mathcal{K}_t^{BZ}$ satisfies the bound $|d_s^k| \leqslant \mathcal{M}_t$.*$\square$

Proof By (3), the coefficient of m_j in z^K is $\frac{1}{2}\alpha_{i_j}^\vee(\gamma_j^K + \gamma_{j+1}^K)$. Here, we recall that $\gamma_j^K, \gamma_{j+1}^K$ are weights of $V(-\varpi_t)$ and that we should interchange roots and coroots. Thus, we are reduced to computing bounds on $\alpha_s(\gamma)$ with γ a weight of $V(-\varpi_t^\vee)$.

Suppose that γ is an extremal weight, thus of the form $-w\varpi_t^\vee$. In this case, we must compute bounds on $-w^{-1}\alpha_s(\varpi_t^\vee)$. This is just the coefficient of α_t in the real root $-w^{-1}\alpha_s$. Hence, the assertion in this case.

For the general case, it suffices to recall a result of Kac [13, Prop. 12.5 b)], which asserts that every weight of an integrable highest (or lowest) weight module is a convex linear combination of extremal weights.

Remark 1 This result can be interpreted and also proved by noting that the length of an α_s-string in $V(-\varpi_t^\vee)$ (or indeed in $V(\varpi_t)$) is bounded by $\mathcal{M}_t$. In particular, in (5), we necessarily have $k_i \leqslant \mathcal{M}_t$, for all $i = 1, 2, \ldots, n$.

Remark 2 It is of interest to describe the extremal elements of $B_J(\lambda)$, particularly when $\lambda = \varpi_t$. This is generally be much more than the extremal vectors $b_{w\lambda} : w \in W$. For example, if $\lambda = \varpi_2$ in type C_3 with J as in Table 1, then the only crystal elements not of the form $b_{w\lambda} : w \in W$ are those of zero weight. Moreover, there are just two, designated by 7 and f in Table 1. Yet 7 is extremal (in $B_J(\lambda)$), whereas the false trail f is not.

Remark 3 In [17], the authors show that $B(\infty)$ is polyhedral under a positivity assumption. This often fails seemingly related to the S-graphs not all being reduced to their pointed chain, contrary to the present case when J is bipartite and $\mathfrak{g}$ is classical. For a particular choice of J being bipartite, Nakashima and Zelevinsky [17] showed that their positivity hypothesis holds in type A. We have so far been unable to show this holds even for the remaining classical Lie algebras.

Remark 4 Under the positivity assumption, Nakashima [16] described the extremal elements $b_{w\lambda} : w \in W$ though a system of linear equations using the functionals described in his work with Zelevinsky [17].

6.2 Corollary of Lemma in Sect. 6.1

Retain the notation of Sect. 6.1. Suppose that $z^K : K \in \mathcal{K}_t^{BZ}$ is a driving function of type s. By the above result, its coefficient set $\mathbf{c} = \{c_i\}_{i=1}^n$ satisfies $c_i \leqslant \mathcal{M}_t$, for all $i = 1, 2, \dots, n$. However, we will obtain a much stronger assertion in the bipartite case based on the following result.

Corollary *Suppose that J is bipartite. Let $z \in Z_t$ be a driving function and* $\mathbf{c}$ *its coefficient set. Then if $b := \Phi(z^K) \in B_J(\varpi_t)$, one has $\sum_{i=1}^n c_i \leqslant \mathcal{M}_t$.*

Proof By Sect. 4.9, one has $f_s^{\sum_{i=1}^n c_i} b \neq 0$. Then, the assertion follows by Remark 1 of Lemma in Sect. 6.1 concerning the length of α_s-strings.

Remark This conclusion fails in the non-periodic case. Indeed already it fails for the example given in Sect. 4.8 in type A_3.

6.3 The Statement of the Main Theorem

The conclusion of Sect. 6.2 gives little choice for a coefficient set in classical type. Thus, either the coefficient set is reduced to one non-zero element that may be either 1 or 2, or it admits two elements equal to 1 separated by a (possibly empty) string of zeros.

This result does not seem to be of much use if we just concentrate on crystals and ignore trails. Simply (and surprisingly), we do not know enough about $B_J(\varpi_t)$ even when J is periodic nor if the latter has special properties because ϖ_t is fundamental.

However, doing a little extra work on the theory of trails developed in [9] to take account of the presence of false trails, the following theorem results. In this, we recall the definition of r given in Sect. 3.1 and of θ given in Sect. 4.5.

Theorem *Let $\mathfrak{g}$ be a simple Lie algebra and J of bipartite type. Suppose $\theta(t) = r$. Then, for all $t \in I$ such that the coefficient $\mathcal{M}_t$ of α_t in the highest root is $\leqslant 2$, one has*

$$\varepsilon_t^\star(b) = \max_{F \in \Phi^{-1}(B_J(\varpi_t))} z^F(b). \tag{18}$$

More precisely, the set of Z_t of dual Kashiwara functions associated to $t \in I$ can be taken to be $\{z^F\}_{F \in \Phi^{-1}(B_J(\varpi_t))}$, that is to say is given by the crystal $B_J(\varpi_t)$. □

Of course, the hypothesis is satisfied for all $t \in I$ for $\mathfrak{g}$ of classical type.

The proof of this result is given in the subsections below. When $\theta(t) \neq r$ and $t > m$, the same analysis will show that Z_t may be replaced by a slightly smaller but easily computable subset of $\Phi^{-1}(B_J(\varpi_t))$. On the other hand by Corollary in Sect. 4.6(i), it follows that Z_t is reduced to the initial driving function z_t^1 if $t \leqslant m$.

Remark Zelikson conjectured that the theorem holds for all $\mathfrak{g}$ simple.

6.4 The Induction

Recall the notation of Sect. 4.1.2 and assume the hypotheses of the theorem in particular that J is of bipartite type. To avoid heavy notation, we take J to be fixed and drop the J subscript on $B(\varpi_t)$.

Our aim is to prove by induction on $j \in \mathbb{N}$:

Proposition *$K \mapsto \Phi(K)$ is a bijection of Φ of $\mathcal{K}_t^j$ onto $B_{w_j}(\varpi_t)$, satisfying the following:*

(i). *If $K + F_s^u \in \mathcal{K}_t^j$ for some $(s, u) \in I \times \mathbb{N}^+$, then $\mathrm{wt}\,\Phi(K + F_s^u) = -\alpha_s + \mathrm{wt}\,\Phi(K)$.*

(ii). *Suppose $K \in \mathcal{K}_t^j$ and $s \in I$. If $\alpha_s^\vee(\mathrm{wt}\,\Phi(K))$ is strictly positive (resp. negative), then there exists an integer $u > 1$ such that $K + F_s^u \in \mathcal{K}_t^j$ (resp. $K - F_s^u \in \mathcal{K}_t^j$) and $\tilde{f}_s\Phi(K) = \Phi(K + F_s^u)$ (resp. $\tilde{e}_s\Phi(K) = \Phi(K - F_s^u)$).*

Remark 1 Continue to assume that J is bipartite. One could hope to show that the proposition holds without restriction on Lie algebra type. Of course for classical Lie algebras, we have the simplifying feature that $\alpha_s^\vee(\mathrm{wt}\,\Phi(K)) \in \{-2, -1, 0, 1, 2\}$.

Remark 2 If $\alpha_s^\vee(\mathrm{wt}\,\Phi(K)) = 0$, it can happen that $\tilde{e}_s\Phi(K) = 0$ (resp. $\tilde{f}_s\Phi(K) = 0$), whilst $K - F_s^u$ (resp. $K + F_s^u$) is a trail for some $u \in \mathbb{N}^+$. Moreover, it can also happen that $K_1 + F_s^{u_1}$, $K_2 + F_s^{u_2}$ coincide as trails even if K_1, K_2 are distinct trails (and necessarily $u_1 \neq u_2$). An example obtains from Fig. 1. Take K_1 to be the trail designated by 7 and K_2 to be the false trail. Then, $K_1 + F_s^{u_1} = K_2 + F_s^{u_2}$ is the trail designated by 8. To obtain an example of not using a false trail, one must go to rank 4. This phenomenon is not reproduced in crystals, yet it should be related to the manner in which the e_s, f_s act on the corresponding global bases—see [15, 3.1]. Indeed the cited result shows that the global basis is *not necessarily* compatible with decomposition into simple $\mathfrak{sl}(2)$ defined by a given $s \in I$, except up to filtration by dimension (see also Remark 2 of Sect. 6.5).

We make a few preliminary steps towards the proof below.

First, if $j < (t, 1)$, then $\mathcal{K}_t^j$ is reduced to K_∞ by Lemma in Sect. 3.6(i), whilst $B_{w_j}(\varpi_t)$ is reduced to b_∞. Then, the assertion holds by definition (Sect. 4.5).

Now take $j = (t, 1)$. Then, $\mathcal{K}_t^{(t,1)} = \{F_t^1\}$, by Lemma in Sect. 3.6. Moreover, $\Phi(F_t^1) = 1_t^1$. Yet as noted in Sect. 4.4, one has $f_t(b_\infty \otimes s_{\varpi_t}) = f_t(b_\infty) \otimes s_{\varpi_t}$, whilst $f_t(b_\infty) = 1_t^1$, by Sect. 3.3. Thus, the Φ is a bijection for $j = (t, 1)$.

Assume that Φ is a bijection of $\mathcal{K}_t^{j-1}$ onto $B_{w_{j-1}}(\varpi_t)$.

Set $s = i_j$. Then, we can write $j = (s, n)$ for some $n \in \mathbb{N}^+$. Define $T_s(\mathbf{e})$ as in Sect. 3.8.2 and define the n-tuple $\mathbf{c}$ with respect to the ℓ-minimal element $L := K_{\ell \min}$ of $T_s(\mathbf{e})$. Since $L \in \mathcal{K}_t^{j-1} \xrightarrow{\sim} B_{w_{j-1}}(\varpi_t) \subseteq B(\varpi_t)$, it follows by Corollary in Sect. 6.2 that $\sum_{i \in [1,n-1]} c_i \leqslant \mathcal{M}_t \leqslant 2$.

Notice we can assume $\sum_{i \in [1,n-1]} c_i > 0$. Otherwise, the S-set to be adjoined is trivial, and on the other hand, $\tilde{f}_s \Phi(L) = 0$.

6.5 False Trails Revisited

Retain the above hypotheses and notation. Recall Sect. 4.1.3. Let $K \in T_s(\mathbf{e})$ be defined by the n-tuple $\mathbf{k}$. Let L be the unique ℓ-minimal trail in $T_s(\mathbf{e})$. Then, we can write $K = L + \sum_{i=1}^n c'_i F_s^{i+1}$, for some $c'_i \in \mathbb{Z}$ and where $c'_i = k^{(i)} - \ell^{(i)}$, as noted in [9, 5.4]. Then by Joseph [9, 7.6], a trail in $T_s(\mathbf{e})$ is false unless

$$c_i \geqslant c'_i \geqslant 0, \forall i \in \hat{N}, \tag{19}$$

together with a more complicated condition [9, (38)], which if the $c_i : i \in N$ are increasing just says that the c'_i are increasing. The latter is all that we need here. (An example of a more complicated condition obtained when the c_i are not increasing is given in the caption to Fig. 2.)

Now let $T_s^-(\mathbf{e})$ denote the subset of trails in $T_s(\mathbf{e})$ that trivialize at w_{j-1}. In the above notation, these are exactly the trails for which $c'_{n-1} = 0$, in particular $L \in T_s^-(\mathbf{e})$. In [9, Sect. 7], we showed that if there are no false trails in $T_s^-(\mathbf{e})$, then there are no false trails in $T_s(\mathbf{e})$ and that the latter set is determined by adjoining the $\mathbb{Z}$ convex hull $K_{\mathbb{Z}}(\mathbf{c})$ of the S-set defined by $\mathbf{c}$. Unfortunately, this does not quite give an inductive argument for determining all trails because the definition of a false trail depends on $s \in I$ defined to be i_j at the jth induction step.

Worse than this, we show here (see Fig. 1) that the second condition on the c'_i : $i \in [1, n-1]$ given above can fail. On the other hand, we still believe that (19) will always hold. Notice it means that L (resp. $L + \sum_{i \in [1,n-1]} c_i F_s^{i+1}$) is then the unique minimal (resp. maximal) element of $T_s(\mathbf{e})$ with respect to the *partial* order defined by adjoining faces.

Recall Sect. 4.1.3, the notion of the $\mathbb{Z}$-skin of an S-set.

Lemma *Let* $\mathbf{c}$ *be the coefficient set of the unique* ℓ*-minimal trail* L *of* $T_s(\mathbf{e})$*. Let* $K \in T_s(\mathbf{e})$ *belong to the* $\mathbb{Z}$*-skin of the* S*-set defined by* $\mathbf{c}$*. Then for all* $j \geqslant 1$, $K + F_s^{j+1}$ *is not a trail if* $c_j = 0$. □

Proof Let $\mathbf{k}$ be the n-tuple defined by K. The assertion follows from Lemma in Sect. 4.1.3(i) if $k_i < \ell_i$, for all $i < j$. We shall reduce to this case by successively removing appropriate multiples of the faces $F_s^2, F_s^3, \ldots, F_s^i$ in that order. The argument is essentially a reversal of the proof of (iii) of Lemma in Sect. 4.1.3.

We can assume $i < j$ minimal such that $k_i > \ell_i$. Then by definition of the $\mathbb{Z}$-skin, one has $k_{i+1} = \ell_{i+1} + c_{i+1} - c, k_i = \ell_i + c$, for some $c \in [1, c_i]$. By (i) of Lemma in Sect. 4.1.3, k_i cannot be increased to be strictly greater than $\ell_i + c_i$, whilst by choice of i and minimality, it cannot be decreased to be strictly less than ℓ_i. On the other hand, $k_{i+1} + k_i - a_{i+1} = \ell_{i+1} + \ell_i - a_{i+1} + c_{i+1} = c_i$. Thus by Joseph [9, Lemma 5.2], we can move $k_i - \ell_i = c$ copies of e_s to the left through e_{-a_i}. This replaces k_i by ℓ_i and increases k_{i+1} by c to $\ell_{i+1} + c_{i+1}$. Repeating this process eventually achieves the required reduction.

Remark 1 This proves a weak version of the upper bound in (19). Proving the upper bound was already very difficult, even given that there are no false trails in $T_s^-(\mathbf{e})$—see [9, Sect. 7].

Remark 2 Decompose $\mathbf{c}$ into connected components so that in every component the entries are all non-zero. Then by Joseph [9, 5.8.1], the S-set (say of type s) defined by $\mathbf{c}$ may be viewed as a union of the S-sets defined by the connected components of $\mathbf{c}$. It is easy to see that the conclusion of the lemma applies when K belongs to the union of the corresponding $\mathbb{Z}$-skins. This can be *larger* than the $\mathbb{Z}$-skin of the S-set itself. We call it the complete $\mathbb{Z}$-skin of S. In this, the only case that will concern us here is when the non-zero values of c_i are separated by a non-empty string of zeros, for example, if $n = 4$ and $c_3 = c_1 = 1, c_2 = 0$. Then, the S-set consists of its $\mathbb{Z}$-skin, forming a 3 element string in $B_J(\varpi_t)$ and a singleton string $\{b\}$ defined by $c'_3 = 0, c'_1 = 1$. The S-set defined by these parameters is given in Fig. 3. Examples occur for J periodic if $\mathfrak{g}$ has rank $\geqslant 4$ and is not of type A.

On the other hand by Kashiwara [15, 3.1], the global basis elements corresponding to the 3 element string span a simple three dimensional submodule V of $V(-\varpi_t)$; but the remaining global basis element $G(b)$ need not form the trivial module and indeed $e_s G(b)$, $f_s G(b)$ can be non-zero elements of V.

One would like to relate these phenomena.

Remark 3 We prove in Sects. 6.6–6.9 under the hypothesis of Theorem in Sect. 6.3 that both bounds in (19) hold. Moreover, all possible solutions to (19) occur. The latter is special to periodic reduced sequences.

6.6 Adjoining Faces to Trails

Retain the hypotheses and notation of Sect. 6.4. Further adopt the hypothesis of Theorem in Sect. 6.3, namely that the coefficient $\mathcal{M}_t$ of α_t in the highest root $\leqslant 2$. By the induction hypothesis, we can assume that $\Phi(\mathbf{F}_t^{j-1}) \subseteq B_{w_{j-1}}(\varpi_t) \subseteq B(\varpi_t)$ and so by Corollary in Sect. 6.2 that $\sum_{i\in[1,n-1]} c_i \leqslant 2$.

6.6.1 One Non-zero Value

Suppose that $c_{u-1} \neq 0$ for just one value of $u \in [2, n]$. Then, in the notation of Sect. 4.1.4, we view $L := K_{\ell\,\min}$ to be given by the subsequence of $\mathbf{l} := (\ell_u, \ell_{u-1})$ corresponding to the vector $\overline{v}_{\mathbf{l}} := e^{\ell_n} e_{-a_n} \cdots e^{\ell_u} e_{-a_u} e^{\ell_{u-1}} v_{-d_{u-1}}$.

By Lemma in Sect. 4.1.4, we may adjoin up to $m \leqslant c_{u-1}$ copies of the face F_s^u to $K_{\ell\,\min}$. This moves m copies of e to the right through e_{-a_u} giving the vector $\overline{v}_{\mathbf{k}} := e^{\ell_n} e_{-a_n} \cdots e^{\ell_u - m} e_{-a_m} e^{\ell_{u-1}+m} v_{-d_{m-1}}$. The latter may be viewed as being given by the subsequence $\mathbf{k} := (\ell_u - m, \ell_{u-1} + m)$.

All these new trails belong to the $\mathbb{Z}$-skin, and so no further faces F_s^v can be adjoined for $v \leqslant u-1$ by minimality or for $v \geqslant u+1$ by Lemma in Sect. 6.5.

6.6.2 Two Consecutive Non-zero Values

Suppose that the c_j take non-zero values for two successive values of $j \in [1, n-1]$. Thus, we can write $c_{u-1} = c_{u-2} = 1$, for some $u \in [3, n]$, with all other coefficients zero.

In this case, we view $K_{\ell\,\min}$ to be given by the subsequence $\mathbf{l} := (\ell_u, \ell_{u-1}, \ell_{u-2})$ corresponding to the vector $\overline{v}_{\mathbf{l}} := e^{\ell_n} e_{-a_n} \cdots e^{\ell_u} e_{-a_u} e^{\ell_{u-1}} e_{a_{u-1}} e^{\ell_{u-2}} v_{-d_{u-2}}$.

By Lemma in Sect. 4.1.4, the $\mathbb{Z}$-skin of the corresponding S-set defines new trails. As in Sect. 6.6.1 to any element of this $\mathbb{Z}$-skin, no faces F_s^v can be adjoined for $v \leqslant u-2$ by minimality or for $v \geqslant u+1$ by Lemma in Sect. 6.5.

Yet we cannot exclude adjoining the face F_s^{u-1} to L, though by Lemma in Sect. 4.1.4, only one copy of F_s^{u-1} may be adjoined. This trail K is defined by the subsequence $\mathbf{k} := (\ell_u, \ell_{u-1} - 1, \ell_{u-2} + 1)$. It is a false trail because $c'_{u-1} = 0, c'_{u-2} = 1$, which are not increasing. It lies in $T_s^-(\mathbf{e})$ and so must arise from a previous induction step.

No further faces F_s^v can be adjoined to $K = L + F_s^{u-1}$ for $v \leqslant u-2$ by minimality, but we cannot immediately deduce the same for $v \geqslant u+1$ using Lemma in Sect. 6.5.

This will be proved below assuming that the second inequality in (19) holds. Notice we only need to assume it holds for $T_s^-(\mathbf{e})$, for then it holds for $T_s(\mathbf{e})$ by Joseph [9, 7.4].

We want to show that a face F_s^v for $n \geqslant v \geqslant u+1$ cannot be adjoined to the false trail $L' := L + F_s^{u-1}$, which we had been unable to do by applying Lemma in Sect. 6.5. Of course this is only of interest if $u < n$.

Suppose otherwise, that is to say $K := L' + F_s^v$ is a trail and deduce a contradiction. By the second inequality in (19) that we are assuming to hold, we cannot remove the face F_s^{v-1} from K. This means that in K we cannot move a factor of e to the left through $e_{-a_{v-1}}$. Now in the n-tuple $\mathbf{k}$ defined by K, we have $k_{v-1} = \ell_{v-1} + 1$ and

$$k_{v-2} = \begin{cases} \ell_{v-2}, & \text{if } v > u+1, \\ \ell_{v-2} - 1, & \text{if } v = u+1. \end{cases}$$

Now recalling that $c_{u-1} = c_{u-2} = 1$ and that all other coefficients vanish, we have

$$a_{v-1} - \ell_{v-1} - \ell_{v-2} = c_{v-1} - c_{v-2} = \begin{cases} 0, & \text{if } v > u+1, \\ -1, & \text{if } v = u+1. \end{cases}$$

Since $v \geqslant u+1$, we obtain in all cases that

$$k_{v-1} + k_{v-2} - a_{v-1} = 1.$$

It follows from [9, Remark 2 of Lemma 5.2] that we can move e^q, for some $q \geqslant 1$, to the right through $e_{-a_{v-1}}$. (We shall henceforth consider just the case $q = 1$ since the general case follows a fortiori by the same reasoning but is just more messy to write down. We may also appeal to Remark 1 of Sect. 6.1 that forces $q \leqslant 1$). Thus, we obtain the trail $K + F_s^{v-1} = L' + F_s^v + F_s^{v-1}$. Repeating this procedure, we obtain the trail $K' = L' + F_s^v + F_s^{v-1} + \cdots + F_s^u = L + F_s^v + F_s^{v-1} + \cdots + F_s^{u-1}$. In the n-tuple $\mathbf{k}'$ defined by K', we have $k'_{u-1} = \ell_{u-1}, k'_{u-2} = \ell_{u-2} + 1, k'_w = \ell_w : w < u-2$. Since $c_{u-2} = 1$, it follows by Lemma in Sect. 4.1.3(i) and the minimality of $\mathbf{l}$ that we cannot increase k'_{u-2}, or decrease it by > 1. On the other hand, $a_{u-1} - \ell_{u-1} - \ell_{u-2} = c_{u-1} - c_{u-2} = 0$ and so $k'_{u-1} + k'_{u-2} - a_{u-1} = 1$. Thus, we can move a copy of e to the left through $e_{-a_{u-1}}$ to obtain the trail $K'' := K' - F_s^{u-1} = L + F_s^v + F_s^{v-1} + \cdots + F_s^u$.

In this, $k''_u = \ell_u, k''_{u-1} = \ell_{u-1} + 1, k''_w = \ell_w : w \leqslant u-2$. Yet k''_{u-1} cannot be increased by Lemma in Sect. 4.1.3 and cannot be decreased by > 1 by the minimality of L. On the other hand, $a_u - k''_u - k''_{u-1} = a_u - \ell_u - (\ell_{u-1} + 1) = c_u - c_{u-1} - 1 = -2$. Together, these contradict the conclusion of [9, Lemma 5.2].

In particular, K'' cannot be a trail. If $v = u+1$, this also follows from Lemma in Sect. 6.5 and in general is a slight addendum to this lemma.

Remark The case $v = n$ is of particular interest being the only case for which we need [9, 7.4]. Indeed in all other cases, $L' + F_s^v \in T_s^-(\mathbf{e})$. Below we give a second proof.

When $v = n$, the vector corresponding to the supposed trail $K := L' + F_s^v$ is a highest weight vector of $M_s(\mathbf{e})$. On the other hand, as noted in [9, 5.3.4] the trails, L, and the false trail L', correspond to lowest weight vectors when ℓ_n is reduced to 0, for which we denoted the corresponding n-tuples by $\bar{\mathbf{l}}'$ and $\bar{\mathbf{l}}$. Then, we can verify that it vanishes by $\mathfrak{sl}(2)$ theory, that is to say by applying [9, Lemma 3.2].

Recall that $n = v > u$. Let $\mathbf{k}$ (resp. $\mathbf{l}$, $\mathbf{l}'$) be the n-tuple corresponding to K (resp. L,L'). All the entries for $i < n$ of $\mathbf{k}$ (resp. $\mathbf{l}'$) coincide with those of $\mathbf{l}$ except $k_{n-1} = \ell_{n-1}+1, k_{u-1} = \ell'_{u-1} = \ell_{u-1}-1, k_{u-2} = \ell'_{u-2} = \ell_{u-2}+1$. This excludes $f\overline{v}_{\mathbf{k}}$ being a multiple of $\overline{v}_{\bar{\mathbf{l}}}$, whilst it is multiple of $\overline{v}_{\bar{\mathbf{l}}'}$ times the right hand side of [9, (7)]. In this, only b_{n-1} is non-zero and indeed equals 1. Then there is only one non-positive factor in [9, Eq. (7)], and this equals $a^{(n-1)}+1-1-k^{(n-2)}-\ell^{(n-1)} = a^{(n-1)} - \ell^{(n-2)} - \ell^{(n-1)} = c_{n-1} = 0$. Thus, $\overline{v}_{\mathbf{k}} = 0$, and so K is not a trail, as required. (Had this factor been positive, we would have obtained a contradiction with our previous result!)

6.6.3 General Case

Finally suppose that $\mathbf{c}$ has two non-consecutive occurrences of 1. By our assumption in Sect. 6.6, we can suppose $c_{u-1} = c_{v-1} = 1$ for some u, $v \in [2, n]$, with $u \geqslant v+2$.

Following Remark 2 of Sect. 6.5, decompose $\mathbf{c}$ into its (two) connected components. Then by Lemma in Sect. 4.1.3, the union of the $\mathbb{Z}$-skins of the corresponding S-set defines trails. Moreover by Lemma in Sect. 6.5, no other trials can be obtained from L by adjoining faces.

Note we obtain no false trails in this case, and this is why the difficulty we met in Sect. 6.6.2 is absent.

Remark Eventually, we show that Sects. 6.6.1–6.6.3 describe all the trails in $T_s(\mathbf{e})$.

6.6.4 A Corollary of Sects. 6.6.1–6.6.3

Take $K \in \mathcal{K}_t^{j-1}$. We can suppose that Sects. 6.6.1–6.6.3 describe all the trails at this previous induction step with s replaced by $s' := i_{j-1}$. Let $c(K)_{s'}^{u'}$, for $(s', u') \in I \times \mathbb{N}^+$ denote the coefficient of $m_{s'}^{u'}$ in z^K. Inspection of coefficients of $m_j : j \in J$ in the trail functions gives the

Corollary $\sum_{u'=1}^{\infty} |c(K)_{s'}^{u'}| \leqslant 2$, *for all* $s' \in I$. □

Remark Again by inspection of coefficients in Sects. 6.6.1–6.6.3, we may observe that the unique minimal trail L' in some $T_{s'}(\mathbf{e}') \subseteq \mathcal{K}_t^{j-1}$ is characterized by all the coefficients of $z^{L'}$ in $m_{s'}^{n'} : n' \in \mathbb{N}^+$ being non-positive. In general, this is of course a necessary condition but is not sufficient—see [10, 5.2] for an example in type D_5 with J non-periodic. This example was obtained by Zelikson using a computer.

6.6.5 Comparison with Crystals

Assume that the second bound in (19) holds at the jth step. We show as a consequence that Proposition in Sect. 6.4 extends from the $(j-1)$th to the jth induction step.

Indeed under this assumption all the *new* trails, that is to say those in $T_s(\mathbf{e}) \setminus T_s^-(\mathbf{e})$, belong to the $\mathbb{Z}$-skin of the corresponding S-set and so by Lemma in Sect. 4.10, lie in $B_J(\varpi_t)$.

Finally, set $\Delta r_s^{u'} = r_s^{u'} - r_s^{u'+1}$. Then, the hypothesis of Question **Q**, namely that $\tilde{f}_s b \notin B_w(\varpi_t)$, implies $\sum_{u' \in \mathbb{N}^+} \Delta r_s^{u'} < 0$. Yet as noted in Sect. 6.4, it follows that $\sum c_i \leqslant 2$. Consequently, the only element in the S-set not satisfying this condition is the driving function itself. Thus, the only new crystal elements are obtained by applying $\tilde{f}_s$ to the image of the driving function, and these in turn belong to the image of the $\mathbb{Z}$-skin, by Corollary in Sect. 4.10.

Comparison of the new trails with the new crystal elements (i) and (ii) of Proposition in Sect. 6.4 results for $K \in T_s(\mathbf{e}) \setminus T_s^-(\mathbf{e})$.

Remark Let us explain why Proposition in Sect. 6.4 does not quite lift to an isomorphism as warned in Remark 2 of Sect. 6.4. This is because $T_{s'}^-(\mathbf{e})$ need *not* be reduced to the unique ℓ-minimal trail. Indeed, it also consists of the false trail L' described in Sect. 6.6.2, which need not be present[2] and the trail L'' not obtained from the $\mathbb{Z}$-skin described in Sect. 6.6.3, which is always present.

Yet from a calculation of $\Delta r_s^{u'}$ on either $\Phi(L')$ or on $\Phi(L'')$, we deduce that they are annihilated by $\tilde{e}_{s'}$ and by $\tilde{f}_{s'}$.

6.7 *The Last Step of the Proof of Theorem in Sect. 6.3*

The proof of Theorem in Sect. 6.3 is concluded by showing that the second inequality in (19) holds for all $K \in \mathcal{K}_t^j$. As before, we set $s = i_j$ and prove the assertion for $K \in T_s^-(\mathbf{e})$, which lies in $K \in \mathcal{K}_t^{j-1}$. Then, it follows for all $K \in T_s(\mathbf{e})$ by Joseph [9, 7.4]. However, the first assertion does *not* follow by the induction hypothesis because that only gives it to hold for $s' = i_{j-1}$. Nevertheless, we can assume that Proposition in Sect. 6.4 and Corollary in Sect. 6.6.4 hold with respect at the previous induction step since these hold independent of the choice of $s' \in I$.

[2] We have no example for J bipartite when this false trail is not present even in type F_4, as noted by Zelikson. Yet already in the example in Sect. 4.8, this false trail is missing. Go figure!

Thus in the notation of (19), we want to show that $c'_i \geqslant 0$ for all $i \in [1, n-1]$. Notice this is *more* than just showing that a face cannot be removed from one of the trails described in Sect. 6.3.

Let L denote the unique ℓ-minimal element of $T_s(\mathbf{e})$ that we recall lies in $T_s^-(\mathbf{e})$. Define n as in Sect. 3.8.2 and as in (5) choose the n-tuple $\mathbf{l}$ so that $\overline{v}_\mathbf{l} = v_{-w_j\varpi_t}$. As a consequence of (9), it follows that $K \in T_s(\mathbf{e})$ lies in $T_s^-(\mathbf{e})$ if and only if its corresponding n-tuple $\mathbf{k}$ satisfies $k_n = \ell_n$.

Consider a possible trail of the form

$$K = L + \sum_{i=1}^{n} d_i F_s^i : d_i \in \mathbb{Z}.$$

Lemma

(i) *If* $d_n > 0$, *then* $K \notin T_s^-(\mathbf{e})$.
(ii) *If* $d_n < 0$, *then* K *is not a trail.* □

Proof Obviously, $k_n = \ell_n - d_n$. Thus, (i) results from the observations proceeding the lemma.

For (ii), set $d'_n = -d_n \in \mathbb{N}^+$. Now if K is a trail

$$\overline{v}_\mathbf{k} = e^{\ell_n + d'_n} v_\gamma,$$

for some non-zero vector $v_{-\gamma} \in V(-\varpi_t)$ of weight $-\gamma$.

We claim that $\overline{v} := \overline{v}_\mathbf{k}$ is zero.

Suppose that $\overline{v} \neq 0$. Then, it lies in the simple module $M_s(\mathbf{e})$ and has the same weight as $\overline{v}_\mathbf{l}$; hence, it must be proportional to the latter, which in turn is the a non-zero multiple of $v_{-w_j\varpi_t}$. On the other hand by (9), we have $w_j\varpi_t + \ell_n\alpha_s = w_{j-1}\varpi_t$.

Yet the expression for $\overline{v}$ means that $-\gamma$ satisfies $\gamma = d'_n\alpha_s + w_{j-1}\varpi_t$. Then, $(\gamma, \gamma) > (\varpi_t, \varpi_t)$, contradicting that $-\gamma$ is a weight of the integrable lowest weight module $V(-\varpi_t)$.

6.8 The Case When $c_{n-1} \neq 0$

Suppose the coefficient set is described as in Sects. 6.6.1 or 6.6.2. Then, the assertion that the second inequality in (19) holds when $c_{n-1} \neq 0$ follows by Lemma in Sect. 6.7.

The case when the coefficient set is described by Sect. 6.6.3 will be considered in Sect. 6.9.5 to avoid repeating the same construction as given in Sect. 6.9.

6.9 Reduction to the Case When $c_{n-1} \neq 0$

Continue to let L denote the unique ℓ-minimal element of $T_s(\mathbf{e})$ with coefficient set $\mathbf{c}$. Recall that the coefficient of $-m_s^{n-1}$ in z^L is c_{n-1}.

Suppose that $c_{n-1} = 0$. This does not quite mean that $T_s(\mathbf{e})$ identifies with the set of trails that trivialize at $w_{(s,n-1)}$.

Indeed, it can happen that there exists $j' \in J$ with $(s, n-1) < j' < (s, n)$ such that the coefficient of $m_{j'}$ in $K_{\ell\,\min}$ is non-zero. Assume j' maximal with this property. Using the notation of Sect. 3.2, we write $j' = (s', n')$, with $s' \in I \setminus \{s\}$, $n' \in \mathbb{N}^+$.

Recall Sect. 4.1.2 that by our inductive construction of trails given by adjoining faces it follows that $m_{j'}$ is a positive integer and we must be able to remove $m_{j'}$ copies of the face $F_{s'}^{n'}$ from L, up to lower order terms (determined by the corresponding S-set).

Then, the resulting trail becomes an element of $L' \in F_s(\mathbf{e}')$, where $\mathbf{e}'$ is defined by moving $m_{j'}$ factors of $e_{s'}$ to the left from its position at $(s', n'-1)$ to position (s', n'). Yet $(s', n'-1) < (s, n-1) < (s', n')$, since J is periodic, so this adds a term $-m_{j'}\alpha_s^\vee(\alpha_{s'})$ to the coefficient of m_s^{n-1}, with a possible change in the remaining coefficients of the $m_s^i : i \in \mathbb{N}^+ \setminus \{n-1\}$ on account of the lower order terms.

6.9.1 Orthogonal Simple Roots

Suppose $\alpha_s^\vee(\alpha_{s'}) = 0$. Then, the coefficients of $m_s^i : i \in [1, n-1]$ for *any* $K \in T_s(\mathbf{e})$ are unchanged by the above operation and become elements of $T_s(\mathbf{e}')$. Conversely, the same applies on adjoining faces $F_{s'}^{n''} : n'' \leqslant n'$.

Thus, $T_s(\mathbf{e})$ having unique ℓ-minimal element L identifies with $T_s(\mathbf{e}')$ having unique ℓ-minimal element L'.

6.9.2 Non-orthogonal Simple Roots

Suppose $\alpha_s^\vee(\alpha_{s'}) \neq 0$ and so $\alpha_{s'}^\vee(\alpha_s) \neq 0$, by definition of the Cartan matrix.

Lemma *Suppose L takes the form given in Sect. 6.6.3 or in Sect. 6.6.2 and the false trail L' defined there exists. Then, $c_{n-1} \neq 0$.* □

Proof Suppose $c_{n-1} = 0$, so the coefficient set $\mathbf{c}$ obtained from L has just non-zero entries $c_u = 1 : 1 \leqslant u \leqslant n-2$ and $c_v = 1 : 1 \leqslant v < u$. Then, the trails described in Sect. 6.6.3 or in Sect. 6.6.2 lie in $T_s^-(\mathbf{e})$. Yet $\alpha_{s'}^\vee(\alpha_s) \neq 0$. Then, one easily checks through (4) that the bound in Sect. 6.6.4 cannot be satisfied with respect to s' for all of them.

6.9.3 A Condition for $K(\mathbf{d})$ and $K'(\mathbf{d})$ to be Trails

Suppose the coefficient set $\mathbf{c}$ obtained from L is as described in Sect. 6.6.2 and that the false trail $L' = L + F_s^{u-1}$ does not exist. Then, $L, L + F_s^u, L + F_s^u + F_s^{u-1}$ are trails. Then, the only new trails we can obtain are $K(\mathbf{d}) = L + F_s^u + \sum_{i>u} d_i F_s^i : d_i \in \mathbb{Z}$ (with some d_i strictly negative) and $K'(\mathbf{d}) = K(\mathbf{d}) + F_s^{u-1}$. By Lemma in Sect. 6.7, they lie in $T_s^-(\mathbf{e})$ if and only if $i < n$ in the sum. By Joseph [9, Lemma 7.4], we need to only consider this case.

Lemma *$K(\mathbf{d})$ and $K'(\mathbf{d})$ are trails (if) and only if d_i=0, for all $i \in [u+1, n-1]$.* □

Proof By Lemma in Sect. 4.1.4(ii) and by reversing its argument as in the proof of Lemma in Sect. 6.5, it follows that $K(\mathbf{d})$ is a trail if and only if $K'(\mathbf{d})$ is a trail. Then one easily checks, as in Sect. 6.9.2, that the bound in Sect. 6.6.4 cannot be satisfied with respect to s' for both $K(\mathbf{d})$, $K'(\mathbf{d})$, whatever (integer) values are given to the d_i unless they are all zero.

6.9.4 The Case When the Coefficient Set Is Described by Sect. 6.6.1

It remains to consider the case when L takes the form given by Sect. 6.6.1. Take m as given there and recall that $m \in \{1, 2\}$. Through Lemma in Sect. 6.7, it is enough to consider the case when $u \leqslant n - 2$. By the above reduction, we are reduced to the case $\alpha_s^\vee(\alpha_{s'}) \neq 0$; moreover, (s', n') is the maximal value of $j \in J$ such that the coefficient of m_j in z^L is non-zero (and hence positive).

We claim that the coefficient of $m_{s',n'-1}$ in z^L is zero. Otherwise, we obtain a contradiction with Corollary in Sect. 6.6.4 with respect to the trail $K' := L + F_s^u$, since $z^{K'}$ has a non-zero coefficient of $m_{s',n''}$, and moreover, by periodicity, $n'' = n' - (n - u) < n' - 1$. Again using Corollary in Sect. 6.6.4, we can even assert that coefficient of $m_{s',n'}$ in z^L is 1.

Recalling the penultimate paragraph of Sect. 4.1.2, the above implies that we can remove one copy of the face $F_{s'}^{n'}$ from L, to obtain a new trail.

Repeating the above process, we obtain $\mathbf{e}'' \in \mathbf{E}_s$ and a trail $L'' \in T_s(\mathbf{e}'')$ for which the coefficients of $m_j : j > (s, n - 1)$ in $z^{L''}$ are all zero. Then, the coefficient of m_s^{n-1} in $z^{L''}$ must be non-negative and the coefficients of the $m_s^i : i < n - 1$ in $z^{L''}$ must be the same as those in z^L. Of the latter, there is exactly one equal to $-m$ and the rest are zero.

Now by Remark in Sect. 6.6.4, it follows that either the coefficient of m_s^{n-1} in $z^{L''}$ is zero (which is necessarily the case if $m = 2$) and $L''' := L''$ is the unique ℓ-minimal trail in $T_s(\mathbf{e}'')$ or it is 1 and $L''' := L'' - F_s^{n-1}$ is the unique ℓ-minimal trail in $T_s(\mathbf{e}'')$.

Now let $K \in T_s^-(\mathbf{e})$ be a trail of the form $L + \sum_{i=u}^{n-1} d_i F_s^i$.

Now the coefficients of $m_{j'} : (s, n-1) < j' < (s, n)$ in z^K are the same as those in z^L because the faces $F_s^v : v \leqslant n - 1$ only change the values of the $m_{s'}^{n''}$, with $n'' < n'$. It follows that we may remove exactly the same faces $F_s^{j'}$ (with the same multiplicities) from K as from L for $(s, n-1) < j' < (s, n)$ to obtain an element $K'' \in T_s(\mathbf{e}'')$. We conclude that $K'' = L''' + \sum_{i=u}^{n-1} d_i F_s^i$.

Yet K'', L''' belong to an earlier induction step, and so the d_i must all be non-negative, as required.

6.9.5 The Case When the Coefficient Set Is Described by Sect. 6.6.3

Finally, we consider the case omitted from Sect. 6.8, that is when the coefficient set is described by Sect. 6.6.3. This follows by removing faces, word for word as in Sect. 6.9.4.

6.10 Concluding Remarks

This concludes the proof of Theorem in Sect. 6.3, and recalling Sect. 6.6.5 shows that Question **Q** has a positive answer under its hypothesis.

We note that by Remark 2 of Sect. 6.5, individual strings in $B_J(\varpi_t)$ do not necessarily form S-sets even for $\mathfrak{g}$ classical—see also the legend to Fig. 3. This precludes the most obvious shortcut to proving that $\{z^F\}_{\Phi(F) \in B_J(\varpi_t)}$ is a t-semi-invariant set in the sense of [10, Prop. 5.1].

Define independent variables $\hat{\alpha}_j : j \in J$. They specialize to the simple roots via the map $\hat{\alpha}_j \mapsto \alpha_{i_j}$. For all $b = (\dots, m_2, m_1) \in B_J$, set $\hat{\text{wt}} b := -\sum_{j \in J} m_j \hat{\alpha}_j$. This defines an extended character $\hat{\text{ch}} B_J(\lambda) := \sum_{b \in B_J}(\lambda) e^{\hat{\text{wt}} b}$, which depends on J, but which specializes to the usual character of $B(\lambda)$ given by the Weyl character formula (as obtained for an arbitrary Kac–Moody algebra by combining [6, 2.4.3,3.1.2,3.1.3]). Similarly for any $F = \sum_{u,v \in I \times \mathbb{N}^+} c_u^v F_u^v \in \mathbf{F}$, define an extended weight $\hat{\text{wt}} F := -\sum_{u,v \in I \times \mathbb{N}^+} c_u^v \hat{\alpha}_{(u,v)}$. In that the elements of $\mathcal{K}_t$ can be expressed as a sum of faces, and this defines an extended character $\hat{\text{ch}} \mathcal{K}_t := \sum_{F \in \mathcal{K}_t} e^{\hat{\text{wt}} F}$. It depends on J and so does it specialization. However, what Theorem in Sect. 6.3 shows is that for J bipartite, $\mathfrak{g}$ classical, and t satisfying $\theta(t) = r$, one has

$$\hat{\text{ch}} \hat{\mathcal{K}}_t = \hat{\text{ch}} B_J(\varpi_t), \tag{20}$$

recalling that $\hat{\mathcal{K}}_t = \mathcal{K}_t \cup \{K_\infty\}$.

When $\theta(t) \neq r$, Lemma in Sect. 4.5 describes how to find a Demazure subcrystal to replace the right hand side of (20).

Obviously, $\mathcal{K}_t$ is determined by its extended character, and one may ask if the latter takes a computable form and to what extent it is determined by its specialization, which for arbitrary J is also rather mysterious.

6.11 Exceptional Types

On the other hand, in say type F_4, the above upper bound becomes $\sum c_i \leqslant 4$. This weaker inequality permits an element of the S-set besides the driving function to satisfy the condition $\sum_{k\in\mathbb{N}^+} \Delta r_s^k < 0$. Indeed, this can even arise for the elements of the extended $\mathbb{Z}$-skin (as defined in Remark 2 of Sect. 4.1.4).

For example, suppose there is a driving function z of type s with the coefficient of $\{m_s^k\}_{k=1}^6$ given by $(0, 0, -1, -1, 0, -1)$. Then the extended $\mathbb{Z}$-skin contains the element with coefficients $\{m_s^k\}_{k=1}^6$ given by $(0, 0, -1, -1, 1, 0)$, for which $\sum_{k\in\mathbb{N}^+} \Delta r_s^k = -1$.

A computer computation of Zelikson shows that this does occur in type F_4 with $J = (1324)^6$. Thus, Question **Q** has a negative answer, but this does not mean that Theorem in Sect. 6.3 must fail in this case.

Indeed, let us write a face sum in the form $\sum_{(s,k)\in I\times\mathbb{N}^+} c_s^k F_s^k$. By definition, its image under Φ is $\sum_{(s,k)\in I\times\mathbb{N}^+} c_s^k 1_s^k$.

Set $F = (0000\,0010\,0131\,2231\,2221\,0100)$ with $c_{s,k}$ above, being the coefficient of the kth block of 1324 and s above, determined as the corresponding entry of that block. For example, F_3^1 occurs with coefficient 1 (so t=3) and F_2^2 occurs with coefficient 2.

Then, the driving face is obtained by removing the face F_1^2 to give the element $F^z = (0000\,0010\,0131\,2231\,1221\,0100)$. One shows that $\Phi(F), \Phi(F^z) \in B(\varpi_3)$ by applying the Kashiwara operators to $B_J(\infty) \otimes S(\varpi_t)$, which is simple enough to do by hand.

6.12 The Difficulty

Can one directly prove that $\Phi^{-1}(B_J(\varpi_t))$ is a t-semi-invariant set in the sense of [10, Prop. 5.1]? Here a severe difficulty is to recognize its S-subsets. Whereas the indices over edges can be expected to be independent of the choice of J, this will certainly be false for the vertices. Besides as pointed out in the legend to Fig. 3, the S-subsets cannot be simply pre-images of *single* $\mathfrak{sl}(2)$-strings.

In type F_4, t-semi-invariance of $\Phi^{-1}(B_J(\varpi_t))$ was checked by Zelikson using a computer in the two remaining cases when $\mathcal{M}_t > 2$, but we have not been able to prove it abstractly. A curious empirical fact is that only the union of complete $\mathbb{Z}$-skins of each S-set is needed because the remaining terms (not present for J bipartite for $\mathfrak{g}$ classical) are redundant being in the convex linear combination of S-sets of some *different* types. Outside the minuscule case and for J not bipartite, this is false even in rank 3.

This leads to a further surprise. Following [9, Sect. 6], define a giant S-set to be subset of X_t, which for all $s \in I$ is a union of S-sets of type s, with the exception that for $s = t$ the set $\{z_t^1\}$ (which is not an S-set of type t) is adjoined. Such a subset is t-semi-invariant, and we may choose it to be Z_t. We had longtime believed [9, 8.8] that for all $t \in I$, a giant S-set exists and indeed given as the extremal points of the set $\{z^F\}_{F\in\mathcal{K}_t}$. For example, by Theorem in Sect. 6.3, this is true for $\mathfrak{g}$ classical with J bipartite, and it is also true for many examples outside J bipartite. However for J bipartite, giant S-sets fail to exist in type F_4 when the coefficient of α_t in the highest root > 2.

7 Examples

Through Table 1 (resp. 2) below, we compute the crystal graph of $B_J(\varpi_2)$ (resp. the set of trails in $V(-\varpi_2)$) for $\mathfrak{g}$ simple of type C_3. the former, we use the Kashiwara functions described in (1) to compute as in Sect. 3.3 the action of the Kashiwara operators $\tilde{f}_i : i \in I$. In the latter, we express the trails as sums of faces using (3) in which roots and coroots have been interchanged. Both modules have dimension 14. Then, we compute Fig. 1, and this illustrates Theorem in Sect. 6.3. Of course, this is a baby example, but nevertheless contains the main feature having an octagon containing the false trail, though in general there will be several such octagons. The simplest case in which more complex diagrams occur is in type F_4, but then the graph has 273 vertices and so difficult to draw!

Table 1 The below table describes the entries of the crystal $B_J(\varpi_2)$ in type C_3 using the Bourbaki convention for the latter. The reduced decomposition of the longest element is given in the first row with s_{i_j} being written as i_j. Below i_j stands for the coefficient of m_{i_j} for the element given at the end of the row. The dimension of the module is 14. In this, b_{ϖ_2} has been omitted, whilst the element f corresponds to a false trail lies between 3 and 6 and has zero weight. In the isomorphism described by Theorem in Sect. 6.3, the latter is a convex linear combination of the trails designated by the same indices (given as superscripts). Thus, the latter as an element determining $\varepsilon_2^\star$ by (18) is redundant

2	1	3	2	1	3	2	1	3	Crystal elements
0	0	0	0	0	0	1	0	0	1
0	0	0	0	0	1	1	0	0	2
0	0	0	0	1	0	1	0	0	3
0	0	0	0	1	1	1	0	0	4
0	0	0	1	0	1	1	0	0	5
0	0	0	1	1	1	1	0	0	f
0	0	0	2	1	1	1	0	0	6
0	1	0	1	0	1	1	0	0	7
0	1	0	1	1	1	1	0	0	8
0	1	0	2	1	1	1	0	0	9
0	0	1	2	1	1	1	0	0	10
0	1	1	2	1	1	1	0	0	11
1	1	1	2	1	1	1	0	0	12

Table 2 The below table describes the computation of Z_2, in type C_3 for the reduced decomposition defined by the first row. This is the same reduced decomposition used in Table 1 above. Then, the entries in the left columns are the coefficients of the coordinate functions m_j occurring in the faces or functions in the right hand columns. The labelling on the $\{z_2^i\}_{i=1}^{12}$ corresponds to the ordering on the Z_2^j. It illustrates the inductive procedure defined in 4.1 and presented above so that one may more easily see how the dual Kashiwara functions are given as a sum of faces.

One trail is missing in the above table, namely the one, which we denote by z_2^f, is a convex linear combination of z_2^3 and z_2^6. Here, one may also remark that z_2^5 is an ℓ-minimal element with respect to $1 \in I$, and the former trail is a "false trail". The corresponding coefficient set of z_2^5 can be read off from the table and is $(0, 1, 1)$. Thus, the S-set associated to z_2^5 is reduced to its pointed chain and consists of $\{z_2^5, z_2^7, z_2^8\}$. In this, z_2^f is a false trail in the sense of Sect. 4.1.3 and is an example of which arises in Sect. 6.6.2. (It may just be a coincidence that the false trail is redundant—here this fact was not used.)

2	1	3	2	1	3	2	1	3	Faces	Functions
0	0	0	0	0	0	1	−1	−1	F_2^1	z_2^1
0	0	0	0	0	1	−2	0	1	F_3^2	
0	0	0	0	0	1	−1	−1	0		z_2^2
0	0	0	1	−1	−1	1	0	0	F_2^2	
0	0	0	1	−1	0	0	−1	0		z_2^5
0	1	0	−1	1	0	0	0	0	F_1^3	
0	1	0	0	0	0	0	−1	0		z_2^7
0	0	0	0	1	0	−1	1	0	F_1^2	
0	1	0	0	1	0	−1	0	0		z_2^8
0	0	0	1	−1	−1	1	0	0	F_2^2	
0	1	0	1	0	−1	0	0	0		z_2^9
0	0	1	−2	0	1	0	0	0	F_3^3	
0	1	1	−1	0	0	0	0	0		z_2^{11}
1	−1	−1	1	0	0	0	0	0	F_2^3	
1	0	0	0	0	0	0	0	0		z_2^{12}
0	0	0	0	0	0	1	−1	−1		z_2^1
0	0	0	0	1	0	−1	1	0	F_1^2	
0	0	0	0	1	0	0	0	−1		z_2^3
0	0	0	0	0	1	−2	0	1	F_3^2	
0	0	0	0	1	1	−2	0	0		z_2^4
0	0	0	2	−2	−2	2	0	0	$2F_2^2$	
0	0	0	2	−1	−1	0	0	0		z_2^6
0	0	1	−2	0	1	0	0	0	F_3^3	
0	0	1	0	−1	0	0	0	0		z_2^{10}
0	1	0	−1	1	0	0	0	0	F_1^3	
0	1	1	−1	0	0	0	0	0		z_2^{11}
0	0	0	0	0	1	−1	−1	0		z_2^2

Table 2 (continued)

2	1	3	2	1	3	2	1	3	Faces	Functions
0	0	0	0	1	0	−1	1	0	F_1^2	
0	0	0	0	1	1	−2	0	0		z_2^4
0	0	0	2	−1	−1	0	0	0		z_2^6
0	1	0	−1	1	0	0	0	0	F_1^3	
0	1	0	1	0	−1	0	0	0		z_2^9

8 Index of Notation

Symbols appearing frequently are given below in the paragraph they are first defined.

1.1. $[1,n], [q], \mathfrak{g}, \ell, I, \mathfrak{h}, \pi, \alpha_i, \pi^\vee, \alpha_i^\vee, \mathfrak{b}, \mathfrak{b}^-, e_s, f_s, h_s, \varpi_i, \varpi_i^\vee, s_i, W$.
1.2. $B(\infty), V(-\varpi_i^\vee)$.
1.3. Z_t.
2.2. $c, \Delta, < \sigma_1, \sigma_2 >, w_0$.
3.1. $J, w_k, r, \hat{J}, V(-\varpi_t)$.
3.2. $B_J, m_j, (s,k), m_s^k, r_s^k$.
3.3. $\mathrm{wt}\, b, \varepsilon_s(b), \varphi_s(b), \tilde{e}_s, \tilde{f}_s$.
3.4. $\tilde{\mathcal{E}}, \tilde{\mathfrak{F}}, b_\infty, B_J(\infty)$.
3.5. $K, \gamma_j^K, e_j^K, \mathbf{e}^K, \varphi(K), \mathcal{K}_t^{BZ}, F_w(-\varpi_t)$.
3.6. K_t^1, F_t^1.
3.7. z^K, δ_j^K, z_t^1.
3.8.1. $F_s^{k+1}, \mathbf{e}^K$.
3.8.2. $J^s, \mathbf{e}, e_{-a_k}, \overline{v}_k, \mathbf{k}, \mathbf{a}, a^{(j)}, \mathbf{E}_s^j, T_s(\mathbf{e})$.
3.8.3. $v_\mathbf{k}, M_s(\mathbf{e})$.
4.1.1. $\mathbf{F}_t, E(\cdot), \mathbf{c}^z, Z(\mathbf{c}^z), K(\mathbf{c}^z), K_\mathbb{Z}(\mathbf{c}^z)$.
4.1.2. $\mathbf{F}_t^j, K_{\ell\,\min}, \mathbf{l}, z^{K_{\ell\,\min}}$.
4.1.4. Z_t, Z_t^j.
4.1.5. $\mathbf{K}_t^{j+1}$.
4.1.6. $\mathcal{K}_t(s)$.
4.1.8. r_s^0.
4.3. F^z, d_u^v.
4.4. $B(\varpi_t), S(\varpi_t), s_{\varpi_t}, B_J(\varpi_t)$.
4.5. $1_{u'}^{v'}, \Phi, \mathrm{Supp}\, b, \mathbf{L}, F_\infty, K_\infty, \hat{\mathcal{K}}_t^{BZ}, \hat{F}_t$.
4.7. $\tilde{d}_u^v$.
4.9. $Z^-(\mathbf{c})$.
5.1. $F_w(\lambda), \mathfrak{F}_w$.
5.2. $\tilde{\mathfrak{F}}_w, B_w(\lambda)$.
6.1. $\mathcal{M}_t$.

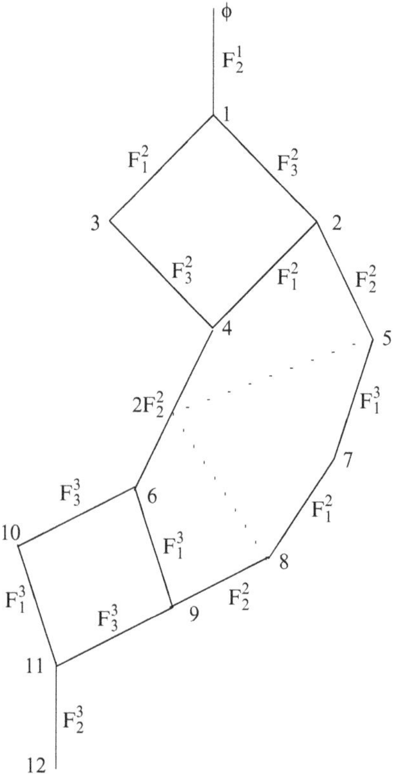

Fig. 1 This describes the graph of the set Z_2 of trails in type C_3, with J chosen as in Table 2. The trivial face F_∞ is designated by ϕ. The vertices are labelled by i designating z_2^i of Table 2. The false trail lies at the end point of the two dotted lines. The trail z_2^5 is minimal for $1 \in I$. Alternatively replacing F_u^v by the Kashiwara operator $\tilde{f}_u$, it describes the crystal graph of $B_j(\varpi_2)$. Then, ϕ designates the highest weight element b_{ϖ_t}. Here, the superscript v determines the place at which $\tilde{f}_u$ enters, that is to say m_u^v increases by 1.

One may remark that there is a unique up to scalars vector $v_f \in V(-\varpi_2)$, namely $f_2 f_1 f_3 f_2 v_{-\varpi_2^\vee}$ through which the false trail f passes. It has zero weight. Again there is a unique crystal element $b_f \otimes s_{\varpi_t} \in B_J(\varpi_2)$ corresponding to the vertex v_2^f. One may check that $\tilde{e}_1(b_f \otimes s_{\varpi_t}) = 0$ and $\tilde{f}_1(b_f \otimes s_{\varpi_t}) = 0$, illustrating the point made in Remark 2 of Sect. 6.4

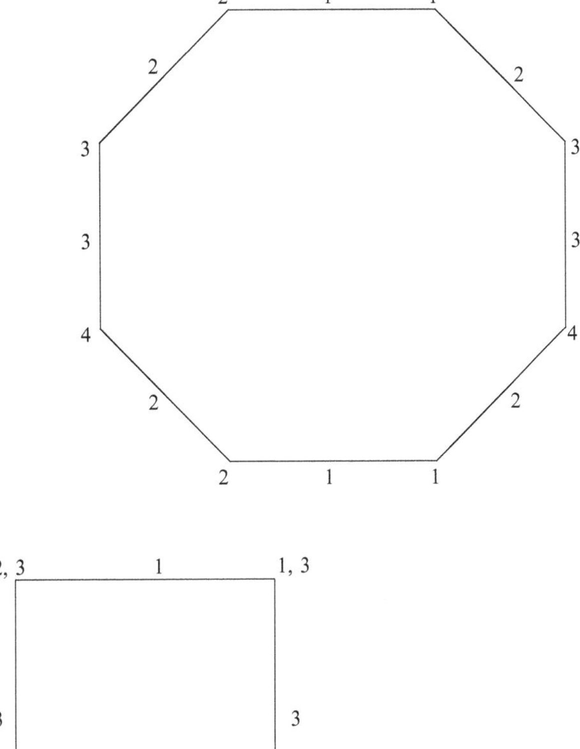

Fig. 2 An S-graph for $n = 4$ and for $c_2 \leqslant c_1, c_3$, computed using the construction of [7, Sect. 7], as exemplified in [12, 5.6.2]. The functions z_v as v runs though the vertices of the graph are the convex set defined by (3) and the relations $c_3' - c_2' \geqslant 0, c_2' - c_1' \geqslant c_2 - c_1$

2, 3 1 1, 3

3 3

2, 4 1 1, 4

Fig. 3 The degeneration of the above S-graph (for $n = 4$) when $c_2 = 0$. This S-graph appears when the coefficient of α_t in the highest root is 2. It is the situation described in Sects. 6.6.3 and 6.9.3. However, unlike the case when $\mathbf{c}$ has two consecutive non-zero entries, it does *not* correspond via Φ to a single crystal string. Rather it corresponds to two crystal strings, one of length 1 coming from the vertex in the bottom right hand corner and a second of length 3. This results in an obstruction to the obvious proof of $\Phi^{-1}(B(\varpi_t))$ being a t-semi-invariant set.[3] In this case, the corresponding convex set is that given by (19)

[3]Indeed, we have a pairing $(b, b') \mapsto (z^{\Phi^{-1}(b)}(b'))$ of $B_J \times B_J$ into $\mathbb{Z}$. It would be pleasant but wrong to believe that this pairing was essentially invariant with respect to the crystal operators for J bipartite.

References

1. A. Berenstein and A. Zelevinsky, Tensor product multiplicities, canonical bases and totally positive varieties. Invent. Math. 143 (2001), no. 1, 77–128.
2. N. Bourbaki, Éléments de mathématique. Fasc. XXXIV. Groupes et algèbres de Lie. Chapitre IV–VI: (French) Actualités Scientifiques et Industrielles, No. 1337 Hermann, Paris 1968.
3. O. Gleizer and A. Postnikov, Littlewood-Richardson coefficients via Yang-Baxter equation. Internat. Math. Res. Notices 2000, no. 14, 741–774.
4. A. Joseph, A decomposition theorem for Demazure crystals. J. Algebra 265 (2003), no. 2, 562–578.
5. A. Joseph, Quantum groups and their primitive ideals. Ergebnisse der Mathematik und ihrer Grenzgebiete (3) [Results in Mathematics and Related Areas (3)], 29. Springer-Verlag, Berlin,1995.
6. A. Joseph, Consequences of the Littelmann path theory for the structure of the Kashiwara $B(\infty)$ crystal. Highlights in Lie algebraic methods, 25–64, Progr. Math., 295, Birkäuser/Springer, New York, 2012.
7. A. Joseph, A Preparation Theorem for the Kashiwara $B(\infty)$ Crystal, Selecta Mathematica, 23, no.2, (2017), 1309–1353.
8. A. Joseph, Convexity properties of the canonical S-graphs, Israel J. Math. 226 (2018), no. 2, 827–849.
9. A. Joseph, Trails, S-graphs and identities in Demazure modules, arXiv:1702.00243.
10. A. Joseph, Dual Kashiwara functions for the $B(\infty)$ crystal. Lie Groups, Geometry, and Represention Theory, Progress in Mathematics 326, 201–233, Birkhauser, Boston, 2018
11. A. Joseph, Trails for minuscule modules and dual Kashiwara functions for the $B(\infty)$ crystal. Quantum theory and symmetries with Lie theory and its applications in physics. Vol. 1, 37–53, Springer Proc. Math. Stat., 263, Springer, Singapore, 2018.
12. A. Joseph and P. Lamprou, A new interpretation of the Catalan numbers, arXiv:1512.00406.
13. V. G Kac, Infinite-dimensional Lie algebras. Second edition. Cambridge University Press, Cambridge, 1985.
14. M. Kashiwara, Global crystal bases of quantum groups. Duke Math. J. 69 (1993), no. 2, 455–485.
15. M. Kashiwara, The crystal base and Littelmann's refined Demazure character formula. Duke Math. J. 71 (1993), no. 3, 839–858.
16. T. Nakashima, Polytopes for crystallized Demazure modules and extremal vectors. Comm. Algebra 30 (2002), no. 3, 1349–1367.
17. T. Nakashima, A. Zelevinsky, Polyhedral realizations of crystal bases for quantized Kac-Moody algebras. Adv. Math. 131 (1997), no. 1, 253–278.
18. D. E. Speyer, Powers of Coxeter elements in infinite groups are reduced. Proc. Amer. Math. Soc. 137 (2009), no. 4, 1295–1302.

Lusztig's t-Analogue of Weight Multiplicity via Crystals

Cédric Lecouvey and Cristian Lenart

Abstract We give a purely combinatorial proof of the positivity of the stabilized forms of the generalized exponents associated with each classical root system. In finite type A_{n-1}, we rederive the description of the generalized exponents in terms of crystal graphs without using the combinatorics of semistandard tableaux or the charge statistic. In finite type C_n, we obtain a combinatorial description of the generalized exponents based on the so-called distinguished vertices in crystals of type A_{2n-1}, which we also connect to symplectic King tableaux. This gives a combinatorial proof of the positivity of Lusztig t-analogues associated with zero weight spaces in the irreducible representations of symplectic Lie algebras. We then present three applications of our combinatorial formula. Our methods are expected to extend to the orthogonal types.

By a result of Lascoux, the type A Kostka–Foulkes polynomials also expand positively in terms of the so-called atomic polynomials. We define, in arbitrary type, a combinatorial version of the atomic decomposition, based on the connected components of a modified crystal graph. We prove this property in type A, as well as in types B, C, and D in a stable range for $t = 1$. We also discuss other cases, applications, and a geometric interpretation. Finally, in classical types, we state the atomic decomposition for stable 1-dimensional sums or, equivalently, for the stable Lusztig t-analogues.

C. Lecouvey
Institut Denis Poisson, Faculté des Sciences et Techniques, Université de Tours, Tours, France
e-mail: cedric.lecouvey@lmpt.univ-tours.fr

C. Lenart (✉)
Department of Mathematics and Statistics, State University of New York at Albany, Albany, NY, USA
e-mail: clenart@albany.edu

J. Greenstein et al. (eds.), *Interactions of Quantum Affine Algebras with Cluster Algebras, Current Algebras and Categorification*, Progress in Mathematics 337,
https://doi.org/10.1007/978-3-030-63849-8_10

1 Introduction

Let $\mathfrak{g}$ be a simple Lie algebra over $\mathbb{C}$ of rank n and G its corresponding Lie group. The group G acts on the symmetric algebra $S(\mathfrak{g})$ of $\mathfrak{g}$, and it was proved by Kostant [13] that $S(\mathfrak{g})$ factors as $S(\mathfrak{g}) = H(\mathfrak{g}) \otimes S(\mathfrak{g})^G$, where $H(\mathfrak{g})$ is the harmonic part of $S(\mathfrak{g})$. The generalized exponents of $\mathfrak{g}$, as defined by Kostant [13], are the polynomials appearing as the coefficients in the expansion of the graded character of $H(\mathfrak{g})$ in the basis of the Weyl characters. It was shown by Hesselink [5] that these polynomials coincide, in fact, with the Lusztig t-analogues $K_{\lambda,0}(t)$ of zero weight multiplicities in the irreducible finite-dimensional representations of $\mathfrak{g}$. In particular, they have nonnegative integer coefficients, because they are affine Kazhdan–Lusztig polynomials.

For $\mathfrak{g} = \mathfrak{sl}_n$, the generalized exponents admit a nice combinatorial description in terms of the Lascoux-Schützenberger charge statistic on semistandard tableaux of zero weight [16]. This statistic is defined via the cyclage operation on tableaux, which is based on the Schensted insertion scheme. This combinatorial description extends, in fact, to any Lusztig t-analogue of type A_{n-1}, that is possibly associated with a nonzero weight (also called Kostka polynomials). Another interpretation of the charge statistic in terms of crystals of type A_{n-1} was given later by Lascoux, Leclerc, and Thibon in [17].

Despite many efforts during the last three decades, no general combinatorial proof of the positivity of the Lusztig t-analogues $K_{\lambda,\mu}(t)$ is known beyond type A. Nevertheless, such proofs have been obtained in some particular cases: [6], [7], [22], and [23].

In type A, the Kostka–Foulkes polynomials $K_{\lambda,\mu}(t)$ are well-known t-analogues of the Kostka numbers $K_{\lambda,\mu}$, i.e., the number of semistandard Young tableaux of shape λ and content μ. Lascoux [18] stated the decomposition of the Kostka–Foulkes polynomials into the so-called *atomic polynomials*. Some arguments of the proof in [18] remained elusive, and it was not until the work of Shimozono [30] that the type A atomic decomposition was completely accepted, this time in larger generality (for the so-called *generalized Kostka–Foulkes polynomials*). However, the latter proof involves several intricate combinatorial arguments and related concepts, such as *plactic monoid*, *cyclage*, and *catabolism*.

This expository paper mainly follows [20, 21], to which we refer for detailed proofs. Nevertheless, the results of Sect. 12 on the atomic decomposition of stable 1-sums are new.

We first give a combinatorial description of the stabilized version of the generalized exponents and a proof of their positivity by using the combinatorics of type $A_{+\infty}$ crystal graphs. This can be regarded as a generalization of results in [17] for the weight zero, and in fact we were able to rederive the latter without any reference to the charge statistic or the combinatorics of semistandard tableaux. Our description is in terms of the so-called distinguished vertices in a crystal of type $A_{+\infty}$, but we show that these vertices are in natural bijection with some generalizations of symplectic King tableaux, which makes the connection with the stable Lusztig t-analogue more natural. Next, we provide a complete combinatorial proof of the positivity of the generalized exponents in the non-stable C_n case. Note that this case is much more involved than the stable one, essentially because we need a combinatorial description of the non-Levi branching from $\mathfrak{gl}_{2n}$ to $\mathfrak{sp}_{2n}$, which is complicated in general. Here one needs in a crucial way recent duality results by Kwon [14, 15], giving a crystal interpretation of the previous branching and a combinatorial model relevant to its study. Our approach has been extended to the orthogonal types in [8].

Next, following Lascoux [18, 19], we formulate the *t-atomic decomposition* property in arbitrary Lie type, as a nonnegative expansion for both a Kostka–Foulkes polynomial $K_{\lambda,\mu}(t)$ (Lusztig's t-analogue of weight multiplicity [26]), and a t-analogue $\chi_{\lambda}^{+}(t)$ of the dominant part of an irreducible character (defined in terms of $K_{\lambda,\mu}(t)$). In fact the underlying t-atomic polynomials are related to the Hall–Littlewood polynomials by the Möbius function for the dominance order on weights [19]. The t-atomic decomposition property is a strengthening of the monotonicity of $K_{\lambda,\mu}(t)$ [1]. As opposed to the above algebraic approach, we define a t-atomic decomposition property at the combinatorial level of the highest weight crystal $B(\lambda)$ [10]. This property involves a partition of the dominant part $B(\lambda)^{+}$ of $B(\lambda)$, and a statistic on $B(\lambda)^{+}$. We prove that the combinatorial t-atomic decomposition holds in type A, thus realizing combinatorially the classical result, while also providing a simple, conceptual proof of it. We also prove this property in types B, C, and D for $t = 1$ in a stable range. Our main ingredients are: the *partial order on dominant weights*, and a *modified crystal graph* structure on $B(\lambda)^{+}$, whose connected components define the needed partition. We conjecture that our result in types B, C, and D holds without specializing t. Furthermore, in type C, we conjecture that this result, together with our combinatorial formula for the corresponding $K_{\lambda,0}(t)$, leads to a statistic which computes any $K_{\lambda,\mu}(t)$. Such a statistic (charge) has been long sought. We also propose a geometric interpretation of the atomic decomposition in terms of the *geometric Satake correspondence*. In classical types, the Lusztig t-analogues have stable forms, which are known to coincide (up to renormalization) with the 1-dimensional sums defined using the coenergy function on finite crystals of classical affine types. We conclude by establishing an atomic decomposition for these 1-dimensional sums.

Section 2 recalls the definition of the generalized exponents. Section 3 is devoted to the combinatorial description of the stabilized form (in classical type) of the generalized exponents in terms of distinguished tableaux, which we define and study here. In Sect. 4, we give the combinatorial description of the generalized exponents in type C_n by using King tableaux [12]. In Sect. 5, we derive three applications of the description in Sect. 4. The concept of a t-atomic decomposition in any type is introduced in Sect. 6. Section 7 is devoted to important properties of the partial order on dominant weights. In Sect. 8, we define the modified crystal operators yielding the desired partition of $B(\lambda)^{+}$. The t-atomic decomposition in type A and the atomic decomposition in types B, C, and D are established in Sect. 9, whereas perspectives and conjectures are proposed in Sect. 10. In Sect. 11, we give a geometric interpretation of the atomic decomposition of characters. Finally, in

Sect. 12 we show that there also exists an atomic decomposition for the stable 1-dimensional sums defined using the coenergy function on finite affine crystals of classical types.

2 Generalized Exponents

2.1 Background

Let $\mathfrak{g}$ be a simple Lie algebra over $\mathbb{C}$ of rank n with triangular decomposition $\mathfrak{g}= \oplus_{\alpha\in R_+} \mathfrak{g}_\alpha \oplus \mathfrak{h} \oplus \oplus_{\alpha\in R_+} \mathfrak{g}_{-\alpha}$,so that $\mathfrak{h}$ is the Cartan subalgebra of $\mathfrak{g}$ and R_+ its set of positive roots. The root system $R = R_+ \sqcup (-R_+)$ of $\mathfrak{g}$ is realized in a real Euclidean space E with inner product $\langle\cdot,\cdot\rangle$. For any $\alpha \in R$, we write $\alpha^\vee = \frac{2\alpha}{\langle\alpha,\alpha\rangle}$ for its coroot. Let $S \subseteq R_+$ be the subset of simple roots and Q_+ the $\mathbb{Z}_+$-cone generated by S. The set P of integral weights for $\mathfrak{g}$ satisfies $\langle\beta,\alpha^\vee\rangle \in \mathbb{Z}$ for any $\beta \in P$ and $\alpha \in R$. We write $P_+ = \{\beta \in P \mid \langle\beta,\alpha^\vee\rangle \geqslant 0 \text{ for any } \alpha \in S\}$ for the cone of dominant weights of $\mathfrak{g}$, and denote by $\omega_1,\dots,\omega_n$ its fundamental weights. Let W be the Weyl group of $\mathfrak{g}$ generated by the reflections s_α with $\alpha \in S$, and write ℓ for the corresponding length function.

By a classical theorem due to Kostant, the graded character of the harmonic part of the symmetric algebra $S(\mathfrak{g})$ satisfies

$$\mathrm{char}_t(H(\mathfrak{g})) = \frac{\prod_{i=1}^n(1-t^{d_i})}{(1-t)^n}\prod_{\alpha\in R}\frac{1}{1-te^\alpha} = \prod_{i=1}^n(1-t^{d_i})\,\mathrm{char}_t(S(\mathfrak{g}))\,,$$

where we have $d_i = m_i+1$, for $i=1,\dots,n$, and $m_1,\dots,m_n$ are the (classical) exponents of $\mathfrak{g}$. In type A_n we have $m_i = i$, in types B_n and C_n $m_i = 2i-1$ and in type D_n $m_i = 2i-1$ for $i=1,\dots,n-1$ with $m_n = n-1$. On the other hand, it is known (see [5]) that $\mathrm{char}_t(H(\mathfrak{g}))$ coincides with the Hall–Littlewood polynomial Q'_0, namely we have

$$\mathrm{char}_t(H(\mathfrak{g})) = Q'_0 = \sum_{\lambda\in P_+} K^{\mathfrak{g}}_{\lambda,0}(t)\, s^{\mathfrak{g}}_\lambda\,,$$

where $s^{\mathfrak{g}}_\lambda$ is the Weyl character associated with the finite-dimensional irreducible representation $V(\lambda)$ of $\mathfrak{g}$ with highest weight λ and $W_0(t) = \prod_{i=1}^n \frac{1-t^{d_i}}{1-t}$. The polynomials $K^{\mathfrak{g}}_{\lambda,0}(t)$ are the generalized exponents of $\mathfrak{g}$, and they coincide with the Lusztig t-analogues associated with the zero weight subspaces in the representations $V(\lambda)$. The classical exponents $m_1,\dots,m_n$ correspond to the adjoint representation of $\mathfrak{g}$, namely we have $K^{\mathfrak{g}_n}_{\widetilde{\alpha},0}(t) = \sum_{i=1}^n t^{m_i}$,where $\widetilde{\alpha}$ is the highest root in R_+.

2.2 Classical Types

In classical types, $\mathrm{char}_t(S(\mathfrak{g}))$ is easy to compute. Let $\mathcal{P}_n$ be the set of partitions with at most n parts, and $\mathcal{P}$ the set of all partitions. The rank of the partition γ is defined as the sum of its parts, and is denoted by $|\gamma|$.

In type A_{n-1}, we start from the Cauchy identity[1]

$$\prod_{1\leqslant i,j\leqslant n} \frac{1}{1-tx_iy_j} = \sum_{\gamma\in\mathcal{P}_n} t^{|\gamma|} s_\gamma(x)s_\gamma(y).$$

By setting $y_i = \frac{1}{x_i}$ for any $i = 1,\dots,n$, and by considering the images of the symmetric polynomials in $R^{A_{n-1}} = \mathrm{Sym}[x_1,\dots,x_n]/(x_1\cdots x_n - 1)$, we get

$$\begin{aligned}\mathrm{char}_t(S(\mathfrak{sl}_n)) &= (1-t)\sum_{\gamma\in\mathcal{P}_n} t^{|\gamma|} s_\gamma(x)s_\gamma(x^{-1}) = (1-t)\sum_{\gamma\in\mathcal{P}_n} t^{|\gamma|} s_\gamma s_{\gamma^*} = \\ &= (1-t)\sum_{\gamma\in\mathcal{P}_n} t^{|\gamma|} \sum_{\lambda\in\mathcal{P}_{n-1}} c^{\lambda}_{\gamma,\gamma^*} s_\lambda(x).\end{aligned} \tag{2.1}$$

Here $\gamma^* = -w_\circ(\gamma)$, where $w_\circ$ is the permutation of maximal length in S_n, and we use the same notation for a symmetric polynomial and its image in $R^{A_{n-1}}$.

For any positive integer m, define $\mathcal{P}^{(2)}_m$ as the set of partitions of the form 2κ with $\kappa \in \mathcal{P}_m$, and $\mathcal{P}^{(1,1)}_m$ as the subset of $\mathcal{P}_m$ containing the partitions of the form $(2\kappa)'$ with $\kappa \in \mathcal{P}$. Moreover, we denote by $s^{\mathfrak{so}_{2n+1}}_\lambda$, $s^{\mathfrak{sp}_{2n}}_\lambda$, and $s^{\mathfrak{so}_{2n}}_\lambda$ the irreducible characters corresponding to the highest weight λ, for the Lie algebras of types B_n, C_n, and D_n, respectively.

In type B_n, we start from the Littlewood identity [25]

$$\prod_{1\leqslant i<j\leqslant 2n+1} \frac{1}{1-ty_iy_j} = \sum_{\nu\in\mathcal{P}^{(1,1)}_{2n+1}} t^{|\nu|/2} s_\nu(y),$$

and we specialize $y_{2n+1} = 1$, $y_{2i-1} = x_i$, and $y_{2i} = \frac{1}{x_i}$, for any $i = 1,\dots,n$. This gives

$$\mathrm{char}_t(S(\mathfrak{so}_{2n+1})) = \sum_{\nu\in\mathcal{P}^{(1,1)}_{2n+1}} t^{|\nu|/2} \sum_{\lambda\in\mathcal{P}_n} c^{\lambda}_{\nu}(\mathfrak{so}_{2n+1}) s^{\mathfrak{so}_{2n+1}}_\lambda,$$

[1] Here $s_\nu(x)$ stands for the ordinary Schur function in the variables $x_1,\dots,x_n$.

where $c_{\nu}^{\lambda}(\mathfrak{so}_{2n+1})$ is the branching coefficient corresponding to the restriction from $\mathfrak{gl}_{2n+1}$ to $\mathfrak{so}_{2n+1}$. Similarly, we get

$$\mathrm{char}_t(S(\mathfrak{sp}_{2n})) = \sum_{\nu \in \mathcal{P}_{2n}^{(2)}} t^{|\nu|/2} \sum_{\lambda \in \mathcal{P}_n} c_{\nu}^{\lambda}(\mathfrak{sp}_{2n}) s_{\lambda}^{\mathfrak{sp}_{2n}} \quad \text{and}$$

$$\mathrm{char}_t(S(\mathfrak{so}_{2n})) = \sum_{\nu \in \mathcal{P}_{2n}^{(1,1)}} t^{|\nu|/2} \sum_{\lambda \in \mathcal{P}_n} c_{\nu}^{\lambda}(O_{2n}) s_{\lambda}^{O_{2n}}.$$

Note that here we considered the character $s_{\lambda}^{O_{2n}}$ of the $O(2n)$-module $V^{O(2n)}(\lambda)$ parametrized by the partition λ.

Proposition 2.1 *We have the following identities.*

(a) In type A_{n-1}, for any $\lambda \in \mathcal{P}_{n-1}$, we have[2]

$$\frac{K_{\lambda,0}^{\mathfrak{sl}_n}(t)}{\prod_{i=1}^{n}(1-t^i)} = \sum_{\gamma \in \mathcal{P}_n} t^{|\gamma|} c_{\gamma,\gamma^*}^{\lambda}.$$

(b) In type B_n, for any $\lambda \in \mathcal{P}_n$, we have[3]

$$\frac{K_{\lambda,0}^{\mathfrak{so}_{2n+1}}(t)}{\prod_{i=1}^{n}(1-t^{2i})} = \sum_{\nu \in \mathcal{P}_{2n+1}^{(1,1)}} t^{|\nu|/2} c_{\nu}^{\lambda}(\mathfrak{so}_{2n+1}).$$

(c) In type C_n, for any $\lambda \in \mathcal{P}_n$, we have

$$\frac{K_{\lambda,0}^{\mathfrak{sp}_{2n}}(t)}{\prod_{i=1}^{n}(1-t^{2i})} = \sum_{\nu \in \mathcal{P}_{2n}^{(2)}} t^{|\nu|/2} c_{\nu}^{\lambda}(\mathfrak{sp}_{2n}).$$

(d) In type D_n, for any $\lambda \in \mathcal{P}_n$, we have

$$\frac{K_{\lambda,0}^{O(2n)}(t)}{(1-t^n)\prod_{i=1}^{n-1}(1-t^{2i})} = \sum_{\nu \in \mathcal{P}_{2n}^{(1,1)}} t^{|\nu|/2} c_{\nu}^{\lambda}(O_{2n}).$$

For type D_n, the dominant weights are not necessarily partitions, whereas this is the case in Assertion 4 of the previous proposition. So here we have in fact to write

[2]The factor $(1-t)$ in (2.1) gives the missing "$d_i = 1$" in type A_{n-1}.

[3]Here that the partition λ can have an odd rank.

$$K_{\lambda,0}^{O(2n)}(t) = K_{\omega(\lambda),0}^{\mathfrak{so}_{2n}}(t) = K_{\iota(\omega(\lambda)),0}^{\mathfrak{so}_{2n}}(t)$$

for any partition $\lambda \in \mathcal{P}_n \setminus \mathcal{P}_{n-1}$ and $\omega(\lambda)$.

We have then by a theorem of Lascoux and Schützenberger [16]

$$K_{\lambda,0}^{\mathfrak{sl}_n}(t) = \sum_{T \in SST(\lambda)_0} t^{\mathrm{ch}_n(T)},$$

where $SST(\lambda)_0$ is the set of semistandard tableaux labeled by letters of $\{1 < \cdots < n\}$ of weight $\mu = (a, \ldots, a) = 0$ (i.e., each letter i appears a times in T) where $a = |\lambda|/n$, and $\mathrm{ch}_n(T)$ is the charge statistic evaluated on T. Recall that this charge statistic is defined by rather involved combinatorial operation such as cyclage on tableaux.

2.3 Stable Versions

When the ranks of the classical root systems considered go to infinity, the previous relations simplify. In particular, for n sufficiently large, we have

$$c_\nu^\lambda(\mathfrak{so}_{2n+1}) = \sum_{\delta \in \mathcal{P}} c_{\lambda,2\delta}^\nu, \quad c_\nu^\lambda(\mathfrak{sp}_{2n}) = \sum_{\delta \in \mathcal{P}} c_{\lambda,(2\delta)'}^\nu, \quad \text{and} \quad c_\nu^\lambda(\mathfrak{so}_{2n}) = \sum_{\delta \in \mathcal{P}} c_{\lambda,2\delta}^\nu.$$

Observe that, for $\mathfrak{g} = \mathfrak{so}_{2n+1}$, this implies in particular that $c_\nu^\lambda(\mathfrak{so}_{2n+1}) = 0$ when the ranks of λ and ν do not have the same parity, which is false in general. Thus when $|\lambda|$ is even, we get the relations

$$\frac{K_{\lambda,0}^{B_\infty}(t)}{\prod_{i=1}^\infty (1-t^{2i})} = \frac{K_{\lambda,0}^{D_\infty}(t)}{\prod_{i=1}^\infty (1-t^{2i})} = \sum_{\nu \in \mathcal{P}^{(1,1)}} \sum_{\delta \in \mathcal{P}^{(2)}} t^{|\nu|/2} c_{\lambda,\delta}^\nu \text{ in type } B_\infty, D_\infty,$$

$$\frac{K_{\lambda,0}^{C_\infty}(t)}{\prod_{i=1}^\infty (1-t^{2i})} = \sum_{\nu \in \mathcal{P}^{(2)}} \sum_{\delta \in \mathcal{P}^{(1,1)}} t^{|\nu|/2} c_{\lambda,\delta}^\nu \quad \text{in type } C_\infty.$$

In particular, this gives

$$K_{\lambda,0}^{B_\infty}(t) = K_{\lambda,0}^{D_\infty}(t) \quad \text{and} \quad K_{\lambda,0}^{B_\infty}(t) = K_{\lambda',0}^{C_\infty}(t). \tag{2.2}$$

All these stabilized forms are in fact formal power series in t equal to zero when the rank of λ is odd.

3 Stabilized Generalized Exponents and Crystal Graphs of Type $A_{+\infty}$

3.1 Crystal of Type $A_{+\infty}$

Recall that crystals of type $A_{+\infty}$ are those associated with the infinite Dynkin diagram

$$\overset{1}{\circ} - \overset{2}{\circ} - \overset{3}{\circ} \cdots$$

The partitions label the dominant weights of $\mathfrak{sl}_{+\infty}$. If we denote by $(\omega_i)_{\geqslant 1}$ the sequence of fundamental weights of $\mathfrak{sl}_{+\infty}$, we have for any partition $\lambda \in \mathcal{P}$, $\lambda = \sum_i a_i \omega_i$,where a_i is the number of columns with height i in the Young diagram of λ.

To each partition λ corresponds the crystal $B(\lambda)$ of the irreducible infinite-dimensional representation of $\mathfrak{sl}_{+\infty}$ parametrized by λ. A classical model for $B(\lambda)$ is that of semistandard tableaux of shape λ on the infinite alphabet $\mathbb{Z}_{>0} = \{1 < 2 < 3 < \cdots\}$. Given $b \in B(\lambda)$, we define $\boldsymbol{\varepsilon}(b) = \sum_{i=1}^{+\infty} \varepsilon_i(b)\omega_i$ and $\boldsymbol{\varphi}(b) = \sum_{i=1}^{+\infty} \varphi_i(b)\omega_i$,where both sums are in fact finite. The weight of $b \in B(\lambda)$ then verifies $\mathrm{wt}(b) = \boldsymbol{\varphi}(b) - \boldsymbol{\varepsilon}(b)$.

3.2 Combinatorial Preliminaries

In the sequel we consider the order $\leqslant$ on $\mathcal{P}$ such that $\lambda \leqslant \mu$ if and only if $\mu - \lambda \in P_+^{\infty}$, that is, $\mu - \lambda$ decomposes in the basis of the ω_i's with nonnegative integer coefficients.

The partitions in $\mathcal{P}^{(2)}$ (resp. in $\mathcal{P}^{(1,1)}$) are those which can be tiled with horizontal (resp. vertical) dominoes. Equivalently, a partition κ belongs to $\mathcal{P}^{(2)}$ (resp. $\mathcal{P}^{(1,1)}$) if and only if the number of columns (resp. rows) of fixed height (resp. length) is even. So

$$\kappa \in \mathcal{P}^{(2)} \iff \kappa = \sum_i 2a_i\omega_i \quad \text{and} \quad \kappa \in \mathcal{P}^{(1,1)} \iff \kappa = \sum_i a_i\omega_{2i} .$$

Set $\mathcal{P}^{\boxplus} = \mathcal{P}^{(2)} \cap \mathcal{P}^{(1,1)}$. It follows that

$$\kappa \in \mathcal{P}^{\boxplus} \iff \kappa = \sum_i 2a_i\omega_{2i} ,$$

that is, λ decomposes in terms of the fundamental weights ω_{2i} with even coefficients. In the general case of a partition $\kappa \in \mathcal{P}$ written as $\kappa = \sum_i a_i\omega_i$ we define

$$\kappa_{\boxplus} = \sum_i (a_{2i} - (a_{2i} \bmod 2))\omega_{2i}$$

and

$$\kappa^{\boxplus} = \kappa - \kappa_{\boxplus} = \sum_i a_{2i+1}\omega_{2i+1} + \sum_i (a_{2i} \bmod 2)\omega_{2i}\,.$$

So $\kappa_{\boxplus}$ and $\kappa^{\boxplus}$ are partitions and $\kappa_{\boxplus} \in \mathcal{P}^{\boxplus}$.

We denote by $P^{\infty}_{(2)}$ and $P^{\infty}_{(1,1)}$ the sublattices of $P = \oplus_{i\geqslant 1}\mathbb{Z}\omega_i$ defined by $P^{\infty}_{(2)} = \oplus_{i\geqslant 1}2\mathbb{Z}\omega_i$ and $P^{\infty}_{(1,1)} = \oplus_{i\geqslant 1}\mathbb{Z}\omega_{2i}$. Observe that $P^{\infty}_{(2)} \cap \mathcal{P} = \mathcal{P}_{(2)}$ and $P^{\infty}_{(1,1)} \cap \mathcal{P} = \mathcal{P}_{(1,1)}$. We have also $P^{\infty}_{\boxplus} = P^{\infty}_{(2)} \cap P^{\infty}_{(1,1)} = \oplus_{i\geqslant 1}2\mathbb{Z}\omega_{2i}$ and $\mathcal{P}^{\boxplus} = \mathcal{P} \cap P^{\infty}_{(2)} \cap P^{\infty}_{(1,1)}$.We define the order $\leqslant_{\boxplus}$ on $\mathcal{P}$ by $\lambda \leqslant_{\boxplus} \mu \iff \mu - \lambda \in \mathcal{P}^{\boxplus}$.

3.3 A Combinatorial Description of the Series $K^{C_\infty}_{\lambda,0}(t)$

Definition 3.3.1 Consider a partition μ. A vertex $b \in B(\lambda)$ is called μ-distinguished if there exists $(\nu, \delta) \in \mathcal{P}^{(2)} \times \mathcal{P}^{(1,1)}$ such that $\boldsymbol{\varphi}(b) = \nu - \mu$ and $\boldsymbol{\varepsilon}(b) = \delta - \mu$.

Definition 3.3.2 Let $D(\lambda)$ be the set of all vertices in $B(\lambda)$ which are μ-distinguished for at least a partition μ.

Clearly, if b is μ-distinguished, then b is $(\mu + \kappa)$-distinguished for any $\kappa \in \mathcal{P}^{\boxplus}$ (change $(\nu, \delta) \in \mathcal{P}^{(2)} \times \mathcal{P}^{(1,1)}$ to $(\nu + \kappa, \delta + \kappa) \in \mathcal{P}^{(2)} \times \mathcal{P}^{(1,1)}$). For any $b \in D(\lambda)$, set

$$S_b = \{\mu \in \mathcal{P} \mid b \text{ is } \mu\text{-distinguished}\}\,.$$

Lemma 3.1 *The set S_b has the form $S_b = \mu_b + \mathcal{P}^{\boxplus}$ and μ_b is minimal for $\leqslant_{\boxplus}$ such that b is μ_b-distinguished. Moreover, for any $\mu \in S_b$, we have $\mu_b = \mu^{\boxplus}$.* □

The following proposition makes more explicit the structure of the distinguished tableaux.

Proposition 3.2 *Let b be a vertex of $B(\lambda)$ with $\lambda \in \mathcal{P}$. Then b is distinguished if and only if $\varepsilon_i(b) = 0$ for any odd i and $\varphi_i(b)$ is even for any odd i. Moreover, we then have*

$$\mu_b = \sum_i (\varphi_{2i}(b) \bmod 2)\omega_{2i} =: \varphi(b) \bmod 2.$$

Theorem 3.3 *We have*

$$K_{\lambda,0}^{C_\infty}(t) = \sum_{b \in D(\lambda)} t^{|\varphi(b)+\mu_b|/2}.$$

3.4 Distinguished Tableaux and Zero Weight King Type Tableaux

To see that the distinguished tableaux we introduced previously are in natural bijection with zero weight tableaux very close to King tableaux, consider the sets $T_{C_\infty}(\lambda)$ of semistandard tableaux of shape λ on the infinite ordered alphabet $\{1 < \overline{1} < 2 < \overline{2} < \cdots\}$. There will be no condition on the position of the barred letters here, contrary to the definition of King tableaux. Recall the notation of Sect. 3.3. For any distinguished vertex b in $D(\lambda)$, set $\boldsymbol{\theta}(b) = \varphi(b) + \mu_b$, and let $\theta_j(b)$ be the coefficient of ω_j in the expansion of $\boldsymbol{\theta}(b)$. Since $\boldsymbol{\theta}(b)$ is a dominant weight for $\mathfrak{sl}_\infty$, it can be regarded as a partition. Recall also that $|\lambda|$ is even, says $|\lambda| = 2\ell$. In the sequel of this section, we shall assume that $B(\lambda)$ is realized as the set of semistandard tableaux on the infinite ordered alphabet $\mathbb{Z}_{>0}$. For any integer $i \geqslant 1$, a reverse lattice skew tableau on $\{2i-1, 2i\}$ is a semistandard filling of a skew Young diagram with columns of height at most 2 by letters $2i-1$ and $2i$ whose Japanese reading is a lattice word (i.e., in each left factor the number of letters $2i$ is less than or equal to that of letters $2i-1$).

Example 3.4 Assume $i = 2$. Then

$$\begin{array}{cccccccccccccc}
 & & & & & & & & & & 3 & 3 & 3 & 3 \\
 & & & & & & 3 & 3 & 4 & 4 & 4 & & & \\
 & & & 3 & 3 & 4 & & & & & & & & \\
3 & 3 & 4 & 4 & & & & & & & & & &
\end{array} \tag{3.1}$$

is a reverse lattice skew tableau on $\{3, 4\}$. □

The following proposition is a reformulation of Proposition 3.2.

Proposition 3.5 *A semistandard tableau T of shape λ is distinguished if and only if for any integer $i \geqslant 1$, the skew tableau obtained by keeping only the letters $2i-1$ and $2i$ in T is a reverse lattice tableau, and the rows of $\boldsymbol{\theta}(T)$ have even lengths.* □

We now explain the correspondence between distinguished tableaux and zero weight King type tableaux. Observe that a tableau T in $T_{C_\infty}(\lambda)$ of weight zero is a juxtaposition of skew tableaux of weight 0 on $\{i, \overline{i}\}$ obtained by keeping only the letters i and $\overline{i}$. So to obtain a bijection between the set of distinguished tableaux of shape λ and the subset $T_{C_\infty}^0(\lambda) \subseteq T_{C_\infty}(\lambda)$ of zero weight tableaux, it suffices to describe a bijection between the set of reverse lattice tableaux on $\{2i-1, 2i\}$ of given shape and weight in $2\omega_i\mathbb{Z}_{\geqslant 0}$, and the set of skew tableaux on $\{i, \overline{i}\}$ with weight 0. Now recall that we have the structure of a $U_q(\mathfrak{sl}_2)$-crystal on the set of all skew semistandard tableaux of fixed skew shape both on $\{2i-1, 2i\}$ and $\{i, \bar{\imath}\}$. By replacing each letter $2i-1$ by i and each letter $2i$ by $\bar{\imath}$, we get a crystal isomorphism f. The distinguished tableaux correspond to the highest weight vertices of weight in $2\omega_i\mathbb{Z}_{\geqslant 0}$ for the $\{2i-1, 2i\}$-structure, whereas the tableaux of weight 0 give the vertices of weight 0 in the $\{i, \bar{\imath}\}$-crystal structure. By observing that only $U_q(\mathfrak{sl}_2)$-crystals with highest weight in $2\omega_i\mathbb{Z}_{\geqslant 0}$ admit a vertex of weight 0, which is then unique, we obtain that the map $\mathcal{C}$ which associates to each zero weight vertex in the $\{i, \bar{\imath}\}$-crystal structure its highest weight vertex in the $\{2i-1, 2i\}$-crystal structure is the bijection we need. More precisely, the map $\mathcal{C}$ (resp. its inverse) is obtained as usual: we start by encoding in the reading of each $\{i, \bar{\imath}\}$-tableau (resp. of each $\{2i-1, 2i\}$-tableau) the letters i by $+$ and the letters $\bar{\imath}$ by $-$ (resp. the letters $2i-1$ by $+$ and the letters $2i-1$ by $-$), and next by recursively deleting all the factors $+-$, thus obtaining a reduced word of the form $-^m+^m$ (resp. $+^{2m}$). It then suffices to change the m letters $\bar{\imath}$ corresponding to the m surviving symbols $-$ into i and to apply the isomorphism f^{-1} (resp. change m letters $2i-1$ corresponding to the rightmost m surviving symbols $+$ into $2i$ and apply the isomorphism f).

Example 3.6 The skew tableau of weight 0 on $\{2, \bar{2}\}$ corresponding to *(3.1)* is

$$
\begin{array}{ccccccccccccc}
 & & & & & & & & & 2 & 2 & 2 & \bar{2} \\
 & & & & & 2 & \bar{2} & \bar{2} & \bar{2} & \bar{2} & & & \\
 & & 2 & 2 & \bar{2} & & & & & & & & \\
2 & 2 & \bar{2} & \bar{2} & & & & & & & & &
\end{array}
$$

□

In the sequel, we will abuse the notation and identify the two crystal structures corresponding up to the isomorphism f.

Example 3.7 Assume $\lambda = (1, 1)$. Then we get

$$T^0_{C_\infty}(\lambda) = \left\{ \begin{array}{|c|}\hline k \\ \hline \bar{k} \\ \hline \end{array} \mid k \in \mathbb{Z}_{\geq 1} \right\} \quad \text{and} \quad K^0_{C_\infty}(\lambda) = \left\{ \begin{array}{|c|}\hline k \\ \hline \bar{k} \\ \hline \end{array} \mid k \in \mathbb{Z}_{\geqslant 2} \right\}.$$

This gives

$$H\left(\begin{array}{|c|}\hline k \\ \hline \bar{k} \\ \hline \end{array} \right) = \begin{array}{|c|}\hline 2k-1 \\ \hline 2k \\ \hline \end{array} \quad \text{and} \quad \varphi\left(\begin{array}{|c|}\hline 2k-1 \\ \hline 2k \\ \hline \end{array} \right) = \omega_{2k} \text{ for any } k \geqslant 1.$$

Therefore

$$\theta\left(\begin{array}{|c|}\hline 2k-1 \\ \hline 2k \\ \hline \end{array} \right) = 2\omega_{2k} \text{ for any } k \geqslant 1.$$

Finally $K^{C_\infty}_{\lambda,0}(t) = \sum_{k\geqslant 1} t^{2k} = \frac{t^2}{1-t^2}$. □

4 Type C_n Generalized Exponents via the Kwon Model

In this section, we refine the results in Sect. 3.4 to the finite type C_n, based on Kwon's model for the corresponding branching coefficients [14, 15]. We also need to use a combinatorial map realizing the conjugation symmetry of Littlewood–Richardson coefficients. It turns out that Kwon's model, the version of the conjugation symmetry map used here, and the distinguished tableaux in Sect. 3.3 fit together in a beautiful way. This allows us to express the related statistic in terms of a natural combinatorial labeling of the vertices of weight 0 in the corresponding type C_n crystal of highest weight λ, namely the corresponding tableaux due to King [12].

Given a King tableau T of weight 0, we replace the entries i and $\bar{\imath}$ with $2i-1$ and $2i$, respectively. Then we map the resulting filling to the lowest weight element with respect to the corresponding $U_q(\mathfrak{sl}_2 \oplus \ldots \oplus \mathfrak{sl}_2)$-crystal structure. We denote this map by $T \mapsto L(T)$. The tableaux $L(T)$ is not distinguished but its image under the Schützenberger involution is. In fact one can reformulate the results of Sect. 3.3 in terms of the image of the distinguished vertices under the Schützenberger involution. This flipped version has the advantage to match well with the definition of King tableaux.

Theorem 4.1 *We have*

$$K_{\lambda,0}^{C_n}(t) = \sum_{T \in K_{C_n}^{0}(\lambda)} t^{\mathrm{ch}_{C_n}(L(T))},$$

where

$$\mathrm{ch}_{C_n}(L(T)) = \sum_{i=1}^{2n-1} (2n-i) \left\lceil \frac{\varepsilon_i(L(T))}{2} \right\rceil.$$

Remarks 4.2

(1) There does not seem to be a simple way to express the related statistic above directly in terms of T. However, the map $T \mapsto L(T)$ is a simple one.
(2) Theorem 4.1 gives a statistic for computing the Kostka–Foulkes polynomial on King tableaux, rather than on the Kashiwara–Nakashima (KN). A natural question is whether the statistic above can be translated to the KN tableaux and moreover if it is related to the charge statistic constructed in [22] (which conjecturally computes the Kostka–Foulkes polynomials). This question is addressed in [3]. □

We will now continue Example 3.7.

Example 4.3 Assume $\lambda = (1, 1)$ in type C_n. Then we get

$$K^0_{C_n}(\lambda) = \left\{ \begin{array}{|c|}\hline k \\ \hline \overline{k} \\ \hline \end{array} \mid k = 2, \ldots, n \right\}.$$

This gives

$$L\left(\begin{array}{|c|}\hline k \\ \hline \overline{k} \\ \hline \end{array} \right) = \begin{array}{|c|}\hline 2k-1 \\ \hline 2k \\ \hline \end{array} \quad \text{and} \quad \boldsymbol{\varepsilon}^*\left(\begin{array}{|c|}\hline 2k-1 \\ \hline 2k \\ \hline \end{array} \right) = \omega_{2(n-k+1)} \text{ for any } k = 2, \ldots, n.$$

Therefore

$$\boldsymbol{\theta}^*_n\left(\begin{array}{|c|}\hline 2k-1 \\ \hline 2k \\ \hline \end{array} \right) = 2\omega_{2(n-k+1)} \text{ for any } k = 2, \ldots, n.$$

Finally

$$K^{C_n}_{\lambda,0}(t) = \sum_{k=1}^{n-1} t^{2k} = \frac{t^2 - t^{2n}}{1 - t^2}.$$

□

5 Three Applications of Theorem 4.1

5.1 Growth of Generalized Exponents

First we analyze the growth of the generalized exponents of type C_n with respect to the rank n. The (weight 0) symplectic King tableaux of type C_n embed into those of type C_{n+1} by changing the entries $k, \overline{k}$ to $k + 1, \overline{k+1}$, for all k, respectively. Moreover, it is easy to see that this map preserves the statistic in Theorem 4.1. So we obtain the following result, which to our knowledge is new.

Theorem 5.1 *For any integer n and any partition λ with at most n parts, we have*

$$K^{C_{n+1}}_{\lambda,0}(t) - K^{C_n}_{\lambda,0}(t) \in \mathbb{Z}_{\geqslant 0}[t].$$

5.2 *Reducing a Type C Generalized Exponent to One of Type A*

We now prove a conjecture of the first author [22]. This conjecture is the first step in the construction of the type C_n charge statistic in [22], and proves the conjecture that this charge computes the corresponding Kostka–Foulkes polynomials in the case of column shapes; see Remark 4.2 (2). We now label the Dynkin diagram of type C_n such that the special node is n. Consider the fundamental weight ω_{2p}, where $p \in \{1, \dots, \lfloor n/2 \rfloor\}$. All the zero weight vertices in the crystal $B(\omega_{2p})$ belong to the same type A_{n-1} component, which has highest weight $\gamma_p := \varepsilon_1 + \dots + \varepsilon_p - \varepsilon_{n-p+1} - \dots - \varepsilon_n$, where ε_i are the coordinate vectors in $\mathbb{R}^n$. In type A_{n-1}, this weight corresponds to the partition $(1^{n-2p}, 2^p)$.

Theorem 5.2 *We have*

$$K^{C_n}_{\omega_{2p},0}(t) = K^{A_{n-1}}_{\gamma_p,0}(t^2)\,.$$

Remark 5.3 *Theorem 5.2 also permits to establish the conjecture of [22] for Lusztig t-analogues associated with any fundamental weight.* □

5.3 *The Smallest Power of t in* $K^{C_n}_{\lambda,0}(t)$

The largest power of t in $K^{C_n}_{\lambda,0}(t)$ is well-known to be $\langle \lambda, \rho^\vee \rangle$, where $\rho^\vee$ is half the sum of the positive coroots. Furthermore, it is also known that the smallest power is greater than or equal to $|\lambda|/2$, see [23]. As the third application of our formula for $K^{C_n}_{\lambda,0}(t)$, we will determine this smallest power. Let $\lambda \in \mathcal{P}_n$ be such that $|\lambda|$ is even, and write $\lambda = \sum_{i=1}^n a_i\, \omega_{n+1-i}$. Define

$$s_k := \sum_{i=1}^{k} a_i\,, \qquad b_i := \begin{cases} a_i + 1 & \text{if } a_i \text{ odd and } s_i \text{ odd} \\ a_i - 1 & \text{if } a_i \text{ odd and } s_i \text{ even} \\ a_i & \text{if } a_i \text{ even}\,. \end{cases}$$

Also let $s_0 := 0$ and $S := s_n$.

Theorem 5.4 *The smallest power of t in* $K^{C_n}_{\lambda,0}(t)$, *for* $|\lambda|$ *even, is*

$$\frac{1}{2}\sum_{i=1}^{n}(n+1-i)b_i = \frac{|\lambda|}{2} + \frac{1}{2}\sum_{i\,:\,a_i \text{ odd}} (-1)^{s_i-1}(n+1-i)\,. \tag{5.1}$$

6 The Atomic Decomposition: Definitions and Basic Facts

6.1 *Characters and t-Deformations*

Recall that the dominance order $\leqslant$ on P^+ is defined by $\alpha < \beta$ if and only if $\beta - \alpha$ decomposes as a sum of positive roots (or equivalently, simple roots) with nonnegative integer coefficients.

Let χ_λ be the character of the finite-dimensional irreducible representation $V(\lambda)$ of $\mathfrak{g}$ with highest weight $\lambda \in P^+$. Let $K_{\lambda,\gamma}$ denote the multiplicity of the weight γ in $V(\lambda)$. Let $P(\lambda)$ be the set of weights of $V(\lambda)$, i.e., the set of γ such that $K_{\lambda,\gamma} > 0$. Set $P^+(\lambda) = P(\lambda) \cap P^+$, and note that $P^+(\lambda) = \{\mu \in P^+ \mid \mu \leqslant \lambda\}$. Since $K_{\lambda,\gamma} = K_{\lambda,w(\gamma)}$ for any $w \in W$, the character χ_λ is determined by its dominant part

$$\chi_\lambda^+ := \sum_{\mu \in P^+(\lambda)} K_{\lambda,\mu}\, e^\mu .$$

Lusztig [26] defined a remarkable t-analogue $K_{\lambda,\mu}(t)$ of $K_{\lambda,\mu}$ (a polynomial in t satisfying $K_{\lambda,\mu}(1) = K_{\lambda,\mu}$) by introducing a variable t in the Weyl character formula for χ_λ:

$$\frac{\sum_{w \in W} (-1)^{\ell(w)} e^{w(\lambda+\rho)-\rho}}{\prod_{\alpha \in R^+} (1 - t e^{-\alpha})} = \sum_{\gamma \in P^+(\lambda)} K_{\lambda,\gamma}(t)\, e^\mu ; \tag{6.1}$$

here ρ is the half sum of positive roots. We define the t-analogue of χ_λ^+ by

$$\chi_\lambda^+(t) := \sum_{\mu \in P^+(\lambda)} K_{\lambda,\mu}(t)\, e^\mu . \tag{6.2}$$

When $\gamma = \mu$ is dominant, the polynomial $K_{\lambda,\mu}(t)$ is known as a *Kostka–Foulkes polynomial.*

This polynomial has remarkable properties, such as being an *affine Kazhdan–Lusztig polynomial*, which implies that it has nonnegative integer coefficients. More precisely,

$$K_{\lambda,\mu}(t) = t^{\langle \lambda-\mu, \rho^\vee \rangle} P_{w_\mu, w_\lambda}(t^{-1}) , \tag{6.3}$$

where w_λ denotes the longest element of $W t_\lambda W$, and t_λ is the translation by λ in the extended affine Weyl group [11]; note that $\langle \lambda - \mu, \rho^\vee \rangle$ is the number of simple roots in the decomposition of $\lambda - \mu$, counted with multiplicity. Based on (6.3), we let

$$\widetilde{K}_{\lambda,\mu}(t) := t^{\langle \lambda-\mu, \rho^\vee \rangle} K_{\lambda,\mu}(t^{-1}) , \qquad \text{so } \widetilde{K}_{\lambda,\mu}(t) = P_{w_\mu, w_\lambda}(t) . \tag{6.4}$$

To each irreducible representation $V(\lambda)$ is associated with an abstract *Kashiwara crystal* $B(\lambda)$ [10]. This is a colored directed graph which encodes the action of certain modified versions of the Chevalley generators, upon passing to the quantum group of $\mathfrak{g}$ and letting the quantum parameter go to 0. We have

$$\chi_\lambda = \sum_{b \in B(\lambda)} e^{\mathrm{wt}(b)}\,, \tag{6.5}$$

where $\mathrm{wt}(b)$ is the weight of the vertex $b \in B(\lambda)$.

6.2 The Definition of the Atomic Decomposition

For any dominant weight μ, define the *layer sum polynomials* by

$$w_\mu := \sum_{\gamma \in P(\mu)} e^\gamma \quad \text{and} \quad w_\mu^+ := \sum_{\nu \in P^+(\mu)} e^\nu = \sum_{\nu \leqslant \mu} e^\nu\,. \tag{6.6}$$

Similarly, define

$$w_\mu^+(t) := \sum_{\nu \in P^+(\mu)} t^{\langle \mu-\nu, \rho^\vee \rangle} e^\nu = \sum_{\nu \leqslant \mu} t^{\langle \mu-\nu, \rho^\vee \rangle} e^\nu\,. \tag{6.7}$$

Consider the expansion

$$\chi_\lambda = \sum_{\mu \in P^+(\lambda)} A_{\lambda,\mu}\, w_\mu\,, \quad \text{or equivalently } \chi_\lambda^+ = \sum_{\mu \in P^+(\lambda)} A_{\lambda,\mu}\, w_\mu^+\,. \tag{6.8}$$

Similarly, consider the polynomials $A_{\lambda,\mu}(t)$ defined by

$$\chi_\lambda^+(t) = \sum_{\mu \in P^+(\lambda)} A_{\lambda,\mu}(t)\, w_\mu^+(t)\,. \tag{6.9}$$

We have $A_{\lambda,\lambda}(t) = 1$. It is not hard to prove that the expansion (6.9) is equivalent to

$$K_{\lambda,\nu}(t) = \sum_{\nu \leqslant \mu \leqslant \lambda} t^{\langle \mu-\nu, \rho^\vee \rangle} A_{\lambda,\mu}(t)\,, \qquad \text{for all } \nu \in P^+(\lambda)\,. \tag{6.10}$$

Definition 6.2.1 The character χ_λ admits an *atomic decomposition* if $A_{\lambda,\mu} \in \mathbb{Z}_{\geqslant 0}$. Similarly, we say that $\chi_\lambda^+(t)$ admits a t-atomic decomposition if $A_{\lambda,\mu}(t) \in \mathbb{Z}_{\geqslant 0}[t]$. These polynomials are called *atomic polynomials*.

We state a property equivalent to the t-atomic decomposition. To this end, by analogy with (6.2) and using (6.4), we define

$$\widetilde{\chi}_\lambda^+(t) := \sum_{\mu \in P^+(\lambda)} \widetilde{K}_{\lambda,\mu}(t)\, e^\mu\,.$$

Like in (6.9), consider the polynomials $\widetilde{A}_{\lambda,\mu}(t)$ defined by

$$\widetilde{\chi}_\lambda^+(t) = \sum_{\mu \in P^+(\lambda)} \widetilde{A}_{\lambda,\mu}(t)\, w_\mu^+\,, \tag{6.11}$$

where we recall that $w_\mu^+ := w_\mu^+(1)$. Like above, we can then prove that the expansion (6.11) is equivalent to

$$\widetilde{K}_{\lambda,\nu}(t) = \sum_{\nu \leqslant \mu \leqslant \lambda} \widetilde{A}_{\lambda,\mu}(t)\,, \qquad \text{for all } \nu \in P^+(\lambda)\,. \tag{6.12}$$

Proposition 6.1 *The polynomials $A_{\lambda,\mu}(t)$ and $\widetilde{A}_{\lambda,\mu}(t)$ satisfy*

$$\widetilde{A}_{\lambda,\mu}(t) = t^{\langle \lambda-\mu, \rho^\vee\rangle} A_{\lambda,\mu}(t^{-1})\,.$$

Thus, the t-atomic decomposition is equivalent to the fact that $\widetilde{A}_{\lambda,\mu}(t) \in \mathbb{Z}_{\geqslant 0}[t]$. □

Remarks 6.2

(1) Lascoux [18] stated the atomic decomposition in type A as in (6.10). However, there is a slight difference in the definition of $\widetilde{K}_{\lambda,\mu}(t)$, for given partitions λ, μ; namely, Lascoux defines $\widetilde{K}_{\lambda,\mu}(t) := t^{n(\mu)}\, K_{\lambda,\mu}(t^{-1})$, where $n(\mu) := \sum_i (i-1)\mu_i$.

(2) The t-atomic decomposition, as stated in Definition 6.2.1, implies the monotonicity of the Kostka–Foulkes polynomials, which holds in the full generality of Kazhdan–Lusztig polynomials for finite and affine Weyl groups [1, Corollary 3.7], cf. (6.4). Indeed, this property says that, for $x \leqslant y \leqslant z$ in such a Weyl group, the difference of Kazhdan–Lusztig polynomials $P_{x,z}(t) - P_{y,z}(t)$ is in $\mathbb{Z}_{\geqslant 0}[t]$.

(3) We will give a simpler, conceptual proof of the t-atomic decomposition in type A, cf. Sect. 9. However, even the atomic decomposition (i.e., the positivity in (6.8)) might fail beyond type A; but this failure seems limited to small ranks, as explained below. □

6.3 Atomic Decomposition of Finite Crystals

Let $B(\lambda)^+$ be the subset of $B(\lambda)$ of vertices with dominant weights.

Definition 6.3.1 An *atomic decomposition* of the crystal $B(\lambda)$ is a partition

$$B(\lambda)^+ = \bigsqcup_{h \in H(\lambda)} \mathbb{B}(\lambda, h)\,, \tag{6.13}$$

where $H(\lambda) \subseteq B(\lambda)^+$, $h \in \mathbb{B}(\lambda, h)$ is a distinguished vertex, and each component $\mathbb{B}(\lambda, h)$ consists of exactly one vertex of weight ν for each $\nu \leqslant \mathrm{wt}(h)$.

An atomic decomposition of $B(\lambda)$ clearly gives the atomic decomposition (6.8) of χ_λ^+, where $A_{\lambda,\mu}$ is the number of vertices of weight μ in $H(\lambda)$.

Definition 6.3.2 A *t-atomic decomposition* of the crystal $B(\lambda)$ is an atomic decomposition together with a statistic $\mathrm{c} : H(\lambda) \to \mathbb{Z}_{\geqslant 0}$ such that the following polynomials satisfy (6.9):

$$A_{\lambda,\mu}(t) = \sum_{\substack{h \in H(\lambda) \\ \mathrm{wt}(h)=\mu}} t^{\mathrm{c}(h)}\,. \tag{6.14}$$

As we see, the t-atomic decomposition of $\chi_\lambda^+(t)$ is part of Definition 6.3.2. Assuming that $B(\lambda)$ has a t-atomic decomposition, one can extend the statistic c to $B(\lambda)^+$ by setting

$$\mathrm{c}(b) := \mathrm{c}(h) + \langle \mathrm{wt}(h) - \mathrm{wt}(b), \rho^\vee \rangle\,, \qquad \text{for any } b \in \mathbb{B}(\lambda, h)\,. \tag{6.15}$$

The t-analogue of the combinatorial formula (6.2) follows from Definition 6.3.2:

$$\chi_\lambda^+(t) = \sum_{b \in B(\lambda)^+} t^{\mathrm{c}(b)} e^{\mathrm{wt}(b)}\,. \tag{6.16}$$

Moreover, by comparing (6.16) with (6.2), we obtain the following combinatorial formula for Kostka–Foulkes polynomials:

$$K_{\lambda,\mu}(t) = \sum_{\substack{b \in B(\lambda) \\ \mathrm{wt}(b)=\mu}} t^{\mathrm{c}(b)}\,. \tag{6.17}$$

To summarize, the existence of a t-atomic decomposition of a crystal is highly desirable because: (1) it gives the t-atomic decomposition of $\chi_\lambda^+(t)$ and of $K_{\lambda,\mu}(t)$, which are now realized combinatorially; (2) it leads to combinatorial formulas for both $K_{\lambda,\mu}(t)$ and the atomic polynomials $A_{\lambda,\mu}(t)$, namely (6.17) and (6.14), respectively.

7 The Partial Order on Dominant Weights

Before we consider the atomic decomposition of finite crystals, we need some information about the dominant weight poset defined in Sect. 6.1. In full generality, this poset was first studied in [32], so we will recall some results from this paper.

The components of the dominant weight poset are lattices. Each cocover is of the form $\mu > \mu - \alpha$, where α is a positive root, so we represent it as a downward edge in the Hasse diagram labeled by α. The cocovers were completely described in [32, Theorem 2.8]. Fixing a dominant weight λ, we will consider the lower order ideal determined by λ. This is an interval $[\widehat{0}, \lambda]$, with $\widehat{0}$ a minimal element of the dominant weight poset.

7.1 Type A_{n-1}

Now λ is a partition $(\lambda_1 \geqslant \ldots \geqslant \lambda_{n-1} \geqslant 0)$, and let $N = |\lambda| := \sum_i \lambda_i$. We denote a partition with p parts a, q parts b ($a \geqslant b$), etc. by $(a^p b^q \ldots)$. In the interval $[\widehat{0}, \lambda]$ mentioned above, $\widehat{0}$ is the partition $\omega_p = (1^p)$, where $p := N \bmod n$ and $N := \sum \lambda_i$. The cocovers $\mu > \mu - \alpha_{ij}$, where $\alpha_{ij} = \varepsilon_i - \varepsilon_j$ is a positive root (i.e., $i < j$), are labeled by (i, j). It turns out that there are only two types of cocovers, namely:

$$(\ldots ab \ldots) > (\ldots (a-1)(b+1) \ldots), \quad \text{and} \quad (\ldots (a+1)a^p(a-1) \ldots) > (\ldots a^{p+2} \ldots), \tag{7.1}$$

where in the first case $a \geqslant b + 2$ and the cocover is labeled by a simple root. These types are referred to as (*) and (**), respectively, while a cocover of type (**) which is not of type (*) is called *proper*.

An important result in [2] concerns the structure of short intervals in the dominance order. To state it, we need some more definitions. Consider two distinct cocovers $\mu > \nu$ and $\mu > \pi$ of a partition μ, which are labeled (i, j) and (k, l), where we assume $i < k$. These cocovers can only have one of the following relative positions (in terms of their labels): (1) *nonoverlapping* if $j < k$; (2) *partially overlapping* if $j = k$; (3) *fully overlapping* if $k = j - 1$. By [2, Proposition 3.2], the interval $[\nu \wedge \pi, \mu]$ can only have one of the following structures; the two cocovers above are shown in the diagrams below in bold.

Case A1: cocovers which are (a) nonoverlapping; (b) partially overlapping and both of type (*); (c) fully overlapping and both proper of type (**). As subcase (a) is easy, only subcases (b) and (c) are represented in the diagrams below.
In subcase (b), we have $a \geqslant c + 2$ and $c \geqslant e + 2$, while i is the position of a in μ.

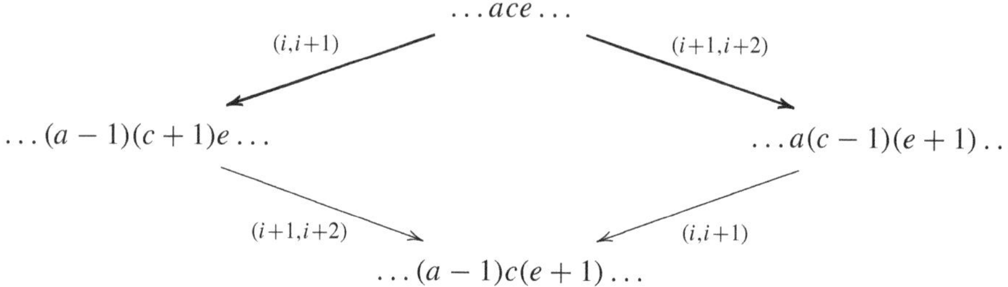

In subcase (c), we have $b = a - 1$, $c = b - 1$, $d = c - 1$, $p, q \geq 1$, while i is the position of a, $j = i + p + 1$ is the position of the first c, and $k = j + q$ is the position of d.

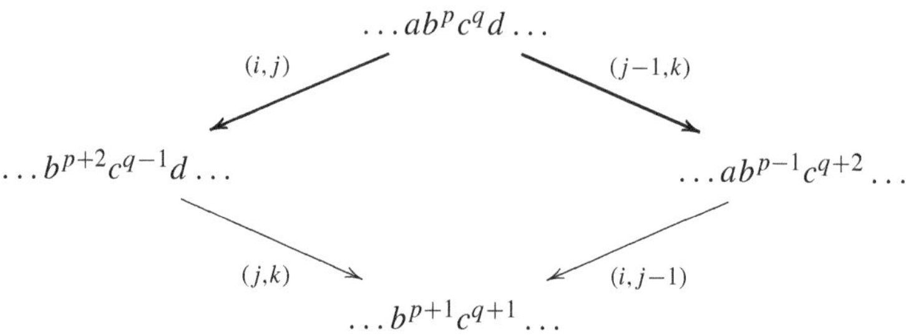

Case A2: partially overlapping cocovers, where (a) the first is of type (*) and the second proper of type (**); (b) vice versa.
In subcase (a), we have $a \geqslant c + 2$, $d = c - 1$, $e = d - 1$, $p \geqslant 1$, while i is the position of a in the partition μ and $j = i + p + 2$ is the position of e.

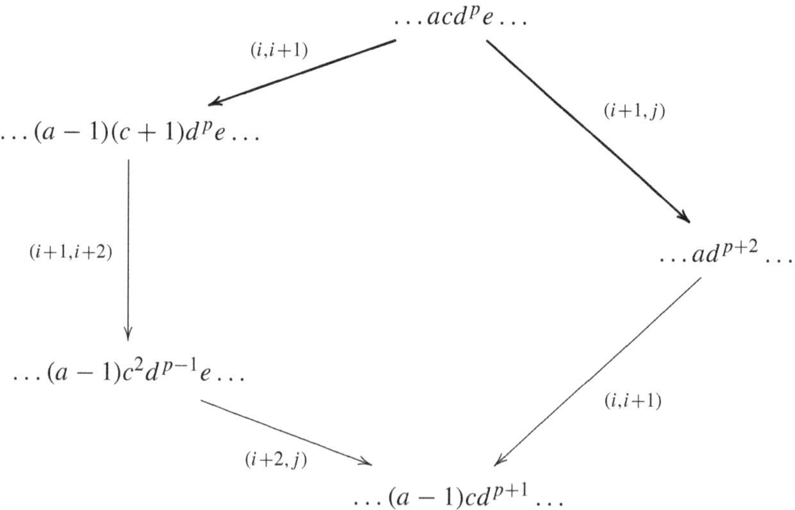

In subcase (b), we have a similar pentagon.

Case A3: partially overlapping cocovers, both proper of type (**). Here $b = a - 1, c = b - 1, d = c - 1, e = d - 1, p, q \geqslant 1$, while i is the position of a in the partition μ, $j = i + p + 1$ is the position of c, and $k = j + q + 1$ is the position of e.

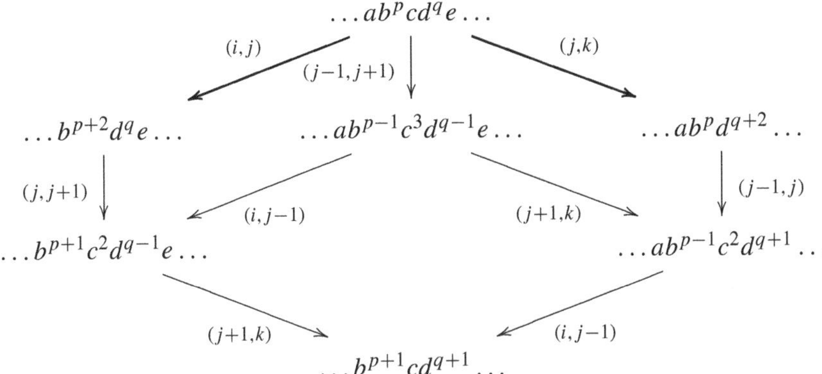

Given two distinct covers $\mu \lessdot \nu$ and $\mu \lessdot \pi$ of μ, the isomorphism type of the interval $[\mu, \nu \vee \pi]$ is always given by one of the above graphs turned upside down.

7.2 *Types B_n, C_n, and D_n*

We start with type C_n. Now λ is a partition $(\lambda_1 \geqslant \ldots \geqslant \lambda_n \geqslant 0)$. It is easy to see that the minimal element $\widehat{0}$ mentioned above (i.e., the unique minimal element below λ) is either 0 or $\omega_1 = (10^{n-1})$, depending on $|\lambda|$ being even or odd, respectively.

Proposition 7.1 *If $n > (|\lambda| + 1)/2$, then the cocover $(\ldots 1^2 0^k) \gtrdot (\ldots 0^{k+2})$ is the only one which can appear in the Hasse diagram of the interval $[\widehat{0}, \lambda]$ beside the type A cocovers in* (7.1). □

We turn to type B_n, and assume that λ is a partition $(\lambda_1 \geqslant \ldots \geqslant \lambda_n \geqslant 0)$, where $\lambda_i \in \mathbb{Z}$. The unique minimal element below λ is clearly always 0.

Proposition 7.2 *If $n > |\lambda|/2$, then the cocover $(\ldots 10^{n-k}) \gtrdot (\ldots 0^{n-k+1})$ is the only one which can appear in the Hasse diagram of the interval $[\widehat{0}, \lambda]$ beside the type A cocovers in* (7.1). □

We conclude with type D_n. Now λ is a sequence $(\lambda_1 \geqslant \ldots \geqslant \lambda_n)$ with $\lambda_i \in \frac{1}{2}\mathbb{Z}$, all congruent mod $\mathbb{Z}$, such that $\lambda_{n-1} + \lambda_n \geqslant 0$. We will now assume that $\lambda_i \in \mathbb{Z}$. This implies that the interval $[\widehat{0}, \lambda]$ only contains weights $\mu = (\mu_1 \geqslant \ldots \geqslant \mu_n)$ with $\mu_i \in \mathbb{Z}$. Note that, in this case, there are the same possibilities for the minimal element $\widehat{0}$ as in type C.

Proposition 7.3 *If $n > |\lambda|$, then the cocover $(\dots 1^2 0^k) \gtrdot (\dots 0^{k+2})$ is the only one which can appear in the Hasse diagram of the interval $[\widehat{0}, \lambda]$ beside the type A cocovers in* (7.1). □

We will work under the assumptions of Propositions 7.1, 7.2, and 7.3, and we call this the *stable range*. In each classical type, we derived a classification of the short intervals, consisting of the same cases as in type A, and a few extra ones involving the new cover.

8 Modified Crystal Operators on Classical Crystals

Now consider a classical Lie algebra, with Dynkin diagram labeled in the standard way.

8.1 Definition of the Modified Crystal Operators

Given a positive root α, consider the shortest length Weyl group element satisfying $w(\alpha_1) = \alpha$. We define the modified crystal operators f_α and e_α as the conjugations

$$\mathrm{f}_\alpha := w\tilde{f}_1 w^{-1}, \qquad \mathrm{e}_\alpha := w\tilde{e}_1 w^{-1} \tag{8.1}$$

of the ordinary crystal operators $\tilde{f}_1$ and $\tilde{e}_1$ by the Kashiwara action of w on $B(\lambda)$ [10]. This means that $\mathrm{f}_\alpha(b) = 0$ precisely when $\tilde{f}_1$ applied to $w^{-1}(b)$ is 0. Clearly, f_α and e_α are inverses to one another. Moreover, for any $b \in B(\lambda)$, we have $\mathrm{wt}(\mathrm{f}_\alpha(b)) = \mathrm{wt}(b) - \alpha$. We endow the vertices of $B(\lambda)$ with the structure of a colored directed graph $\mathbb{B}(\lambda)$ with edges $b \overset{\alpha}{\dashrightarrow} b'$ when $b' = \mathrm{f}_\alpha(b)$. The graph $\mathbb{B}(\lambda)$ is different from the Kashiwara crystal $B(\lambda)$ and, unlike the latter, the former is not connected in general.

In type B_n, in addition to the operators f_α for $\alpha \in W\alpha_1$, i.e., a long root, we need such operators indexed by short roots. They are defined completely similarly to (8.1). Namely, given a short root α, consider *any* $w \in W$ satisfying $w(\alpha_n) = \alpha$. We then define

$$\mathrm{f}_\alpha := w\tilde{f}_n w^{-1}, \qquad \mathrm{e}_\alpha := w\tilde{e}_n w^{-1}. \tag{8.2}$$

Note that, in this case, the definition does not depend on the choice of w. All the basic properties of the modified crystal operators indexed by long roots extend to the additional operators. Moreover, to the modified crystal graph $\mathbb{B}(\lambda)$ constructed before, we add the extra edges $b \overset{\alpha}{\dashrightarrow} b'$ when $b' = \mathrm{f}_\alpha(b)$ and α is a short positive root.

8.2 Properties of the Modified Crystal Operators

We start with some properties of the modified crystal operators indexed by roots $\alpha \in W\alpha_1$, in any classical type.

Lemma 8.1 *For $b \in \mathbb{B}(\lambda)$ and a positive root $\alpha \in W\alpha_1$, if $\langle \mathrm{wt}(b), \alpha \rangle > 0$, then $\mathrm{f}_\alpha(b) \neq 0$.* □

Theorem 8.2 *Consider two positive roots α and β in $W\alpha_1$ and a vertex b in $\mathbb{B}(\lambda)$ such that $\langle \mathrm{wt}(b), \alpha \rangle > 0$ and $\langle \mathrm{wt}(b), \beta \rangle > 0$.*

(a) Assume that (α, β) satisfies: (i) it is $(\varepsilon_i - \varepsilon_j,\ \varepsilon_j \pm \varepsilon_k)$ or $(\varepsilon_j \pm \varepsilon_k,\ \varepsilon_i - \varepsilon_j)$, for $i < j < k$; (ii) it is $(\varepsilon_{j-1} + \varepsilon_j,\ \varepsilon_i - \varepsilon_j)$ for $i < j-1$, and $\langle \mathrm{wt}(b) - \beta, \varepsilon_{j-1} - \varepsilon_j \rangle = 0$. Then we have $\mathrm{f}_\alpha \mathrm{f}_\beta(b) = \mathrm{f}_{\alpha+\beta}(b) \neq 0$.
(b) Assume that the pair (α, β) is in the W-orbit of (α_1, α_3). Then $\mathrm{f}_\alpha \mathrm{f}_\beta(b) = \mathrm{f}_\beta \mathrm{f}_\alpha(b) \neq 0$. □

We have an analogous result to Theorem 8.2 for the e. operators.

Theorem 8.3 *Consider two positive roots α and β in $W\alpha_1$ and a vertex b in $\mathbb{B}(\lambda)$ such that $\langle \mathrm{wt}(b), \alpha \rangle \geqslant 0$ and $\langle \mathrm{wt}(b), \beta \rangle \geqslant 0$. Assume also that $\mathrm{e}_\alpha(b) \neq 0$ and $\mathrm{e}_\beta(b) \neq 0$.*

(a) Assume that (α, β) satisfies: (i) it is $(\varepsilon_i - \varepsilon_j,\ \varepsilon_j \pm \varepsilon_k)$ or $(\varepsilon_j \pm \varepsilon_k,\ \varepsilon_i - \varepsilon_j)$, for $i < j < k$; (ii) it is $(\varepsilon_i - \varepsilon_j,\ \varepsilon_{j-1} + \varepsilon_j)$ for $i < j-1$, and $\langle \mathrm{wt}(b), \varepsilon_{j-1} - \varepsilon_j \rangle = 0$. Then we have $\mathrm{e}_\alpha \mathrm{e}_\beta(b) = \mathrm{e}_{\alpha+\beta}(b) \neq 0$.
(b) Assume that the pair (α, β) is in the W-orbit of (α_1, α_3), and w is a shortest length element satisfying $w(\alpha_1, \alpha_3) = (\alpha, \beta)$. Let $\gamma := w(\alpha_2)$, and also assume that $\langle \mathrm{wt}(b), \gamma \rangle > 0$. Then $\mathrm{e}_\alpha \mathrm{e}_\beta(b) = \mathrm{e}_\beta \mathrm{e}_\alpha(b) \neq 0$. □

We conclude with some properties of the modified crystal operators f_α indexed by short roots α, in type B_n; we will not need analogues of these properties for the operators e_α.

Theorem 8.4

(a) Given $i < j < k$, assume that $\mathrm{f}_{\varepsilon_i - \varepsilon_j}(b) \neq 0$ and $\mathrm{f}_{\varepsilon_k}(b) \neq 0$. Then we have

$$\mathrm{f}_{\varepsilon_i - \varepsilon_j} \mathrm{f}_{\varepsilon_k}(b) = \mathrm{f}_{\varepsilon_k} \mathrm{f}_{\varepsilon_i - \varepsilon_j}(b) \neq 0 .$$

The same is true with the f. operators replaced by the e. operators.
(b) Consider a vertex b with $\langle \mathrm{wt}(b), \varepsilon_{j-1} \rangle = 1$ and $\langle \mathrm{wt}(b), \varepsilon_j \rangle = 0$. Then we have

$$\mathrm{f}_{\varepsilon_j} \mathrm{f}_{\varepsilon_{j-1} - \varepsilon_j}(b) = \mathrm{f}_{\varepsilon_{j-1}}(b) \neq 0 .$$

(c) Assume that, in addition to the conditions in (1), we have $i < j-1$ and $\langle \mathrm{wt}(b), \varepsilon_i \rangle > 1$. Then we have

$$\mathrm{f}_{\varepsilon_j} \mathrm{f}_{\varepsilon_i - \varepsilon_j}(b) = \mathrm{f}_{\varepsilon_i - \varepsilon_{j-1}} \mathrm{f}_{\varepsilon_{j-1}}(b) \neq 0 .$$

9 Atomic Decomposition of Crystals in Classical Types

Fix a dominant weight λ for a classical Lie algebra. Consider the subgraph of $\mathbb{B}(\lambda)$ consisting of the vertices of dominant weight, and the edges $b \overset{\alpha}{\dashrightarrow} \mathrm{f}_\alpha(b)$ for which $\mathrm{wt}(b) \gtrdot \mathrm{wt}(\mathrm{f}_\alpha(b))$ is a cocover in the dominant weight poset. This new colored directed graph on the vertices of $B(\lambda)^+$ will be denoted by $\mathbb{B}(\lambda)^+$. It can also be viewed as a poset (with cocovers given by the above edges), and the weight function is a poset projection to the interval $[\widehat{0}, \lambda]$ in the dominant weight poset. The two points of view will be used interchangeably.

The main goal is to identify situations in which the components of the poset $\mathbb{B}(\lambda)^+$ define an atomic, respectively, t-atomic decomposition, cf. Definitions 6.3.1 and 6.3.2.

9.1 Type A_{n-1}

Lemma 9.1

(a) *Consider two distinct edges $b \dashrightarrow b'$ and $b \dashrightarrow b''$ in $\mathbb{B}(\lambda)^+$. The vertices b' and b'' have a lower bound in this poset.*

(b) *Consider two distinct edges $b' \dashrightarrow b$ and $b'' \dashrightarrow b$ in $\mathbb{B}(\lambda)^+$. The vertices b' and b'' have an upper bound in this poset.* □

The proof of this lemma relies on the structure of the short intervals in the dominant weight poset, discussed in Sect. 7.1. More precisely, we consider one by one all the types of short intervals, and for each of them, we show that we obtain the same structure in the poset $\mathbb{B}(\lambda)^+$. This is achieved by using the commutation relations between the modified crystal operators discussed in Sect. 8.1, namely Theorems 8.2 and 8.3

Theorem 9.2 *The components of $\mathbb{B}(\lambda)^+$ define a t-atomic decomposition. These components are isomorphic to intervals of the form $[\widehat{0}, \mu]$ in the dominant weight poset via the weight projection, and the distinguished vertex $h \in H(\lambda)$ in each of them is chosen to be the respective maximum.* □

The proof has two parts. First, the atomic decomposition is proved (i.e., the $t = 1$ case), by using Lemma 9.1 to derive the existence of a maximum and a minimum in each interval. Using the realization of $B(\lambda)$ in terms of semistandard tableaux, we show that we can choose the statistic $\mathrm{c}(\cdot)$ in Definition 6.3.2 to be the Lascoux-Schützenberger *charge* [16], which expresses the type A Kostka–Foulkes polynomials combinatorially; only some basic properties of charge are needed.

Example 9.3 Consider $\lambda = (3, 2, 1)$ in type A_3. The modified crystal graph $\mathbb{B}(\lambda)^+$ is shown in Fig. 1. Its vertices are labeled by semistandard Young tableaux whose

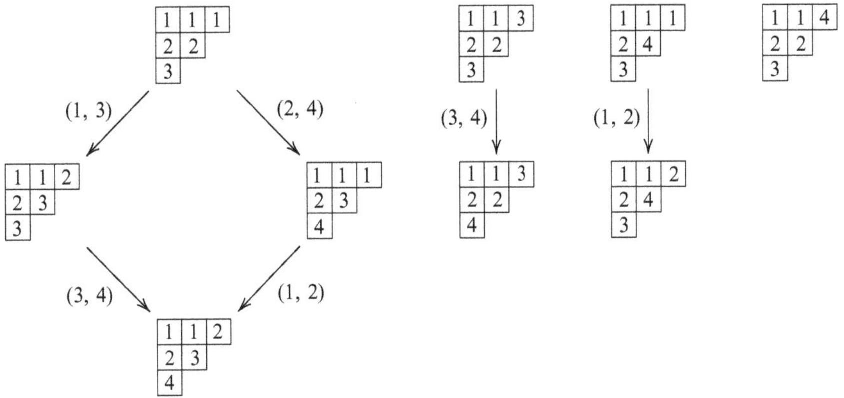

Fig. 1 The modified crystal graph $\mathbb{B}(\lambda)^+$ in Example 9.3

content is a partition, and its edges are labeled as above. In particular, this graph gives the following atomic decomposition of the character:

$$\chi_\lambda = w_{(3,2,1)} + w_{(2,2,2)} + w_{(3,1,1,1)} + w_{(2,2,1,1)} .$$

9.2 *Types B_n, C_n, and D_n*

This section refers to the stable ranges in types B_n, C_n, and D_n, namely to a corresponding graph/poset $\mathbb{B}(\lambda)^+$. The approach is completely similar to that in Sect. 9.1.

Theorem 9.4 *The components of $\mathbb{B}(\lambda)^+$ define an atomic decomposition. These components are isomorphic to intervals of the form $[\widehat{0}, \mu]$ in the dominant weight poset via the weight projection.* □

10 Atomic Decomposition: Additional Facts and Perspectives

We conjecture that the atomic decompositions in type B, C, D, as stated in Theorem 9.4, are t-atomic decompositions, for appropriate choices of the statistic $c(\cdot)$ in Definition 6.3.2. Let us describe a possible such choice in type C, for partitions λ of even rank. In [20], we gave a combinatorial formula $K_{\lambda,0}(t)$ based on a new statistic on King tableaux of zero weight (indexing the crystal vertices of zero weight). We can translate this statistic via the *Sheats bijection* [29] between King and Kashiwara–Nakashima tableaux (the crystal structure is only known on the latter), and then we can extend it recursively on $\mathbb{B}(\lambda)^+$ via (6.15). By (6.17), such a statistic would also give a combinatorial formula for all $K_{\lambda,\mu}(t)$.

Other cases in which we derive a t-atomic decomposition are: (1) the crystal $B(\infty)$ in types $A - D$ and G_2; (2) the crystal of the adjoint representation of $\mathfrak{g}$ in any type.

11 Geometric Interpretation of the Atomic Decomposition

We give an interpretation of the combinatorial atomic decomposition in terms of the *geometric Satake correspondence*. For a reductive group G, this important theory exhibits a geometric realization of the irreducible representation $V(\lambda)$ of highest weight λ of the Langlands dual group, as the *intersection cohomology* $IH^*(\overline{Gr_\lambda})$ of the *Schubert variety* denoted $\overline{Gr_\lambda}$ in the *affine Grassmannian* Gr_G for G; there is also a geometric basis of *MV-cycles* [27]. However, it is hard to give concrete formulas for the MV-cycles and the action. We will show how one can understand the combinatorics of the geometric Satake correspondence via our combinatorial atomic decomposition.

The module $IH^*(\overline{Gr^\lambda})$ has the *truncation filtration* (or *standard Grothendieck filtration*), which gives the Kostka–Foulkes polynomials when restricted to the weight spaces [4]. The degree 0 piece in this filtration is the cohomology of the constant sheaf $H^*(\overline{Gr^\lambda})$, so

$$IH^*(\overline{Gr^\lambda}) \simeq H^*(\overline{Gr^\lambda}) \oplus \text{other summands}\,. \tag{11.1}$$

On another hand, $H^*(\overline{Gr^\lambda})$ has the basis of classes of Schubert varieties inside $\overline{Gr^\lambda}$, which are indexed by the weights of $V(\lambda)$ considered without multiplicity, as recorded by the layer sum polynomials. In this language, the atomic decomposition decomposition (6.9), cf. Definition 6.2.1, is expressing the fact that there is a refinement of the truncation filtration (with the $H^*(Gr_G)$-action), whose successive quotients are isomorphic to $H^*(\overline{Gr^\mu})$ for $\mu \in P^+(\lambda)$. These quotients correspond precisely to the blocks of the partition in the combinatorial atomic decomposition, cf. Definition 6.3.1.

12 Atomic Decomposition for 1-Dimensional Sums

For this section, we refer the reader to [24] and the references therein. As explained in Remark 6.2, Lascoux's original atomic decomposition was established for a slight renormalization of the Kostka polynomials. This renormalization coincides in fact with the 1-dimensional sums, defined via the coenergy function on finite crystals of affine type A. The aim of this section is to establish that this atomic decomposition has an analogue in any classical affine root system of sufficiently large rank.

12.1 Background on 1-Dimensional Sums

Let $\mathfrak{g}$ be an affine algebra of nonexceptional type, and $I = \{0, 1, \ldots, n\}$ the index set of its Dynkin nodes. Let $0 \in I$ as specified in [9], and set $I_0 = I \backslash \{0\}$. For a pair (r, s) ($r \in I_0$, $s \in \mathbb{Z}_{>0}$), there exists a crystal $B^{r,s}$ called the *Kirillov–Reshetikhin* (KR) *crystal*. It is a crystal base (in the sense of Kashiwara) of the Kirillov–Reshetikhin module $W^{r,s}(a)$, for a suitable parameter a, over the quantum affine algebra $U_q'(\mathfrak{g})$ without the degree operator q^d. Let B be a tensor product of KR crystals $B = B^{r_1,s_1} \otimes B^{r_2,s_2} \otimes \cdots \otimes B^{r_l,s_l}$, and for a subset J of I, set $\mathrm{hw}_J(B) = \{b \in B \mid \tilde{e}_i b = 0 \text{ for any } i \in J\}$, where $\tilde{e}_i$ is the Kashiwara operator acting on B. We set

$$|B| := \sum_{i=1}^{l} r_i s_i \quad \text{and} \quad \|B\| := \sum_{1 \leqslant i < j \leqslant l} \min(r_i, r_j) \min(s_i, s_j) .$$

We call an element of $\mathrm{hw}_J(B)$ J-highest. For an I_0-weight λ, we define the 1-dimensional sum $\overline{X}_{\lambda,B}(t)$ by

$$\overline{X}_{\lambda,B}(t) := \sum_{\substack{b \in \mathrm{hw}_{I_0}(B) \\ \mathrm{wt}(b) = \lambda}} t^{\overline{D}(b)} .$$

Here $\overline{D} : B \to \mathbb{Z}$ is the intrinsic coenergy function, and $\mathrm{wt}(b)$ is the I_0-weight of b. Assume now that $n = |I_0|$ is sufficiently large. Then it can be shown that $\overline{X}_{\lambda,B}(t)$ depends only on the attachment of the node 0 to the rest of the Dynkin diagram of $\mathfrak{g}$, so one can assume that $\mathfrak{g}$ is of type $A_n^{(1)}$, $D_n^{(1)}$, $C_n^{(1)}$, or $D_{n+1}^{(2)}$, and thus admits the Dynkin diagram automorphism flipping the nodes i and $n - i$. To each of these affine root systems we attach a partition symbol $\diamondsuit$, as follows:

	$A_n^{(1)}$	$D_n^{(1)}$	$C_n^{(1)}$	$D_{n+1}^{(2)}$
$\diamondsuit$	$\emptyset$	$(1, 1)$	(2)	(1)

Hence, we have four kinds of "stable" 1-dimensional sums denoted by $\overline{X}^{\diamondsuit}_{\lambda,B}(t)$, and the non-spin I_0-weight $\lambda = (\lambda_1, \lambda_2, \ldots)$ is identified with a partition in the standard way. Let $\mathcal{P}_N^{\diamondsuit}$ be the set of partitions of N tiled by $\diamondsuit$. As before, we write $|\lambda|$ for the rank of the partition λ, i.e., the sum of its parts. The following theorem proved in [24] reduces the computation of all stable 1-dimensional sums to type A.

Theorem 12.1 ([24])

(a) For $\diamondsuit \neq \varnothing$, we have

$$\overline{X}^{\diamondsuit}_{\lambda,B}(t) = t^{\frac{|B|-|\lambda|}{|\diamondsuit|}} \sum_{\substack{\mu \in \mathcal{P}^{\diamondsuit}_{|B|-|\lambda|} \\ \nu \in \mathcal{P}^{(1)}_{|B|}}} c^{\nu}_{\lambda\mu}\, \overline{X}^{\varnothing}_{\nu,B}(t^{\frac{2}{|\diamondsuit|}}),$$

where $c^{\nu}_{\lambda\mu}$ stands for the Littlewood–Richardson coefficient.

(b) We also have

$$\overline{X}^{\diamondsuit^t}_{\lambda^t,B^t}(t) = t^{\frac{2(\|B\|+|B|-|\lambda|)}{|\diamondsuit|}}\, \overline{X}^{\diamondsuit}_{\lambda,B}(t^{-1}), \tag{12.1}$$

where t is the conjugation of partitions. □

We will not use this theorem in the sequel, but we will need some properties which are crucial ingredients of its proof. First, for each affine algebra $\mathfrak{g}^{\diamondsuit}$ (recall that $\mathfrak{g}^{\diamondsuit}$ is of type $A^{(1)}_{n}$, $D^{(1)}_{n}$, $C^{(1)}_{n}$, or $D^{(2)}_{n+1}$), the Dynkin diagram automorphism $i \mapsto n-i$ $(i \in I)$ yields an automorphism F on the KR crystal $B^{r,s}$ for $\mathfrak{g}^{\diamondsuit}$ satisfying

$$\mathrm{F}(\tilde{e}_i b) = \tilde{e}_{n-i}\mathrm{F}(b),$$

for any $i \in I$ and $b \in B^{r,s}$. This automorphism F is extended to B by $\mathrm{F}(b) := \mathrm{F}(b_1) \otimes \mathrm{F}(b_2) \otimes \cdots \otimes \mathrm{F}(b_l)$. Write $\max(B) := \bigoplus_{\gamma} B(\gamma)$, where $B(\gamma)$ is the $U_q(\mathfrak{g}^{\diamondsuit}_{I_0})$-crystal of highest weight γ, and γ runs over all weights with $|\gamma| = |B|$ such that $B(\gamma)$ appears in the restriction of B. Namely, $\max(B)$ is the disjoint union of classical highest weight crystals of maximal highest weights. We remark that $\mathfrak{g}_{I\setminus\{0,n\}}$ is of type A_{n-1}, and set $\overline{\lambda} := (-\lambda_n, \ldots, -\lambda_1)$ for $\lambda = (\lambda_1, \ldots, \lambda_n)$. The following proposition is a crucial step in the proof of the previous theorem.

Proposition 12.2

(a) F *restricts to the following bijection:*

$$\left\{\begin{array}{c} I_0\text{-highest elements} \\ \text{in } B \text{ of wt}\,\lambda \end{array}\right\} \xrightarrow{\ \mathrm{F}\ } \left\{\begin{array}{c} I \setminus \{0,n\}\text{-highest elements} \\ \text{in } \max(B) \text{ of wt}\,\overline{\lambda} \end{array}\right\}.$$

(b) $\overline{D}(b) = \overline{D}(\mathrm{F}(b)) + (|B| - |\mathrm{wt}\,(b)|)/|\diamondsuit|$ *for* $b \in \mathrm{hw}_{I_0}(B)$.

(c) For any vertex b in $\max(B)$, *we have* $\overline{D}^{\diamondsuit}(b) = \frac{2}{|\diamondsuit|}\overline{D}^{\varnothing}(b)$. □

From now on, we assume that $r_1 = \cdots = r_l = 1$ (that is, each KR crystal has row shape,[4]) and set $\sigma = (s_1, \ldots, s_l)$. We then write for short

[4]Nevertheless, we will establish that the atomic decomposition also holds in the column shape case, by (12.1).

$$B^{1,s_1} \otimes \cdots \otimes B^{1,s_l} = B^{s_1} \otimes \cdots \otimes B^{s_l} =: B_\sigma .$$

Let μ be the partition obtained by reordering σ, and set $\|\mu\| = \sum_{1 \leqslant i \leqslant n} (i-1)\mu_i$. Assuming n large enough, we set $\widehat{\lambda} = (n - \lambda_n, \ldots, n - \lambda_1)$ and $\widehat{\mu} := (n - \mu_n, \ldots, n - \mu_1)$. By Lecouvey et al. [24, Theorem 10.9], we then have

$$\overline{X}^{\diamondsuit}_{\lambda,\sigma}(t) = t^{\frac{2(\|\mu\|+|\mu|-|\lambda|}{|\diamondsuit|}} K^{\mathrm{stab}}_{\widehat{\lambda},\widehat{\mu}}(t^{-1}) , \tag{12.2}$$

where $K^{\mathrm{stab}}_{\lambda,\mu}(t)$ is the stable Lusztig t-analogue of classical type with Dynkin diagram indexed by I_0, that is, $K^{\mathrm{stab}}_{\lambda,\mu}(t) = K_{\lambda+k\omega_n,\mu+k\omega_n}(t)$, with $\omega_n = (1^n)$ and $k \in \mathbb{Z}_{>0}$ sufficiently large. In particular, we obtain $\overline{X}^{\emptyset}_{\lambda,\sigma}(t) = t^{\|\mu\|} K_{\widehat{\lambda},\widehat{\mu}}(t^{-1}) = t^{\|\mu\|} K_{\lambda,\mu}(t^{-1})$ for $\diamondsuit = \emptyset$. For any $\sigma = (s_1, \ldots, s_l) \in \mathbb{Z}^l_{\geqslant 0}$, set $|\sigma| := s_1 + \cdots + s_l$. Given a positive integer s, define

$$B_s := \bigoplus_{\substack{\sigma=(s_1,\ldots,s_l)\in\mathbb{Z}^l_{\geqslant 0} \\ |\sigma|=s}} B_\sigma \tag{12.3}$$

as the direct sum of all tensor products of row KR crystals with length sum s. For each vertex $b \in B_s$, let $\Sigma(b)$ be the unique l-tuple in $\mathbb{Z}^l_{\geqslant 0}$ such that $b \in B_{\Sigma(b)}$.

12.2 Bi-Crystal Structure and Atomic Decomposition in Type A

In this section, we assume $\diamondsuit = \emptyset$. The vertices of the KR crystals B^{s_k}, for $k = 1, \ldots, l$, can be identified with the row tableaux of length s_k on the ordered alphabet $\mathcal{A}_n = \{1 < \cdots < n\}$. It is well-known (see [31]) that the RSK correspondence yields a bijection

$$b \in B^{s_1} \otimes \cdots \otimes B^{s_l} \overset{1:1}{\longleftrightarrow} (P(b), Q(b)) ,$$

where P and Q are semistandard tableaux with the same shape on $\mathcal{A}_n$ and $\mathcal{A}_l$, respectively. Given two vertices b_1 and b_2 of B_s, we then have:

(a) $P(b_1) = P(b_2)$ if and only if b_1 and b_2 belong to isomorphic connected I_0-components of B_s, and are matched by the associated isomorphism;
(b) $Q(b_1) = Q(b_2)$ if and only if b_1 and b_2 belong to the same connected I_0-component.

By considering the $A_{n-1} \times A_{l-1}$ crystal action on the pairs of semistandard tableaux (P, Q), one gets a $A_{n-1} \times A_{l-1}$ bi-crystal structure on B_s (i.e., the actions of the crystal operators for A_{n-1} and A_{l-1} commute). The A_{n-1}-action is just the

I_0-action on B_s. It was proved in [31] that for the Weyl group A_{l-1}-action (that is, of the symmetric group S_l) on B_s, each elementary transposition $(i, i+1)$, for $i = 1, \ldots, l-1$, acts as the combinatorial R-matrix

$$B^{s_1} \otimes \cdots \otimes B^{s_i} \otimes B^{s_{i+1}} \otimes \cdots \otimes B^{s_l} \xrightarrow{\simeq} B^{s_1} \otimes \cdots \otimes B^{s_{i+1}} \otimes B^{s_i} \otimes \cdots \otimes B^{s_l}\,. \tag{12.4}$$

Thus, for any $b \in B_s$ and $w \in S_l$, we have

$$\overline{D}(w(b)) = \overline{D}(b)\,. \tag{12.5}$$

In the sequel, we will need the action of the A_{l-1}-crystal operator $\tilde{F}_1$ on B_s. In fact, it suffices to describe its action on the A_{n-1}-highest weight (i.e., I_0) vertices. Observe that each such vertex has the form $b = R_1 \otimes R_2 \otimes \cdots \otimes R_l$ where, in particular, $R_1 = (1^{s_1})$ contains only letters 1, and $R_2 = (1^{s_2-a}2^a)$ contains only letters 1 and 2 with $a \leqslant \max(s_1, s_2)$.

Proposition 12.3 *Assume $s_1 > s_2$. Then for any A_{n-1}-highest vertex $b \in B^{s_1} \otimes B^{s_2} \otimes \cdots \otimes B^{s_l}$, $\tilde{F}_1(b)$ is the vertex of $B^{s_1-1} \otimes B^{s_2+1} \otimes \cdots \otimes B^{s_l}$ obtained by moving one letter* 1 *in b from R_1 to R_2. Moreover $\overline{D}(\tilde{F}_1(b)) = \overline{D}(b)$.* □

Proof Set $\mathrm{RSK}(b) = (T_\lambda, Q)$. Then $\mathrm{RSK}(\tilde{F}_1(b)) = (T_\lambda, \tilde{F}_1 Q)$, and $\tilde{F}_1 Q$ is obtained by changing in Q the rightmost letter 1 into a letter 2. By the definition of the RSK correspondence based on the Schensted bumping algorithm, this exactly means that $\tilde{F}_1(b)$ is obtained by moving one letter 1 in b from R_1 to R_2. It is also known that $\mathrm{coch}_n(Q) = \overline{D}(b)$ (see [31]) and $\mathrm{coch}_n(\tilde{F}_1 Q) = \mathrm{coch}_n(Q)$. Here coch_n is the cocharge statistics of Lascoux and Schützenberger such that $\mathrm{coch}_n(T) + \mathrm{ch}_n(T) = n(\mu) = \sum_i (i-1)\mu_i$ for any semistandard tableau T of evaluation μ. We therefore get $\mathrm{coch}_{\mathrm{n}}(\tilde{F}_1(Q)) = \overline{D}(\tilde{F}_1(b))$ and $\overline{D}(\tilde{F}_1(b)) = \overline{D}(b)$.

We call the action of $\tilde{F}_1(b)$ the splitting of the highest weight vertex b. Assume that μ is a partition. Then, by the RSK correspondence, each vertex b in $\mathrm{hw}_{I_0}(B_\mu)$ of highest weight λ is matched with a pair (P_λ, Q), such that P_λ is the semistandard tableau of shape λ with letters i in row i, and Q has shape λ and dominant evaluation μ. In the semistandard realization of the A_{l-1}-crystals, the atomic decomposition yields a graph structure on the set of semistandard tableaux with s boxes and dominant evaluation (that is, with an evaluation equal to a partition μ of s). By the RSK correspondence, these tableaux are in bijection with the highest weight vertices of the crystals $B_\mu \subseteq B_s$, where μ is a partition of s. This leads us to defining

$$B_{s,\emptyset}^{\mathrm{part}} := \bigoplus_{\mu \text{ partition of } s} B_\mu\,.$$

Each component of $\mathrm{hw}_{I_0}(B_{s,\emptyset}^{\mathrm{part}})$ is isomorphic to an interval for the dominance order on partitions, and in particular has a minimum and a maximum. More precisely, for each positive root $\varepsilon_i - \varepsilon_j$, $1 \leqslant i < j \leqslant l$, of type A_{l-1}, one can define the operator L_α (resp. R_α) on $\mathrm{hw}_{I_0}(B_{s,\emptyset}^{\mathrm{part}}) \cup \{0\}$ by $L_\alpha := \mathrm{RSK}^{-1} \circ \mathrm{f}_\alpha \circ \mathrm{RSK}$

(resp. $R_\alpha := \mathrm{RSK}^{-1} \circ \mathrm{e}_\alpha \circ \mathrm{RSK}$), where f_α (resp. e_α) is the modified crystal operator associated with α. Then, under the action of the operators L_α and R_α corresponding to the covering relations for the dominance order on partitions, the set $\mathrm{hw}_{I_0}(B^{\mathrm{part}}_{s,\emptyset})$ decomposes into connected components, each of which is isomorphic to a component $\mathbb{B}(\lambda, h)$ defined in Sect. 6.3, and thus, under the map Σ, to an interval in the dominance order on partitions. This is to say that the mentioned connected components are defined by considering arrows $b \dashrightarrow b'$ between two vertices of $\mathrm{hw}_{I_0}(B^{\mathrm{part}}_{s,\emptyset})$ when $b' = L_\alpha(b)$ for a root $\alpha = \varepsilon_i - \varepsilon_j$, and the partition $\Sigma(b)$ covers $\Sigma(b')$. Observe that, by (12.4) and Proposition 12.3, the action of each operator L_α can be obtained as the conjugation of the splitting by combinatorial R-matrices. Therefore, it can be defined independently of the RSK correspondence. Furthermore, for each partition λ, the operators L_α and R_α stabilize the set

$$\mathrm{hw}_{I_0}(B^{\mathrm{part}}_{s,\emptyset})_\lambda := \{b \in \mathrm{hw}_{I_0}(B^{\mathrm{part}}_{s,\emptyset}) \mid \mathrm{wt}(b) = \lambda\} \cup \{0\}.$$

Now set

$$\mathcal{H}^\lambda_{s,\emptyset} := \{h \in \mathrm{hw}_{I_0}(B^{\mathrm{part}}_{s,\emptyset})_\lambda \mid R_\alpha(h) = 0 \text{ for any } \alpha = \varepsilon_i - \varepsilon_j, 1 \leqslant i < j \leqslant l\},$$

and for each vertex $h \in \mathcal{H}^\lambda_s$, let $\mathbb{B}^\emptyset_{(h,\lambda)}$ be the connected component (for the arrows $\dashrightarrow$) of b in $\mathrm{hw}_{I_0}(B^{\mathrm{part}}_{s,\emptyset})_\lambda$. Then the map Σ is a bijection between $\mathbb{B}^\emptyset_{(h,\lambda)}$ and the set of partitions μ such that $\mu \leqslant \Sigma(h)$. For any partition ν, we define the atomic 1-dimensional sum

$$\overline{Y}^\emptyset_{\lambda,\nu}(t) := \sum_{\substack{h \in \mathcal{H}^\lambda_{s,\emptyset} \\ \Sigma(h)=\nu}} t^{\overline{D}(h)}.$$

Theorem 12.4 *For any partition μ, we have the atomic decomposition*

$$\overline{X}^\emptyset_{\lambda,\mu}(t) = \sum_{\mu \leqslant \nu \leqslant \lambda} \overline{Y}^\emptyset_{\lambda,\nu}(t).$$

Proof We have

$$\overline{X}^\emptyset_{\lambda,\mu}(t) = \sum_{b \in \mathrm{hw}_{I_0}(B^\lambda_\mu)} t^{\overline{D}(b)},$$

where $\mathrm{hw}_{I_0}(B^\lambda_\mu) = \{b \in \mathrm{hw}_{I_0}(B_\mu) \mid \mathrm{wt}(b) = \lambda\}$. Since $\mathrm{hw}_{I_0}(B^\lambda_\mu)$ is a subset of $\mathrm{hw}_{I_0}(B^{\mathrm{part}}_{s,\emptyset})_\lambda$, we can write

$$\mathrm{hw}_{I_0}(B^\lambda_\mu) = \bigsqcup_{h \in \mathcal{H}^\lambda_{s,\emptyset}} \mathbb{B}^\emptyset_{(h,\lambda)} \cap B_\mu.$$

Moreover, each set $\mathbb{B}^{\emptyset}_{(h,\lambda)} \cap B_\mu$ is nonempty if and only if $\mu \leqslant \Sigma(h)$, in which case it reduces to one vertex. This gives

$$\overline{X}^{\emptyset}_{\lambda,\mu}(t) = \sum_{h\in\mathcal{H}^{\lambda}_{s,\emptyset}} \sum_{b\in\mathbb{B}^{\emptyset}_{(h,\lambda)}\cap B_\mu} t^{\overline{D}(b)} = \sum_{\mu\leqslant\nu} \sum_{\substack{h\in\mathcal{H}^{\lambda}_{s,\emptyset}\\ \Sigma(h)=\nu}} \sum_{b\in\mathbb{B}^{\emptyset}_{(h,\lambda)}\cap B_\mu} t^{\overline{D}(b)}.$$

But for any b in $\mathbb{B}^{\emptyset}_{(h,\lambda)} \cap B_\mu$, we must have $\overline{D}(b) = \overline{D}(h)$, by (12.5) and Proposition 12.3. Finally, we get

$$\overline{X}^{\emptyset}_{\lambda,\mu}(t) = \sum_{\mu\leqslant\nu} \sum_{\substack{h\in\mathcal{H}^{\lambda}_{s,\emptyset}\\ \Sigma(h)=\nu}} t^{\overline{D}(h)} = \sum_{\mu\leqslant\nu\leqslant\lambda} \overline{Y}^{\emptyset}_{\lambda,\nu}(t).$$

Remark 12.5 *By (12.2), the previous decomposition coincides with the original atomic decomposition of the renormalized Kostka polynomials obtained by Lascoux from the cocharge statistics.* □

12.3 Atomic Decomposition of the Stable 1-Dimensional Sums

We now assume that $\diamondsuit \in \{\emptyset, (1,1), (2), (1)\}$, and for any positive integer s, we consider the affine crystal B_s defined in (12.3). Then $\max(B_s)$ is the union of the I_0-components whose highest weight vertices have weight a partition λ of rank s. This means that their highest weight vertices are exactly those appearing in B_s when $\diamondsuit = \emptyset$. In the previous section, we defined operators L_α and R_α on the set $\mathrm{hw}_{I_0}(B^{\mathrm{part}}_{s,\emptyset})$, for each positive root α of type A_{l-1}. We will first extend this definition to the subset

$$\max(B^{\mathrm{part}}_s) := \max(B_s) \cap \bigoplus_{\mu \text{ partition of } s} B_\mu.$$

Since L_α and R_α are defined on the highest weight vertices of $\max(B^{\mathrm{part}}_s)$ and preserve these vertices, it suffices to define these operators such that they commute with the actions of the crystal operators $\tilde{f}_i, \tilde{e}_i$, $i = 1, \ldots, n-1$. Then L_α and R_α correspond to I_0-crystal isomorphisms. In particular, we have for any b in $\max(B^{\mathrm{part}}_s)$:

$$\overline{D}(L_\alpha(b)) = \overline{D}(b) \quad\text{and}\quad \overline{D}(R_\alpha(b)) = \overline{D}(b)\,; \tag{12.6}$$

indeed, the coenergy $\overline{D}$ is constant on the I_0-connected components, and both relations hold on the highest weight vertices of $\max(B^{\mathrm{part}}_s)$, since $\overline{D}$ does not depend on $\diamondsuit$ on $\max(B_s)$. Observe also that L_α and R_α stabilize $\mathrm{hw}_{I\setminus\{0,n\}}(B^{\mathrm{part}}_s) \cap \max(B_s)$.

Letting

$$B_{s,\diamondsuit}^{\text{part}} := \bigoplus_{\mu \text{ partition of } s} B_\mu \,,$$

we can now define L_α and R_α on $\text{hw}_{I_0}(B_{s,\diamondsuit}^{\text{part}})$ by

$$L_\alpha := \text{F} \circ L_\alpha \circ \text{F} \quad \text{and} \quad R_\alpha := \text{F} \circ R_\alpha \circ \text{F} \,.$$

Indeed, assume $b \in \text{hw}_{I_0}(B_{s,\diamondsuit}^{\text{part}})$ is of highest weight λ in B_μ. Then, by Proposition 12.2, $\text{F}(b)$ belongs to $\text{hw}_{I\setminus\{0,n\}}(B_{s,\diamondsuit}^{\text{part}}) \cap \max(B_s) \cap B_\mu$ with weight $\overline{\lambda}$. When $L_\alpha \circ \text{F}(b) \neq 0$, it belongs to $\text{hw}_{I\setminus\{0,n\}}(B_{s,\diamondsuit}^{\text{part}}) \cap \max(B_s) \cap B_{\mu-\alpha}$ with weight $\overline{\lambda}$. Therefore, $\text{F} \circ L_\alpha \circ \text{F}(b)$ is well defined and belongs to $\text{hw}_{I_0}(B_{s,\diamondsuit}^{\text{part}}) \cap B_{\mu-\alpha}$ with weight λ. By Proposition 12.2, we also have for $\diamondsuit = (1,1)$ or $\diamondsuit = (2)$:

$$\overline{D}(L_\alpha(b)) = \overline{D}(\text{F}(L_\alpha(\text{F}(b)))) = \frac{|\mu| - |\lambda|}{2} + \overline{D}(L_\alpha(\text{F}(b))) \,.$$

But $\overline{D}(L_\alpha(\text{F}(b))) = \overline{D}(\text{F}(b))$ by (12.6). Thus we get

$$\overline{D}(L_\alpha(b)) = \frac{|\mu| - |\lambda|}{2} + \overline{D}(\text{F}(b)) = \overline{D}(b) \,.$$

The proof is similar for $\diamondsuit = (1)$ and for R_α. So we established the following result.

Proposition 12.6 *For any positive root α of type A_{l-1}, the operators L_α and R_α are well defined on $\text{hw}_{I_0}(B_{s,\diamondsuit}^{\text{part}})$ and preserve the coenergy.* □

For any partition λ, set

$$\text{hw}_{I_0}(B_{s,\diamondsuit}^{\text{part}})_\lambda := \{b \in \text{hw}_{I_0}(B_{s,\diamondsuit}^{\text{part}}) \mid \text{wt}(b) = \lambda\} \cup \{0\} \,,$$

$$\mathcal{H}_{s,\diamondsuit}^{\lambda} := \{h \in \text{hw}_{I_0}(B_{s,\diamondsuit}^{\text{part}})_\lambda \mid R_\alpha(h) = 0 \text{ for any } \alpha = \varepsilon_i - \varepsilon_j, 1 \leqslant i < j \leqslant l\} \,.$$

For each vertex $h \in \mathcal{H}_{s,\diamondsuit}^{\lambda}$, let $\mathbb{B}_{(h,\lambda)}^{\diamondsuit}$ be the connected component of $\text{hw}_{I_0}(B_{s,\diamondsuit}^{\text{part}})_\lambda$ obtained by applying operators L_α to h. Then the map Σ is a bijection between $\mathbb{B}_{(h,\lambda)}^{\diamondsuit}$ and the set of partitions μ such that $\mu \leqslant \Sigma(h)$. Indeed, if we denote by ν the highest weight vertex of $\text{F}(h) \in \max(B_s^{\text{part}})$, the map $\text{F} : \mathbb{B}_{(h,\lambda)}^{\diamondsuit} \to \mathbb{B}_{(\text{F}(h),\nu)}^{\diamondsuit}$ commutes with the operators L_α and R_α. For any partition ν, define the atomic 1-dimensional sum

$$\overline{Y}_{\lambda,\nu}^{\diamondsuit}(t) := \sum_{\substack{h \in \mathcal{H}_{s,\diamondsuit}^{\lambda} \\ \Sigma(h) = \nu}} t^{\overline{D}(h)} \,,$$

where $s = |\nu|$. The proof of the following theorem is similar to that of Theorem 12.4.

Theorem 12.7 *For any $\diamondsuit \in \{\emptyset, (1,1), (2), (1)\}$ and any partition μ, we have the atomic decomposition of the stable 1-dimensional sum*

$$\overline{X}^{\diamondsuit}_{\lambda,\mu}(t) = \sum_{\mu \leqslant \nu \leqslant \lambda} \overline{Y}^{\diamondsuit}_{\lambda,\nu}(t)\,.$$

Remark 12.8 *The operators L_α, R_α defined on $\mathrm{hw}_{I_0}(B^{\mathrm{part}}_{s,\diamondsuit})$ can be extended to $\mathrm{hw}_{I_0}(B_{s,\diamondsuit}) = \mathrm{hw}_{I_0}(B_s)$ as follows. Given b of highest weight in B_σ, let θ be the unique isomorphism of affine crystals $\theta : B_\sigma \to B_\mu$, where μ is the reordering of σ (θ is unique because B_σ and B_μ are connected affine crystals). Then one sets $L_\alpha(b) := \theta^{-1} \circ L_\alpha \circ \theta(b)$ and $R_\alpha(b) := \theta^{-1} \circ R_\alpha \circ \theta(b)$. It is then also possible to define L_α and R_α on the whole crystal B_s by requiring that they are isomorphisms of I_0-crystals.* □

Acknowledgments The second author Cristian Lenart gratefully acknowledges the partial support from the NSF grant DMS–1362627 and the Simons grant #584738. Both authors are grateful to Arthur Lubovsky and Adam Schultze for the computer tests (based on the Sage [28] system) related to this work; they also received support from the NSF grant mentioned above.

References

1. T. Braden and R. MacPherson. From moment graphs to intersection cohomology. *Math. Ann.* 321:533–551, 2001.
2. T. Brylawski. The lattice of integer partitions. *Discrete Math.* 6:201–219, 1973.
3. M. Dolega, T. Gerber, and J. Torres. A positive combinatorial formula for symplectic Kostka-Foulkes polynomials I: Rows, 2019, *J. Algebra* 560:1253–1296, 2020. arXiv:1911.06732.
4. V. Ginzburg. Perverse sheaves on a Loop group and Langlands' duality. arXiv:math.alg-geom/9511007, 1995.
5. W-H. Hesselink. Characters of the nullcone. *Math. Ann.* 252:179–182, 1980.
6. B. Ion. Generalized exponents of small representations I. *Represent. Theory* 13:401–426, 2009.
7. B. Ion. Generalized exponents of small representations II. *Represent. Theory* 15:433–493, 2011.
8. I.-S. Jang and J.-H. Kwon. Flagged Littlewood-Richardson tableaux and branching rule for classical groups, 2019, *J. Combin. Theory Ser. A* 181:105419, 51, 2021. arXiv:1908.11041.
9. V. G. Kac. Infinite Dimensional Lie Algebras 3rd ed., Cambridge Univ. Press, Cambridge, UK, 1990.
10. M. Kashiwara. Crystal bases of modified quantized enveloping algebra. *Duke Math. J.* 73:383–413, 1994.
11. S. Kato. Spherical functions and a q-analogue of Kostant's weight multiplicity formula. *Invent. Math.* 66:461–468, 1982.
12. R. C. King. Weight multiplicities for the classical groups. *Lectures Notes in Phys.* 50:490–499, 1976.
13. B. Kostant. Lie groups representations on polynomial rings. *Amer. J. Math.* 85:327–404, 1963.
14. J.-H. Kwon. Combinatorial extension of stable branching rules for classical groups. *Trans. Amer. Math. Soc.* 370:6125–6152, 2018.

15. J.-H. Kwon. Lusztig data of Kashiwara-Nakashima tableaux in types B and C. *J. Algebra* 503:222–264, 2018.
16. A. Lascoux and M-P. Schützenberger. Sur une conjecture de H. O. Foulkes. *C. R. Acad. Sci. Paris* 288:95–98, 1979.
17. A. Lascoux, B. Leclerc, and J-Y. Thibon. Crystal graphs and q-analogue of weight multiplicities for the root system A_n. *Lett. Math. Phys.* 35:359–374, 1995.
18. A. Lascoux. Cyclic permutations on words, tableaux and harmonic polynomials. In *Proceedings of the Hyderabad Conference on Algebraic Groups (Hyderabad, 1989)*, pages 323–347. Manoj Prakashan, Madras, 1991.
19. A. Lascoux. Polynomials, 2013, available at http://phalanstere.u-perm.fr/ al/ARTICLES/CoursYGKM.pdf.
20. C. Lecouvey and C. Lenart. Combinatorics of generalized exponents. *Int. Math. Res. Not.* 16:4942–4992, 2020.
21. C. Lecouvey and C. Lenart. Atomic decomposition of characters and crystals, *Adv. Math.*, 376:107453, 51, 2021.
22. C. Lecouvey. Kostka-Foulkes polynomials, cyclage graphs and charge statistic for the root system C_n. *J. Algebraic Combin.* 21:203–240, 2005.
23. C. Lecouvey. Combinatorics of crystal graphs and Kostka-Foulkes polynomials for the root systems B_n, C_n, and D_n. *European J. Combin.* 27:526–557, 2006.
24. C. Lecouvey, M. Okado and M. Shimozono. One-dimensional sums and parabolic Lusztig q-analogues. *Math. Z.* 271:819–865, 2012.
25. D.-E. Littlewood. *The theory of group characters and matrix representations of groups.* Oxford University Press, second edition, 1958.
26. G. Lusztig. Singularities, character formulas, and a q-analog of weight multiplicities. Analyse et topologie sur les espaces singuliers (II-III), *Astérisque* 101–102:208–227, 1983.
27. I. Mirković and K. Vilonen. Geometric Langlands duality and representations of algebraic groups over commutative rings. *Ann. of Math. (2)* 166:95–143, 2007.
28. The Sage-Combinat community. Sage-Combinat: enhancing Sage as a toolbox for computer exploration in algebraic combinatorics, 2008. http://combinat.sagemath.org.
29. J. Sheats. A symplectic jeu de taquin bijection between the tableaux of King and De Concini, *Trans. Amer. Math.*, 351:3569–3607, 1999.
30. M. Shimozono. Multi-atoms and monotonicity of generalized Kostka polynomials. *European J. Combin.* 22:395–414, 2001.
31. M. Shimozono. Crystal for dummies. Available at https://aimath.org/WWN/kostka/crysdumb.pdf.
32. J. Stembridge. The partial order of dominant weights. *Adv. Math.* 136:340–364, 1998.

Conormal Varieties on the Cominuscule Grassmannian

V. Lakshmibai and Rahul Singh

Dedicated to Professor Vyjayanthi Chari on her 60th birthday

Abstract Let G be a simply connected, almost simple group over an algebraically closed field $\mathbf{k}$ of characteristic p, where either $p = 0$ or p is an odd prime which is also a good prime for G. Let P be a maximal parabolic subgroup corresponding to omitting a cominuscule root. We construct a compactification $\phi : T^*G/P \to X(u)$, where $X(u)$ is a Schubert variety in a partial affine flag variety associated with the loop group $G\left(\mathbf{k}[t, t^{-1}]\right)$. Let $N^*X(w) \subseteq T^*G/P$ be the conormal variety of some Schubert variety $X(w)$ in G/P; hence we obtain that the closure of $\phi(N^*X(w))$ in $X(u)$ is a B-stable compactification of $N^*X(w)$. We further show that this compactification is a Schubert subvariety of $X(u)$ if and only if $X(w_0w) \subseteq G/P$ is smooth, where w_0 is the longest element in the Weyl group of G. This result is applied to compute the conormal fibre at the zero matrix in any determinantal variety.

1 Introduction

Given a quiver Q, let $\overline{Q} = Q \sqcup Q^{op}$ be the *double* of Q, the quiver that has the same vertex set as Q and whose set of edges is a disjoint union of the sets of edges of Q and of Q^{op}, the *opposite* quiver. Thus, for any edge $e \in Q$, there is also a reverse edge $e^* \in Q^{op} \subseteq \overline{Q}$ with the same endpoints as e, but in the opposite direction. For $\mathbf{d}$ a dimension vector, the quiver variety $Rep_{\mathbf{d}}\overline{Q}$ is naturally identified as the cotangent bundle of $Rep_{\mathbf{d}}Q$ (cf. [8]).

V. Lakshmibai
Department of Mathematics, Northeastern University, Boston, MA, USA

R. Singh (✉)
Department of Mathematics, Virginia Tech, Blacksburg, VA, USA

J. Greenstein et al. (eds.), *Interactions of Quantum Affine Algebras with Cluster Algebras, Current Algebras and Categorification*, Progress in Mathematics 337,
https://doi.org/10.1007/978-3-030-63849-8_11

Consider the quiver A_2, with dimension vector $\mathbf{d} = (n, m)$. Orbit closures in $Rep_{\mathbf{d}} A_2$ (see Table 1) are called *determinantal varieties*. Lakshmibai and Seshadri [18] have identified determinantal varieties as open subsets of certain *Schubert varieties* in the type A Grassmannian. Consider the following subvariety of $T^* Rep_{\mathbf{d}} A_2 = Rep_{\mathbf{d}} \overline{A}_2$,

$$Z = \left\{ (x, y) \in Hom(\mathbf{k}^m, \mathbf{k}^n) \times Hom(\mathbf{k}^n, \mathbf{k}^m) \,\middle|\, xy = 0, yx = 0 \right\}.$$

Strickland [24] has identified the conormal varieties of determinantal varieties as the irreducible components of Z.

We denote by $\widetilde{A}_n$ the extended Dynkin diagram corresponding to A_n (see Table 1). A point $(x_1, \cdots, x_n) \in Rep_{\mathbf{d}} \widetilde{A}_n$ is called *nilpotent* if the composite map

$$x_n x_{n-1} \cdots x_1 : V_1 \to V_1$$

is nilpotent, and the orbit of such a point is called a *nilpotent orbit*. It is easily seen that $\overline{A}_2 = \widetilde{A}_2$, and that each irreducible component of Z is the closure of a nilpotent orbit in $Rep_{\mathbf{d}} \overline{A}_2$, where $\mathbf{d} = (n, m)$.

On the other hand, Lusztig [19] has identified each nilpotent orbit closure in $Rep_{\mathbf{d}} \widetilde{A}_n$ as an open subset of some Schubert variety in the type A affine Grassmannian. Inspired by these results, Lakshmibai [14] has suggested an exploration of the relationship between the conormal varieties of Schubert varieties and the corresponding affine Schubert varieties.

Recall that a prime p is called a *good prime* for G, if p is co-prime to all the coefficients of the highest root of G written in terms of the simple roots. Let $\mathbf{k}$ be an algebraically closed field of characteristic p, and let G be a simply connected, almost simple algebraic group over $\mathbf{k}$. *We assume that either $p = 0$ or that p is an odd prime which is also a good prime for G.*

We may identify G as a Kac–Moody group corresponding to some irreducible finite type Dynkin diagram $\mathcal{D}_0$. The loop group $LG \stackrel{\text{def}}{=} G\left(\mathbf{k}[t, t^{-1}]\right)$ is then a Kac–Moody group corresponding to the *extended Dynkin diagram* $\mathcal{D}$, obtained by attaching to $\mathcal{D}_0$ the extra root α_0 (see Table 1).

Recall that a simple root $\alpha_d \in \mathcal{D}_0$ is called *cominuscule* if its coefficient in the highest root θ_0 is 1. We see from Table 1 that α_d is cominuscule if and only if there exists an automorphism ι of $\mathcal{D}$ such that $\iota(\alpha_0) = \alpha_d$. Let P be the parabolic subgroup in G corresponding to omitting a cominuscule root α_d, and $\mathcal{P}$ the parabolic subgroup in LG corresponding to omitting both α_d and α_0. For $\mathbf{k} = \mathbb{C}$, Lakshmibai [14], and Lakshmibai et al. [17] have constructed a dense embedding ϕ of T^*G/P into a Schubert variety in $LG/\mathcal{P}$. We use the Kac–Moody functor of Tits [25] to give a definition of ϕ which works in good and odd characteristic (see Proposition 3.14 and Theorem 3.18).

Let $\mathcal{J} = \mathcal{D}_0 \backslash \{\alpha_d\}$, the set of simple roots associated with the parabolic subgroup P as defined above. Further, let $w_{\mathcal{J}}$ (resp. w_0) be the longest element in the Weyl

Table 1 (See [5]) finite type Dynkin diagrams with cominuscule simple roots marked in black (left column), and the corresponding extended Dynkin diagrams with all the cominuscule roots and the additional affine root marked in black (right column)

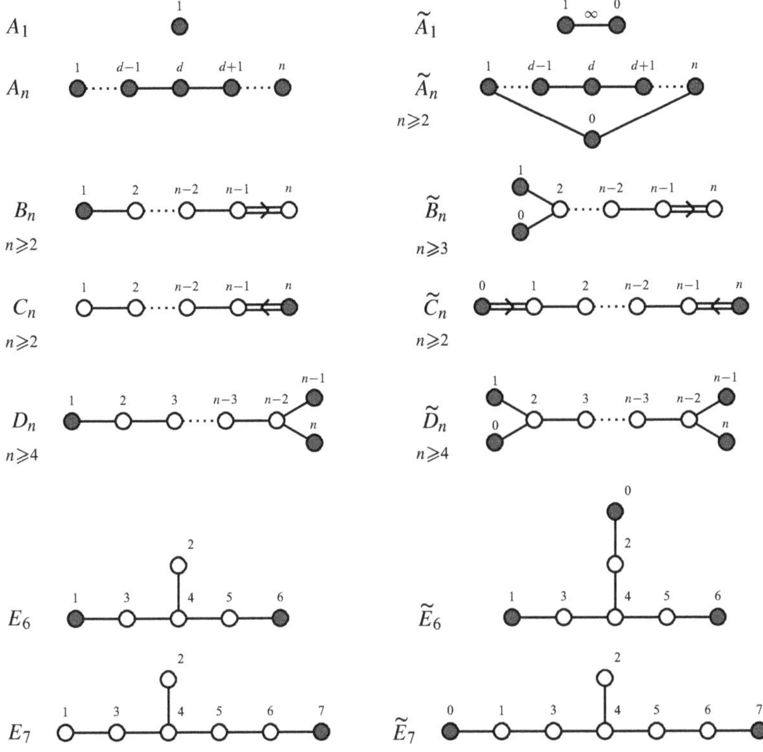

group of P (resp. the Weyl group of G). Our first result, Theorem A (which appears right after Proposition 4.8), is the following:

Theorem A *Let $X_{\mathfrak{J}}(w)$ be a Schubert variety in G/P, and let $N^*X_{\mathfrak{J}}(w)$ be its conormal variety. Then the closure of $\phi(N^*X_{\mathfrak{J}}(w))$ in $LG/\mathcal{P}$ is a Schubert variety if and only if the Schubert variety $X_{\mathfrak{J}}(w_0 w)$ in G/P is smooth.* □

We also obtain (Proposition 4.11) a description of the conormal fibre at the identity for the Schubert varieties to which Theorem A applies.

Lakshmibai and Seshadri [16, 18] have identified determinantal varieties, symmetric determinantal varieties, and skew-symmetric determinantal varieties as open subsets of certain Schubert subvarieties of cominuscule Grassmannians of type A, C, and D, respectively. Using this identification, along with the combinatorial criterion for smoothness of Schubert varieties in cominuscule Grassmannians (see Proposition 4.5) given by Billey and Mitchell [3], we deduce that Theorem A applies to determinantal varieties, symmetric determinantal varieties, and skew-symmetric determinantal varieties.

In our second result, Theorem B (which appears right after Appendix 5.3), we obtain an explicit determination of the conormal fibre at the zero matrix of a skew-symmetric determinantal variety. Similar results are known for determinantal varieties (see [10, 24]) and symmetric determinantal varieties (see [9]), and our proof of Theorem B can be adapted in a straightforward manner to recover these results.

Theorem B *Let* $\overline{\Sigma}_r^{sk,n}$ *denote the* rank r skew-symmetric determinantal variety,

$$\overline{\Sigma}_r^{sk,n} = \left\{ A \in Mat_{n\times n}(\mathbf{k}) \,\middle|\, A = -A^T,\ rank(A) \leqslant r \right\}.$$

The conormal fibre of $\overline{\Sigma}_r^{sk,n}$ *at* 0 *is isomorphic to* $\overline{\Sigma}_{\overline{n}-r}^{sk,n}$, *where*

$$\overline{n} = \begin{cases} n & \text{if } n \text{ is even,} \\ n-1 & \text{if } n \text{ is odd.} \end{cases}$$

The paper is arranged as follows. In Sect. 2, we recall the basics of finite and affine type root systems and Kac–Moody groups. We also describe how extending a finite type root system by attaching an extra root (in the manner prescribed in [11]) corresponds to extending the associated finite type group to its loop group. Finally, we recall some results on Weyl groups and (affine) Schubert varieties.

In Sect. 3, we show that the cotangent bundle of a cominuscule Grassmannian has a compactification ϕ by an affine Schubert variety. Along the way, we construct (Definition 3.6) an involution ι of affine type Dynkin diagrams that exchanges a cominuscule root α_d with the *extra* root α_0. The involution ι, viewed as a linear automorphism of the root lattice, acts on the associated Weyl group by conjugation (Eq. (3.2)).

In Sect. 4, we study the conormal variety $N^*X_{\mathfrak{J}}(w)$ of a Schubert variety $X_{\mathfrak{J}}(w)$ in a cominuscule Grassmannian G/P. We leverage the main result of [3] to develop characterizations of smooth Schubert varieties in G/P, see Proposition 4.5. We then use this to prove that $N^*X_{\mathfrak{J}}(w)$ has a compactification as a Schubert variety via the embedding ϕ if and only if the Schubert variety $X_{\mathfrak{J}}(w_0ww_{\mathfrak{J}})$ is smooth, see Theorem A. This yields powerful results about the geometry of $N^*X_{\mathfrak{J}}(w)$ when $X_{\mathfrak{J}}(w_0ww_{\mathfrak{J}})$ is smooth, see Theorem 4.9. Further, using Littelmann's work [15] on the standard monomial theory of affine Schubert varieties, we can write down the equations defining $N^*X_{\mathfrak{J}}(w)$ as a subvariety of T^*G/P, see Theorem 4.10. In Proposition 4.11, we give a description of the fibre of $N^*X_{\mathfrak{J}}(w)$ at identity as a union of Schubert varieties.

In Sect. 5, we apply the results of Sect. 4 to skew-symmetric determinantal varieties. The (usual, symmetric, skew-symmetric resp.) rank r determinantal varieties can be identified as the opposite cells of certain Schubert varieties $X_{\mathfrak{J}}(w_r)$ in certain cominuscule Grassmannian (of type A, C, D resp.). Working in type D, we first verify that $X_{\mathfrak{J}}(w_0 w w_r)$ is smooth (Eq. (5.5)); hence Proposition 4.11 applies. We then make explicit computations in the Weyl group (Proposition 5.6) to prove that the fibre at the zero matrix of the skew-symmetric determinantal variety is the rank $\overline{n} - r$ skew-symmetric determinantal variety, where $\overline{n}$ is as in Theorem B.

2 Dynkin Diagrams and Weyl Groups

In this section, we recall the basics of the theory of finite type and extended Dynkin diagrams, their root systems, and certain Kac–Moody groups associated with them. Throughout, we assume that the base field $\mathbf{k}$ is algebraically closed. The primary references for the combinatorial results in this section are [5, 13, 25]. For the geometric results, one may refer to [7, 13].

2.1 *Finite Type Dynkin Diagrams*

Let $\mathcal{D}_0$ be an irreducible *finite type* Dynkin diagram, and Δ_0 the *abstract root system* associated with $\mathcal{D}_0$. We shall denote by Δ_0^+, Δ_0^-, $\mathcal{D}_0$, θ_0, and W_0, the positive roots, negative roots, simple roots, highest root, and the Weyl group of Δ_0, respectively.

2.2 *Extended Dynkin Diagram and Root System*

The extended Dynkin diagram $\mathcal{D}$ is obtained by attaching a simple root α_0 to $\mathcal{D}_0$, see [11]. Let Δ, Δ^+, Δ^-, $\mathcal{D}$, and W denote the set of roots, positive roots, negative roots, simple roots, and the Weyl group, respectively, of the abstract root system of $\mathcal{D}$.

A root $\alpha \in \Delta$ is called a *real root* if there exists a $w \in W$ such that $w(\alpha) \in \mathcal{D}$; otherwise α is called an *imaginary root*. The root $\delta \overset{\text{def}}{=} \alpha_0 + \theta_0$ is called the *basic imaginary root*, where θ_0 denotes the highest root in Δ_0. The set of real roots Δ_{re}, and the set of positive roots Δ^+, have the following characterizations in terms of δ:

$$\Delta_{\text{re}} = \left\{\alpha + n\delta \,\middle|\, \alpha \in \Delta_0,\ n \in \mathbb{Z}\right\},$$

$$\Delta^+ = \left\{\alpha + n\delta \,\middle|\, \alpha \in \Delta_0 \sqcup \{0\},\ n > 0\right\} \bigsqcup \Delta_0^+.$$

2.3 Bruhat Order and Reduced Expressions

The Weyl group W is a *Coxeter group* with *simple reflections* $\{s_\alpha \mid \alpha \in \mathcal{D}\}$. The *Bruhat order* $\leq$ on W is the partial order generated by the relations

$$\begin{aligned} ws_\alpha > w &\iff w(\alpha) > 0 \qquad \forall \alpha \in \Delta^+, \\ s_\alpha w > w &\iff w^{-1}(\alpha) > 0 \qquad \forall \alpha \in \Delta^+. \end{aligned} \tag{2.1}$$

We say $w = s_1 \dots s_l$ is a *reduced expression* for w if each s_i is a simple reflection, and any other expression $w = s'_1 \dots s'_k$ satisfies $k \geqslant l$. The *length* $l(w)$ of an element $w \in W$ is the number of simple reflections in a reduced expression for w. The length function satisfies the relation $v < w \implies l(v) < l(w)$.

2.4 The Weyl Involution

The Weyl group W_0 is finite, and has a unique longest element w_0. The element w_0 is an involution, i.e., $w_0^2 = 1$, and further satisfies $w_0(\Delta_0^+) = \Delta_0^-$. It follows that $-w_0$ induces an involution of $\mathcal{D}_0$, called the *Weyl involution* (see [5, pg 158]).

2.5 Semi-Direct Product Decomposition

Let $\Lambda_0^\vee$ be the co-root lattice of Δ_0. There exists (cf. [13, §13.1.7]) a group isomorphism $W \to W_0 \ltimes \Lambda_0^\vee$ given by

$$\begin{aligned} s_\alpha &\mapsto (s_\alpha, 0) \qquad \text{for } \alpha \in \mathcal{D}_0, \\ s_{\alpha_0} &\mapsto (s_{\theta_0}, -\theta_0^\vee), \end{aligned}$$

where θ_0 is the highest root in Δ_0. For $q \in \Lambda_0^\vee$, we write $\tau_q \stackrel{\text{def}}{=} (1, q) \in W_0 \ltimes \Lambda_0^\vee$. The action of τ_q on Δ is determined by the formula $\tau_q(\delta) = \delta$, and

$$\tau_q(\alpha) = \alpha - \alpha(q)\delta \qquad \forall \alpha \in \Delta_0. \tag{2.2}$$

2.6 Support

The *support* of $w \in W$, denoted $Supp(w)$, is the smallest subset $\mathcal{J} \subseteq \mathcal{D}$ satisfying $w \in W_{\mathcal{J}}$. For $\alpha = \sum_{\beta \in \mathcal{D}} a_\beta \beta$, we define the *support* of α to be

$$Supp(\alpha) \stackrel{\text{def}}{=} \left\{\beta \in \mathcal{D} \,\middle|\, a_\beta \neq 0\right\}.$$

Observe that if $\alpha \in \Delta^+$ and $w(\alpha) \in \Delta^-$, then $Supp(\alpha) \subseteq Supp(w)$.

2.7 Minimal Representatives

Let $\mathcal{J}$ be some proper (necessarily finite type) sub-diagram of $\mathcal{D}$. We write $\Delta_{\mathcal{J}}$, $\Delta_{\mathcal{J}}^+$, $\Delta_{\mathcal{J}}^-$, $\mathcal{J}$, and $W_{\mathcal{J}}$ for the set of roots, positive roots, negative roots, simple roots, and the Weyl subgroup, respectively, whose support is contained in $\mathcal{J}$. Given an element $w \in W$, there exists a unique element $w^{\mathcal{J}}$, which is of minimal length in the coset $wW_{\mathcal{J}}$. The element $w^{\mathcal{J}}$ is called the *minimal representative* of w with respect to $\mathcal{J}$. The set of minimal representatives in W with respect to $\mathcal{J}$ is denoted by $W^{\mathcal{J}}$. For $\mathcal{J} = \mathcal{D}_0$, we will also use W^0 to denote the set of minimal representatives $W^{\mathcal{D}_0}$.

It follows from Eq. (2.1) that

$$W^{\mathcal{J}} = \{w \in W \mid w(\alpha) > 0,\ \forall \alpha \in \mathcal{J}\}. \tag{2.3}$$

In particular, we have $W_{\mathcal{J}} \subseteq W^{\mathcal{D}\setminus\mathcal{J}}$ for any $\mathcal{J} \subseteq \mathcal{D}$.

2.8 The Group G

Let G be the simply connected, almost simple algebraic group over $\mathbf{k}$ whose Dynkin diagram is $\mathcal{D}_0$. We fix a *torus* $T \subseteq G$, and a pair of *Borel subgroups* B and B^- in G, which are opposite with respect to T, i.e., we have $B \cap B^- = T$. We identify the root system of (G, B, T) with the abstract root system Δ_0, and the Weyl group W_0 with N/T, where N is the normalizer of T in G.

2.9 The Loop Group

Let $\mathcal{O} \stackrel{\text{def}}{=} \mathbf{k}[t]$, $\mathcal{O}^- \stackrel{\text{def}}{=} \mathbf{k}[t^{-1}]$, and $\mathcal{K} \stackrel{\text{def}}{=} \mathbf{k}[t, t^{-1}]$. The *loop group* $LG \stackrel{\text{def}}{=} G(\mathcal{K})$ is a Kac–Moody group with Dynkin diagram $\mathcal{D}$, and is ind-representable by an affine scheme over $\mathbf{k}$. Throughout, we shall identify the Weyl group W of $\mathcal{D}$ with $N(\mathcal{K})/T$.

Let $\mathfrak{g}$ be the Lie algebra of G. We identify the Lie algebra $L\mathfrak{g}$ of LG with $\mathfrak{g} \otimes \mathcal{K}$. Let U_α denote the *root subgroup* corresponding to a real root $\alpha \in \Delta_{re}$ (see [4, 20, 22]). We can identify G as the subgroup of LG generated by T and $\left\{U_\alpha \,\middle|\, \alpha \in \Delta_0\right\}$.

Let $L^+G \stackrel{\text{def}}{=} G(\mathcal{O})$, $L^-G \stackrel{\text{def}}{=} G(\mathcal{O}^-)$, and consider the *evaluation maps*

$$\pi : L^+G \to G, \quad t \mapsto 0, \qquad \pi_- : L^-G \to G, \quad t^{-1} \mapsto 0.$$

The subgroups $\mathcal{B} \stackrel{\text{def}}{=} \pi^{-1}(B)$ and $\mathcal{B}^- \stackrel{\text{def}}{=} \pi_-^{-1}(B^-)$ are called Borel subgroups of LG. The Borel subgroups $\mathcal{B}$ and $\mathcal{B}^-$ are opposite subgroups with respect to T, i.e., $\mathcal{B} \cap \mathcal{B}^- = T$.

2.10 The Adjoint Group $\mathcal{G}$

Observe that the adjoint action of G on $\mathfrak{g}$ induces a t-linear G-action on $L\mathfrak{g}$, and hence a G-action on $\mathrm{End}(L\mathfrak{g})$. Following [20, §7.54], we have a G-equivariant homomorphism, $Ad : LG \to \mathrm{GL}(L\mathfrak{g})$, induced by the conjugation action of LG on itself.

Let $\mathcal{G}$ (resp. $\mathcal{G}_0$) denote the image of LG (resp. L^+G) under Ad. We have $\ker(Ad) \subseteq T$ (cf. [20, §7.55]); consequently, Ad yields a G-equivariant isomorphism, $LG/L^+G \cong \mathcal{G}/\mathcal{G}_0$.

Finally, the conjugation (resp. left multiplication) action of G on LG (resp. LG/L^+G) yields a G-action on $\mathcal{G}$ (resp. $\mathcal{G}/\mathcal{G}_0$). It is a simple verification that the quotient map $\mathcal{G} \to \mathcal{G}/\mathcal{G}_0$ is G-equivariant with respect to these actions.

2.11 Nilpotent Set of Roots

(See [25]) Let Ψ be a finite set of real roots. We say that Ψ is *pre-nilpotent* if there exist $w, w' \in W$ such that $w(\Psi) \subseteq \Delta^+$ and $w'(\Psi) \subseteq \Delta^-$. We say that Ψ is *closed* if

$$\alpha, \beta \in \Psi, \ \alpha + \beta \in \Delta \implies \alpha + \beta \in \Psi.$$

Finally, we say Ψ is *nilpotent* if it is pre-nilpotent and closed. For $\alpha \in \Psi$, let $\mathfrak{g}^\alpha$ be the associated root space (see, for example, [11]). If Ψ is nilpotent, then so is the Lie sub-algebra

$$\mathfrak{g}^\Psi \stackrel{\text{def}}{=} \bigoplus_{\alpha \in \Psi} \mathfrak{g}^\alpha. \tag{2.4}$$

2.12 Tits' Functor for Kac–Moody Groups

For α a real root, let U_α be the group scheme over $\mathbf{k}$ isomorphic to $\mathbb{G}_a$ with Lie algebra $\mathfrak{g}^\alpha$. To every nilpotent set of roots Ψ, Tits[25] associates a group scheme

U_Ψ that depends only on $\mathfrak{g}^\Psi$, and is naturally a closed subgroup scheme of LG (cf. [25]). For any ordering of Ψ, the product morphism

$$\prod_{\alpha \in \Psi} U_\alpha \to U_\Psi$$

is a scheme isomorphism. Hence, we get a scheme isomorphism $\exp : \mathfrak{g}^\Psi \to U_\Psi$.

2.13 Parabolic Subgroups

Recall the identification $W \cong N(\mathcal{K})/T$. For every $w \in W$, we fix a lift $\widetilde{w}$ of w to $N(\mathcal{K})$. Then, for any subset $\mathcal{J} \subseteq \mathcal{D}$, the subgroup

$$\mathcal{P}_\mathcal{J} \stackrel{\text{def}}{=} \mathcal{B} W_\mathcal{J} \mathcal{B} = \left\{ b\widetilde{w}b' \mid b, b' \in \mathcal{B},\ w \in W_\mathcal{J} \right\}$$

is the *parabolic subgroup* of LG corresponding to $\mathcal{J}$. The parabolic subgroup $\mathcal{P}_{\mathcal{D}_0}$ is precisely L^+G.

For a subset $\mathcal{J} \subseteq \mathcal{D}_0$, the subgroup

$$P_\mathcal{J} \stackrel{\text{def}}{=} B W_\mathcal{J} B = \left\{ b\widetilde{w}b' \mid b, b' \in B,\ w \in W_\mathcal{J} \right\}$$

is the parabolic subgroup of G corresponding to $\mathcal{J} \subseteq \mathcal{D}_0$.

2.14 The Bruhat Decomposition

Consider some subset $\mathcal{J}$ of $\mathcal{D}_0$. The *Bruhat decomposition* of G is

$$G = \bigsqcup_{w \in W_0^\mathcal{J}} B w P_\mathcal{J},$$

where $W_0^\mathcal{J} \stackrel{\text{def}}{=} W_0 \cap W^\mathcal{J}$. The *partial flag variety* $G/P_\mathcal{J}$ has the decomposition

$$G/P_\mathcal{J} = \bigsqcup_{w \in W_0^\mathcal{J}} B w P_\mathcal{J} \ (\text{mod } P_\mathcal{J}).$$

Let $\leq$ denote the *Bruhat order* on W. For $w \in W_0^\mathcal{J}$, the *Schubert variety*

$$X_{\mathcal{J}}(w) \stackrel{\text{def}}{=} \overline{BwP_{\mathcal{J}}} \,(\text{mod}\, P_{\mathcal{J}}) = \bigsqcup_{\substack{v \in W_0^{\mathcal{J}} \\ v \leqslant w}} BvP_{\mathcal{J}} \,(\text{mod}\, P_{\mathcal{J}})$$

is a projective variety of dimension $l(w)$.

2.15 Affine Schubert Varieties

Fix a proper subset $\mathcal{J} \subsetneq \mathcal{D}$. We have the *Bruhat decomposition*

$$LG = \bigsqcup_{w \in W^{\mathcal{J}}} \mathcal{B}w\mathcal{P}_{\mathcal{J}}.$$

The quotient $LG/\mathcal{P}_{\mathcal{J}}$ is an ind-scheme (in the sense of [7]). For $w \in W^{\mathcal{J}}$, the *affine Schubert variety*

$$X_{\mathcal{J}}(w) \stackrel{\text{def}}{=} \overline{\mathcal{B}w\mathcal{P}_{\mathcal{J}}} \,(\text{mod}\, \mathcal{P}_{\mathcal{J}})$$

is a projective variety of dimension $l(w)$, and has the decomposition

$$X_{\mathcal{J}}(w) = \bigsqcup_{\substack{v \leqslant w \\ v \in W^{\mathcal{J}}}} \mathcal{B}v\mathcal{P}_{\mathcal{J}} \,(\text{mod}\, \mathcal{P}_{\mathcal{J}}).$$

Consider a proper subset $\mathcal{L} \subsetneq \mathcal{D}$, and let $w_{\mathcal{L}}$ be the longest element in $W_{\mathcal{L}}$. The Schubert variety $X_{\mathcal{J}}(w_{\mathcal{L}}^{\mathcal{J}})$ is $\mathcal{P}_{\mathcal{L}}$-homogeneous, hence smooth. Indeed, we have $\mathcal{P}_{\mathcal{L}} \cap \mathcal{P}_{\mathcal{J}} = \mathcal{P}_{\mathcal{L} \cap \mathcal{J}}$, and further, $X_{\mathcal{J}}(w_{\mathcal{L}}^{\mathcal{J}}) = \mathcal{P}_{\mathcal{L}}/\mathcal{P}_{\mathcal{L} \cap \mathcal{J}}$.

2.16 The Opposite Cell

Let $\mathcal{J}, \mathcal{L}$ be as above. Let $\boldsymbol{e}$ denote the image of the identity element in the quotient $LG/\mathcal{P}_{\mathcal{J}}$. The *opposite cell* $X_{\mathcal{J}}^{-}(w)$, given by

$$X_{\mathcal{J}}^{-}(w) \stackrel{\text{def}}{=} \mathcal{B}^{-}\boldsymbol{e} \bigcap X_{\mathcal{J}}(w),$$

is an *open affine* subvariety of $X_{\mathcal{J}}(w)$. The opposite cell of the Schubert variety $X_{\mathcal{J}}(w_{\mathcal{L}}^{\mathcal{J}})$ is isomorphic to the *affine space* $\mathbb{A}^k$ for $k = Card(\Delta_{\mathcal{L}}^{-} \backslash \Delta_{\mathcal{J}})$. Indeed, for any enumeration $\alpha_1, \ldots, \alpha_k$ of $\Delta_{\mathcal{L}}^{-} \backslash \Delta_{\mathcal{J}}$, the map

$$U_{\alpha_1} \times \ldots \times U_{\alpha_k} \longrightarrow LG/\mathcal{P}_{\mathcal{J}},$$
$$(u_1, \ldots, u_k) \mapsto u_1 \ldots u_k \,(\mathrm{mod}\, \mathcal{P}_{\mathcal{J}})$$

is an open immersion onto the opposite cell $X^-_{\mathcal{J}}(w^{\mathcal{J}}_{\mathcal{L}})$. We also mention here the *L^-G cell $Y_{\mathcal{J}}(w) \subseteq X_{\mathcal{J}}(w)$*,

$$Y_{\mathcal{J}}(w) \stackrel{\text{def}}{=} L^-Ge \bigcap X_{\mathcal{J}}(w), \tag{2.5}$$

which is also an *open affine* subvariety in $X_{\mathcal{J}}(w)$.

2.17 *The Demazure Product*

The *Demazure product* $\star$ on W is the unique associative product satisfying

$$s_\alpha \star w = \begin{cases} w & \text{if } s_\alpha w < w, \\ s_\alpha w & \text{if } s_\alpha w > w. \end{cases} \tag{2.6}$$

The double coset $\mathcal{B}w\mathcal{B}$ is called a *Bruhat cell* in LG. Suppose $v = s_1 \ldots s_k$ is a reduced presentation for $v \in W$. Then $v = s_1 \star \ldots \star s_k$, and

$$\mathcal{B}v\mathcal{B} = \mathcal{B}s_1\mathcal{B} \ldots \mathcal{B}s_k\mathcal{B}.$$

Consider $v \in W^{\mathcal{J}}$, $w \in W_{\mathcal{J}}$. Then $v \star w = vw$. More generally, we have

$$l(vw) = l(v) + l(w) \iff v \star w = vw. \tag{2.7}$$

Remark The reader may refer to [12, Remark 3.3] for further details, including a justification of the name "Demazure product". □

3 The Cominuscule Grassmannian

Let $\mathcal{D}_0$ be a finite type Dynkin diagram, and $\mathcal{D}$ its associated extended Dynkin diagram. In this section, we first recall the notion of cominuscule roots and cominuscule Grassmannians associated with $\mathcal{D}_0$. We then fix a choice of cominuscule root α_d, and develop the rest of this section (and the next) for that fixed choice of α_d. We introduce in Definition 3.6 a canonical Dynkin diagram involution ι depending only on α_d.

Let G be the almost simple, simply connected algebraic group defined over **k**, with Dynkin diagram $\mathcal{D}_0$, and let $P \subseteq G$ be the maximal parabolic subgroup corresponding to "omitting" the simple root α_d. In the case where char($\mathbf{k}$) $= 0$, Lakshmibai, Ravikumar, and Slofstra [17] have constructed an isomorphism ϕ of T^*G/P with the opposite cell of an affine Schubert variety in $LG/\mathcal{P}$ (see Appendix 3.2 for the definition of $\mathcal{P}$). We give an alternate description of this Schubert variety in Eq. (3.3). Our goal in this section is to extend the results of [17] to the case where char($\mathbf{k}$) is an odd prime which is also a good prime for G, see Proposition 3.14 and Theorem 3.18.

Definition 3.1 A simple root $\alpha_d \in \mathcal{D}_0$ is called *cominuscule* if the coefficient of α_d in the highest root θ_0 (or equivalently, in the basic imaginary root $\delta = \alpha_0 + \theta_0$) is 1. □

Observe from Table 1 that a simple root α_d is cominuscule if and only if there exists an automorphism ι of $\mathcal{D}$ such that $\iota(\alpha_0) = \alpha_d$. For the remainder of this section (and the next), let α_d be some *fixed cominuscule root* in $\mathcal{D}_0$, and set

$$\mathcal{D}_d \stackrel{\text{def}}{=} \mathcal{D} \setminus \{\alpha_d\}, \qquad \mathcal{J} \stackrel{\text{def}}{=} \mathcal{D}_0 \cap \mathcal{D}_d, \qquad \theta_d \stackrel{\text{def}}{=} \delta - \alpha_d.$$

We write Δ_d, Δ_d^+, Δ_d^-, and W_d for the set of roots, positive roots, negative roots, and Weyl group, respectively, of the root system associated with $\mathcal{D}_d$. Further, we denote by W^d the set of minimal representatives in W with respect to W_d. Observe that θ_d is the highest root of the finite type root system Δ_d.

3.2 The Cominuscule Grassmannian

Let $\mathcal{P}_d$ be the parabolic subgroup corresponding to the set of simple roots $\mathcal{D}_d$. We will write P, $\mathcal{P}$ for the parabolic subgroups $P_{\mathcal{J}} \subseteq G$ and $\mathcal{P}_{\mathcal{J}} \subseteq LG$, respectively. Observe that $\mathcal{P} = L^+G \cap \mathcal{P}_d$ and $P = G \cap \mathcal{P}_d$. The variety G/P is called a *cominuscule Grassmannian* of type $\mathcal{D}_0$.

3.3 The Cotangent Bundle

Let $\mathfrak{g}$, $\mathfrak{p}$, $\mathfrak{h}$ denote the Lie algebras of G, P, T, respectively. We have the *root space decompositions*

$$\mathfrak{g} = \mathfrak{h} \oplus \bigoplus_{\alpha \in \Delta_0} \mathfrak{g}^\alpha,$$

$$\mathfrak{p} = \mathfrak{h} \oplus \bigoplus_{\alpha \in \Delta_0^+ \sqcup \Delta_{\mathcal{J}}^-} \mathfrak{g}^\alpha,$$

$$\mathfrak{u}_P = \bigoplus_{\alpha \in \Delta_0^+ \setminus \Delta_{\mathfrak{J}}^+} \mathfrak{g}^\alpha.$$

Following [2, §5.1.11], we have

$$T^*G/P = G \times^P \mathfrak{u}_P = G \times \mathfrak{u}_P / \sim,$$

where $\sim$ is the equivalence relation given by $(gp, u) \sim (g, Ad(p)u)$ for $g \in G,\ u \in \mathfrak{u}_P,\ p \in P$.

Lemma 3.4 *Let $w_{\mathfrak{J}}$ be the longest element of $W_{\mathfrak{J}}$. We have $w_{\mathfrak{J}}(\alpha_d) = \theta_0$ and $w_{\mathfrak{J}}(\alpha_0) = \theta_d$.* □

Proof To show $w_{\mathfrak{J}}(\alpha_d) = \theta_0$, it is enough to show that $w_{\mathfrak{J}}(\alpha_d)$ is maximal in $\Delta_0^+ \setminus \Delta_{\mathfrak{J}}^+$. Observe first that $w_{\mathfrak{J}}(\Delta_0) = \Delta_0$, $w_{\mathfrak{J}}(\Delta_{\mathfrak{J}}) = \Delta_{\mathfrak{J}}$, and

$$\left\{\alpha \in \Delta_0^+ \,\middle|\, w_{\mathfrak{J}}(\alpha) < 0\right\} = \Delta_{\mathfrak{J}}^+.$$

Consequently, $w_{\mathfrak{J}}(\Delta_0^+ \setminus \Delta_{\mathfrak{J}}^+) \subseteq \Delta_0^+ \setminus \Delta_{\mathfrak{J}}^+$. Consider $\alpha \in \Delta_0^+ \setminus \Delta_{\mathfrak{J}}^+$, and let $\gamma = \alpha - \alpha_d$. Observe that $\alpha_d \leqslant \alpha$; hence $\gamma \geqslant 0$. Further, since α_d is cominuscule, we have $2\alpha_d \not\leq \alpha$. It follows that $\alpha_d \notin Supp(\gamma)$, and so $Supp(\gamma) \subseteq \mathfrak{J}$. Hence

$$\begin{aligned} w_{\mathfrak{J}}(\gamma) \leqslant 0 &\implies w_{\mathfrak{J}}(\alpha) = w_{\mathfrak{J}}(\alpha_d) + w_{\mathfrak{J}}(\gamma), \\ &\implies w_{\mathfrak{J}}(\alpha) \leqslant w_{\mathfrak{J}}(\alpha_d). \end{aligned}$$

We see that $w_{\mathfrak{J}}(\alpha_d)$ is maximal in $\Delta_0^+ \setminus \Delta_{\mathfrak{J}}^+$, hence $w_{\mathfrak{J}}(\alpha_d) = \theta_0$. The formula $w_{\mathfrak{J}}(\alpha_0) = \theta_d$ follows similarly, by showing that $w_{\mathfrak{J}}(\alpha_0)$ is maximal in $\Delta_d^+ \setminus \Delta_{\mathfrak{J}}^+$.

3.5 *Bilinear Form*

Let V denote the real vector space with basis $\mathcal{D}$. There exists a W-invariant symmetric bilinear form $(\ |\ )$ on V (cf. [11, §3.7]) such that

$$s_\alpha(\beta) = \beta - 2\frac{(\alpha \mid \beta)}{(\alpha \mid \alpha)}\alpha. \tag{3.1}$$

Definition 3.6 (The Involution ι) Let ι be the linear involution of V given by

$$\iota(\alpha) = \begin{cases} \alpha_d & \text{for } \alpha = \alpha_0, \\ \alpha_0 & \text{for } \alpha = \alpha_d, \\ -w_{\mathfrak{J}}(\alpha) & \text{for } \alpha \in \mathfrak{J}. \end{cases}$$

Lemma 3.7 *The form $(\ |\ )$ is invariant under $\iota \in GL(V)$.* □

Proof Recall that $(\delta\,|\ ) = 0$ (cf. [11, §5.2]). It is sufficient to verify the invariance on the basis $\mathfrak{J} \cup \{\alpha_0, \alpha_d\}$. For $\alpha, \beta \in \mathfrak{J}$, we have

$$\begin{aligned}(\iota(\alpha)\,|\,\iota(\beta)) &= \big(-w_{\mathfrak{J}}(\alpha)\,\big|\,-w_{\mathfrak{J}}(\beta)\big) = (\alpha\,|\,\beta)\,,\\ (\iota(\alpha_0)\,|\,\iota(\beta)) &= \big(\alpha_d\,\big|\,-w_{\mathfrak{J}}(\beta)\big) = \big(w_{\mathfrak{J}}(\theta_0)\,\big|\,-w_{\mathfrak{J}}(\beta)\big)\\ &= (-\theta_0\,|\,\beta) = (\alpha_0 - \delta\,|\,\beta) = (\alpha_0\,|\,\beta)\,,\\ (\iota(\alpha_d)\,|\,\iota(\beta)) &= \big(\alpha_0\,\big|\,-w_{\mathfrak{J}}(\beta)\big) = \big(w_{\mathfrak{J}}(\theta_d)\,\big|\,-w_{\mathfrak{J}}(\beta)\big)\\ &= (-\theta_d\,|\,\beta) = (\alpha_d - \delta\,|\,\beta) = (\alpha_d\,|\,\beta)\,,\\ (\iota(\alpha_0)\,|\,\iota(\alpha_d)) &= (\alpha_d\,|\,\alpha_0) = (\alpha_0\,|\,\alpha_d)\,.\end{aligned}$$

Proposition 3.8 *The map ι induces an involution of the Dynkin diagram $\mathcal{D}$.* □

Proof It is clear from the definition that ι is an involution. Further, since $-w_{\mathfrak{J}}$ induces an involution of $\mathfrak{J}$ that preserves its Dynkin diagram structure (cf. [5, pg 158]), it follows that ι preserves the set of simple roots $\mathcal{D}$. Now, it follows from

$$\alpha_i^\vee(\alpha_j) = \frac{(\alpha_i\,|\,\alpha_j)}{(\alpha_i\,|\,\alpha_i)}$$

that the *Cartan matrix* $\big(\alpha_i^\vee(\alpha_j)\big)_{ij}$ is preserved under ι, and so ι preserves the Dynkin diagram structure on $\mathcal{D}$.

Corollary 3.9 *We have the equality $\iota(\delta) = \delta$.* □

Proof Consider $\gamma \in \mathbb{Z}\mathcal{D}$ satisfying $(\gamma|\ ) = 0$. Then $\gamma = k\delta$ for some $k \in \mathbb{Z}$ (cf. [11, §5.6]). Now, it follows from Lemma 3.7 and the fact $(\delta\,|\ ) = 0$, that $(\iota(\delta)\,|\ ) = 0$. Hence $\iota(\delta) = k\delta$ for some $k \in \mathbb{Z}$. Further it follows from Proposition 3.8 that $k > 0$ and $k^2 = 1$. We deduce that $k = 1$, i.e., $\iota(\delta) = \delta$.

3.10 Action on W

We also define an involution ι of W given by

$$^{\iota}s_\alpha \overset{\text{def}}{=} s_{\iota(\alpha)} \qquad \text{for } \alpha \in \mathcal{D}.$$

It is clear that ι preserves the length and the Bruhat order on W, and

$$\begin{aligned}&{}^{\iota}W_{\mathfrak{J}} = W_{\mathfrak{J}}, &&{}^{\iota}W^{\mathfrak{J}} = W^{\mathfrak{J}}, &&{}^{\iota}w_{\mathfrak{J}} = w_{\mathfrak{J}},\\ &{}^{\iota}W_0 = W_d, &&{}^{\iota}W^0 = W^d, &&{}^{\iota}w_d = w_0.\end{aligned}$$

Next, we show that the action of ι on W is the same as conjugation by ι, i.e.,

$$ {}^{\iota}w = \iota \circ w \circ \iota, \tag{3.2} $$

where both w and ι are viewed as elements of $GL(V)$. Using Eq. (3.1), Definition 3.6, and Lemma 3.7, we have for any $\beta \in \mathcal{D}$,

$$ \begin{aligned} (\iota \circ s_\alpha \circ \iota)(\beta) = \iota(s_\alpha(\iota(\beta))) &= \iota\left(\iota(\beta) - 2\frac{(\alpha \mid \iota(\beta))}{(\alpha \mid \alpha)}\alpha\right) \\ &= \beta - 2\frac{(\iota(\alpha) \mid \beta)}{(\iota(\alpha) \mid \iota(\alpha))}\iota(\alpha) \\ &= s_{\iota(\alpha)}(\beta). \end{aligned} $$

Note also that since Schubert varieties depend only on the underlying Dynkin diagrams, there exists an isomorphism $X_{\mathfrak{J}}({}^{\iota}w) \cong X_{\mathfrak{J}}(w)$ for any $w \in W$.

3.11 The Co-Weight q

Let w_0, w_d be the maximal elements in W_0, W_d, respectively, and let $\varpi_d^\vee$ be the fundamental co-weight dual to α_d. Set

$$ q \overset{\text{def}}{=} w_0(\varpi_d^\vee) - \varpi_d^\vee \in \Lambda_0^\vee, \tag{3.3} $$

and let $\tau_q \in W$ be the element corresponding to $q \in \Lambda_0^\vee$ (see Appendix 2.5).

Proposition 3.12 *We have the equality* $\tau_q = w_0 w_{\mathfrak{J}} w_d w_{\mathfrak{J}} = w_0^{\mathfrak{J}} w_d^{\mathfrak{J}}$. □

Proof The set of roots Δ is contained in the $\mathbb{Z}$-span of the set $\mathcal{D}_0 \cup \{\delta\}$. Since the action of W on Δ is faithful, it is enough to verify

$$ w_0 w_{\mathfrak{J}} w_d w_{\mathfrak{J}}(\alpha) = \tau_q(\alpha) \qquad \forall \alpha \in \mathcal{D}_0 \cup \{\delta\}. \tag{3.4} $$

Further, since δ is fixed under the action of W (cf. Appendix 2.5), it is sufficient to verify Eq. (3.4) for $\alpha \in \mathcal{D}_0$.

Recall from Appendix 2.4 $-w_0$ induces an involution on $\mathcal{D}_0$. Set $\beta = -w_0(\alpha_d)$, so that $-w_0(\varpi_d^\vee) = \varpi_\beta^\vee$, the fundamental co-weight dual to β. It follows from Eqs. (2.2) and (3.3) that

$$ \begin{aligned} \tau_q(\alpha) &= \alpha + \alpha(\varpi_d^\vee)\delta - \alpha(w_0(\varpi_d^\vee))\delta \\ &= \alpha + \alpha(\varpi_d^\vee + \varpi_\beta^\vee)\delta. \end{aligned} \tag{3.5} $$

Hence we can rewrite Eq. (3.4) as

$$w_0 w_{\mathfrak{J}} w_d w_{\mathfrak{J}}(\alpha) = \alpha + \alpha(\varpi_d^\vee + \varpi_\beta^\vee)\delta \qquad \forall\, \alpha \in \mathcal{D}_0. \tag{3.6}$$

Next, it follows from $\beta = -w_0(\alpha_d)$ that

$$\begin{aligned} \iota(\beta) &= -\iota(w_0(\alpha_d)) = -{}^\iota w_0(\iota(\alpha_d)) && \text{using Eq. (3.2),} \\ \Longrightarrow \iota(\beta) &= -w_d(\alpha_0) && \text{using Appendix 3.10.} \end{aligned} \tag{3.7}$$

Further, we have

$$\begin{aligned} w_d w_{\mathfrak{J}}(\alpha_d) &= w_d(\theta_0) = w_d(\delta - \alpha_0) && \text{using Lemma 3.4} \\ &= \delta - w_d(\alpha_0) = \delta + \iota(\beta) && \text{using Eq. (3.7),} \end{aligned} \tag{3.8}$$

$$\begin{aligned} w_0 w_{\mathfrak{J}}(\alpha_0) &= {}^\iota(w_d w_{\mathfrak{J}})(\iota(\alpha_d)) && \text{using Appendix 3.10} \\ &= \iota(\delta + \iota(\beta)) = \delta + \beta && \text{using Corollary 3.9.} \end{aligned} \tag{3.9}$$

We are now ready to prove that Eq. (3.6) holds for $\alpha \in \{\alpha_d, \beta\}$.

Case 1 Suppose $\beta = \alpha_d$. Then $\iota(\beta) = \alpha_0$, $\varpi_d^\vee = \varpi_\beta^\vee$, and $q = \tau_{2\varpi_d^\vee}$. We have

$$\begin{aligned} & w_0 w_{\mathfrak{J}} w_d w_{\mathfrak{J}}(\alpha_d) \\ &\quad = w_0 w_{\mathfrak{J}}(\delta + \iota(\beta)) && \text{using Eq. (3.8)} \\ &\quad = w_0 w_{\mathfrak{J}}(\delta + \alpha_0) && \text{using } \beta = \alpha_d \\ &\quad = w_0(\delta + \theta_d) && \text{using Lemma 3.4} \\ &\quad = w_0(2\delta - \alpha_d) = 2\delta + \beta && \text{using } \beta = -w_0(\alpha_d) \\ &\quad = \alpha_d + 2\delta = \tau_q(\alpha_d) && \text{using } \beta = \alpha_d \text{ and Eq. (3.5).} \end{aligned}$$

Case 2 Suppose $\beta \neq \alpha_d$. Then $\beta, \iota(\beta) \in \mathfrak{J}$, and $\varpi_\beta^\vee(\alpha_d) = \varpi_d^\vee(\beta) = 0$. It follows from Definition 3.6 that $\iota(\beta) = -w_{\mathfrak{J}}(\beta)$, hence $w_{\mathfrak{J}}(\iota(\beta)) = -\beta$. We have

$$\begin{aligned} w_0 w_{\mathfrak{J}} w_d w_{\mathfrak{J}}(\alpha_d) &= w_0 w_{\mathfrak{J}}(\delta + \iota(\beta)) && \text{using Eq. (3.8)} \\ &= w_0(\delta - \beta) = \delta - w_0(\beta) \\ &= \delta + \alpha_d = \tau_q(\alpha_d) && \text{using Eq. (3.5)} \\ w_0 w_{\mathfrak{J}} w_d w_{\mathfrak{J}}(\beta) &= w_0 w_{\mathfrak{J}} w_d(-\iota(\beta)) \\ &= w_0 w_{\mathfrak{J}} w_d w_d(\alpha_0) && \text{using Eq. (3.7)} \\ &= w_0 w_{\mathfrak{J}}(\alpha_0) = \delta + \beta && \text{using Eq. (3.9)} \\ &= \tau_q(\beta) && \text{using Eq. (3.5).} \end{aligned}$$

Finally, we prove that Eq. (3.6) holds for any $\alpha \in \mathfrak{D}_0 \backslash \{\alpha_d, \beta\} = \mathfrak{J} \backslash \{\beta\}$. Since $\alpha \in \mathfrak{D}_0$, we have $-w_0(\alpha) \in \mathfrak{D}_0$. Observe further that since $\alpha \neq \beta$, we have $-w_0(\alpha) \neq \alpha_d$, and so $-w_0(\alpha) \in \mathfrak{J}$. Applying Definition 3.6, we get

$$\begin{aligned} \iota(-w_0(\alpha)) &= w_{\mathfrak{J}} w_0(\alpha) \\ \Longrightarrow w_{\mathfrak{J}} \iota w_0(\alpha) &= -w_0(\alpha). \end{aligned} \tag{3.10}$$

We see from Eq. (3.5) that $\tau_q(\alpha) = \alpha$. We compute

$$\begin{aligned} w_0 w_{\mathfrak{J}} w_d w_{\mathfrak{J}}(\alpha) &= -w_0 w_{\mathfrak{J}}{}' w_0 \iota(\alpha) && \text{using } \alpha \notin \mathfrak{J}, \text{ and Definition 3.6} \\ &= -w_0 w_{\mathfrak{J}} \iota w_0(\alpha) && \text{using Eq. (3.2)} \\ &= w_0^2(\alpha) = \alpha = \tau_q(\alpha) && \text{using Eq. (3.10).} \end{aligned}$$

Lemma 3.13 *Let* $\Psi \stackrel{\text{def}}{=} \Delta_d^- \backslash \Delta_{\mathfrak{J}}$, *and consider some* $\gamma \in \mathfrak{D}_0 \cup \pm\mathfrak{J}$. *Then*

(a) All subsets of Ψ *are nilpotent (see Appendix 2.11).*
(b) The set $\Psi \cup \{\gamma\}$ *is nilpotent.* □

Proof First observe that $\Psi = \Delta_d^- \backslash \Delta_{\mathfrak{J}} = \{\alpha \in \Delta \,|\, -\theta_d \leqslant \alpha \leqslant -\alpha_0\}$. Consider $\alpha, \beta \in \Psi$. Then $Supp(\alpha + \beta) \subseteq \mathfrak{D}_d$. Further, we have

$$\alpha, \beta \leqslant -\alpha_0 \Longrightarrow \alpha + \beta \leqslant -2\alpha_0.$$

Now, since α_0 is cominuscule in $\mathfrak{D}_d$, it follows that $\alpha + \beta \notin \Delta$. Consequently, every subset of Ψ is closed. Next, we prove that $\Psi \cup \{\gamma\}$ is closed.

(a) Suppose $\gamma = \alpha_d$. Consider $\alpha \in \Psi$. The coefficient of α_0 in $\alpha + \gamma$ is -1, and the coefficient of α_d is 1. Hence, $\alpha + \gamma$ is not a root.
(b) Suppose $\gamma \in \mathfrak{J}$. Suppose further that $\alpha + \gamma \in \Delta$ for some $\alpha \in \Psi$. Since $Supp(\alpha + \gamma) \subseteq \mathfrak{D}_d$, we see that $\alpha + \gamma \in \Delta_d^-$. Further, since the coefficient of α_0 in $\alpha + \gamma$ is -1, we have $\alpha + \gamma \notin \Delta_{\mathfrak{J}}$.
(c) Suppose $\gamma \in -\mathfrak{J}$. Suppose further that $\alpha + \gamma \in \Delta$ for some $\alpha \in \Psi$. Then $\alpha + \gamma \leqslant -\alpha_0$ and $\alpha + \gamma \in \Delta_d^-$. It follows that $\alpha + \gamma \in \Psi$.

Finally, consider $u_+, u_- \in W$ given by

$$u_+ = \begin{cases} w_d & \text{if } \gamma = \alpha_d, \\ w_d s_\gamma & \text{if } \gamma \in \mathfrak{J}, \\ w_d & \text{if } \gamma \in -\mathfrak{J}, \end{cases} \qquad u_- = \begin{cases} s_{\alpha_d} & \text{if } \gamma = \alpha_d, \\ s_\gamma & \text{if } \gamma \in \mathfrak{J}, \\ 1 & \text{if } \gamma \in -\mathfrak{J}. \end{cases}$$

It is easy to verify that $u_\pm(\Psi \cup \{\gamma\}) \subseteq \Delta^\pm$.

Recall from Eq. (2.4) the Lie sub-algebra $\mathfrak{g}^\Psi = \sum_{\alpha \in \Psi} \mathfrak{g}^\alpha$. Recall also from Appendix 2.12 the isomorphism, $\exp : \mathfrak{g}^\Psi \to U_\Psi$, for some ordering on Ψ.

Proposition 3.14 *Recall the opposite cell $X^-_{\mathfrak{z}}(w^{\mathfrak{z}}_d)$ from Appendix 2.16. There exists a P-equivariant isomorphism $\phi : \mathfrak{u}_P \to X^-_{\mathfrak{z}}(w^{\mathfrak{z}}_d)$ given by*

$$\phi(X) = \exp(t^{-1}X) \bmod \mathcal{P}.$$

Proof The map $X \mapsto t^{-1}X$ is G-equivariant (hence also P-equivariant), and takes the root space $\mathfrak{g}^\alpha$ to $\mathfrak{g}^{\alpha-\delta}$. Observe that $\mathfrak{u}_P = \mathfrak{g}^{\Delta_0^+\backslash\Delta_{\mathfrak{z}}}$, and further

$$\begin{aligned} \Delta_0^+\backslash\Delta_{\mathfrak{z}} - \delta &= \{\alpha - \delta \mid \alpha \in \Delta_0\backslash\Delta_{\mathfrak{z}}\} \\ &= \{\alpha - \delta \mid \alpha \in \Delta,\ \alpha_d \leqslant \alpha \leqslant \theta_0\} \\ &= \{\alpha \in \Delta \mid -\theta_d \leqslant \alpha \leqslant -\alpha_0\} = \Delta_d^-\backslash\Delta_{\mathfrak{z}}. \end{aligned} \tag{3.11}$$

Hence, we have a map $t^{-1} : \mathfrak{u}_P \to \mathfrak{g}^\Psi$, where $\Psi \stackrel{\text{def}}{=} \Delta_d^-\backslash\Delta_{\mathfrak{z}}$. It follows from Appendix 2.16 that $\exp \circ t^{-1}$ is an isomorphism from $\mathfrak{u}_P$ to $Y_{\mathfrak{z}}(w^{\mathfrak{z}}_d)$. It remains to show that exp is P-equivariant. For $\gamma \in \mathcal{D}_0 \cup -\mathfrak{z}$, it follows from Lemma 3.13 that the group scheme $U_{\Psi\cup\{\gamma\}}$ is well-defined. The action of U_γ on U_Ψ (resp. $\mathfrak{g}^\Psi$) being the restriction of the adjoint action of $U_{\Psi\cup\{\gamma\}}$ on itself (resp. $\mathfrak{g}^{\Psi\cup\{\gamma\}}$), the map exp is U_γ-equivariant. It follows that exp is P-equivariant, since $P = \langle T,\ U_\gamma \mid \gamma \in \mathcal{D}_0 \cup -\mathfrak{z}\rangle$, and exp is T-equivariant by construction.

Lemma 3.15 *For $X \in \mathfrak{g}$, let $\mathrm{ad}_X : L\mathfrak{g} \to \mathfrak{g}$ be given by $\mathrm{ad}_X(Y) = [X, Y]$. For $X \in \mathfrak{u}_P$, we have $\mathrm{ad}_X{}^3 = 0$.* □

Proof Since the ad_X is t-linear, it is sufficient to show that $\mathrm{ad}^3_X(Y) = 0$ for all $Y \in \mathfrak{g}$. Consider $\gamma \in \Delta_0$, $\beta_1, \beta_2, \beta_3 \in \Delta_0^+\backslash\Delta_{\mathfrak{z}}$, and set $\gamma' = \gamma + \beta_1 + \beta_2 + \beta_3$. We have $\beta_1, \beta_2, \beta_3 \geqslant \alpha_d$. Suppose

$$\gamma = \sum_{\alpha\in\mathcal{D}_0} a_\alpha \alpha.$$

Since α_d is cominuscule, we have $a_{\alpha_d} \geqslant -1$. Consequently, the coefficient of α_d in γ' is $\geqslant 2$, and so γ' is not a root. In particular, we have

$$[\mathfrak{g}^{\beta_1}, [\mathfrak{g}^{\beta_2}, [\mathfrak{g}^{\beta_3}, \mathfrak{g}^\gamma]]] \subseteq \mathfrak{g}^{\gamma'} = 0.$$

The lemma now follows from the observation that the map $(X, Y) \to [X, Y]$ is linear in X and Y, and from the equality $\mathfrak{u}_P = \mathfrak{g}^{\Delta_0^+\backslash\Delta_{\mathfrak{z}}}$.

Proposition 3.16 *Let $\boldsymbol{\theta} : T^*G/P \to \mathcal{N}$ be the* Springer map, *see [23], and let $C \subseteq \mathcal{N}$ be the G-orbit such that $\boldsymbol{\theta}(T^*G/P) = \overline{C}$. The map ϕ from Proposition 3.14 extends to a G-equivariant map, $\psi : \overline{C} \to LG/L^+G$.* □

Proof Recall from Appendix 2.10 the groups $\mathcal{G}$ and $\mathcal{G}_0$, the G-equivariant isomorphism, $LG/L^+G \cong \mathcal{G}/\mathcal{G}_0$, and the G-equivariant quotient map $\mathrm{pr} : \mathcal{G} \to \mathcal{G}/\mathcal{G}_0$.

Consider first the map, $\widetilde{\psi} : \overline{C} \to \mathrm{End}(L\mathfrak{g})$, given by

$$\widetilde{\psi}(X) = 1 + t^{-1}\mathrm{ad}_X + \frac{t^{-2}}{2}\mathrm{ad}_X{}^2. \tag{3.12}$$

Since the map $X \mapsto \mathrm{ad}_X$ is G-equivariant, the same is true for $\widetilde{\psi}$. Note that our assumption that $\mathrm{char}(\mathbf{k}) \neq 2$ ensures that $\widetilde{\psi}(X)$ is well-defined.

Following Lemma 3.15, we have $\mathrm{ad}_X^3 \,|\mathfrak{g} = 0$ for all $X \in \mathfrak{u}_P$, and hence,

$$\widetilde{\psi}|\mathfrak{u}_P = Ad \circ \exp . \tag{3.13}$$

In particular, we have $\widetilde{\psi}(\mathfrak{u}_P) \subseteq \mathcal{G}$. The G-equivariance of $\widetilde{\psi}$ now yields the inclusion $\widetilde{\psi}(\overline{C}) \subseteq \mathcal{G}$. We define $\psi \stackrel{\mathrm{def}}{=} \mathrm{pr} \circ \widetilde{\psi}$. The map ψ is G-equivariant because $\widetilde{\psi}$ and pr are G-equivariant, and agrees with ϕ on $\mathfrak{u}_P$, as can be seen from Eq. (3.13).

Proposition 3.17 *Recall the map $\psi : \overline{C} \to LG/L^+G$ from Proposition 3.16 and the co-weight $q = w_0(\varpi_d^\vee) - \varpi_d^\vee$ from Eq. (3.3). We have $\psi(\overline{C}) = Y_{\mathcal{D}_0}(\tau_q)$, and further, the induced map, $\psi : \overline{C} \to Y_{\mathcal{D}_0}(\tau_q)$, is an isomorphism.* □

Proof Since ψ is G-equivariant and $\psi(\mathfrak{u}_P) = \phi(\mathfrak{u}_P) = X_{\mathfrak{Z}}^-(w_d^{\mathfrak{Z}})$, we have

$$\psi(\overline{C}) = GX_{\mathcal{D}_0}^-(w_d^{\mathfrak{Z}}) = G\mathcal{B}^-X_{\mathcal{D}_0}^-(w_d^{\mathfrak{Z}}).$$

Next, $\tau_q = w_0^{\mathfrak{Z}} w_d^{\mathfrak{Z}}$ (see Proposition 3.12) is the maximal element in the coset $W_0 w_d^{\mathfrak{Z}}$ (cf. [17]), and hence we have

$$G\mathcal{B}^-X_{\mathcal{D}_0}^-(w_d^{\mathfrak{Z}}) = Y_{\mathcal{D}_0}(w_0^{\mathfrak{Z}} w_d^{\mathfrak{Z}}) = Y_{\mathcal{D}_0}(\tau_q).$$

The induced map, $\psi : \overline{C} \to Y_{\mathcal{D}_0}(\tau_q)$, is an isomorphism because it admits an inverse map, which we describe below. In [1], Achar and Henderson study a G-equivariant finite map π from a subset of the affine Grassmannian to a subset of the nilpotent cone. The co-weight q satisfies the equality $q = -w_0(q)$, and hence we can apply [1, Proposition 6.6], to obtain an induced isomorphism, $Y_{\mathcal{D}_0} \xrightarrow{\sim} \overline{C}$. That this isomorphism is inverse to ψ can be seen by comparing the definition of the map π (see [1, (2.1),(2.2)]) with Eqs. (3.12) and (3.13).

Theorem 3.18 *Let θ, C, ψ, and q be as in Proposition 3.16. The map ϕ from Proposition 3.14 extends to a G-equivariant isomorphism, $\phi : T^*G/P \to Y_{\mathfrak{Z}}(\tau_q)$,*

$$\phi(g, X) = g\,\phi(X) \qquad \textit{for } g \in G,\ X \in \mathfrak{u}_P.$$

Further, we have a commutative diagram,

$$\begin{array}{ccc} T^*G/P & \xrightarrow{\phi} & Y_{\mathcal{J}}(\tau_q) \\ \downarrow \theta & & \downarrow \mathrm{pr} \\ \overline{C} & \xrightarrow{\psi} & Y_{\mathcal{D}_0}(\tau_q), \end{array}$$

where pr *denotes the quotient map* $LG/\mathcal{P} \to LG/L^+G$. □

Proof The proof of [17, Theorem 1.3] applies to show that ϕ is well-defined and G-equivariant, and that $\phi(T^*G/P) \subseteq X_{\mathcal{J}}(w_0^{\mathcal{J}} w_d^{\mathcal{J}})$. The commutativity of the diagram follows from the G-equivariance of ϕ and ψ, and the observation that

$$\mathrm{pr}(\phi(1, X)) = \phi(X) \,(\mathrm{mod}\, L^+G) = \psi(\theta(X)), \qquad \forall\, X \in \mathfrak{u}_P.$$

Next, we show that ϕ is injective. Let $\pi : T^*G/P \to G/P$ denote the structure map defining the cotangent bundle. The product map,

$$(\pi, \theta) : T^*G/P \to G/P \times \overline{C},$$

is injective. Let $\mathcal{Q}$ denote the parabolic subgroup of LG corresponding to the subset $\mathcal{D}_d \subseteq \mathcal{D}$, and let $\mathrm{pr}' : LG/\mathcal{P} \to LG/\mathcal{Q}$ denote the quotient map. We have an identification $\mathrm{id} : G/P \xrightarrow{\sim} X_{\mathcal{D}_d}(w_0^{\mathcal{J}})$. Consider the commutative diagram,

$$\begin{array}{ccc} T^*G/P & \xrightarrow{\phi} & X_{\mathcal{J}}(\tau_q) \\ \downarrow (\pi,\theta) & & \downarrow (\mathrm{pr}',\mathrm{pr}) \\ G/P \times \overline{C} & \xrightarrow{\mathrm{id}\times\psi} & X_{\mathcal{D}_d}(w_0^{\mathcal{J}}) \times X_{\mathcal{D}_0}(\tau_q). \end{array}$$

The map $(\mathrm{pr}', \mathrm{pr})$ is injective because it is the restriction to $X_{\mathcal{J}}(\tau_q)$ of the map

$$LG/\mathcal{P} \hookrightarrow LG/L^+G \times LG/\mathcal{Q}.$$

Since $(\mathrm{pr}', \mathrm{pr}) \circ \phi = (\mathrm{id} \times \psi) \circ (\pi, \theta)$ is injective, the same is true of ϕ.

Further, $X_{\mathcal{J}}(\tau_q)$ is a normal variety, and it follows from Zariski's Main Theorem that ϕ is an open embedding. We use the G-equivariance of ϕ, along with Proposition 3.14 to compute,

$$\phi(T^*G/P) = GX^-_{\mathcal{J}}(w_d^{\mathcal{J}}) = G\mathcal{B}^- X^-_{\mathcal{J}}(w_d^{\mathcal{J}}) = Y_{\mathcal{J}}(\tau_q).$$

Hence, we have an induced isomorphism, $\phi : T^*G/P \to Y_{\mathcal{J}}(\tau_q)$, as claimed.

Remark The G-orbits $\mathcal{N}_\lambda \subseteq \mathcal{N}$ are naturally indexed by certain co-weights, see, for example, [6]. As a consequence of Theorem 3.18, we see that the orbit $C \subseteq \mathcal{N}$ is precisely the orbit corresponding to the co-weight q. □

4 The Conormal Variety of a Schubert variety

4.1 The Conormal Variety

Consider the Schubert variety $X_{\mathfrak{J}}(w)$, viewed as a subvariety of G/P. For a smooth point $x \in X_{\mathfrak{J}}(w)$, the *conormal fibre* $N_x^* X_{\mathcal{P}}(w)$ is the annihilator of $T_x X_{\mathfrak{J}}(w)$ in $T_x^* G/P$, i.e.,

$$N_x^* X_{\mathcal{P}}(w) = \left\{ f \in T_x^* G/P \mid f(v) = 0 \ \forall v \in T_x X_{\mathcal{P}}(w) \right\}.$$

The *conormal variety* $N^* X_{\mathfrak{J}}(w)$ of $X_{\mathfrak{J}}(w) \hookrightarrow G/P$ is the closure in T^*G/P of the conormal bundle of the smooth locus of $X_{\mathfrak{J}}(w)$.

Let $\mathcal{D}_0$ be a finite type Dynkin diagram and $\mathcal{D}$ the associated extended diagram. Fix a cominuscule root $\alpha_d \in \mathcal{D}_0$ and let $\mathfrak{J}, \iota, G, LG, P, \mathcal{P}$, and $\phi : T^*G/P \hookrightarrow LG/\mathcal{P}$ be as in the previous section. We fix $w \in W_0^{\mathfrak{J}} \overset{\text{def}}{=} W_0 \cap W^{\mathfrak{J}}$ and set

$$v \overset{\text{def}}{=} \iota(w_0 w w_{\mathfrak{J}}). \tag{4.1}$$

It follows from Theorem 3.18 that the closure of $\phi(N^* X_{\mathfrak{J}}(w))$ in $X_{\mathfrak{J}}(\tau_q)$ is a B-stable compactification of $N^* X_{\mathfrak{J}}(w)$. The primary goal of this section is Theorem A: This compactification of $\phi(N^* X_{\mathfrak{J}}(w))$ is a Schubert subvariety of $X_{\mathfrak{J}}(\tau_q)$ if and only if $X_{\mathfrak{J}}(w_0 w w_{\mathfrak{J}})$ is smooth. This yields powerful results (Theorems 4.9 and 4.10) regarding the geometry of $N^* X_{\mathfrak{J}}(w)$ in the case where $X_{\mathfrak{J}}(w_0 w w_{\mathfrak{J}})$ is smooth.

We first give a combinatorial description of $N^* X_{\mathfrak{J}}(w)$ in Proposition 4.2, and then develop a sequence of lemmas leading to the proof of Theorem A. Finally, we show in Proposition 4.11 that the fibre at identity of $N^* X_{\mathfrak{J}}(w)$ can be identified as the union of opposite cells of certain Schubert varieties.

Proposition 4.2 *Let* $R \overset{\text{def}}{=} \left\{ \alpha \in \Delta_0^+ \mid \alpha \geqslant \alpha_d,\ w(\alpha) > 0 \right\}$ *and* $\mathfrak{g}^R = \sum_{\alpha \in R} \mathfrak{g}^\alpha$. *The conormal variety* $N^* X_{\mathfrak{J}}(w)$ *is the closure in* T^*G/P *of*

$$\left\{ (bw, X) \in G \times^P \mathfrak{u}_P \,\middle|\, b \in B,\ X \in \mathfrak{g}^R \right\}.$$

Proof The tangent space of G/P at identity is $\mathfrak{g}/\mathfrak{p}$. Consider the action of P on $\mathfrak{g}/\mathfrak{p}$ induced from the adjoint action of P on $\mathfrak{g}$. The tangent bundle $T\,G/P$ is the fibre bundle over G/P associated with the principal P-bundle $G \to G/P$, for the aforementioned action of P on $\mathfrak{g}/\mathfrak{p}$, i.e., $T\,G/P = G \times^P \mathfrak{g}/\mathfrak{p}$. Let

$$R' = \Delta^+ \setminus \left(\Delta_{\mathfrak{J}}^+ \cup R \right) = \left\{ \alpha \in \Delta^+ \mid \alpha \geqslant \alpha_d,\ w(\alpha) < 0 \right\},$$

$$U_w = \left\langle U_\alpha \mid -\alpha \in R' \right\rangle.$$

For any point $b \in B$, we have (see, for example, [4]):

$$\begin{aligned} BwP \text{ (mod } P) &= bBwP \text{ (mod } P) \\ &= b(wU_w w^{-1})wP \text{ (mod } P) \\ &= bwU_w P \text{ (mod } P). \end{aligned}$$

It follows that the tangent subspace at bw of the big cell BwP (mod P) is given by

$$T_w BwP \text{ (mod } P) = \left\{ (bw, X) \in G \times^P \mathfrak{g}/\mathfrak{p} \;\middle|\; X \in \bigoplus_{-\alpha \in R'} \mathfrak{g}^\alpha/\mathfrak{p} \right\},$$

where $\mathfrak{g}^\alpha/\mathfrak{p}$ denotes the image of a root space $\mathfrak{g}^\alpha$ under the map $\mathfrak{g} \to \mathfrak{g}/\mathfrak{p}$.

Recall from Appendix 3.3 the symmetric bilinear G-invariant form on $\mathfrak{g}$ identifying the dual of a root space $\mathfrak{g}^\alpha$ with the root space $\mathfrak{g}^{-\alpha}$. We see that a root space $\mathfrak{g}^\alpha \subseteq \mathfrak{u}$ annihilates $T_{bw} BwP$ (mod P) if and only if $\alpha \in \Delta_0^+ \setminus \Delta_\partial^+$ and $\alpha \notin R'$, or equivalently, $\alpha \in R$. The result now follows from the observation that BwP (mod P) is a dense open subset of $X_\partial(w)$, and is contained in the smooth locus of $X_\partial(w)$.

Lemma 4.3 *Recall from Eq. (4.1) that $v = {}^\iota(w_0 w w_\partial)$. We have:*

(a) $W_0 \cap W^\partial = W_0^d \overset{\text{def}}{=} W_0 \cap W^d$ *and* $W_d \cap W^\partial = W_d^0 \overset{\text{def}}{=} W_d \cap W^0$.
(b) $v \in W_d^0$.
(c) $l(wv) = l(w) + l(v) = \dim G/P$. □

Proof It follows from Appendix 2.6 that $W_0 \subseteq W^{\{\alpha_0\}}$, hence

$$W_0 \cap W^\partial \subseteq W_0 \cap W^{\{\alpha_0\}} \cap W^\partial = W_0^d.$$

Conversely, since $W^d \subseteq W^\partial$, we have $W_0^d = W_0 \cap W^d \subseteq W_0 \cap W^\partial$. Consequently,

$$W_0 \cap W^\partial = W_0^d.$$

Applying ι to this equality, we have $W_d \cap W^\partial = W_d^0$.

Next, we prove $v \in W^\partial$. It is sufficient to verify $v(\Delta_\partial^+) \subseteq \Delta^+$, see Eq. (2.3).

$$\begin{aligned} & v(\Delta_\partial^+) = \iota w_0 w w_\partial (\iota(\Delta_\partial^+)) && \text{using Eq. (3.2)} \\ \implies & v(\Delta_\partial^+) = \iota w_0 w w_\partial (\Delta_\partial^+) && \text{using Appendix 3.10} \\ \implies & v(\Delta_\partial^+) = \iota w_0 w (\Delta_\partial^-) && \text{using Appendix 2.4.} \end{aligned} \tag{4.2}$$

Now, since $Supp(w) \subseteq \mathcal{D}_0$, we see from Appendix 2.6 that $w(\Delta_{\mathfrak{J}}^-) \subseteq \Delta_0$. Further, since $w \in W^{\mathfrak{J}}$, it follows from Eq. (2.3) that $w(\Delta_{\mathfrak{J}}^-) \subseteq \Delta_0^-$. Applying Eq. (4.2), we have $v(\Delta_{\mathfrak{J}}^+) \subseteq \iota w_0(\Delta_0^-) = \iota(\Delta_0^+) \subseteq \Delta^+$. This proves $v \in W^{\mathfrak{J}}$.

Further, since $w_0, w, w_{\mathfrak{J}} \in W_0$, we have $w_0 w w_{\mathfrak{J}} \in W_0$. It follows from Appendix 3.10 that $v = {}^{\iota}(w_0 w w_{\mathfrak{J}}) \in W_d$. Combining with $v \in W^{\mathfrak{J}}$, we deduce (2):

$$v \in W^{\mathfrak{J}} \cap W_d = W_d^0.$$

Finally, since $v \in W^{\mathfrak{J}}$, we have ${}^{\iota}v = w_0 w w_{\mathfrak{J}} \in W^{\mathfrak{J}}$. Consequently,

$$\begin{aligned} l(w_0 w) &= l(w_0 w w_{\mathfrak{J}}) + l(w_{\mathfrak{J}}) \\ \implies \dim G - l(w) &= l(v) + \dim P \\ \implies \quad \dim G/P &= l(w) + l(v) = l(wv), \end{aligned}$$

where the last equality follows from the observations $w \in W_0$ and $v \in W^0$.

Lemma 4.4 *Let $u \in W_d^0$. Then $Supp(u)$ is a connected sub-graph of $\mathcal{D}_d$.* □

Proof Suppose $Supp(u)$ is not connected. Let $\mathcal{L}_1$ be the connected component of $Supp(u)$ containing α_d, and let $\mathcal{L}_2 \stackrel{\text{def}}{=} Supp(u) \backslash \mathcal{L}_1$. Now, since $\mathcal{L}_1$ and $\mathcal{L}_2$ are disconnected, we have

$$s_\alpha s_\beta = s_\beta s_\alpha \qquad\qquad \forall\, s_\alpha \in \mathcal{L}_1,\ s_\beta \in \mathcal{L}_2.$$

Let $s_1 \ldots s_l$ be a reduced word for u, and let k be the largest index such that $s_k \in \mathcal{L}_2$. Then $s_k s_m = s_m s_k$ for all $m > k$. It follows that

$$\begin{aligned} u s_k = (s_1 \ldots s_l) s_k &= (s_1 \ldots s_{k-1} s_{k+1} \ldots s_l s_k) s_k \\ &= s_1 \ldots s_{k-1} s_{k+1} \ldots s_l. \end{aligned}$$

Now, since $Supp(u) \subseteq \mathcal{D}_d$, we have $\alpha_0, \alpha_d \notin \mathcal{L}_2$. In particular, $s_k \in W_{\mathfrak{J}}$. Hence $u s_k = u \pmod{W_{\mathfrak{J}}}$ and $u s_k < u$, contradicting the assumption $u \in W_d^0 \subseteq W^{\mathfrak{J}}$.

Proposition 4.5 *For $u \in W_d^0$, the following are equivalent:*

(a) $X_{\mathfrak{J}}(u)$ *is smooth.*
(b) $X_{\mathfrak{J}}(u)$ *is $\mathcal{P}_{\mathcal{L}}$-homogeneous, i.e., $X_{\mathfrak{J}}(u) = \mathcal{P}_{\mathcal{L}}/(\mathcal{P}_{\mathcal{L}} \cap \mathcal{P})$ for some connected sub-graph $\mathcal{L} \subseteq \mathcal{D}_d$.*
(c) $l(u^{-1} \star u w_{\mathfrak{J}}) = l(u w_{\mathfrak{J}})$, *where $\star$ is the Demazure product, see Appendix 2.17.*
(d) $(u w_{\mathfrak{J}})^{-1}(\alpha) < 0$ *for all $\alpha \in Supp(u)$.*
(e) $\{\alpha \in \Delta^+ \mid u(\alpha) < 0\} = \Delta^+_{Supp(u)} \backslash \Delta^+_{\mathfrak{J}}$.
(f) $u = w_{\mathcal{L}} w_{\mathcal{L} \cap \mathfrak{J}}$, *where $\mathcal{L} = Supp(u)$, and $w_{\mathcal{L}}, w_{\mathcal{L} \cap \mathfrak{J}}$ denote the maximal elements in $W_{\mathcal{L}}, W_{\mathcal{L} \cap \mathfrak{J}}$, respectively.* □

Proof The claim (1) $\iff$ (2) is [3, Theorem 1.1]. Suppose (2) holds, i.e., $X_{\mathcal{J}}(u) = \mathcal{P}_{\mathcal{L}}/\mathcal{P}_{\mathcal{L}\cap\mathcal{J}}$. We have the following Cartesian square:

$$\begin{array}{ccc} X_{\mathcal{B}}(uw_{\mathcal{J}}) & \hookrightarrow & LG/\mathcal{B} \\ \downarrow & & \downarrow \\ \mathcal{P}_{\mathcal{L}}/\mathcal{P}_{\mathcal{L}\cap\mathcal{J}} = X_{\mathcal{J}}(u) & \hookrightarrow & LG/\mathcal{P}. \end{array}$$

Since $X_{\mathcal{J}}(u)$ is $\mathcal{P}_{\mathcal{L}}$-stable, the same is true of its pull-back $X_{\mathcal{B}}(uw_{\mathcal{J}})$. Further, since $u \in W_{\mathcal{L}}$, any lift of u^{-1} to $N(\mathcal{K})$ (see Appendix 2.9) is in $\mathcal{P}_{\mathcal{L}}$. Consequently, $X_{\mathcal{B}}(uw_{\mathcal{J}})$ is u^{-1} stable, and so $u^{-1} \star uw_{\mathcal{J}} = uw_{\mathcal{J}}$. Hence we obtain (2) $\implies$ (3).

It is clear from Eq. (2.6) that (3) is equivalent to $u^{-1} \star uw_{\mathcal{J}} = uw_{\mathcal{J}}$, which holds if and only if $s_\alpha \star uw_{\mathcal{J}} = uw_{\mathcal{J}}$ for all $\alpha \in \mathcal{L}$. This is equivalent to (4) from Eq. (2.1). Hence we obtain (3) $\iff$ (4).

Suppose (4) holds. Then $(uw_{\mathcal{J}})^{-1}(\alpha) < 0$ for all $\alpha \in \Delta^+_{Supp(u)}$. It follows from Appendix 2.6 and Eq. (2.3) that

$$\{\alpha \in \Delta^+ \mid u(\alpha) < 0\} \subseteq \Delta^+_{Supp(u)} \backslash \Delta^+_{\mathcal{J}}.$$

Consider $\alpha \in \Delta^+_{Supp(u)} \backslash \Delta^+_{\mathcal{J}}$ satisfying $u(\alpha) > 0$. Since u preserves $\Delta_{Supp(u)}$, we have $u(\alpha) \in \Delta^+_{Supp(u)}$. Applying (4) to $u(\alpha)$, we get $w_{\mathcal{J}}u^{-1}(u(\alpha)) = w_{\mathcal{J}}(\alpha) < 0$. It follows that $\alpha \in \Delta^+_{\mathcal{J}}$, contradicting the assumption $\alpha \notin \Delta^+_{\mathcal{J}}$. Hence we obtain the implication (4) $\implies$ (5).

Suppose (5) holds. We verify that

$$\{\alpha \in \Delta^+ \mid w_{\mathcal{L}}w_{\mathcal{L}\cap\mathcal{J}}(\alpha) < 0\} = \Delta^+_{\mathcal{L}} \backslash \Delta^+_{\mathcal{J}}.$$

Now since $u \in W$ is uniquely determined by the set $\{\alpha \in \Delta^+ \mid u(\alpha) < 0\}$ (cf. [13, §1.3.14]), we get $u = w_{\mathcal{L}}w_{\mathcal{L}\cap\mathcal{J}}$. Hence we obtain (5) $\implies$ (6).

Finally, suppose (6) holds. Since $w_{\mathcal{L}\cap\mathcal{J}} \subseteq \mathcal{J}$, we have $u = w_{\mathcal{L}} \pmod{W_{\mathcal{J}}}$. It follows that $X_{\mathcal{J}}(u) = X_{\mathcal{J}}(w_{\mathcal{L}}) = \mathcal{P}_{\mathcal{L}}/(\mathcal{P}_{\mathcal{L}\cap\mathcal{J}})$. Hence we obtain (6) $\implies$ (2). □

Lemma 4.6 *For $\alpha \in \Delta_0^+ \backslash \Delta_{\mathcal{J}}$, we have $v(\alpha - \delta) = -\iota w_0 w(\alpha)$.*

Proof Since $\alpha \in \Delta_0^+ \backslash \Delta_{\mathcal{J}}$, we have $\alpha \geqslant \alpha_d$. Set $\gamma = \alpha - \alpha_d$. Since α_d is cominuscule, $Supp(\gamma) \subseteq \mathcal{J}$. In particular, $\iota(\gamma) = -w_{\mathcal{J}}(\gamma)$. We compute

$$\begin{aligned} \iota(\alpha) &= \iota(\alpha_d) + \iota(\gamma) = \alpha_0 - w_{\mathcal{J}}(\gamma) \\ \implies \iota(\alpha - \delta) &= -\theta_0 - w_{\mathcal{J}}(\gamma). \end{aligned} \tag{4.3}$$

Recall from Eq. (4.1) that $v = {}^{\iota}(w_0ww_{\mathcal{J}})$. We now compute $v(\alpha - \delta)$:

$$\begin{aligned} v(\alpha - \delta) &= {}^{\iota}(w_0ww_{\mathcal{J}})(\alpha - \delta) = \iota w_0ww_{\mathcal{J}}(\iota(\alpha - \delta)) && \text{using Appendix 3.10} \\ &= -\iota w_0ww_{\mathcal{J}}(\theta_0 + w_{\mathcal{J}}(\gamma)) && \text{using Eq. (4.3)} \end{aligned}$$

$$= -\iota w_0 w(\alpha_d) - \iota w_0 w(\gamma) \qquad \text{using Lemma 3.4}$$
$$= -\iota w_0 w(\alpha_d + \gamma) = -\iota w_0 w(\alpha).$$

Lemma 4.7 *The map* $\alpha \mapsto \alpha - \delta$ *induces a bijection*

$$\left\{\alpha \in \Delta_0^+ \mid \alpha \geqslant \alpha_d,\ w(\alpha) > 0\right\} \xrightarrow{\sim} \left\{\alpha \in \Delta_d^- \mid v(\alpha) > 0\right\}.$$

Proof Observe that

$$-\iota w_0(\Delta_0^{\pm}) = \iota(-w_0\Delta_0^{\pm}) = \iota(\Delta_0^{\pm}) \qquad \text{using Appendix 2.4}$$
$$= \Delta_d^{\pm} \qquad \text{using Appendix 3.10.}$$

Now, since $v(\alpha - \delta) = -\iota w_0 w(\alpha)$ (see Lemma 4.6), it follows that for $\alpha \in \Delta_0^+ \backslash \Delta_{\mathfrak{z}}$, $w(\alpha) > 0$ is equivalent to $v(\alpha - \delta) > 0$. The result now follows from Eq. (3.11), which states that $\alpha \mapsto \alpha - \delta$ induces a bijection from $\Delta_0^+ \backslash \Delta_{\mathfrak{z}}$ to $\Delta_d^- \backslash \Delta_{\mathfrak{z}}$.

Proposition 4.8 *Recall the map* ϕ *from Proposition 3.14. Let*

$$R = \left\{\alpha \in \Delta_0^+ \mid \alpha \geqslant \alpha_d,\ w(\alpha) > 0\right\}.$$

Then $\phi(\mathfrak{g}^R)$ *is dense in* $v^{-1} X_{\mathfrak{z}}(v)$. □

Proof Following the proof of Proposition 3.14, we see that $\exp(t^{-1}\mathfrak{g}^R) = U_\Phi$, where $\Phi \overset{\text{def}}{=} \{\alpha - \delta \mid \alpha \in R\}$. It follows from Lemma 4.7 that

$$\Phi = \left\{\alpha \in \Delta_d^- \mid v(\alpha) > 0\right\} = v^{-1}(\Delta^+) \cap \Delta^-.$$

In particular, $\#\Phi = l(v)$ (see [13, §1.3.14]), and so,

$$\dim \mathfrak{g}^R = \dim U_\Phi = \#\Phi = l(v) = \dim X_{\mathfrak{z}}(v) = \dim v^{-1} X_{\mathfrak{z}}(v).$$

Further, observe that

$$\phi(\mathfrak{g}^R) = \exp(t^{-1}\mathfrak{g}^R) \ (\mathrm{mod}\ \mathcal{P}) \subseteq v^{-1}\mathcal{B}v \ (\mathrm{mod}\ \mathcal{P}) \subseteq v^{-1} X_{\mathfrak{z}}(v).$$

The result follows from the injectivity of ϕ and the irreducibility of $v^{-1} X_{\mathfrak{z}}(v)$.

Theorem A *The closure of* $\phi(N^* X_{\mathfrak{z}}(w))$ *in* $LG/\mathcal{P}$ *is a Schubert variety if and only if the variety* $X_{\mathfrak{z}}(w_0 w w_{\mathfrak{z}})$ *is smooth.* □

Proof For (bw, X) be a generic point in $N^* X_{\mathfrak{z}}(w)$, we have

$$\phi(bw, X) = bw\, \phi(X) \in \mathcal{B}wv^{-1}\mathcal{B}v\mathcal{P} \ (\mathrm{mod}\ \mathcal{P})$$

Hence the minimal Schubert variety containing $\phi(N^* X_{\mathcal{J}}(w))$ is $X_{\mathcal{J}}(wv^{-1} \star v)$. Consequently, the closure $\overline{\phi(N^* X_{\mathcal{J}}(w))}$ is a Schubert variety if and only if

$$\dim X_{\mathcal{J}}(wv^{-1} \star v) = \dim N^* X_{\mathcal{J}}(w) = \dim G/P. \tag{4.4}$$

Consider the following Cartesian diagram,

$$\begin{array}{ccc} X_{\mathcal{B}}(wv^{-1} \star v \star w_{\mathcal{J}}) & \hookrightarrow & LG/\mathcal{B} \\ \downarrow & & \downarrow \\ X_{\mathcal{J}}(wv^{-1} \star v) & \hookrightarrow & LG/\mathcal{P}. \end{array}$$

The dimension of the generic fibre for the right projection is $\dim \mathcal{P}/\mathcal{B}$. Observe that $X_{\mathcal{B}}(wv^{-1} \star v \star w_{\mathcal{J}})$ is the pull-back of $X_{\mathcal{J}}(wv^{-1} \star v)$ to $LG/\mathcal{B}$. It follows that Eq. (4.4) is equivalent to

$$\dim X_{\mathcal{B}}(wv^{-1} \star v \star w_{\mathcal{J}}) = \dim G/P + \dim \mathcal{P}/\mathcal{B} = \dim G/B. \tag{4.5}$$

We see from Lemma 4.3 and Eq. (2.7) that $wv^{-1} = w \star v^{-1}$ and $v \star w_{\mathcal{J}} = vw_{\mathcal{J}}$. Hence, we have

$$wv^{-1} \star v \star w_{\mathcal{J}} = w \star v^{-1} \star vw_{\mathcal{J}}. \tag{4.6}$$

Observe that since $v, w_{\mathcal{J}} \in W_d$, we have $v^{-1} \star vw_{\mathcal{J}} \in W_d$. Recall also from Lemma 4.3 that $w \in W^{\mathcal{J}} \cap W_0 \subseteq W^d$. It follows that

$$\begin{aligned} w \star v^{-1} \star vw_{\mathcal{J}} &= w(v^{-1} \star vw_{\mathcal{J}}) && \text{using Eq. (2.7)} \\ \implies l(wv^{-1} \star v \star w_{\mathcal{J}}) &= l(w(v^{-1} \star vw_{\mathcal{J}})) && \text{using Eq. (4.6)} \\ &= l(w) + l(v^{-1} \star vw_{\mathcal{J}}) && \text{using } w \in W^d, \\ & && v^{-1} \star vw_{\mathcal{J}} \in W_d. \end{aligned} \tag{4.7}$$

Now $l(v^{-1} \star vw_{\mathcal{J}}) \geqslant l(vw_{\mathcal{J}})$, with equality holding if and only if $v^{-1} \star vw_{\mathcal{J}} = vw_{\mathcal{J}}$. Continuing Eq. (4.7), we have

$$\begin{aligned} \dim X_{\mathcal{B}}(wv^{-1} \star v \star w_{\mathcal{J}}) &\geqslant l(w) + l(vw_{\mathcal{J}}) \\ &= l(w) + l(v) + l(w_{\mathcal{J}}) && v \in W^{\mathcal{J}},\ w_{\mathcal{J}} \in W_{\mathcal{J}} \\ &= l(wv) + l(w_{\mathcal{J}}) && w \in W^d,\ v \in W_d \\ &= \dim G/B. \end{aligned}$$

Hence, Eq. (4.5) holds if and only if $v^{-1} \star vw_{\mathcal{J}} = vw_{\mathcal{J}}$, which is equivalent to $X_{\mathcal{J}}(v)$ being smooth, see Proposition 4.5. Finally, the Schubert varieties $X_{\mathcal{J}}(v)$ and

$X_{\mathfrak{J}}(w_0ww_{\mathfrak{J}})$ being isomorphic (since $v = {}^{\iota}(w_0ww_{\mathfrak{J}})$), we deduce that the closure $\overline{\phi(N^*X_{\mathfrak{J}}(w))}$ is a Schubert variety if and only if $X_{\mathfrak{J}}(w_0ww_{\mathfrak{J}})$ is smooth.

Theorem 4.9 *Let $w \in W_0^{\mathfrak{J}}$ be such that $X_{\mathfrak{J}}(w_0ww_{\mathfrak{J}})$ is smooth. Then $N^*X_{\mathfrak{J}}(w)$ is normal, Cohen–Macaulay, and has a resolution via Bott–Samelson varieties. Further, the family*

$$\left\{N^*X_{\mathfrak{J}}(w) \,\middle|\, X_{\mathfrak{J}}(w_0ww_{\mathfrak{J}}) \text{ is smooth}\right\}$$

is compatibly Frobenius split in good and odd positive characteristic. □

Proof These are standard results for Schubert varieties. One can find details in [7, 15, 21].

Let ϖ_i denote the fundamental weight associated with the simple co-root α_i. Consider a dominant weight $\lambda = \sum a_i\varpi_i,\ a_i \in \mathbb{Z}_{\geqslant 0}$, satisfying $a_i = 0$ if and only if $\alpha_i \in \mathfrak{J}$. Let $L(\lambda)$ be the line bundle on G/P associated with the weight λ, and set

$$V(\lambda) = H^0(G/P, L(\lambda)),$$

the dual Weyl module (see [26]). Recall from [15, Theorem 2], the monomial basis of $V(\lambda)$, whose elements u_π are indexed by LS paths π of shape λ. For an LS path π, we shall denote its initial point by $i(\pi)$, as in [15].

Theorem 4.10 *Let $w \in W_0^{\mathfrak{J}}$ be such that $X_{\mathfrak{J}}(w_0ww_{\mathfrak{J}})$ is smooth. Then the ideal sheaf of $N^*X_{\mathfrak{J}}(w)$ in T^*G/P is $\phi^{-1}\mathfrak{I}$, where $\mathfrak{I}$ is the ideal generated by the monomials*

$$\left\{u_\pi \,\middle|\, i(\pi) \leqslant \tau_q,\ i(\pi) \nleqslant wv\right\}.$$

Proof The ideal sheaf of $X_{\mathfrak{J}}(wv)$ in $X_{\mathfrak{J}}(\tau_q)$ is generated by the monomials

$$\left\{u_\pi \,\middle|\, i(\pi) \leq \tau_q,\ i(\pi) \nleqslant wv\right\},$$

see [15, Theorem 6]. Since $N^*X_{\mathfrak{J}}(w)$ is closed in T^*G/P, we have

$$\begin{aligned}\phi(N^*X_{\mathfrak{J}}(w)) &= \overline{\phi(N^*X_{\mathfrak{J}}(w))} \cap \phi(T^*G/P)\\ &= X_{\mathfrak{J}}(wv) \cap L^-Ge \cap X_{\mathfrak{J}}(\tau_q) = Y_{\mathfrak{J}}(wv).\end{aligned}$$

It follows that the ideal sheaf of $N^*X_{\mathfrak{J}}(w)$ in T^*G/P is the pull-back (via ϕ) of the restriction of $\mathfrak{I}$ to $Y_{\mathfrak{J}}(\tau_q)$, i.e., the ideal sheaf is $\phi^{-1}(\mathfrak{I})$.

Proposition 4.11 *Let $w \in W_0^{\mathfrak{J}}$ be such that $X_{\mathfrak{J}}(w_0ww_{\mathfrak{J}})$ is smooth, and let $N_0^*X_{\mathfrak{J}}(w)$ denote the fibre at identity of the conormal variety $N^*X_{\mathfrak{J}}(w)$. Then*

$$\phi(N_0^* X_{\mathcal{J}}(w)) = \bigcup_{u \in \mathcal{S}} X_{\mathcal{J}}^-(u),$$

where $\mathcal{S} = \{u \in W_d^0 \mid u \leqslant (wv)^{\mathcal{D}_0}\}$, *and* $(wv)^{\mathcal{D}_0}$ *is the minimal representative of* wv *with respect to* $\mathcal{D}_0$. □

Proof Recall from Proposition 3.14 that $\phi(T_e^* G/P) = X_{\mathcal{J}}^-(w_d^{\mathcal{J}})$. It follows that

$$\phi(N_0^* X_{\mathcal{J}}(w)) = X_{\mathcal{J}}^-(w_d^{\mathcal{J}}) \cap X_{\mathcal{J}}(wv) = \bigcup X_{\mathcal{J}}^-(u),$$

where the union runs over $\left\{u \in W^{\mathcal{J}} \,\middle|\, u \leqslant w_d^{\mathcal{J}},\ u \leqslant wv\right\}$. Since $w_d^{\mathcal{J}}$ is maximal in $W_d^{\mathcal{J}}$, the condition $\left\{u \in W^{\mathcal{J}},\ u \leqslant w_d^{\mathcal{J}}\right\}$ is equivalent to $u \in W_d \cap W^{\mathcal{J}} = W_d^0$, see Lemma 4.3. Hence,

$$\left\{u \in W^{\mathcal{J}} \,\middle|\, u \leqslant w_d^{\mathcal{J}},\ u \leqslant wv\right\} = \left\{u \in W_d^0 \,\middle|\, u \leqslant w_d^{\mathcal{J}},\ u \leqslant wv\right\}.$$

Finally, consider $u \in W^0$. If $u \leqslant wv$, then $u \leqslant (wv)^{\mathcal{D}_0}$. It follows that

$$\begin{aligned}\left\{u \in W^{\mathcal{J}} \,\middle|\, u \leqslant w_d^{\mathcal{J}},\ u \leqslant wv\right\} &= \left\{u \in W_d^0 \,\middle|\, u \leqslant w_d^{\mathcal{J}},\ u \leqslant wv\right\} \\ &= \left\{u \in W_d^0 \,\middle|\, u \leqslant (wv)^{\mathcal{D}_0}\right\} = \mathcal{S}.\end{aligned}$$

5 Determinantal Varieties

In this section, we use the results of Appendix 4 to prove the following: The conormal fibre at the zero matrix of the rank r (usual, symmetric, skew-symmetric resp.) determinantal variety is the co-rank r (usual, symmetric, skew-symmetric resp.) determinantal variety.

Consider a rank r (usual, symmetric, skew-symmetric resp.) determinantal variety Σ. There exists a simply connected, almost simple group G (of type A, C, D resp.) and a cominuscule Grassmannian G/P such that Σ is naturally identified as the opposite cell of some Schubert variety $X_{\mathcal{J}}(w) \subseteq G/P$, see [16, 18]. For such w, we verify that the Schubert variety $X_{\mathcal{J}}(w_0 w w_{\mathcal{J}})$ is smooth. This allows us to apply Proposition 4.11. Finally, we show that the union of the various Schubert varieties in Proposition 4.11 is equal to a single Schubert variety, which we further verify to be isomorphic to the co-rank r (usual, symmetric, skew-symmetric resp.) determinantal variety.

We carry out the proof in detail only for the skew-symmetric determinantal varieties. The other two cases are completely analogous. For the usual determinantal varieties, this result has been proved by Strickland [24]. Further, the conormal fibres at the zero matrix of the usual determinantal varieties and symmetric determinantal varieties have been studied by Gaffney and Rangachev (cf. [10]) and Gaffney and Molina (cf. [9]).

5.1 The Weyl Group of D_n

Let $\mathcal{D}_0 = D_n$, $\mathcal{D} = \widetilde{D}_n$, and W_0 (resp. W) the Weyl group of $\mathcal{D}_0$ (resp. $\mathcal{D}$). For $1 \leqslant i, j \leqslant n-1$, the relations in W_0 are

$$\begin{aligned} s_i s_j &= s_j s_i && \text{if } |i-j| \geqslant 2, \\ s_i s_j s_i &= s_j s_i s_j && \text{if } |i-j| = 1. \end{aligned}$$

The latter are called braid relations. The remaining relations are $s_n s_i = s_i s_n$ for $i \neq n-2$, and the braid relation

$$s_n s_{n-2} s_n = s_{n-2} s_n s_{n-2}.$$

Let μ be the involution on $\{1, \ldots, 2n\}$ given by $\mu(i) \stackrel{\text{def}}{=} 2n+1-i$. Following [16, 18], we embed the Weyl group W_0 into the symmetric group S_{2n} via the homomorphism given by

$$\begin{aligned} s_i &\mapsto r_i r_{2n-i} && \text{for } i \neq n, \\ s_n &\mapsto r_n r_{n-1} r_{n+1} r_n, \end{aligned}$$

where r_i denotes the transposition $(i\ i+1)$ in S_{2n}. Under this embedding, we have

$$W_0 = \{w \in S_{2n} \mid w\mu = \mu w,\ sgn(w) = 1\}. \tag{5.1}$$

It is clear that $w \in W_0$ is uniquely determined by its value on $1, \ldots, n$. Accordingly, we represent w by the string $[w(1), \ldots, w(n)]$.

5.2 The Involution ι for D_n

The simple root α_n is cominuscule in $\mathcal{D}_0 = D_n$, and $\mathcal{J} \stackrel{\text{def}}{=} \mathcal{D}_0 \backslash \{\alpha_n\} = \{\alpha_1, \ldots, \alpha_{n-1}\}$ is isomorphic to A_{n-1}. Let ι be the involution defined in Definition 3.6. Recall that the action of the *Weyl involution* $-w_{\mathcal{J}}$ on $\mathcal{J} \cong A_{n-1}$ is given by $-w_{\mathcal{J}}(\alpha_i) = \alpha_{n-i}$ (cf. [5, Ch.VI§4.7]), and so $\iota(\alpha_i) = \alpha_{n-i}$ for $1 \leqslant i \leqslant n-1$. Further, since ι interchanges α_0 and α_n, we have

$$\iota(\alpha_i) = \alpha_{n-i} \qquad \forall \alpha_i \in \mathcal{D}. \tag{5.2}$$

5.3 Skew-Symmetric Determinantal Varieties

Let Mat_n^{sk} be the variety of skew-symmetric $n \times n$ matrices. The *rank r skew-symmetric determinantal variety* $\overline{\Sigma}_r^{sk,n}$ is the subvariety of Mat_n^{sk} given by

$$\overline{\Sigma}_r^{sk,n} = \left\{ A \in Mat_n^{sk} \,\middle|\, rank(A) \leqslant r \right\}.$$

Recall that the rank of a skew-symmetric matrix is necessarily even. Hence, we assume without loss of generality that *r is even.*

Let G be the simply connected, almost simple group of type D_n, and let $P \subseteq G$ be the parabolic group corresponding to $\mathcal{J} = \{\alpha_1, \ldots, \alpha_{n-1}\}$, see Table 1. Following [18], we identify Mat_n^{sk} with the opposite cell in G/P. Under this identification, the zero matrix corresponds to $\boldsymbol{e} \in G/P$, and $\overline{\Sigma}_r^{sk,n} = X_{\mathcal{J}}^-(w_r)$, where

$$w_r \stackrel{\text{def}}{=} [r+1, \ldots, n, 2n-r+1, \ldots, 2n], \tag{5.3}$$

in the sense of Eq. (5.1). Observe that

$$w_r \in W^{\mathcal{J}} \cap W_0 = W_0^{\mathcal{J} \cup \{\alpha_0\}}. \tag{5.4}$$

The last equality is Lemma 4.3, (1).

Remark In [18], the skew-symmetric variety is identified with a Schubert variety corresponding to the group $SO(2n)$, which is not simply connected. This is not a problem, however, since Schubert varieties depend only on the underlying Dynkin diagram, and not on the group per se. □

Theorem B *The conormal fibre of $\overline{\Sigma}_r^{sk,n}$ at 0 is isomorphic to $\overline{\Sigma}_{\overline{n}-r}^{sk,n}$, where*

$$\overline{n} = \begin{cases} n & \text{if } n \text{ is even,} \\ n-1 & \text{if } n \text{ is odd.} \end{cases}$$

Proof Let $\mathcal{D}_0$ (resp. $\mathcal{D}$, resp. $\mathcal{J}$) be the Dynkin diagram D_n (resp. $\widetilde{D}_n$, $D_n \backslash \{\alpha_n\}$), and W_0 (resp. W, $W_{\mathcal{J}}$) its Weyl group. Recall that $\mathcal{J} \cong A_{n-1}$. For $\mathcal{L}$ a sub-diagram of $\mathcal{D}_0$, we write $w_{\mathcal{L}}$ for the longest element in W supported on $\mathcal{L}$, and $w_{\mathcal{L}}^{\mathcal{J}}$ for its minimal representative with respect to $\mathcal{J}$. The longest elements $w_0 \in W_0$ and $w_{\mathcal{J}} \in W_{\mathcal{J}}$ are given by

$$w_0 = w_{\overline{n}} = [2n, \ldots, n+2, \overline{n}+1],$$

$$w_{\mathfrak{I}} = [n, \ldots, 1],$$

in the sense of Eq. (5.1), see [16]. Let w_r be as defined in Eq. (5.3), and set $v_r \stackrel{\text{def}}{=} {}^{\iota}(w_0 w_r w_{\mathfrak{I}})$. We have

$$w_0 w_r w_{\mathfrak{I}} = [1, \ldots, r, \overline{n}+1, n+2, \ldots, 2n-r] = w_{\mathcal{L}}^{\mathfrak{I}}, \tag{5.5}$$

where $\mathcal{L} = \{\alpha_{r+1}, \ldots, \alpha_n\}$. Hence $X_{\mathfrak{I}}(w_0 w_r w_{\mathfrak{I}})$ is smooth, see Proposition 4.5. It now follows from Proposition 4.11 and Proposition 5.6 that

$$\phi(N_0^* X_{\mathfrak{I}}(w_r)) = X_{\mathfrak{I}}^-({}^{\iota}w_{\overline{n}-r}) \cong X_{\mathfrak{I}}^-(w_{\overline{n}-r}) \cong \overline{\Sigma}_{\overline{n}-r}^{sk,n}.$$

It remains to prove Proposition 5.6. The proof is obtained as a consequence of the following two lemmas.

Lemma 5.4 *Consider $x_i \in W_0$ defined inductively as*

$$x_i = \begin{cases} s_n & \text{for } i = n-1, \\ s_{i+1} s_i x_{i+1} & \text{for } 1 \leqslant i < n-1. \end{cases} \tag{5.6}$$

Then $s_{i+2} s_{i+3} x_i = x_i s_i s_{i+1}$ for $1 \leqslant i \leqslant n-4$. □

Proof Observe that for $j \leqslant i-2$, we have $s_j x_i = x_i s_j$, from which we deduce the equality $s_i s_{i+1} x_{i+3} = x_{i+3} s_i s_{i+1}$. Now,

$$\begin{aligned}
& s_{i+2} s_{i+3} x_i \\
&\quad = s_{i+2} s_{i+3} s_{i+1} s_i s_{i+2} s_{i+1} s_{i+3} s_{i+2} x_{i+3} && \text{using Eq. (5.6)} \\
&\quad = s_{i+2} s_{i+1} s_i (s_{i+3} s_{i+2} s_{i+3}) s_{i+1} s_{i+2} x_{i+3} \\
&\quad = (s_{i+2} s_{i+1} s_{i+2}) s_{i+3} s_i (s_{i+2} s_{i+1} s_{i+2}) x_{i+3} && \text{using Braid relations} \\
&\quad = s_{i+1} s_{i+2} s_{i+3} (s_{i+1} s_i s_{i+1}) s_{i+2} s_{i+1} x_{i+3} && \text{using Braid relations} \\
&\quad = s_{i+1} s_{i+2} s_i s_{i+1} s_{i+3} s_{i+2} s_i s_{i+1} x_{i+3} && \text{using Braid relations} \\
&\quad = s_{i+1} s_i s_{i+2} s_{i+1} s_{i+3} s_{i+2} x_{i+3} s_i s_{i+1} \\
&\quad = x_i s_i s_{i+1} && \text{using Eq. (5.6).}
\end{aligned}$$

Lemma 5.5 *Let x_i be given by Eq. (5.6). For $3 \leqslant j, k \leq \overline{n}-1$, we have*

$${}^{\iota}x_{\overline{n}-k}\, x_k = x_{k-2}\, {}^{\iota}x_{\overline{n}-k+2},$$

$${}^{\iota}x_{\overline{n}-k}\, x_k x_{k-2} \ldots x_j = x_{k-2} x_{k-4} \ldots x_{j-2}\, {}^{\iota}x_{\overline{n}-j}.$$

Proof The second equality follows from repeated applications of the first. Observe first that $x_k \in \langle s_j \mid j \geqslant k \rangle$, or equivalently, ${}^{\iota}x_{n-k} \in \langle s_j \mid j \leqslant k \rangle$. Consequently, the braid relations yields ${}^{\iota}x_i x_j = x_j {}^{\iota}x_i$ whenever $i + j \geqslant n + 2$. Now,

$$\begin{aligned} &{}^{\iota}x_{n-k}\, x_k \\ &\quad = {}^{\iota}s_{n-k+1} {}^{\iota}s_{n-k} {}^{\iota}s_{n-k+2} {}^{\iota}s_{n-k+1} {}^{\iota}x_{n-k+2}\, x_k && \text{using Eq. (5.6)} \\ &\quad = s_{k-1}s_k s_{k-2}s_{k-1} {}^{\iota}x_{n-k+2}x_k && \text{using Eq. (5.2)} \\ &\quad = s_{k-1}s_{k-2}s_k s_{k-1} x_k {}^{\iota}x_{n-k+2} \\ &\quad = x_{k-2} {}^{\iota}x_{n-k+2} && \text{using Eq. (5.6).} \end{aligned}$$

This proves the claim when n is even. Suppose n is odd, so that $\overline{n} = n - 1$ and $k \leqslant n - 2$. Then,

$$\begin{aligned} &{}^{\iota}x_{n-k-1}\, x_k \\ &\quad = {}^{\iota}s_{n-k} {}^{\iota}s_{n-k-1} {}^{\iota}x_{n-k}\, x_k && \text{using Eq. (5.6)} \\ &\quad = s_k s_{k+1}\, x_{k-2} {}^{\iota}x_{n-k+2} && \text{using Eq. (5.2)} \\ &\quad = x_{k-2} s_{k-2} s_{k-1} {}^{\iota}x_{n-k+2} && \text{using Lemma 5.4} \\ &\quad = x_{k-2} {}^{\iota}x_{n-k+1} && \text{using Eq. (5.6).} \end{aligned}$$

This proves the claim when n is odd.

Proposition 5.6 *For w_r given by Eq. (5.3), and $v_r = {}^{\iota}(w_0 w_r w_{\mathfrak{J}})$, we have*

$$(w_r v_r)^{\mathcal{D}_0} = w_r v_r {}^{\iota}v_{\overline{n}-r}^{-1} = {}^{\iota}w_{\overline{n}-r} \in W^0_{\mathfrak{J}\cup\{\alpha_0\}}.$$

Consequently, ${}^{\iota}w_{\overline{n}-r}$ is the unique maximal element in

$$\left\{ u \in W^0_{\mathfrak{J}\cup\{\alpha_0\}} \,\middle|\, u \leqslant (w_r v_r)^{\mathcal{D}_0} \right\}.$$

Proof Let x_i be as in Eq. (5.6). We have the following formulae, which are easily verified inductively,

$$\begin{aligned} x_i &= [1, \ldots, i-1, i+2, \ldots, n-2, 2n-i, 2n-i+1] \\ w_r &= x_{r-1}x_{r-3}\ldots x_1 \\ w_0 w_r w_{\mathfrak{J}} &= x_{\overline{n}-1}x_{\overline{n}-3}\ldots x_{r+1}. \end{aligned}$$

Now,

$$\begin{aligned}{}^{\iota}w_r{}^{\iota}v_r &= {}^{\iota}x_{r-1}{}^{\iota}x_{r-3}\dots{}^{\iota}x_1x_{\overline{n}-1}x_{\overline{n}-3}\dots x_{r+1}\\ &= x_{\overline{n}-r-1}x_{\overline{n}-r-3}\dots x_1\,{}^{\iota}x_{\overline{n}-1}{}^{\iota}x_{\overline{n}-3}\dots{}^{\iota}x_{\overline{n}-r+1} \qquad \text{using Lemma 5.5.}\\ &= w_{\overline{n}-r}v_{\overline{n}-r}.\end{aligned}$$

It follows that ${}^{\iota}w_r{}^{\iota}v_r v_{\overline{n}-r}^{-1} = w_{\overline{n}-r}$, hence $w_r v_r{}^{\iota}v_{\overline{n}-r}^{-1} = {}^{\iota}w_{\overline{n}-r}$. Next, Eq. (5.4) yields

$$w_{\overline{n}-r} \in W_0^{\mathcal{J}\cup\{\alpha_0\}} \implies {}^{\iota}w_{\overline{n}-r} \in W^0_{\mathcal{J}\cup\{\alpha_0\}} \subseteq W^{\mathcal{D}_0}.$$

Further, since ${}^{\iota}v_{\overline{n}-r} \in W_0$ (see Eq. (5.5)), we have $w_r v_r = {}^{\iota}w_{\overline{n}-r} \pmod{W_0}$. Together, we deduce $(w_r v_r)^{\mathcal{D}_0} = {}^{\iota}w_{\overline{n}-r}$.

Acknowledgments We thank Terence Gaffney for fruitful discussions that pointed us towards the results in Appendix 5. We also thank the referee for pointing out inconsistencies in our assumptions on the characteristic of **k** in an earlier version of this paper.

References

1. Pramod N. Achar and Anthony Henderson, *Geometric Satake, Springer correspondence and small representations*, Selecta Math. (N.S.) **19** (2013), no. 4, 949–986. MR3131493
2. Michel Brion and Shrawan Kumar, *Frobenius splitting methods in geometry and representation theory*, Progress in Mathematics, vol. 231, Birkhäuser Boston, Inc., Boston, MA, 2005. MR2107324
3. Sara C. Billey and Stephen A. Mitchell, *Smooth and palindromic Schubert varieties in affine Grassmannians*, J. Algebraic Combin. **31** (2010), no. 2, 169–216. MR2592076
4. Armand Borel, *Linear algebraic groups*, second ed., Graduate Texts in Mathematics, vol. 126, Springer-Verlag, New York, 1991. MR1102012
5. Nicolas Bourbaki, *éléments de mathématique. Fasc. XXXIV. Groupes et algèbres de Lie. Chapitre IV: Groupes de Coxeter et systèmes de Tits. Chapitre V: Groupes engendrés par des réflexions. Chapitre VI: systèmes de racines*, Actualités Scientifiques et Industrielles, No. 1337, Hermann, Paris, 1968. MR0240238
6. Roger W. Carter, *Finite groups of Lie type*, Pure and Applied Mathematics (New York), John Wiley & Sons, Inc., New York, 1985, Conjugacy classes and complex characters, A Wiley-Interscience Publication. MR794307
7. Gerd Faltings, *Algebraic loop groups and moduli spaces of bundles*, J. Eur. Math. Soc. (JEMS) **5** (2003), no. 1, 41–68. MR1961134
8. Victor Ginzburg, *Lectures on Nakajima's quiver varieties*, preprint arXiv:0905.0686 (2009).
9. Terence Gaffney and Michelle Molino, *Determinantal symmetric singularities and Whitney equisingularity*, Ph.D. thesis, Universidade Federal Fluminense, 2018.
10. Terence Gaffney and Antoni Rangachev, *Pairs of modules and determinantal isolated singularities*, arXiv preprint arXiv:1501.00201 (2014).
11. Victor G. Kac, *Infinite-dimensional Lie algebras*, third ed., Cambridge University Press, Cambridge, 1990. MR1104219
12. Allen Knutson and Ezra Miller, *Subword complexes in Coxeter groups*, Advances in Mathematics **184** (2004), no. 1, 161–176.

13. Shrawan Kumar, *Kac-Moody groups, their flag varieties and representation theory*, Progress in Mathematics, vol. 204, Birkhäuser Boston, Inc., Boston, MA, 2002. MR1923198
14. V. Lakshmibai, *Cotangent bundle to the Grassmann variety*, Transform. Groups **21** (2016), no. 2, 519–530. MR3492046
15. Peter Littelmann, *Bases for representations, LS-paths and Verma flags*, 323–345. MR2017591
16. V. Lakshmibai and K. N. Raghavan, *Standard monomial theory*, Encyclopaedia of Mathematical Sciences, vol. 137, Springer-Verlag, Berlin, 2008, Invariant theoretic approach, Invariant Theory and Algebraic Transformation Groups, 8. MR2388163
17. V. Lakshmibai, Vijay Ravikumar, and William Slofstra, *The cotangent bundle of a cominuscule Grassmannian*, Michigan Math. J. **65** (2016), no. 4, 749–759. MR3579184
18. V. Lakshmibai and C. S. Seshadri, *Geometry of G/P. II. The work of de Concini and Procesi and the basic conjectures*, vol. 87, 1978, pp. 1–54. MR490244
19. G. Lusztig, *Canonical bases arising from quantized enveloping algebras*, J. Amer. Math. Soc. **3** (1990), no. 2, 447–498. MR1035415
20. Timothée Marquis, *An introduction to Kac-Moody groups over fields*, EMS Textbooks in Mathematics, European Mathematical Society (EMS), Zürich, 2018. MR3838421
21. Vikram B. Mehta and A. Ramanathan, *Frobenius splitting and cohomology vanishing for Schubert varieties*, Ann. of Math. (2) **122** (1985), no. 1, 27–40. MR799251
22. Bertrand Rémy, *Groupes de Kac-Moody déployés et presque déployés*, vol. 277, Société mathématique de France, 2002.
23. T. A. Springer, *The unipotent variety of a semi-simple group*, Algebraic Geometry (Internat. Colloq., Tata Inst. Fund. Res., Bombay, 1968), Oxford Univ. Press, London, 1969, pp. 373–391. MR0263830
24. Elisabetta Strickland, *On the conormal bundle of the determinantal variety*, J. Algebra **75** (1982), no. 2, 523–537. MR653906
25. Jacques Tits, *Uniqueness and presentation of Kac-Moody groups over fields*, J. Algebra **105** (1987), no. 2, 542–573. MR873684
26. Jerzy Weyman, *Cohomology of vector bundles and syzygies*, Cambridge Tracts in Mathematics, vol. 149, Cambridge University Press, Cambridge, 2003. MR1988690

Evaluation Modules for Quantum Toroidal $\mathfrak{gl}_n$ Algebras

Boris Feigin, Michio Jimbo, and Evgeny Mukhin

To Vyjayanthi Chari on the occasion of her 60th birthday

Abstract The affine evaluation map is a surjective homomorphism from the quantum toroidal $\mathfrak{gl}_n$ algebra $\mathcal{E}'_n(q_1, q_2, q_3)$ to the quantum affine algebra $U'_q\widehat{\mathfrak{gl}}_n$ at level κ completed with respect to the homogeneous grading, where $q_2 = q^2$ and $q_3^n = \kappa^2$. We discuss $\mathcal{E}'_n(q_1, q_2, q_3)$ evaluation modules. We give highest weights of evaluation highest weight modules. We also obtain the decomposition of the evaluation Wakimoto module with respect to a Gelfand–Zeitlin-type subalgebra of a completion of $\mathcal{E}'_n(q_1, q_2, q_3)$, which describes a deformation of the coset theory $\widehat{\mathfrak{gl}}_n/\widehat{\mathfrak{gl}}_{n-1}$.

1 Introduction

For an arbitrary complex Lie algebra $\mathfrak{g}$ and a non-zero constant u, we have the evaluation map

$$\mathfrak{g} \otimes \mathbb{C}[t, t^{-1}] \to \mathfrak{g}, \qquad g \otimes t^k \mapsto u^k g.$$

B. Feigin
National Research University Higher School of Economics, Russian Federation, International Laboratory of Representation Theory and Mathematical Physics, Moscow, Russia

Landau Institute for Theoretical Physics, Chernogolovka, Russia

M. Jimbo
Department of Mathematics, Rikkyo University, Toshima-ku, Tokyo, Japan
e-mail: jimbomm@rikkyo.ac.jp

E. Mukhin (✉)
Department of Mathematics, Indiana University-Purdue University-Indianapolis, Indianapolis, IN, USA
e-mail: emukhin@iupui.edu

J. Greenstein et al. (eds.), *Interactions of Quantum Affine Algebras with Cluster Algebras, Current Algebras and Categorification*, Progress in Mathematics 337,
https://doi.org/10.1007/978-3-030-63849-8_12

The evaluation map is a surjective homomorphism of Lie algebras. It plays a prominent role in representation theory of current algebras and various constructions in mathematics and physics.

In type A, a quantum version of the evaluation map $\overline{ev}_u : U'_q\widehat{\mathfrak{sl}}_n \to U_q\mathfrak{gl}_n$ was introduced in [6], see (2.4) below. This map is used to construct simplest possible representations of the quantum affine algebra $U_q\widehat{\mathfrak{sl}}_n$ called evaluation modules. The evaluation modules are central for many studies. For example, the R matrix of the celebrated six vertex model is an intertwiner for tensor products of two evaluation modules. It is well-known that the quantum evaluation map does not exist for simple Lie algebras $\mathfrak{g}$ other than in type A.

The affine analog of the quantum evaluation map was discovered in [9], which we now recall. Let $\mathcal{E}'_n = \mathcal{E}'_n(q_1, q_2, q_3)$ be the quantum toroidal algebra associated with $\mathfrak{gl}_n$. It depends on complex parameters q_1, q_2, q_3 such that $q_1q_2q_3 = 1$ and a central elements C (we set the second central element to 1), see Sect. 3. Let $U'_q\widehat{\mathfrak{gl}}_n$ be the quantum affine algebra associated with $\mathfrak{gl}_n$, see Sect. 2.2. It depends on a complex parameter q, and it has a central element C. We always assume $q^2 = q_2$. An easy well-known fact is that there is a homomorphism of algebras $v : U'_q\widehat{\mathfrak{gl}}_n \to \mathcal{E}'_n(q_1, q_2, q_3)$ such that $v(C) = C$, see (3.12).

We consider $U'_q\widehat{\mathfrak{gl}}_n$ in the Drinfeld new realization, and we denote by $\widetilde{U}'_q\widehat{\mathfrak{gl}}_n$ its completion with respect to the homogeneous grading, see (2.5). We also impose the following key relation for the central elements of $\mathcal{E}'_n$ and $\widetilde{U}'_q\widehat{\mathfrak{gl}}_n$ with parameter q_3:

$$C^2 = q_3^n. \tag{1.1}$$

Then by [9] there exists a surjective algebra homomorphism depending on a non-zero complex number u:

$$\mathrm{ev}_u^{(3)} : \mathcal{E}'_n \to \widetilde{U}'_q\widehat{\mathfrak{gl}}_n$$

such that $\mathrm{ev}_u^{(3)} \circ v = id$. We call the homomorphism $\mathrm{ev}_u^{(3)}$ the quantum affine evaluation map.

We give formulas for the quantum affine evaluation map in Sect. 4 and provide the proof in Sect. 4.2. We supply a number of details that were omitted in [9].

The completed algebra $\widetilde{U}'_q\widehat{\mathfrak{gl}}_n$ acts on highest weight $U'_q\widehat{\mathfrak{gl}}_n$ modules, and therefore every such representation becomes an evaluation representation of the quantum toroidal algebra with the parameters satisfying (1.1). Moreover, this representation of $\mathcal{E}'_n$ is also a highest weight module. We discuss the highest weights of evaluation modules in Sect. 5.2. It turns out that some evaluation representations appeared already in [5].

The algebra $\mathcal{E}'_n(q_1, q_2, q_3)$ has mutually commuting subalgebras $\mathcal{E}'_{1,n-1} \simeq \mathcal{E}'_1(q_1^n, q_2, q_3q_1^{-n+1})$ and $\mathcal{E}^{0|n-1}_{n-1} \simeq \mathcal{E}'_{n-1}(q_1q_3^{-1/(n-1)}, q_2, q_3q_3^{1/(n-1)})$. Under the affine evaluation map, the image of $\mathcal{E}'_{1,n-1}$ is a deformation of the coset W algebra

$\widehat{\mathfrak{gl}}_n/\widehat{\mathfrak{gl}}_{n-1}$. Therefore it is important to describe the decomposition of evaluation modules with respect to these subalgebras.

For that purpose, we recall the n remarkable pairwise commuting subalgebras $\mathcal{E}'_{1,m}(q_{1,m}, q_2, q_{3,m})$ corresponding to the diagonal inclusions of $\mathfrak{gl}_1$ to $\mathfrak{gl}_n$, see [4]. We consider the Wakimoto $U'_q\widehat{\mathfrak{gl}}_n$ modules generated from Gelfand–Zeitlin modules of $U_q\mathfrak{gl}_n$ and the corresponding evaluation module $\widehat{\mathrm{GZ}}_{\lambda^0}(u)$ of $\mathcal{E}'_n(q_1, q_2, q_3)$, see Sect. 5.3. We describe the decomposition of $\widehat{\mathrm{GZ}}_{\lambda^0}(u)$ with respect to algebra $\otimes_{m=0}^{n-1}\mathcal{E}'_{1,m}(q_{1,m}, q_2, q_{3,m})$, see Theorem 5.3.

This result is also important for applications, see [3]. In fact, the screening operators commuting with a copy of quantum affine $\mathfrak{gl}_2$ in [3] are obtained from currents of $\mathcal{E}'_{1,2}$ acting in an $\mathcal{E}'_2$ evaluation Wakimoto module.

The paper is constructed as follows. We give definitions of various quantum algebras in Sects. 2 and 3.1. In Sect. 3.2 we recall the fused currents of [4] and study their commutation relations with other currents. In Sect. 3.3 we recall the definition of the subalgebras $\mathcal{E}'_{1,m}(q_{1,m}, q_2, q_{3,m})$. We give the evaluation map in Sect. 4 and prove that it is well-defined in Sect. 4.2. We discuss evaluation modules in Sect. 5. We discuss evaluation highest weight modules in Sect. 5.2 and evaluation Wakimoto modules in Sect. 5.3. We give a proof of the result on evaluation Wakimoto modules in Appendix 6.

2 Quantum Groups

In this section we set up the notation for various quantum groups.

Let n be a positive integer. For $n \geqslant 2$, let $(a_{i,j})_{i,j\in\mathbb{Z}/n\mathbb{Z}}$ be the Cartan matrix of type $A^{(1)}_{n-1}$.

Let $\mathbb{C}P = \oplus_{i\in\mathbb{Z}/n\mathbb{Z}}\mathbb{C}\varepsilon_i$ be an n-dimensional vector space with the chosen basis and a non-degenerate form such that $(\varepsilon_i, \varepsilon_j) = \delta_{i,j}$. Let $P = \oplus_{i\in\mathbb{Z}/n\mathbb{Z}}\mathbb{Z}\varepsilon_i$ be the lattice.

Set $\alpha_i = \varepsilon_{i-1} - \varepsilon_i$ and $\Lambda_i = \sum_{j=0}^{i-1}\varepsilon_j$, $1 \leqslant i \leqslant n-1$. We have $(\alpha_i, \Lambda_j) = \delta_{i,j}$, $(\alpha_i, \alpha_j) = a_{i,j}$.

Let $\mathbb{C}\bar{P} = (\sum_{i\in\mathbb{Z}/n\mathbb{Z}}\varepsilon_i)^\perp \subseteq \mathbb{C}P$. For $p \in \mathbb{C}P$, denote by $\bar{p} \in \mathbb{C}\bar{P}$ the projection along $\mathbb{C}(\sum_{i\in\mathbb{Z}/n\mathbb{Z}}\varepsilon_i)$. Then $\bar{\alpha}_i$, $\bar{\Lambda}_i$, $1 \leqslant i \leqslant n-1$, are simple roots and fundamental weights of $\mathfrak{sl}_n$, respectively.

Fix $\log q, \log d \in \mathbb{C}$ and set $q = e^{\log q}$, $d = e^{\log d}$, $q_1 = q^{-1}d$, $q_2 = q^2$, $q_3 = q^{-1}d^{-1}$, so that $q_1q_2q_3 = 1$. We assume that, for rational numbers a, b, c, the equality $q_1^a q_2^b q_3^c = 1$ holds if and only if $a = b = c$.

We use the standard notation $[A, B]_p = AB - pBA$ and $[r] = (q^r - q^{-r})/(q - q^{-1})$.

2.1 Quantum Algebra $U_q\mathfrak{gl}_n$

The quantum $\mathfrak{gl}_n$ algebra $U_q\mathfrak{gl}_n$ has generators e_i, f_i, q^h, $1 \leqslant i \leqslant n-1$, $h \in P$, with the defining relations

$$q^h q^{h'} = q^{h+h'}, \qquad q^0 = 1, \qquad q^h e_i = q^{(h,\alpha_i)} e_i q^h, \qquad q^h f_i = q^{-(h,\alpha_i)} f_i q^h,$$

$$[e_i, f_j] = \delta_{i,j} \frac{K_i - K_i^{-1}}{q - q^{-1}},$$

$$[e_i, e_j] = [f_i, f_j] = 0 \qquad (|i-j| \geqslant 2),$$

$$[e_i, [e_i, e_j]_{q^{-1}}]_q = [f_i, [f_i, f_j]_{q^{-1}}]_q = 0 \qquad (|i-j| = 1),$$

where $K_i = q^{\alpha_i}$.

The quantum $\mathfrak{sl}_n$ algebra $U_q\mathfrak{sl}_n$ is the subalgebra of $U_q\mathfrak{gl}_n$ generated by e_i, f_i, K_i, $1 \leqslant i \leqslant n-1$.

The element $\mathsf{t} = q^{\varepsilon_0+\varepsilon_1+\cdots+\varepsilon_{n-1}} \in U_q\mathfrak{gl}_n$ is central and split.

2.2 Quantum Affine Algebra $U_q'\widehat{\mathfrak{gl}}_n$

The quantum affine algebra $U_q'\widehat{\mathfrak{sl}}_n$ in the Drinfeld new realization is defined by generators $x^{\pm}_{i,k}$, $h_{i,r}$, q^h, C, where $1 \leqslant i \leqslant n-1$, $k \in \mathbb{Z}$, $r \in \mathbb{Z} \setminus \{0\}$, $h \in P$, with the defining relations

$$C \text{ is central}, \quad q^h q^{h'} = q^{h+h'}, \quad q^0 = 1,$$

$$q^h x_i^{\pm}(z) q^{-h} = q^{\pm(h,\alpha_i)} x_i^{\pm}(z), \quad [q^h, h_{j,r}] = 0,$$

$$[h_{i,r}, h_{j,s}] = \delta_{r+s,0} \frac{[r a_{i,j}]}{r} \frac{C^r - C^{-r}}{q - q^{-1}},$$

$$[h_{i,r}, x_j^{\pm}(z)] = \pm \frac{[r a_{i,j}]}{r} C^{-(r\pm|r|)/2} z^r x_j^{\pm}(z),$$

$$[x_i^+(z), x_j^-(w)] = \frac{\delta_{i,j}}{q - q^{-1}} \Big(\delta\big(C\frac{w}{z}\big)\phi_i^+(w) - \delta\big(C\frac{z}{w}\big)\phi_i^-(z) \Big),$$

$$(z - q^{\pm a_{ij}} w) x_i^{\pm}(z) x_j^{\pm}(w) + (w - q^{\pm a_{ij}} z) x_j^{\pm}(w) x_i^{\pm}(z) = 0,$$

$$[x_i^{\pm}(z), x_j^{\pm}(w)] = 0 \quad \text{if } a_{ij} = 0,$$

$$\mathrm{Sym}_{z_1,z_2} [x_i^{\pm}(z_1), [x_i^{\pm}(z_2), x_j^{\pm}(w)]_q]_{q^{-1}} = 0 \quad \text{if } a_{ij} = -1.$$

Here we set $x_i^{\pm}(z) = \sum_{k\in\mathbb{Z}} x_{i,k}^{\pm} z^{-k}$, $\phi_j^{\pm}(z) = K_j^{\pm 1} \exp\left(\pm(q-q^{-1}) \sum_{r>0} h_{j,\pm r} z^{\mp r}\right)$, where $K_j = q^{\alpha_j}$, $1 \leqslant j \leqslant n-1$.

The quantum affine algebra $U_q'\widehat{\mathfrak{gl}}_n$ is the algebra $U_q'\widehat{\mathfrak{sl}}_n$ with additional Heisenberg generators Z_r, $r \in \mathbb{Z} \setminus \{0\}$, such that

$$[Z_k, U_q'\widehat{\mathfrak{sl}}_n] = 0, \quad [Z_r, Z_s] = -\delta_{r+s,0}[nr]\frac{1}{r}\frac{C^r - C^{-r}}{q-q^{-1}}. \tag{2.1}$$

The prime in the notation $U_q'\widehat{\mathfrak{sl}}_n$, $U_q'\widehat{\mathfrak{gl}}_n$ indicates that we do not consider the degree operator.

The element $\mathsf{t} = q^{\varepsilon_0+\cdots+\varepsilon_{n-1}}$ is central and split in both $U_q'\widehat{\mathfrak{sl}}_n$ and $U_q'\widehat{\mathfrak{gl}}_n$.

The subalgebra of $U_q'\widehat{\mathfrak{gl}}_n$ generated by $x_{i,0}^{\pm}$, q^h, $1 \le i \le n-1$, $h \in P$, is isomorphic to $U_q\mathfrak{gl}_n$.

For $n > 2$, the algebra $U_q'\widehat{\mathfrak{gl}}_n$ can be described by the same relations as above by allowing $h_{0,r}$ and setting $\alpha_0 = \varepsilon_{n-1} - \varepsilon_0$. Then we have

$$Z_r = \sum_{i=0}^{n-1} \frac{q^{ir} + q^{(n-i)r}}{q^r - q^{-r}} h_{i,r}.$$

Define the Chevalley generators of $U_q'\widehat{\mathfrak{sl}}_n$:

$$\begin{aligned} e_i &= x_{i,0}^+, \quad f_i = x_{i,0}^- \qquad (1 \leqslant i \leqslant n-1), \\ e_0^{(r)} &= [x_{n-1,0}^-, \cdots [x_{2,0}^-, x_{1,-1}^-]_q \cdots]_q\, q^{\alpha_1+\cdots+\alpha_{n-1}}, \\ f_0^{(r)} &= q^{-\alpha_1-\cdots-\alpha_{n-1}}[\cdots [x_{1,1}^+, x_{2,0}^+]_{q^{-1}}, \cdots x_{n-1,0}^+]_{q^{-1}}. \end{aligned} \tag{2.2}$$

The Chevalley generators e_j, f_j, $1 \leqslant j \leqslant n-1$, $e_0^{(r)}$, $f_0^{(r)}$ generate $U_q'\widehat{\mathfrak{sl}}_n$.

We will also use the other set of Chevalley generators, which we call left Chevalley generators:

$$\begin{aligned} e_0^{(l)} &= q^{-\alpha_1-\cdots-\alpha_{n-1}}[\cdots [x_{1,1}^-, x_{2,0}^-]_q \cdots, x_{n-1,0}^-]_q, \\ f_0^{(l)} &= [x_{n-1,0}^+, \cdots [x_{2,0}^+, x_{1,-1}^+]_{q^{-1}} \cdots]_{q^{-1}} q^{\alpha_1+\cdots+\alpha_{n-1}}. \end{aligned} \tag{2.3}$$

The elements e_j, f_j, $1 \le j \le n-1$, $e_0^{(l)}$, $f_0^{(l)}$ generate $U_q'\widehat{\mathfrak{sl}}_n$ as well.

For $u \in \mathbb{C}^{\times}$, we have the evaluation homomorphism, $\overline{\mathrm{ev}}_u : U_q'\widehat{\mathfrak{sl}}_n \to U_q\mathfrak{gl}_n$, given in Chevalley generators by (see [6])

$$\begin{aligned} \overline{\mathrm{ev}}_u(e_i) &= e_i, \qquad \overline{\mathrm{ev}}_u(f_i) = f_i \qquad (1 \leqslant i \leqslant n-1), \\ \overline{\mathrm{ev}}_u(e_0) &= u^{-1} q^{-\Lambda_1+\Lambda_{n-1}}[\cdots [f_1, f_2]_{q^{-1}}, \cdots, f_{n-1}]_{q^{-1}}, \end{aligned} \tag{2.4}$$

$$\overline{\mathrm{ev}}_u(f_0) = u\,[e_{n-1}, \cdots [e_2, e_1]_q, \cdots]_q\, q^{\Lambda_1-\Lambda_{n-1}}\,.$$

Note that $\overline{\mathrm{ev}}_u(C) = 1$.

We have the homogeneous grading given by

$$\begin{aligned} &\deg e_i = -\deg f_i = 0 \qquad (1 \leqslant i \leqslant n-1), && \deg q^h = 0 \qquad (h \in P), \\ &\deg\, e_0^{(l)} = -\deg\, f_0^{(l)} = 1, && \deg Z_r = r \qquad (r \neq 0)\,. \end{aligned} \tag{2.5}$$

We also have the principal degree given by

$$\begin{aligned} &\mathrm{pdeg}\, e_i = -\mathrm{pdeg}\, f_i = 1 \qquad (1 \leqslant i \leqslant n-1), && \mathrm{pdeg}\, q^h = 0 \qquad (h \in P), \\ &\mathrm{pdeg}\, e_0^{(l)} = -\mathrm{pdeg}\, f_0^{(l)} = 1, && \mathrm{pdeg}\, Z_r = nr \qquad (r \neq 0)\,. \end{aligned} \tag{2.6}$$

Then, in particular, $\deg x^{\pm}_{i,k} = k$ and $\mathrm{pdeg}\, x^{\pm}_{i,k} = nk \pm 1$.

For $\kappa \in \mathbb{C}^{\times}$, we denote by $U'_{q,\kappa}\widehat{\mathfrak{gl}}_n$ the quotient of $U'_q\widehat{\mathfrak{gl}}_n$ by the relation $C = \kappa$. We denote by $\widetilde{U}'_{q,\kappa}\widehat{\mathfrak{gl}}_n$ the completion of the algebra $U'_{q,\kappa}\widehat{\mathfrak{gl}}_n$ with respect to the homogeneous grading in the positive direction. Elements of $\widetilde{U}'_{q,\kappa}\widehat{\mathfrak{gl}}_n$ have the form $\sum_{r=r_0}^{\infty} g_r$, where $g_r \in U'_{q,\kappa}\widehat{\mathfrak{gl}}_n$, $\deg g_r = r$.

We denote by $\mathfrak{b}_{q,\kappa}$ the subalgebra of $U'_{q,\kappa}\widehat{\mathfrak{gl}}_n$ generated by $x^{+}_{i,k}$, $h_{j,r}$, Z_r, q^h, where $1 \leqslant i \leqslant n-1$, $1 \leqslant j \leqslant n-1$, $k \geqslant 0$, $r \geqslant 1$, $h \in P$.

3 Quantum Toroidal $\mathfrak{gl}_n$

In this section we recall the definition of quantum toroidal algebras and some facts we will use.

3.1 Definition of $\mathcal{E}'_n$

For $i, j \in \mathbb{Z}/n\mathbb{Z}$ and $r \neq 0$, we set

$$a_{i,j}(r) = \frac{[r]}{r} \times \left((q^r + q^{-r})\delta^{(n)}_{i,j} - d^r \delta^{(n)}_{i,j-1} - d^{-r}\delta^{(n)}_{i,j+1} \right),$$

where $\delta^{(n)}_{i,j}$ is Kronecker's delta modulo n: $\delta^{(n)}_{i,j} = 1$ (if $i \equiv j \bmod n$) and $\delta^{(n)}_{i,j} = 0$ (otherwise).

Define further functions $g_{i,j}(z, w)$ by

$$n \geqslant 3: \quad g_{i,j}(z,w) = \begin{cases} z - q_1 w & (i \equiv j-1), \\ z - q_2 w & (i \equiv j), \\ z - q_3 w & (i \equiv j+1), \\ z - w & (i \not\equiv j, j \pm 1), \end{cases}$$

$$n = 2: \quad g_{i,j}(z,w) = \begin{cases} z - q_2 w & (i \equiv j), \\ (z - q_1 w)(z - q_3 w) & (i \not\equiv j), \end{cases}$$

$$n = 1: \quad g_{0,0}(z,w) = (z - q_1 w)(z - q_2 w)(z - q_3 w),$$

and set $d_{i,j} = d^{\mp 1}$ $(i \equiv j \mp 1, n \geqslant 3)$, $d_{i,j} = -1$ $(i \not\equiv j, n = 2)$, and $d_{i,j} = 1$ (otherwise).

The quantum toroidal algebra of type $\mathfrak{gl}_n$, which we denote by $\mathcal{E}'_n = \mathcal{E}'_n(q_1, q_2, q_3)$, is a unital associative algebra generated by $E_{i,k}$, $F_{i,k}$, and $H_{i,r}$ and invertible elements q^h, C, where $i \in \mathbb{Z}/n\mathbb{Z}$, $k \in \mathbb{Z}$, $r \in \mathbb{Z}\backslash\{0\}$, $h \in P$. As always, we set $K_i = q^{\alpha_i}$ for $i \in \mathbb{Z}/n\mathbb{Z}$. We have

$$K_0 = \prod_{i=1}^{n-1} K_i^{-1}. \tag{3.1}$$

We present below the defining relations in terms of generating series:

$$E_i(z) = \sum_{k \in \mathbb{Z}} E_{i,k} z^{-k}, \quad F_i(z) = \sum_{k \in \mathbb{Z}} F_{i,k} z^{-k},$$

$$K_i^{\pm}(z) = K_i^{\pm 1} \bar{K}_i^{\pm}(z), \quad \bar{K}_i^{\pm}(z) = \exp\Big(\pm (q - q^{-1}) \sum_{r>0} H_{i,\pm r} z^{\mp r} \Big).$$

The relations are as follows.

C, q^h Relations

$$C \text{ is central}, \quad q^h q^{h'} = q^{h+h'}, \quad q^0 = 1, \tag{3.2}$$

$$\begin{aligned} &q^h E_i(z) q^{-h} = q^{(h,\alpha_i)} E_i(z), \quad q^h F_i(z) q^{-h} = q^{-(h,\alpha_i)} F_i(z), \\ &q^h K_i^{\pm}(z) = K_i^{\pm}(z) q^h. \end{aligned} \tag{3.3}$$

K-K, K-E, and K-F Relations

$$K_i^{\pm}(z) K_j^{\pm}(w) = K_j^{\pm}(w) K_i^{\pm}(z), \tag{3.4}$$

$$\frac{g_{i,j}(C^{-1}z,w)}{g_{i,j}(Cz,w)}K_i^-(z)K_j^+(w)=\frac{g_{j,i}(w,C^{-1}z)}{g_{j,i}(w,Cz)}K_j^+(w)K_i^-(z), \tag{3.5}$$

$$d_{i,j}g_{i,j}(z,w)K_i^{\pm}(C^{-(1\pm1)/2}z)E_j(w)+g_{j,i}(w,z)E_j(w)K_i^{\pm}(C^{-(1\pm1)/2}z)=0, \tag{3.6}$$

$$d_{j,i}g_{j,i}(w,z)K_i^{\pm}(C^{-(1\mp1)/2}z)F_j(w)+g_{i,j}(z,w)F_j(w)K_i^{\pm}(C^{-(1\mp1)/2}z)=0\,. \tag{3.7}$$

E-F Relations

$$[E_i(z),F_j(w)]=\frac{\delta_{i,j}}{q-q^{-1}}(\delta(C\frac{w}{z})K_i^+(w)-\delta(C\frac{z}{w})K_i^-(z))\,. \tag{3.8}$$

E-E and F-F Relations

$$[E_i(z),E_j(w)]=0\,,\quad [F_i(z),F_j(w)]=0\quad (i\not\equiv j,j\pm1)\,,$$
$$d_{i,j}g_{i,j}(z,w)E_i(z)E_j(w)+g_{j,i}(w,z)E_j(w)E_i(z)=0,$$
$$d_{j,i}g_{j,i}(w,z)F_i(z)F_j(w)+g_{i,j}(z,w)F_j(w)F_i(z)=0.$$

Serre Relations For $n\geqslant 3$,

$$\mathrm{Sym}_{z_1,z_2}[E_i(z_1),[E_i(z_2),E_{i\pm1}(w)]_q]_{q^{-1}}=0\,,$$
$$\mathrm{Sym}_{z_1,z_2}[F_i(z_1),[F_i(z_2),F_{i\pm1}(w)]_q]_{q^{-1}}=0\,.$$

For $n=2$, $i\not\equiv j$,

$$\mathrm{Sym}_{z_1,z_2,z_3}[E_i(z_1),[E_i(z_2),[E_i(z_3),E_j(w)]_{q^2}]]_{q^{-2}}=0\,,$$
$$\mathrm{Sym}_{z_1,z_2,z_3}[F_i(z_1),[F_i(z_2),[F_i(z_3),F_j(w)]_{q^2}]]_{q^{-2}}=0\,.$$

For $n=1$,

$$\mathrm{Sym}_{z_1,z_2,z_3}z_2z_3^{-1}[E_0(z_1),[E_0(z_2),E_0(z_3)]]=0\,,$$
$$\mathrm{Sym}_{z_1,z_2,z_3}z_2z_3^{-1}[F_0(z_1),[F_0(z_2),F_0(z_3)]]=0\,.$$

In the above we set

$$\mathrm{Sym}\,f(x_1,\ldots,x_N)=\frac{1}{N!}\sum_{\pi\in\mathfrak{S}_N}f(x_{\pi(1)},\ldots,x_{\pi(N)})\,.$$

Under the C, q^h relations (3.3), the K-K, K-E, and K-F relations (3.4)–(3.7) are equivalently written as

H-E, H-F, and H-H Relations For $r \neq 0$,

$$[H_{i,r}, E_j(z)] = a_{i,j}(r)C^{-(r+|r|)/2}\, z^r E_j(z)\,,$$

$$[H_{i,r}, F_j(z)] = -a_{i,j}(r)C^{-(r-|r|)/2}\, z^r F_j(z)\,,$$

$$[H_{i,r}, H_{j,s}] = \delta_{r+s,0} \cdot a_{i,j}(r)\eta_r\,, \qquad \eta_r = \frac{C^r - C^{-r}}{q - q^{-1}}\,.$$

The E-E and F-F relations with $j \equiv i \pm 1$ in Fourier components read for $n \geqslant 3$

$$[E_{i,k+1}, E_{i+1,l}]_{q^{-1}} = q_1[E_{i,k}, E_{i+1,l+1}]_q\,, \tag{3.9}$$

$$[F_{i,k+1}, F_{i+1,l}]_q = q_3^{-1}[F_{i,k}, F_{i+1,l+1}]_{q^{-1}}\,,$$

and for $n = 2$

$$[E_{i,k+1}, E_{i+1,l-1}]_{q^{-2}} - (q_1 + q_3)[E_{i,k}, E_{i+1,l}] + q_1q_3[E_{i,k-1}, E_{i+1,l+1}]_{q^2} = 0\,, \tag{3.10}$$

$$[F_{i+1,l-1}, F_{i,k+1}]_{q^{-2}} - (q_1 + q_3)[F_{i+1,l}, F_{i,k}] + q_1q_3[F_{i+1,l+1}, F_{i,k-1}]_{q^2} = 0\,.$$

The algebra $\mathcal{E}'_n$ considered here is obtained from [4] by setting the second central element $\prod_{i=0}^{n-1} K_i$ to 1, dropping the scaling elements D, $D^\perp$ and adding the split central element $\mathsf{t} = q^{\varepsilon_0+\cdots+\varepsilon_{n-1}}$. Our generators $E_i(z)$, $F_i(z)$, $K_i^\pm(z)$, C correspond to the perpendicular generators $E_i^\perp(z)$, $F_i^\perp(z)$, $K_i^{\pm,\perp}((q^c)^\perp z)$, $(q^c)^\perp$ of [4].

Algebra $\mathcal{E}'_n$ is $\mathbb{Z}$-graded by

$$\deg E_{i,k} = \deg F_{i,k} = k, \qquad \deg H_{i,r} = r, \qquad \deg C = \deg q^h = 0. \tag{3.11}$$

We denote by $\widetilde{\mathcal{E}}'_n$ the completion of $\mathcal{E}'_n$ with respect to this grading in the positive direction.

We have a graded embedding $v:\ U'_q\widehat{\mathfrak{gl}}_n \to \mathcal{E}'_n$ given by

$$x_i^+(z) \mapsto E_i(d^{-i}z)\,, \quad x_i^-(z) \mapsto F_i(d^{-i}z)\,, \quad \phi_i^\pm(z) \mapsto K_i^\pm(d^{-i}z)\,, \tag{3.12}$$

and $C \mapsto C$, $q^h \mapsto q^h$. We call the image of v the vertical subalgebra and denote it by $U_q^v\widehat{\mathfrak{gl}}_n$.

We call the subalgebra of $\mathcal{E}'_n$ generated by $E_{i,0}, F_{i,0}, 0 \leqslant i \leqslant n-1$, the horizontal subalgebra. The horizontal subalgebra, which we denote by $U_q^h\widehat{\mathfrak{sl}}_n$, is isomorphic to the quotient of $U'_q\widehat{\mathfrak{sl}}_n$ by the relation $C = 1$.

There exists an isomorphism of algebras

$$\iota : \mathcal{E}'_n(q_1, q_2, q_3) \to \mathcal{E}'_n(q_3, q_2, q_1),$$

given by

$$\iota : E_i(z) \mapsto E_{n-i}(z), \quad F_i(z) \mapsto F_{n-i}(z), \quad K_i^{\pm}(z) \mapsto K_{n-i}^{\pm}(z), \tag{3.13}$$

and $\iota(C) = C$.

We have also the Miki automorphism θ that interchanges vertical and horizontal subalgebras, see [8]. We fix the definition of θ as in [4]. We remark that θ is defined for algebra $\mathcal{E}_n$, not $\mathcal{E}'_n$, but expressions such as $\theta^{-1}(K_i^{\pm}(z))$ have a well-defined meaning in $\mathcal{E}'_n$ as well. In particular, we have for $n \geqslant 2$,

$$\theta^{-1}(H_{i,1}) = -(-d)^{-i}[[\cdots[[\cdots[F_{0,0}, F_{n-1,0}]_q, \cdots, F_{i+1,0}]_q, F_{1,0}]_q, \cdots,$$
$$F_{i-1,0}]_q, F_{i,0}]_{q^2}, \tag{3.14}$$

$$\theta^{-1}(H_{i,-1}) = -(-d)^i[E_{i,0}, [E_{i-1,0}, \cdots, [E_{1,0}, [E_{i+1,0}, \cdots,$$
$$[E_{n-1,0}, E_{0,0}]_{q^{-1}}, \cdots]_{q^{-1}}]_{q^{-1}}, \cdots]_{q^{-1}}]_{q^{-2}} \tag{3.15}$$

for $1 \leqslant i \leqslant n-1$, and

$$\theta^{-1}(H_{0,1}) = -(-d)^{-n+1}[[\cdots[F_{1,1}, F_{2,0}]_q, \cdots, F_{n-1,0}]_q, F_{0,-1}]_{q^2}, \tag{3.16}$$

$$\theta^{-1}(H_{0,-1}) = -(-d)^{n-1}[E_{0,1}, [E_{n-1,0}, \cdots, [E_{2,0}, E_{1,-1}]_{q^{-1}} \cdots]_{q^{-1}}]_{q^{-2}}. \tag{3.17}$$

In addition,

$$\theta(E_{0,0}) = d^{-1}[F_{n-1,0}, \ldots, [F^-_{2,0}, F^-_{1,-1}]_q \ldots]_q K_1 \cdots K_{n-1}, \tag{3.18}$$

$$\theta(F_{0,0}) = d(K_1 \cdots K_{n-1})^{-1}[\ldots[E_{1,1}, E_{2,0}]_{q^{-1}}, \ldots, E_{n-1,0}]_{q^{-1}}. \tag{3.19}$$

We define $U_q^h\widehat{\mathfrak{gl}}_n$ to be the subalgebra generated by $U_q^h\widehat{\mathfrak{sl}}_n$ and $\theta^{-1}(K_0^{\pm}(z))$.

3.2 Fused Currents

We recall the fused currents of [4] and compute the commutation relations with generators of $U'_q\widehat{\mathfrak{gl}}_n$.

It is convenient to use series of generators of $U'_q\widehat{\mathfrak{gl}}_n$ defined by $E_i(z) = x_i^+(d^i z)$, $F_i(z) = x_i^-(d^i z)$, $K_i^\pm(z) = \phi_i^\pm(d^i z)$, $1 \leqslant i \leqslant n-1$, cf. (3.12).

Following [4], let us introduce the following elements of $\widetilde{U}'_q\widehat{\mathfrak{gl}}_n$:

$$\mathsf{E}(z) = \prod_{i=1}^{n-2}(1 - \frac{z_i}{z_{i+1}}) \cdot E_{n-1}(q_3^{n/2-1} z_{n-1}) \cdots \times E_2(q_3^{-n/2+2} z_2) E_1(q_3^{-n/2+1} z_1)\Big|_{z_1=\cdots=z_{n-1}=z}, \tag{3.20}$$

$$\mathsf{F}(z) = \prod_{i=1}^{n-2}(1 - \frac{z_{i+1}}{z_i}) \cdot F_1(q_3^{-n/2+1} z_1) F_2(q_3^{-n/2+2} z_2) \cdots \times F_{n-1}(q_3^{n/2-1} z_{n-1})\Big|_{z_1=\cdots=z_{n-1}=z}, \tag{3.21}$$

$$\mathsf{K}^\pm(z) = \prod_{i=1}^{n-1} K_i^\pm(q_3^{-n/2+i} z). \tag{3.22}$$

When $n = 2$, we have $\mathsf{E}(z) = E_1(z)$, $\mathsf{F}(z) = F_1(z)$, $\mathsf{K}^\pm(z) = K_1^\pm(z)$.

The following result is a special case of the construction in [4].

Proposition 3.1 ([4]) *For $n \geqslant 2$, the currents* (3.20)–(3.22) *satisfy*

$$[\mathsf{E}(z), \mathsf{F}(w)] = \frac{1}{q - q^{-1}}\Big(\delta(C\frac{w}{z})\mathsf{K}^+(z) - \delta(C\frac{z}{w})\mathsf{K}^-(z)\Big), \tag{3.23}$$

$$(z - q_2 w)\mathsf{E}(z)\mathsf{E}(w) + (w - q_2 z)\mathsf{E}(w)\mathsf{E}(z) = 0, \tag{3.24}$$

$$(w - q_2 z)\mathsf{F}(z)\mathsf{F}(w) + (z - q_2 w)\mathsf{F}(w)\mathsf{F}(z) = 0, \tag{3.25}$$

$$[\mathsf{E}(z), E_i(w)] = [\mathsf{E}(z), F_i(w)] = 0 \quad (2 \leqslant i \leqslant n-2), \tag{3.26}$$

$$[E_i(z), \mathsf{F}(w)] = [F_i(z), \mathsf{F}(w)] = 0 \quad (2 \leqslant i \leqslant n-2). \tag{3.27}$$

In addition we calculate the other commutation relations.

Proposition 3.2 *For $n \geqslant 3$, we have*

$$(z - q_3^{-n/2} q_1^{-1} w) E_1(z)\mathsf{E}(w) = q(z - q_3^{-n/2+1} w)\mathsf{E}(w) E_1(z), \tag{3.28}$$

$$(z - q_3^{n/2-1} w) E_{n-1}(z)\mathsf{E}(w) = q(z - q_3^{n/2} q_1^{-1} w)\mathsf{E}(w) E_{n-1}(z), \tag{3.29}$$

$$(z - q_3^{-n/2+1} w) F_1(z)\mathsf{F}(w) = q^{-1}(z - q_3^{-n/2} q_1^{-1} w)\mathsf{F}(w) F_1(z), \tag{3.30}$$

$$(z - q_3^{n/2} q_1 w) F_{n-1}(z)\mathsf{F}(w) = q^{-1}(z - q_3^{n/2-1} w)\mathsf{F}(w) F_{n-1}(z), \tag{3.31}$$

$$(z - C^{-1}q_3^{n/2-1}w)[\mathsf{E}(z), F_1(w)] = (z - Cq_3^{-n/2+1}w)[\mathsf{E}(z), F_{n-1}(w)] = 0\,, \tag{3.32}$$

$$(z - Cq_3^{-n/2+1}w)[E_1(z), \mathsf{F}(w)] = (z - C^{-1}q_3^{n/2-1}w)[E_{n-1}(z), \mathsf{F}(w)] = 0\,. \tag{3.33}$$

Proof As an example, we consider (3.32). From the defining relations (3.8), we have

$$\begin{aligned}
&(q-q^{-1})[E_{n-1}(q_3^{n/2-1}z_{n-1})\cdots E_1(q_3^{-n/2+1}z_1), F_1(w)] \\
&= E_{n-1}(q_3^{n/2-1}z_{n-1})\cdots E_2(q_3^{-n/2+2}z_2)\Big(\delta\big(Cq_3^{n/2-1}\frac{w}{z_1}\big)K_1^+(w) \\
&\qquad\qquad - \delta\big(Cq_3^{-n/2+1}\frac{z_1}{w}\big)K_1^-(q_3^{-n/2+1}z_1)\Big)\,.
\end{aligned}$$

Upon multiplying by $z_1 - C^{-1}q_3^{n/2-1}w$, the second term in the right-hand side vanishes. The first term does not have a pole at $z_1 = z_2$ because $K_1^+(w)$ is a power series in w^{-1} placed at the rightmost. Multiplying further by $\prod_{i=1}^{n-2}(1 - z_i/z_{i+1})$ and setting $z_1 = \cdots = z_{n-1} = z$, we find

$$(z - C^{-1}q_3^{n/2-1}w)[\mathsf{E}(z), F_1(w)] = 0$$

as desired.

The rest of the relations can be shown in a similar manner.

3.3 *The Subalgebra* $\mathcal{A}$

We also recall from [4] the fusion construction of subalgebras

$$\begin{aligned}
&\mathcal{E}'_{1,m} = \mathcal{E}'_1(q_{1,m}, q_2, q_{3,m}) \subseteq \widetilde{\mathcal{E}}'_n(q_1, q_2, q_3) \quad (0 \leqslant m \leqslant n-1)\,, \\
&q_{1,m} = q_1^{m+1}q_3^{-n+m+1}\,, \quad q_{3,m} = q_3^{n-m}q_1^{-m}\,,
\end{aligned}$$

which mutually commute and intersect only by the central element C.

The algebra $\mathcal{E}'_{1,m}$ is generated by the fused currents

$$\begin{aligned}
&-\frac{q_{3,m}}{1-q_{3,m}}\mathsf{E}_m(z) \\
&\qquad = \prod_{i=0}^{n-2}\big(1 - \frac{z_i}{z_{i+1}}\big)\cdot E_0(q_3^n z_0)E_{n-1}(q_3^{n-1}z_{n-1})\cdots E_{m+1}(q_3^{m+1}z_{m+1})
\end{aligned}$$

$$\times\, E_1(q_3^n q_1^{-1} z_1)\cdots E_m(q_3^n q_1^{-m} z_m)\Big|_{z_0=\cdots=z_{n-1}=z}\,,$$

$$(1-q_{1,m})\mathsf{F}_m(z)$$
$$=\prod_{i=0}^{n-2}\Big(1-\frac{z_{i+1}}{z_i}\Big)\cdot F_m(q_3^n q_1^{-m} z_m)\cdots F_1(q_3^n q_1^{-1} z_1)$$
$$\times\, F_{m+1}(q_3^{m+1} z_{m+1})\cdots F_{n-1}(q_3^{n-1} z_{n-1})F_0(q_3^n z_0)\Big|_{z_0=\cdots=z_{n-1}=z}\,,$$

$$\mathsf{K}_m^{\pm}(z)=\prod_{i=m+1}^{n-1} K_i^{\pm}(q_3^i z)\prod_{i=0}^{m} K_i^{\pm}(q_1^{-i} q_3^n z).$$

Let $\mathcal{A}$ be the algebra generated by subalgebras $\mathcal{E}'_{1,m}$, $0\leqslant m\leqslant n-1$.

We have

$$\mathcal{A}=\mathcal{E}'_{1,0}\otimes\mathcal{E}'_{1,1}\otimes\cdots\otimes\mathcal{E}'_{1,n-1}\subseteq\widetilde{\mathcal{E}}'_n(q_1,q_2,q_3)\,. \tag{3.34}$$

The algebra $\mathcal{A}$ can be considered as an analog of the Gelfand–Zeitlin subalgebra in $\widetilde{\mathcal{E}}'_n$.

4 Quantum Affine Evaluation Map

In this section we define and prove the evaluation map in the quantum toroidal setting.

4.1 *The Definition of the Quantum Affine Evaluation Map*

From now on, we consider the quotient algebra of $\mathcal{E}'_n$ by the relation

$$C=q_3^{n/2}\,.$$

Denote this quotient by $\mathcal{E}_n^{(3)}$.

Introduce currents $A_\pm(z)=\sum_{r>0}A_{\pm r}z^{\mp r}$, $B_\pm(z)=\sum_{r>0}B_{\pm r}z^{\mp r}$ in $U'_q\widehat{\mathfrak{gl}}_n$ by setting

$$A_{-r}=\eta_r^{-1}C^{-r}(H_{0,-r}+\sum_{i=1}^{n-1}q_3^{ir}H_{i,-r})\,,$$

$$A_r = -\eta_r^{-1}(H_{0,r} + \sum_{i=1}^{n-1} q_3^{(n-i)r} H_{i,r}),$$

$$B_{-r} = -\eta_r^{-1}(H_{0,-r} + \sum_{i=1}^{n-1} q_3^{-(n-i)r} H_{i,-r}),$$

$$B_r = \eta_r^{-1} C^r (H_{0,r} + \sum_{i=1}^{n-1} q_3^{-ir} H_{i,r}).$$

Set further

$$\mathcal{K} = q^{-\Lambda_1 + \Lambda_{n-1}}.$$

Note that $\mathcal{K}$ commutes with $\mathsf{E}(z)$, $\mathsf{F}(z)$.

Theorem 4.1 ([9]) *Let* $u \in \mathbb{C}^\times$, *and set* $\kappa = q_3^{n/2}$. *The following assignment gives a homomorphism of algebras* $\mathrm{ev}_u^{(3)} : \mathcal{E}_n^{(3)} \to \widetilde{U}'_{q,\kappa}\widehat{\mathfrak{gl}}_n$ *such that* $\mathrm{ev}_u^{(3)} \circ \upsilon = \mathrm{id}$*:*

$$E_0(z) \mapsto u^{-1} e^{A_-(z)} \mathsf{F}(z) e^{A_+(z)} \mathcal{K}, \quad F_0(z) \mapsto u\, e^{B_-(z)} \mathsf{E}(z) e^{B_+(z)} \mathcal{K}^{-1}, \tag{4.1}$$

$$E_i(z) \mapsto E_i(z), \quad F_i(z) \mapsto F_i(z) \quad (i = 1, \ldots, n-1), \tag{4.2}$$

$$K_i^{\pm}(z) \mapsto K_i^{\pm}(z) \quad (i = 0, 1, \ldots, n-1), \quad q^h \mapsto q^h \quad (h \in P). \tag{4.3}$$

Remark When $n = 1$, we set formally $\mathsf{E}(z) = \mathsf{F}(z) = \mathcal{K} = 1$. Then (4.1) is nothing but the known vertex operator realization of $\mathcal{E}'_1$ for $C = q_3^{1/2}$.

A proof of Theorem 4.1 is provided in Sect. 4.2.

In the above theorem we have chosen the currents $E_0(z)$, $F_0(z)$ to play a special role. In view of the cyclic symmetry of $\mathcal{E}'_n$ that sends $E_i(z) \mapsto E_{i+1}(z)$, $F_i(z) \mapsto F_{i+1}(z)$, $K_i^{\pm}(z) \mapsto K_{i+1}^{\pm}(z)$, we could have started with $E_i(z)$, $F_i(z)$ for any i.

Using the isomorphism ι (3.13) that interchanges q_1 with q_3, it is easy to write another evaluation homomorphism $\mathrm{ev}_u^{(1)}$ when

$$C = q_1^{n/2}.$$

This is parallel to the fact that there are two evaluation homomorphisms $U'_q\widehat{\mathfrak{sl}}_n \to U_q\mathfrak{gl}_n$.

We also remark that the evaluation map is clearly graded with respect to the homogeneous degree, see (2.5), (3.11), and commutes with the automorphism that rescales the spectral parameter, see (2.7) in [4].

4.2 Proof

In this section, we prove Theorem 4.1. To simplify the notation, we consider $\mathrm{ev} = \mathrm{ev}_1^{(3)}$.

We shall need commutation relations between $A_\pm(z)$, $B_\pm(z)$ and $E_i(w)$, $F_i(w)$. First we have

$$[A_\pm(z), E_i(w)] = [A_\pm(z), F_i(w)] = [B_\pm(z), E_i(w)] \\ = [B_\pm(z), F_i(w)] = 0 \qquad (2 \leqslant i \leqslant n-2).$$

Other relations are given as follows:

$$e^{A_+(z)} E_1(w) e^{-A_+(z)} = \frac{z - q_3^{-1} w}{z - q_1 w} E_1(w)\,, \\ e^{A_+(z)} F_1(w) e^{-A_+(z)} = \frac{z - C q_1 w}{z - C q_3^{-1} w} F_1(w)\,, \tag{4.4}$$

$$e^{B_+(z)} E_{n-1}(w) e^{-B_+(z)} = \frac{z - C^{-1} q_1^{-1} w}{z - C^{-1} q_3 w} E_{n-1}(w)\,, \\ e^{B_+(z)} F_{n-1}(w) e^{-B_+(z)} = \frac{z - q_3 w}{z - q_1^{-1} w} F_{n-1}(w)\,, \tag{4.5}$$

$$e^{-A_-(z)} E_{n-1}(w) e^{A_-(z)} = \frac{w - q_3^{-1} z}{w - q_1 z} E_{n-1}(w)\,, \\ e^{-A_-(z)} F_{n-1}(w) e^{A_-(z)} = \frac{w - C q_1 z}{w - C q_3^{-1} z} F_{n-1}(w)\,, \tag{4.6}$$

$$e^{-B_-(z)} E_1(w) e^{B_-(z)} = \frac{w - C^{-1} q_1^{-1} z}{w - C^{-1} q_3 z} E_1(w)\,, \\ e^{-B_-(z)} F_1(w) e^{B_-(z)} = \frac{w - q_3 z}{w - q_1^{-1} z} F_1(w)\,. \tag{4.7}$$

We also have

$$e^{A_+(z)} e^{A_-(w)} = \frac{(z-w)^2}{(z - q_2 w)(z - q_2^{-1} w)} e^{A_-(w)} e^{A_+(z)}\,, \tag{4.8}$$

$$e^{A_+(z)} e^{B_-(w)} \frac{(z - C q_2 w)(z - C^{-1} q_2^{-1} w)}{(z - C w)(z - C^{-1} w)} e^{B_-(w)} e^{A_+(z)}\,, \tag{4.9}$$

$$e^{B_+(z)}e^{A_-(w)} = \frac{(z - Cq_2w)(z - C^{-1}q_2^{-1}w)}{(z - Cw)(z - C^{-1}w)} e^{A_-(w)}e^{B_+(z)}, \tag{4.10}$$

$$e^{B_+(z)}e^{B_-(w)} = \frac{(z-w)^2}{(z - q_2w)(z - q_2^{-1}w)} e^{B_-(w)}e^{B_+(z)}. \tag{4.11}$$

Let us verify the relations involving $E_0(z)$, $F_0(z)$ case-by-case.

C, q^h Relations These are easy to check.

H-E and H-F Relations The relation

$$[\mathrm{ev}(H_{i,r}), \mathrm{ev}(E_0(z))] = a_{i,0}(r)C^{-r}z^r\mathrm{ev}(E_0(z)) \quad (r \geqslant 1)$$

follows from $C = q_3^{n/2}$ and

$$[H_{i,r}, e^{A_-(z)}] = z^r e^{A_-(z)} \times C^{-r}\big(a_{i,0}(r) + \sum_{j=1}^{n-1} a_{i,j}(r)q_3^{jr}\big),$$

$$[H_{i,r}, \mathsf{F}(z)] = -z^r\mathsf{F}(z) \times \sum_{j=1}^{n-1} a_{i,j}(r)\big(q_3^{-n/2+j}\big)^r.$$

The relations for $r \leqslant -1$ and $[\mathrm{ev}(H_{i,r}), \mathrm{ev}(F_0(z))]$ can be verified in a similar manner.

E-F Relations First consider the relation

$$[\mathrm{ev}(E_0(z)), \mathrm{ev}(F_i(w))] = 0 \quad (i \neq 0).$$

We have

$$\mathrm{ev}(E_0(z))\mathrm{ev}(F_i(w)) = e^{A_-(z)}\mathsf{F}(z)F_i(w)e^{A_+(z)}\mathcal{K}\times \begin{cases} q(z - Cq_1w)/(z - Cq_3^{-1}w) & (i = 1) \\ q^{-\delta_{i,n-1}} & (2 \leqslant i \leqslant n-1) \end{cases},$$

$$\mathrm{ev}(F_i(w))\mathrm{ev}(E_0(z)) = e^{A_-(z)}F_i(w)\mathsf{F}(z)e^{A_+(z)}\mathcal{K}\times \begin{cases} 1 & (1 \leqslant i \leqslant n-2) \\ (w - Cq_1z)/(w - Cq_3^{-1}z) & (i = n-1) \end{cases},$$

so that the relations reduce to (3.30), (3.31), and (3.27). The case $[\mathrm{ev}(E_i(z)), \mathrm{ev}(F_0(w))] = 0$ $(i \neq 0)$ is similar.

Using (4.4), (4.7), and (4.9), we obtain

$$\begin{aligned}\mathrm{ev}(E_0(z))\mathrm{ev}(F_0(w)) &= e^{A_-(z)}\mathsf{F}(z)e^{A_+(z)}\mathcal{K}\cdot e^{B_-(z)}\mathsf{E}(z)e^{B_+(z)}\mathcal{K}^{-1} \\ &= e^{A_-(z)+B_-(w)}\mathsf{F}(z)\mathsf{E}(w)e^{A_+(z)+B_+(w)}\,.\end{aligned}$$

Computing $\mathrm{ev}(F_0(w))\mathrm{ev}(E_0(z))$ similarly and using further (3.23), we find

$$\begin{aligned}&[\mathrm{ev}(E_0(z)), \mathrm{ev}(F_0(w))] = \frac{1}{q-q^{-1}} \\ &\times e^{A_-(z)+B_-(w)}\Bigl(-\delta\bigl(C\frac{z}{w}\bigr)\mathsf{K}^+(z)+\delta\bigl(C\frac{w}{z}\bigr)\mathsf{K}^-(w)\Bigr)e^{A_+(z)+B_+(w)}\,.\end{aligned}$$

Noting that

$$\begin{aligned}&e^{A_-(z)+B_-(Cz)} = \bar{K}_0^-(z)\,, \quad e^{A_+(Cw)+B_+(w)} = \bar{K}_0^+(w)\,, \\ &e^{A_-(Cw)+B_-(w)} = K_0\mathsf{K}^-(w)^{-1}\,, \quad e^{A_+(z)+B_+(Cz)} = K_0^{-1}\mathsf{K}^+(z)^{-1}\,,\end{aligned}$$

we obtain the desired result.

E-E and F-F Relations To check the quadratic relations

$$d_{0,j}g_{0,j}(z,w)\mathrm{ev}(E_0(z))\mathrm{ev}(E_j(w)) + g_{j,0}(w,z)\mathrm{ev}(E_j(w))\mathrm{ev}(E_0(z)) = 0,$$

we proceed in the same way as above; using (4.4) and (4.6), we bring $A_+(z)$ to the right, $A_-(z)$ to the left, and apply (3.28) and (3.29). Verification of the F-F relations is entirely similar.

Serre Relations Let us check the Serre relations assuming $n \geqslant 3$. We have

$$\begin{aligned}&e^{-A_-(w)}[\mathrm{ev}(E_1(z_1)), [\mathrm{ev}(E_1(z_2)), \mathrm{ev}(E_0(w))]_q]_{q^{-1}}\, e^{-A_+(w)} \\ &= E_1(z_1)E_1(z_2)\mathsf{F}(w) - (q+q^{-1})q^{-1}\frac{w-q_3^{-1}z_2}{w-q_1z_2}E_1(z_1)\mathsf{F}(w)E_1(z_2) \\ &+q^{-2}\frac{w-q_3^{-1}z_1}{w-q_1z_1}\frac{w-q_3^{-1}z_2}{w-q_1z_2}\mathsf{F}(w)E_1(z_1)E_1(z_2)\,.\end{aligned}$$

In view of (3.33), we can move $\mathsf{F}(w)$ to the right without producing delta functions. After simplification, the right-hand side becomes

$$-\frac{(1-q_2^{-1})w}{(w-q_1z_1)(w-q_1z_2)}(z_1-q_2z_2)E_1(z_1)E_1(z_2)\mathsf{F}(w)\,.$$

Symmetrizing in z_1 and z_2, we obtain 0 due to the quadratic relations for $E_1(z)$.

Likewise we compute

$$e^{-A_-(z_1)-A_-(z_2)}[\mathrm{ev}(E_0(z_1)), [\mathrm{ev}(E_0(z_2)), \mathrm{ev}(E_1(w))]_q]_{q^{-1}}\, e^{-A_+(z_1)-A_+(z_2)}$$

$$= \frac{z_1 - q_2^{-1} z_2}{z_1 - q_2 z_2}\Big(q^{-2}\frac{z_1 - q_3^{-1} w}{z_1 - q_1 w}\frac{z_2 - q_3^{-1} w}{z_2 - q_1 w}\mathsf{F}(z_1)\mathsf{F}(z_2)E_1(w)$$

$$-(q+q^{-1})q^{-1}\frac{z_1 - q_3^{-1} w}{z_1 - q_1 w}\mathsf{F}(z_1)E_1(w)\mathsf{F}(z_2) + E_1(w)\mathsf{F}(z_1)\mathsf{F}(z_2)\Big)$$

$$= \frac{q_2^{-1}(q_1 - q_3^{-1})w}{(z_1 - q_1 w)(z_2 - q_1 w)}(z_1 - q_2^{-1} z_2)E_1(w)\mathsf{F}(z_1)\mathsf{F}(z_2)\,.$$

Due to (3.25), the last line vanishes after symmetrization.

Serre relations in the remaining cases (including the case $n = 2$) can be verified by the same argument. We omit further details.

The proof is over.

5 Evaluation Modules

In this section we define and discuss the evaluation modules.

5.1 Evaluation Modules

Recall the grading of $U'_q\widehat{\mathfrak{gl}}_n$ given by (2.5). We say that a $U'_q\widehat{\mathfrak{gl}}_n$ module V is admissible if for any $v \in V$ there exists an N such that $xv = 0$ holds for any $x \in U'_q\widehat{\mathfrak{gl}}_n$ with $\deg x > N$. Algebra $\widetilde{U}'_{q,\kappa}\widehat{\mathfrak{gl}}_n$ has a well-defined action on admissible modules of level κ.

The quantum affine evaluation map $\mathrm{ev}_u^{(3)}$ goes from the quantum toroidal $\mathfrak{gl}_n$ algebra $\mathcal{E}'_n(q_1, q_2, q_3)$ to the (completed) quantum affine $\mathfrak{gl}_n$ algebra, provided that q_3 has the special value related to the central charge by $C = q_3^{n/2}$. Note that for the quantum affine algebra the value for the central element is completely arbitrary as q and q_3 are independent variables.

It follows that any admissible representation V of $U'_q\widehat{\mathfrak{gl}}_n$ on which C acts as an arbitrary scalar κ can be pulled back by $\mathrm{ev}_u^{(3)}$ to a representation of $\mathcal{E}_n^{(3)}$, by choosing q_3 so that $\kappa = q_3^{n/2}$. We call the resulting $\mathcal{E}_n^{(3)}$ module the evaluation module and denote it by $V(u)$.

5.2 Highest Weight Evaluation Modules

An example of admissible modules is given by highest weight modules.

A $U_q'\widehat{\mathfrak{gl}}_n$ module V is called a highest weight module of highest weight $(\kappa_0, \dots, \kappa_{n-1}) \in (\mathbb{C}^\times)^n$ with highest weight vector v if v is a cyclic vector in V satisfying

$$e_j v = 0 \quad (1 \leqslant j \leqslant n-1), \qquad e_0^{(l)} v = 0, \qquad Z_r v = 0 \quad (r \geqslant 1),$$

$$K_i v = \kappa_i v \quad (1 \leqslant i \leqslant n-1), \qquad C v = \prod_{i=0}^{n-1} \kappa_i \cdot v,$$

where Z_r is defined in (2.1) and $e_0^{(l)}$ is the left Chevalley generator (2.3).

Highest weight $U_q'\widehat{\mathfrak{gl}}_n$ modules are principally graded, see (2.6).

An $\mathcal{E}_n'$ module V is called a highest weight module with highest weight vector v if v is a cyclic vector in V satisfying

$$\theta^{-1}(E_i(z))v = 0, \qquad \theta^{-1}(K_i^{\pm}(z))v = P_i^{\pm}(z)v \qquad (0 \leqslant i \leqslant n-1).$$

Here $P_i^{\pm} \in \mathbb{C}[[z^{\mp 1}]]$ and θ is the Miki automorphism, see Sect. 3.2.

We set the degree of the highest weight vectors to zero. Then all highest weight $\mathcal{E}_n'$ modules are graded, see (3.11): $V = \oplus_{k \leqslant 0} V_k$.

Let V be an irreducible highest weight $\mathcal{E}_n'$ module such that $\dim V_k < \infty$ for all k. Then in the terminology of [5] the module V is quasi-finite. Moreover, by Theorem 2.3 in [5], the series $P_i^{\pm}(z)$ are expansions of a rational function $P_i(z)$. Moreover, the rational function $P_i(z)$ is regular at $z^{\pm 1} = \infty$ and satisfies $P_i(0)P_i(\infty) = 1$.

Denote $\mathbf{P} = (P_0(z), \dots, P_{n-1}(z))$. We call the n-tuple of rational functions $\boldsymbol{P}$ the highest weight of V.

Highest weight modules of $\mathcal{E}_n'$ were studied in detail in [5].

Since the evaluation map is graded, the evaluation highest weight $U_q'\widehat{\mathfrak{gl}}_n$ module with highest weight vector v is a highest weight $\mathcal{E}_n'$ module with highest vector v. Indeed, using Lemma 2.4 in [4], we find the principal degrees

$$\operatorname{pdeg} \operatorname{ev}_u^{(3)}(\theta^{-1}(E_{i,k})) = 1, \qquad \operatorname{pdeg} \operatorname{ev}_u^{(3)}(\theta^{-1}(H_{i,r})) = 0,$$

$0 \leqslant i \leqslant n-1, k \in \mathbb{Z}, r \in \mathbb{Z} \setminus \{0\}$.

The following proposition describes the corresponding highest weight.

Theorem 5.1 *Let V be a highest weight $U_q'\widehat{\mathfrak{gl}}_n$ module with highest weight $(\kappa_0, \dots, \kappa_{n-1})$. Let $V(u)$ be the evaluation $\mathcal{E}_n^{(3)}$ module. Then $q_3^n = \prod_{i=0}^{n-1} \kappa_i^2$ and $V(u)$ is a highest weight module with highest weight*

$$P = \Big(\kappa_0 \frac{1-\kappa_0^{-2}u'/z}{1-u'/z},\ \kappa_1 \frac{1-q_3(\kappa_0\kappa_1)^{-2}u'/z}{1-q_3\kappa_0^{-2}u'/z},\ \dots,\ \kappa_{n-1}\frac{1-q_3^{n-1}(\prod_{i=0}^{n-1}\kappa_i^{-2})u'/z}{1-q_3^{n-1}(\prod_{i=0}^{n-2}\kappa_i^{-2})u'/z}\Big),$$

for an appropriate choice of $u' \in \mathbb{C}^\times$. □

Proof The proposition is proved similarly to Theorem 5.7 in [10].

In the case of evaluation modules defined for $C = q_1^{n/2}$, the highest weight reads

$$P = \Big(\kappa_0 \frac{1-u'/z}{1-\kappa_0^2 u'/z},\ \kappa_1 \frac{1-q_1^{-1}\kappa_0^2 u'/z}{1-q_1^{-1}\kappa_0^2\kappa_1^2 u'/z},\ \dots,\ \kappa_{n-1}\frac{1-q_1^{-n+1}(\prod_{i=0}^{n-2}\kappa_i^{2})u'/z}{1-q_1^{-n+1}(\prod_{i=0}^{n-1}\kappa_i^{2})u'/z}\Big).$$

It follows from Theorem 5.1 that the modules $\mathcal{H}^{(k)}(u_1, \dots, u_n)$ and $\mathcal{G}^{(k)}_{\mu,\nu}$ in [5] are evaluation Verma- and Weyl-type modules, respectively.

5.3 Wakimoto Evaluation Modules

A Gelfand–Zeitlin pattern for $\mathfrak{gl}_n$ is an array of complex numbers

$$\boldsymbol{\lambda} = \begin{array}{ccccccccc} \lambda_{0,n-1} & & \lambda_{1,n-1} & & \cdots & \lambda_{n-2,n-1} & & \lambda_{n-1,n-1} \\ & \lambda_{0,n-2} & & \lambda_{1,n-2} & \cdots & & \lambda_{n-2,n-2} & \\ & & \ddots & & \ddots & & & \\ & & & \lambda_{0,1} & & \lambda_{1,1} & & \\ & & & & \lambda_{0,0} & & & \end{array}$$

Fix generic complex numbers $\lambda^0_{i,j} \in \mathbb{C}$ and consider the linear space $\mathrm{GZ}_{\boldsymbol{\lambda}^0}$ with basis $\{|\boldsymbol{\lambda}\rangle\}$, where $\boldsymbol{\lambda}$ runs over Gelfand–Zeitlin patterns for $\mathfrak{gl}_n$ such that $\lambda_{i,j} \in \lambda^0_{i,j} + \mathbb{Z}$ for $0 \leqslant i \leqslant j \leqslant n-2$ and $\lambda_{i,n-1} = \lambda^0_{i,n-1}$ for $0 \leqslant i \leqslant n-1$:

$$\mathrm{GZ}_{\boldsymbol{\lambda}^0} = \mathrm{Span}_{\mathbb{C}}\{|\boldsymbol{\lambda}\rangle \mid \lambda_{i,j} \in \lambda^0_{i,j} + \mathbb{Z}\ (0 \leqslant i \leqslant j \leqslant n-2),\ \lambda_{i,n-1} = \lambda^0_{i,n-1}\ (0 \leqslant i \leqslant n-1)\}. \tag{5.1}$$

Proposition 5.2 ([7],[12]) *The following formulas give a representation of* $U_q\mathfrak{gl}_n$ *on* $\mathrm{GZ}_{\boldsymbol{\lambda}^0}$.

$$e_i|\boldsymbol{\lambda}\rangle = \sum_{k=0}^{i-1} c^+_{k,i-1}(\boldsymbol{\lambda})|\boldsymbol{\lambda} + \mathbf{1}_{k,i-1}\rangle\,,$$

$$f_i|\boldsymbol{\lambda}\rangle = \sum_{k=0}^{i-1} c^-_{k,i-1}(\boldsymbol{\lambda})|\boldsymbol{\lambda} - \mathbf{1}_{k,i-1}\rangle\,,$$

$$q^{\varepsilon_r}|\boldsymbol{\lambda}\rangle = q^{h_r(\boldsymbol{\lambda})-h_{r-1}(\boldsymbol{\lambda})}|\boldsymbol{\lambda}\rangle\,.$$

Here $1 \leqslant i \leqslant n-1$, $0 \leqslant r \leqslant n-1$, $\mathbf{1}_{i,j}$ *stands for the vector* $(\delta_{r,i}\delta_{s,j})_{0\leqslant r\leqslant s\leqslant n-1}$, *and*

$$c^+_{k,i-1}(\boldsymbol{\lambda}) = \frac{\prod_{l=0}^{i-2}[\lambda_{l,i-2} - \lambda_{k,i-1} - l + k - 1]}{\prod_{l=k+1}^{i-1}[\lambda_{l,i-1} - \lambda_{k,i-1} - l + k - 1][\lambda_{l,i-1} - \lambda_{k,i-1} - l + k]}\,,$$

$$c^-_{k,i-1}(\boldsymbol{\lambda}) = -\frac{\prod_{l=0}^{i}[\lambda_{l,i} - \lambda_{k,i-1} - l + k + 1]}{\prod_{l=0}^{k-1}[\lambda_{l,i-1} - \lambda_{k,i-1} - l + k][\lambda_{l,i-1} - \lambda_{k,i-1} - l + k + 1]}\,,$$

$$h_r(\boldsymbol{\lambda}) = \sum_{k=0}^{r} \lambda_{k,r}\,.$$

□

We call the $U_q\mathfrak{gl}_n$ module $\mathrm{GZ}_{\boldsymbol{\lambda}^0}$ the Gelfand–Zeitlin module.

Now we choose an arbitrary $\kappa \in \mathbb{C}^\times$ and define the $U'_{q,\kappa}\widehat{\mathfrak{gl}}_n$ module $\widehat{\mathrm{GZ}}_{\boldsymbol{\lambda}^0}$ of level κ as follows.

Consider the $U_q\mathfrak{gl}_n$ Gelfand–Zeitlin module $\mathrm{GZ}_{\boldsymbol{\lambda}^0}$ described in Proposition 5.2. We trivially extend it to the action of the subalgebra $\mathfrak{b}_{q,\kappa}$ by setting

$$x^\pm_{i,r}|\boldsymbol{\lambda}\rangle = h_{j,r}|\boldsymbol{\lambda}\rangle = 0 \qquad (1 \leqslant i \leqslant n-1, \quad 0 \leqslant j \leqslant n-1, \quad r \geqslant 1, \quad |\boldsymbol{\lambda}\rangle \in \mathrm{GZ}_{\boldsymbol{\lambda}^0}).$$

Then we induce

$$\widehat{\mathrm{GZ}}_{\boldsymbol{\lambda}^0} = \mathrm{Ind}_{\mathfrak{b}_{q,\kappa}}^{U'_{q,\kappa}\widehat{\mathfrak{gl}}_n} \mathrm{GZ}_{\boldsymbol{\lambda}^0}.$$

We call $\widehat{\mathrm{GZ}}_{\boldsymbol{\lambda}^0}$ the Wakimoto module. Clearly, any Wakimoto module is an admissible $U'_{q,\kappa}\widehat{\mathfrak{gl}}_n$ module.

Finally, we choose $u \in \mathbb{C}^\times$, set $q_3^{n/2} = \kappa$, and, using the evaluation homomorphism $\mathrm{ev}_u^{(3)}$, we view $\widehat{\mathrm{GZ}}_{\boldsymbol{\lambda}^0}$ as an $\mathcal{E}'_n(q_1, q_2, q_3)$ module. We denote this evaluation module by $\widehat{\mathrm{GZ}}_{\boldsymbol{\lambda}^0}(u)$.

Recall the subalgebras $\mathcal{E}_{1,m} = \mathcal{E}_1(q_{1,m}, q_2, q_{3,m}) \subseteq \mathcal{E}'_n$, $0 \leqslant m \leqslant n-1$, and the algebra $\mathcal{A}$, see (3.34).

For each $i \in \{1, 2, 3\}$, the Fock module $\mathcal{F}_i(u)$ for $\mathcal{E}'_1$ is defined to be the highest weight $\mathcal{E}'_1(q_1, q_2, q_3)$ module with highest weight

$$\boldsymbol{P} = q_i^{1/2}\,\frac{1-q_i^{-1}u/z}{1-u/z}\,.$$

The Fock modules are well-known and understood, see, for example, [2] and [5]. For $\mathcal{E}'_1$, the action of $E_0(z)$, $F_0(z)$, $K_0^{\pm}(z)$ can be described explicitly via one free boson and vertex operators. In particular the character $\sum_{r\geq 0}\dim(\mathcal{F}_i(u))_{-r}x^r$ of the Fock module $\mathcal{F}_i(u)$ is given by $1/(x)_\infty = \prod_{r>0}(1-x^r)^{-1}$.

Our main result of this section is the following statement. For this statement, we make a technical assumption that q is generic, meaning we exclude a set of values of q that is at most countable.

Theorem 5.3 *We have the decomposition of $\mathcal{A}$ modules*

$$\widehat{\mathrm{GZ}}_{\lambda^0}(u) = \oplus_{\boldsymbol{\lambda}}\, W(\boldsymbol{\lambda})\,,\qquad W(\boldsymbol{\lambda}) = \mathcal{A}|\boldsymbol{\lambda}\rangle = W_0(\boldsymbol{\lambda})\boxtimes W_1(\boldsymbol{\lambda})\boxtimes\cdots\boxtimes W_{n-1}(\boldsymbol{\lambda})\,,$$

where the sum is over all GZ patterns as in (5.1), *and the $\mathcal{E}_{1,m}$ module $W_m(\boldsymbol{\lambda})$ is given by*

$$W_m(\boldsymbol{\lambda}) = \overbrace{\mathcal{F}_3(u_{0,m}(\boldsymbol{\lambda}))\otimes\cdots\otimes\mathcal{F}_3(u_{m,m}(\boldsymbol{\lambda}))}^{m+1}\otimes$$
$$\overbrace{\mathcal{F}_1(v_{0,m-1}(\boldsymbol{\lambda}))\otimes\cdots\otimes\mathcal{F}_1(v_{m-1,m-1}(\boldsymbol{\lambda}))}^{m}\,,$$

where

$$u_{l,m}(\boldsymbol{\lambda}) = q_2^{\lambda_{l,m}-l+m}q_3^n\tilde{u}\,,\qquad v_{l,m-1}(\boldsymbol{\lambda}) = q_2^{\lambda_{l,m-1}-l-1}\tilde{u}\,,$$
$$\tilde{u} = (-1)^n q^{-\sum_{r=0}^{n-1}\lambda^0_{r,n-1}+n-1}u\,.$$

Theorem 5.3 is proved in Appendix 6.

The $\mathcal{E}_1$ modules $W_m(\boldsymbol{\lambda})$ also appeared in [1].

6 Proof of Theorem 5.3

6.1 The Plan of the Proof

Our logic is the following.

Denote $K_i^{\pm,=}(z) = \theta^{-1}(K_i(z))$. Denote $\mathsf{K}_m^{\pm,=}(z) = \theta_m^{-1}(\mathsf{K}_m(z))$, where θ_m is the Miki automorphism of $\mathcal{E}_{1,m}$.

Since the degrees of $K_i^{\pm,=}(z)$, $\mathsf{K}_m^{\pm,=}(z)$ are zero, these operators preserve the space $\mathrm{GZ}_{\lambda^0}\subseteq\widehat{\mathrm{GZ}}_{\lambda^0}$. We show that these operators are diagonal in the basis $\{|\boldsymbol{\lambda}\rangle\}$ and give their eigenvalues.

First, we compute the eigenvalues of $K_i^{\pm,=}(z)$ using the fact that these operators belong to the horizontal algebra, which acts on $\mathrm{GZ}_{\boldsymbol{\lambda}^0}$ through the standard evaluation map $\overline{\mathrm{ev}}_u$. Then we calculate the eigenvalues of $\mathsf{K}_m^{\pm,=}(z)$ by finding the projection of these operators to the algebra generated by $K_i^{\pm,=}(z)$ along annihilating operators. This is a long calculation. We first compute the projection of the first components of $\mathsf{K}_m^{\pm,=}(z)$ explicitly and then argue that this is sufficient.

We observe that the eigenvalue of $\mathsf{K}_m^{\pm,=}(z)$ on eigenvector $|\boldsymbol{\lambda}\rangle$ coincides with the highest weight of module $W_m(\boldsymbol{\lambda})$. It follows that the character of $\mathcal{A}|\boldsymbol{\lambda}\rangle$ is at least $((x)_\infty)^{-n^2}$. We show that the joint spectrum of $\mathsf{K}_m^{\pm,=}(z)$, $0 \leqslant m \leqslant n-1$, is simple in $\oplus_{\boldsymbol{\lambda}} W(\boldsymbol{\lambda})$.

Comparing to the character of $\widehat{\mathrm{GZ}}_{\boldsymbol{\lambda}^0}$, we obtain the theorem.

6.2 Projection

Let $\mathcal{E}_n^{\pm} \subseteq \mathcal{E}_n'$ be the subalgebras generated by $E_{i,r}, F_{i,r}, H_{i,r}$ with $0 \leqslant i \leqslant n-1$ and $\pm r > 0$, and let $\mathcal{E}_n^0 = U_q^h\widehat{\mathfrak{gl}}_n$. Let $\widetilde{\mathcal{E}}_n^+ \subseteq \widetilde{\mathcal{E}}_n'$ be the completion of $\mathcal{E}_n^+$. We use the triangular decomposition

$$\widetilde{\mathcal{E}}_n' = \mathcal{E}_n^- \otimes \mathcal{E}_n^0 \otimes \widetilde{\mathcal{E}}_n^+ .$$

While we expect this to hold for all q, we have been unable to find it in the literature. For generic q, it can be shown by taking the limit $q \to 1$ and using the result of [11].

Consider the projection to the middle factor in the triangular decomposition

$$\mathrm{pr} : \widetilde{\mathcal{E}}_n' \longrightarrow \mathcal{E}_n^0 .$$

We shall write $x \equiv y$ if $\mathrm{pr}\,(x) = \mathrm{pr}\,(y)$ for $x, y \in \widetilde{\mathcal{E}}_n'$. Similarly to the usual Harish-Chandra map, this projection is a homomorphism when restricted to the subalgebra $(\widetilde{\mathcal{E}}_n')_0 \subseteq \widetilde{\mathcal{E}}_n'$ consisting of elements of homogeneous degree 0.

The algebras $\mathcal{E}_{1,m}$ are obtained by taking $(n-1)$ simple fusions. It is sufficient to consider a single simple fusion. For that purpose, following [5], we consider the subalgebra

$$\mathcal{E}_{n-1}^{0|1} = \mathcal{E}_{n-1}'(\tilde{q}_1, q_2, \tilde{q}_3) \subseteq \widetilde{\mathcal{E}}_n'(q_1, q_2, q_3), \quad \tilde{q}_1 = q_1 q_1^{\frac{1}{n-1}}, \tilde{q}_3 = q_3 q_1^{-\frac{1}{n-1}} .$$

It is generated by the following currents:

$$\tilde{E}_i(z) = E_{i+1}(q_1^{\frac{i}{n-1}} z)\,, \quad \tilde{F}_i(z) = F_{i+1}(q_1^{\frac{i}{n-1}} z)\,, \quad \tilde{K}_i^{\pm}(z) = K_{i+1}^{\pm}(q_1^{\frac{i}{n-1}} z)$$
$$(1 \leqslant i \leqslant n-2)\,,$$
$$\tilde{E}_0(z) = E_{0|1}(z) = \Big(1 - \frac{z}{z'}\Big) E_0(q_1 z') E_1(z)\Big|_{z'=z}\,,$$
$$\tilde{F}_0(z) = F_{0|1}(z) = \Big(1 - \frac{z'}{z}\Big) F_1(z) F_0(q_1 z')\Big|_{z'=z}\,,$$
$$\tilde{K}_0^{\pm}(z) = K_0^{\pm}(q_1 z) K_1^{\pm}(z)\,,$$

and for $n = 2$

$$-\frac{q q_1^{-1} q_3}{1 - q_1^{-1} q_3} \tilde{E}_0(z) = E_{0|1}^{(1)}(z) = \Big(1 - \frac{z}{z'}\Big) E_0(q_1 z') E_1(z)\Big|_{z'=z}\,,$$
$$(1 - q_1^2)\tilde{F}_0(z) = F_{0|1}^{(1)}(z) = \Big(1 - \frac{z'}{z}\Big) F_1(z) F_0(q_1 z')\Big|_{z'=z}\,,$$
$$\tilde{K}_0^{\pm}(z) = K_0^{\pm}(q_1 z) K_1^{\pm}(z)\,.$$

We also set $\tilde{K}_i^{\pm}(z) = \tilde{K}_i^{\pm 1} \exp(\pm(q - q^{-1}) \sum_{r>0} \tilde{H}_{i,\pm r} z^{\mp 1})$.

We will use also another subalgebra $\mathcal{E}_{n-1}^{0|n-1}$ obtained from $\mathcal{E}_{n-1}^{0|1}$ by applying the isomorphism ι (3.13). In what follows we study projections of elements of $\mathcal{E}_{n-1}^{0|1}$, $\mathcal{E}_{n-1}^{0|n-1}$. We will write formulas only for the former, while those for the latter are easily obtained by applying ι.

Let $\mathcal{N}_{\geqslant r}$ denote the left ideal of $\widetilde{\mathcal{E}}_n'$ generated by elements of homogeneous degree $\geqslant r$.

Lemma 6.1 *Let $n \geqslant 3$. Then,*

$$\tilde{E}_{0,0}, \equiv -q[E_{1,0}, E_{0,0}]_{q^{-1}} \bmod \mathcal{N}_{\geqslant 1}\,, \quad \tilde{F}_{0,0} \equiv -q^{-1}[F_{0,0}, F_{1,0}]_q \bmod \mathcal{N}_{\geqslant 1}\,, \tag{6.1}$$

$$\tilde{E}_{0,1} \equiv q_1^{-1}[E_{0,1}, E_{1,0}]_q \bmod \mathcal{N}_{\geqslant 2}\,, \quad \tilde{F}_{0,-1} \equiv -q_1 q^{-1}[F_{0,-1}, F_{1,0}]_q \bmod \mathcal{N}_{\geqslant 1}\,. \tag{6.2}$$

For $n = 2$ and by setting $a = q(1 - q_1^2)(1 - q_1^{-1} q_3)$ and $b = -q q_1^2$, we have

$$b^{-1}\tilde{E}_{0,0} \equiv E_{0,0}E_{1,0} - (1 + q_1^{-1}q_3 - q_1^{-2})E_{1,0}E_{0,0} + q_1^{-1}E_{1,1}E_{0,-1} \bmod \mathcal{N}_{\geqslant 1}\,,$$
$$b^{-1}\tilde{E}_{0,1} \equiv q_1^{-1}E_{0,1}E_{1,0} - q_1^{-2}(q_1 + q_3 - q_1^{-1})E_{1,0}E_{0,1} - q_1^{-1}q_3 E_{0,0}E_{1,0}$$
$$+ q_1^{-2}E_{1,1}E_{0,0} \bmod \mathcal{N}_{\geqslant 2}\,,$$

$$ab\tilde{F}_{0,0} \equiv F_{1,0}F_{0,0} - (1 + q_1q_3^{-1} - q_1^2)F_{0,0}F_{1,0} + q_1F_{0,1}F_{1,-1} \bmod \mathcal{N}_{\geqslant 1}\,,$$

$$ab\tilde{F}_{0,-1} \equiv q_1F_{1,0}F_{0,-1} - q_1^2(q_1^{-1} + q_3^{-1} - q_1)F_{0,-1}F_{1,0} - q_1q_3^{-1}F_{1,-1}F_{0,0} + q_1^2F_{0,0}F_{1,-1} \bmod \mathcal{N}_{\geqslant 1}\,.$$

Proof Let $n \geqslant 3$. By the definition along with (3.9), we obtain

$$E_{0|1,l} = \sum_{j\geqslant 0} q_1^{j-l}\Big(E_{0,-j+l}E_{1,j} - q_1E_{0,-j-1+l}E_{1,j+1}\Big) + \sum_{j\geqslant 1} q^{-1}q_1^{-j-l}\Big(E_{1,-j}E_{0,j+l} - q_3^{-1}E_{1,-j+1}E_{0,j+l-1}\Big)\,,$$

$$F_{0|1,l} = \sum_{j\geqslant 0} q_1^{-j-l}\Big(F_{1,-j}F_{0,l+j} - q_1^{-1}F_{1,-j-1}F_{0,l+j+1}\Big) + \sum_{j\geqslant 1} qq_1^{j-l}\Big(F_{0,l-j}F_{1,j} - q_3F_{0,l-j+1}F_{1,j-1}\Big)\,,$$

from which the assertion follows.

For $n = 2$, we have

$$E^{(1)}_{0|1,l} = \sum_{j\in\mathbb{Z}} q_1^{j-l}(E_{0,-j+l}E_{1,j} - q_1^{-1}E_{0,-j+l-1}E_{1,j+1})\,.$$

Rewriting it as

$$(1 - q_1^{-1}q_3)E^{(1)}_{0|1,l} = \sum_{j\in\mathbb{Z}} q_1^{j-l}(E_{0,-j+l}E_{1,j} - (q_1 + q_3)E_{0,-j+l-1}E_{1,j+1} + q_1q_3E_{0,-j+l+2}E_{1,j+2})$$

and applying (3.10), we obtain

$$(1 - q_1^{-1}q_3)E^{(1)}_{0|1,l} = \sum_{j\geqslant 0} q_1^{j-l}\Big(E_{0,l-j}E_{1,j} - (q_1 + q_3)E_{0,l-j-1}E_{1,j+1} + q_1q_3E_{0,l-j-2}E_{1,j+2}\Big) + \sum_{j\geqslant 1} q_1^{-j-l}\Big(q_1q_3E_{1,-j}E_{0,j+l} - (q_1 + q_3)E_{1,-j+1}E_{0,l+j-1} + E_{1,-j+2}E_{0,j+l-2}\Big)\,,$$

Proceeding similarly with $F^{(1)}_{0|1}(z)$, we get

$$(1-q_1q_3^{-1})F^{(1)}_{0|1,l}$$
$$=\sum_{j\geqslant 0} q_1^{-j-l}\Big(F_{1,-j}F_{0,l+j}-(q_1^{-1}+q_3^{-1})F_{1,-j-1}F_{0,l+j+1}$$
$$+q_1^{-1}q_3^{-1}F_{1,-j-2}F_{0,l+j+2}\Big)$$
$$+\sum_{j\geqslant 1} q_1^{j-l}\Big(q_1^{-1}q_3^{-1}F_{0,l-j}F_{1,j}-(q_1^{-1}+q_3^{-1})F_{0,l-j+1}F_{1,j-1}$$
$$+F_{0,l-j+2}F_{1,j-2}\Big),$$

which imply the relations stated in Lemma.

6.3 Action of $\mathbf{K}_i^{\pm,=}(z)$ *on* $\mathbf{GZ}_{\boldsymbol{\lambda}^0}$

Recall that the horizontal subalgebra $U_q^h\widehat{\mathfrak{sl}}_n$ of $\mathcal{E}'_n$ is given in Chevalley generators. The next lemma describes the corresponding Drinfeld generators under the identification (2.2).

Note that we use a different identification (2.3) for Chevalley generators of the vertical algebra $U_q^v\widehat{\mathfrak{sl}}_n$.

Lemma 6.2 *The Drinfeld generators of the horizontal subalgebra* $U_q^h\widehat{\mathfrak{sl}}_n$ *are given by*

$$x_i^+(z)=\theta^{-1}(E_i(d^{-i}z)),\quad x_i^-(z)=\theta^{-1}(F_i(d^{-i}z)),$$
$$\phi_i^\pm(z)=\theta^{-1}(K_i^\pm(d^{-i}z))\quad (1\leqslant i\leqslant n-1).$$

Proof Define the currents $x_i^\pm(z)$, $\phi_i^\pm(z)$ by the above formulas. Then they belong to the horizontal subalgebra and satisfy the relations for the Drinfeld currents of $U'_q\widehat{\mathfrak{sl}}_n$. We have $x_{i,0}^+=E_{i,0}$, $x_{i,0}^-=F_{i,0}$ for $1\leqslant i\leqslant n-1$. Moreover, from (3.18) and (3.19) we obtain

$$E_{0,0}=[x^-_{n-1,0},\ldots,[x^-_{2,0},x^-_{1,-1}]_q\ldots]_qK_1\cdots K_{n-1},$$
$$F_{0,0}=(K_1\cdots K_{n-1})^{-1}[\ldots[x^+_{1,1},x^+_{2,0}]_{q^{-1}},\ldots,x^+_{n-1,0}]_{q^{-1}}.$$

Hence the assertion follows from the identification (2.2) between the Chevalley and the Drinfeld generators.

Recall the standard evaluation map $\overline{\mathrm{ev}}_u$ (2.4).

Lemma 6.3 *On* $\mathrm{GZ}_{\boldsymbol{\lambda}^0}$, *the action of horizontal algebra* $U_q^h\widehat{\mathfrak{sl}}_n$ *is given by* $\overline{\mathrm{ev}}_u$. *Namely, for any* $x \in U_q^h\widehat{\mathfrak{sl}}_n$, *we have*

$$\mathrm{ev}_u^{(3)}(x)|\boldsymbol{\lambda}\rangle = \overline{\mathrm{ev}}_u(x)|\boldsymbol{\lambda}\rangle\,.$$

Proof It suffices to check the statement for $x = e_0, f_0$. By the definition of $\mathrm{ev}_u^{(3)}$, we have

$$\mathrm{ev}_u^{(3)}\big(E_0(z)\big)|\boldsymbol{\lambda}\rangle = u^{-1}\mathcal{K}e^{A_-(z)}\mathsf{F}(z)|\boldsymbol{\lambda}\rangle\,,$$

and in particular

$$\mathrm{ev}_u^{(3)}(E_{0,0})|\boldsymbol{\lambda}\rangle = u^{-1}\mathcal{K}\mathsf{F}_0|\boldsymbol{\lambda}\rangle\,.$$

Using the q_3 version of Lemma 6.1 Repeatedly, we find that the right-hand side becomes

$$u^{-1}q^{-\Lambda_1+\Lambda_{n-1}}[\ldots[F_{1,0}, F_{2,0}]_{q^{-1}}, \ldots, F_{n-1,0}]_{q^{-1}}|\boldsymbol{\lambda}\rangle$$

and hence coincides with $\overline{\mathrm{ev}}_u(e_0)|\boldsymbol{\lambda}\rangle$.

The case of f_0 is entirely similar.

From Lemma 6.3, we obtain the explicit action of Drinfeld generators of horizontal algebra $U_q^h\widehat{\mathfrak{sl}}_n$ on the "top level" $\mathrm{GZ}_{\boldsymbol{\lambda}^0}$.

Lemma 6.4 *On* $\mathrm{GZ}_{\boldsymbol{\lambda}^0}$, *the Drinfeld generators of horizontal algebra* $U_q^h\widehat{\mathfrak{sl}}_n$ *act as follows:*

$$\mathrm{ev}_u^{(3)}\big(x_i^{\pm}(z)\big)|\boldsymbol{\lambda}\rangle = \sum_{k=0}^{i-1} c_{k,i-1}^{\pm}(\boldsymbol{\lambda})\delta\big(q^{2\lambda_{k,i-1}+i-2k\pm1}\bar{u}/z\big)|\boldsymbol{\lambda} \pm \mathbf{1}_{k,i-1}\rangle\,,$$

$$\mathrm{ev}_u^{(3)}\big(\phi_i^{\pm}(z)\big)|\boldsymbol{\lambda}\rangle = q^{2h_{i-1}(\boldsymbol{\lambda})-h_{i-2}(\boldsymbol{\lambda})-h_i(\boldsymbol{\lambda})}\times$$
$$\frac{\prod_{l=0}^{i-2}(1-q^{2\lambda_{l,i-2}+i-2l-1}\bar{u}/z)\prod_{l=0}^{i}(1-q^{2\lambda_{l,i}+i-2l+1}\bar{u}/z)}{\prod_{l=0}^{i-1}(1-q^{2\lambda_{l,i-1}+i-2l-1}\bar{u}/z)(1-q^{2\lambda_{l,i-1}+i-2l+1}\bar{u}/z)}|\boldsymbol{\lambda}\rangle\,,$$

where $1 \leqslant i \leqslant n-1$ *and*

$$\bar{u} = (-1)^n q^{-\sum_{r=0}^{n-1}\lambda^0_{r,n-1}+n-2}u\,. \tag{6.3}$$

Proof Define operators $\mathrm{ev}'_{\bar{u}}\big(x_i^{\pm}(z)\big)$, $\mathrm{ev}'_{\bar{u}}\big(\phi_i^{\pm}(z)\big)$ acting on $|\boldsymbol{\lambda}\rangle$ by the right-hand sides of the above formulas. A direct computation shows that they satisfy the defining relations for $U'_q\widehat{\mathfrak{sl}}_n$. It suffices to check that $\mathrm{ev}'_{\bar{u}}(x) = \mathrm{ev}^{(3)}_u(x)$ for $x = e_0, f_0$.

We have

$$\mathrm{ev}'_{\bar{u}}(x^-_{1,-1})|\boldsymbol{\lambda}\rangle = \bar{u}^{-1} f_1 q^{-2\varepsilon_0}|\boldsymbol{\lambda}\rangle\,, \quad \mathrm{ev}'_{\bar{u}}(x^-_{i,0})|\boldsymbol{\lambda}\rangle = f_i|\boldsymbol{\lambda}\rangle \quad (2 \leqslant i \leqslant n-1).$$

Noting that q^{ε_0} commutes with f_i for $i \geqslant 2$ and using (2.2), we obtain

$$\begin{aligned}\mathrm{ev}'_{\bar{u}}(e_0) &= \mathrm{ev}'_{\bar{u}}\big([x^-_{n-1,0}, \ldots, [x^-_{2,0}, x^-_{1,-1}]_q \cdots]_q q^{\alpha_1+\cdots+\alpha_{n-1}}\big)|\boldsymbol{\lambda}\rangle \\ &= \bar{u}^{-1}[f_{n-1}, \ldots, [f_2, f_1]_q \cdots]_q q^{-2\varepsilon_0} q^{\varepsilon_0-\varepsilon_{n-1}}|\boldsymbol{\lambda}\rangle \\ &= (-q)^{n-2}\bar{u}^{-1}[\ldots[f_1, f_2]_{q^{-1}}, \ldots, f_{n-1}]_{q^{-1}} q^{-\Lambda_1+\Lambda_{n-1}}\mathfrak{t}^{-1}|\boldsymbol{\lambda}\rangle \\ &= \mathrm{ev}^{(3)}_u(e_0)\,.\end{aligned}$$

Similarly we check $\mathrm{ev}'_{\bar{u}}(f_0) = \mathrm{ev}^{(3)}_u(f_0)$.

Now we are in a position to compute the action of $\bar{K}^{\pm,=}_i(z) = K_i^{\mp 1} K^{\pm,=}_i(z)$ on $\mathrm{GZ}_{\boldsymbol{\lambda}^0}$.

Denote by $\bar{K}_i(z, \boldsymbol{\lambda})$ the eigenvalues of $\bar{K}^{\pm,=}_i(z)$ on $|\boldsymbol{\lambda}\rangle$, $0 \leqslant i \leqslant n-1$.

Proposition 6.5 *With the definition* (6.3), *we have*

$$\bar{K}_i(z, \boldsymbol{\lambda}) = \frac{\prod_{l=0}^{i-2}(1 - q_3^i q_2^{\lambda_{l,i-2}+i-l-1} q\bar{u}/z) \prod_{l=0}^{i}(1 - q_3^i q_2^{\lambda_{l,i}+i-l} q\bar{u}/z)}{\prod_{l=0}^{i-1}(1 - q_3^i q_2^{\lambda_{l,i-1}+i-l-1} q\bar{u}/z)(1 - q_3^i q_2^{\lambda_{l,i-1}+i-l} q\bar{u}/z)} \qquad (1 \leqslant i \leqslant n-1),$$

$$\bar{K}_0(z, \boldsymbol{\lambda}) = \frac{1 - q_2^{\lambda_{0,0}} q\bar{u}/z}{1 - q_3^n q_2^{\lambda_{n-1,n-1}} q\bar{u}/z} \prod_{l=0}^{n-2} \frac{1 - q_3^n q_2^{\lambda_{l,n-2}+n-l-1} q\bar{u}/z}{1 - q_3^n q_2^{\lambda_{l,n-1}+n-l-1} q\bar{u}/z}\,.$$

Proof The formulas for $1 \leqslant i \leqslant n-1$ follow from Lemmas 6.2 and 6.4.

Consider the special case where $\boldsymbol{\lambda}$ is dominant. Then the Wakimoto module has a highest weight submodule, and the eigenvalue of $K^{\pm,=}_0(z)$ on highest weight vector can be determined by Theorem 5.1 along with the knowledge of $\bar{K}_i(z, \boldsymbol{\lambda})$ for $1 \leqslant i \leqslant n-1$. In this case general eigenvalues can then be obtained by acting with $E_i(z), F_i(z)$.

Since the eigenvalues of $K^{\pm}_{i,r}$ are polynomial functions of the parameters $q^{\lambda_{j,r}}$, the formula for $K^{\pm,=}_0(z)$ in the general case follows by "analytic continuation."

Denote by $\Psi_m(z, \boldsymbol{\lambda})$ the highest weight of $\mathcal{E}_{1,m}$ module $W_m(\boldsymbol{\lambda})$ and let $\bar{\Psi}_m(z, \boldsymbol{\lambda}) = \Psi_m(z, \boldsymbol{\lambda})/\Psi_m(\infty, \boldsymbol{\lambda})$.

Corollary 6.6 *For $0 \leqslant m \leqslant n-1$, the highest weight of $\mathcal{E}_{1,m}$ module $W_m(\boldsymbol{\lambda})$ is given by*

$$\bar{\Psi}_m(z,\boldsymbol{\lambda}) = \prod_{i=m+1}^{n-1} \bar{K}_i(q_3^{-n+i}z,\boldsymbol{\lambda}) \prod_{i=0}^{m} \bar{K}_i(q_1^{-i}z,\boldsymbol{\lambda})$$

$$= \prod_{l=0}^{m} \frac{1-q_2^{\lambda_{l,m}-l} q\bar{u}/z}{1-q_2^{\lambda_{l,m}-l+m} q_3^n q\bar{u}/z} \prod_{l=0}^{m-1} \frac{1-q_2^{\lambda_{l,m-1}-l+m} q_3^n q\bar{u}/z}{1-q_2^{\lambda_{l,m-1}-l-1} q\bar{u}/z},$$

where $\bar{u}$ is given by (6.3). □

6.4 Action of $\mathsf{K}_m^{\pm,=}(z)$ on GZ_{λ^0}: The First Components

We show that the action of $\mathsf{K}_m^{\pm,=}(z)$ on the top level is diagonal in the basis of $|\boldsymbol{\lambda}\rangle$ with eigenvalues given in Corollary 6.6. More precisely, we show the following formulas for the horizontal currents:

$$\mathsf{K}_m^{\pm,=}(z) \equiv \prod_{i=m+1}^{n-1} \bar{K}_i^{\pm,=}(q_3^{-n+i}z) \prod_{i=0}^{m} \bar{K}_i^{\pm,=}(q_1^{-i}z). \tag{6.4}$$

In this section we do it for the first components using an explicit computation.

Recall the subalgebra $\mathcal{E}_{n-1}^{0|1}$ and the currents $\tilde{K}_i^{\pm}(z)$ in Sect. 6.2. Let $\tilde{\theta}$ be the Miki automorphism for $\mathcal{E}_{n-1}^{0|1}$.

To show (6.4), it is enough to prove the following statement.

Proposition 6.7 *For all $0 \leqslant i \leqslant n-2$, we have*

$$\tilde{\theta}^{-1}(\tilde{K}_i^{\pm}(z)) \equiv \theta^{-1}\big(K_{i+1}^{\pm}(q_1^{\frac{i}{n-1}-1}z)\big) \quad (1 \leqslant i \leqslant n-2), \tag{6.5}$$

$$\tilde{\theta}^{-1}(\tilde{K}_0^{\pm}(z)) \equiv \theta^{-1}\big(K_0^{\pm}(z)K_1^{\pm}(q_1^{-1}z)\big). \tag{6.6}$$

In this section first we explicitly compute the projections of $\tilde{\theta}^{-1}(\tilde{H}_{i,\pm 1})$. The general case is done in Sect. 6.5.

Proposition 6.8 *Let $n \geqslant 3$. We have* mod $\mathcal{N}_{\geqslant 1}$,

$$\tilde{\theta}^{-1}\big(\tilde{H}_{i,1}\big) \equiv q_1^{1-\frac{i}{n-1}}\, \theta^{-1}\big(H_{i+1,1}\big),$$

$$\tilde{\theta}^{-1}\big(\tilde{H}_{i,-1}\big) \equiv q_1^{-1+\frac{i}{n-1}}\, \theta^{-1}\big(H_{i+1,-1}\big) \quad (1 \leqslant i \leqslant n-2),$$

$$\tilde{\theta}^{-1}\big(\tilde{H}_{0,1}\big) \equiv \theta^{-1}\big(H_{0,1}\big) + q_1\theta^{-1}\big(H_{1,1}\big),$$

$$\tilde{\theta}^{-1}\big(\tilde{H}_{0,-1}\big) \equiv \theta^{-1}\big(H_{0,-1}\big) + q_1^{-1}\theta^{-1}\big(H_{1,-1}\big).$$

Proof We prove the case $\tilde{\theta}^{-1}(H_{i,1})$. Set $\tilde{d} = q_1^{\frac{1}{n-1}} d$.

Suppose $1 \leqslant i \leqslant n-2$. By the definition

$$\begin{aligned}
&\tilde{\theta}^{-1}(\tilde{H}_{i,1}) \\
&= -(-\tilde{d})^{-i}[[\cdots[[\cdots[\tilde{F}_{0,0}, \tilde{F}_{n-2,0}]_q, \cdots, \tilde{F}_{i+1,0}]_q, \tilde{F}_{1,0}]_q, \cdots, \tilde{F}_{i-1,0}]_q, \tilde{F}_{i,0}]_{q^2} \\
&= -(-\tilde{d})^{-i}[[\cdots[[\cdots[\tilde{F}_{0,0}, \tilde{F}_{n-2,0}]_q, \cdots, \tilde{F}_{i+1,0}]_q, \tilde{F}_{1,0}]_q, \cdots, \tilde{F}_{i-1,0}]_q, \tilde{F}_{i,0}]_{q^2} .
\end{aligned}$$

Substituting (6.1) and noting that

$$\begin{aligned}
[\cdots[F_{0|1,0}, \tilde{F}_{n-2,0}]_q \cdots \tilde{F}_{i+1,0}]_q &\equiv -q^{-1}[\cdots[[F_{0,0}, F_{1,0}]_q, F_{n-1,0}]_q \cdots F_{i+2,0}]_q \\
&= -q^{-1}[\cdots[[F_{0,0}, F_{n-1,0}]_q \cdots F_{i+2,0}]_q F_{1,0}]_q ,
\end{aligned}$$

we obtain the result.

Next let $i = 0$. From (3.16), we have

$$\begin{aligned}
\tilde{\theta}^{-1}(\tilde{H}_{0,1}) &\equiv -(-\tilde{d})^{-n+2}[\cdots[\tilde{F}_{1,1}, \tilde{F}_{2,0}]_q, \cdots, \tilde{F}_{n-2,0}]_q \cdot \tilde{F}_{0,-1} \\
&= -(-d)^{-n+1} q_1 A \cdot \left(F_{0,-1}F_{1,0} - qF_{1,0}F_{0,-1}\right),
\end{aligned}$$

with

$$A = [[\cdots[F_{2,1}, F_{3,0}]_q, \cdots, F_{n-2,0}]_q, F_{n-1,0}]_q .$$

Since A has degree 1, we have

$$\begin{aligned}
A \cdot \left(F_{0,-1}F_{1,0} - qF_{1,0}F_{0,-1}\right) &\equiv [A, F_{0,-1}]_q F_{1,0} \\
&\qquad - q\left([A, F_{1,0}]_q F_{0,-1} + qF_{1,0}AF_{0,-1}\right) \\
&\equiv [[A, F_{0,-1}]_q, F_{1,0}]_{q^2} - q[[A, F_{1,0}]_q, F_{0,-1}]_{q^2} .
\end{aligned}$$

Noting that $[[A, B]_p, C]_p = [[A, C]_p, B]_p$ if $BC = CB$, we find

$$\begin{aligned}
[A, F_{1,0}]_q &= [[\cdots[[F_{2,1}, F_{1,0}]_q, F_{3,0}]_q, \cdots, F_{n-2,0}]_q, F_{n-1,0}]_q \\
&= (-d)^{-1}[[\cdots[[F_{1,1}, F_{2,0}]_q, F_{3,0}]_q, \cdots, F_{n-2,0}]_q, F_{n-1,0}]_q .
\end{aligned}$$

In the last line we use the quadratic relations. By the same token, we have

$$\begin{aligned}
A &= (-d)[[\cdots[F_{3,1}, F_{2,0}]_q, \cdots, F_{n-2,0}]_q, F_{n-1,0}]_q \\
&= (-d)[[[\cdots[F_{3,1}, F_{4,0}]_q, \cdots, F_{n-2,0}]_q, F_{n-1,0}]_q, F_{2,0}]_q ,
\end{aligned}$$

And by repeating this we get

$$A=(-d)^{n-3}[[\cdots[F_{n-1,1},F_{n-2,0}]_q,\cdots,F_{3,0}]_q,F_{2,0}]_q\,.$$

Hence,

$$\begin{aligned}[A,F_{0,-1}]_q &\equiv (-d)^{n-3}[[\cdots[F_{n-1,1},F_{n-2,0}]_q,\cdots,F_{2,0}]_q,F_{0,-1}]_q\\ &=(-d)^{n-3}[\cdots[[F_{n-1,1},F_{0,-1}]_q,F_{n-2,0}]_q,\cdots,F_{2,0}]_q\\ &=(-d)^{n-2}[\cdots[[F_{0,0},F_{n-1,0}]_q,F_{n-2,0}]_q,\cdots,F_{2,0}]_q\,.\end{aligned}$$

We obtain after simplification

$$\tilde{\theta}^{-1}\big(\tilde{H}_{0,1}\big)=\theta^{-1}(H_{0,1})+q_1\theta^{-1}(H_{1,1})\,.$$

The case $\tilde{H}_{0,-1}$ can be handled in the same way.

Proposition 6.9 *Let $n=2$. Then we have* mod $\mathcal{N}_{\geqslant 1}$

$$\begin{aligned}\tilde{\theta}^{-1}(\tilde{H}_{0,1})&\equiv\theta^{-1}(H_{0,1})+q_1\theta^{-1}(H_{1,1}),\\ \tilde{\theta}^{-1}(\tilde{H}_{0,-1})&\equiv\theta^{-1}(H_{0,-1})+q_1^{-1}\theta^{-1}(H_{1,-1})\,.\end{aligned}$$

Proof We recall that the Miki automorphism for $\mathcal{E}_1^{0|1}$ is given by

$$\begin{aligned}&\tilde{\theta}^{-1}\big(\tilde{H}_{0,1}\big)=-a\tilde{K}_0^{-1}\tilde{F}_{0,0}\,,\quad \tilde{\theta}^{-1}\big(\tilde{H}_{0,-1}\big)=-\tilde{K}_0\tilde{E}_{0,0}\,,\\ &\tilde{\theta}^{-1}\big(\tilde{E}_{0,0}\big)=-\tilde{C}^{-1}\tilde{H}_{0,1}\,,\quad \tilde{\theta}^{-1}\big(\tilde{F}_{0,0}\big)=-a^{-1}\tilde{C}\tilde{H}_{0,-1}\,,\\ &\tilde{\theta}^{-1}\big(\tilde{C}\big)=\tilde{K}_0^{-1}\,,\quad \tilde{\theta}^{-1}\big(\tilde{K}_0\big)=\tilde{C}\,,\end{aligned}$$

where $a=q(1-\tilde{q}_1)(1-\tilde{q}_3)$. As an example, let us check $\tilde{\theta}^{-1}\big(\tilde{H}_{0,1}\big)$. By Lemma 6.1, we have

$$qq_1^2\tilde{\theta}^{-1}\big(\tilde{H}_{0,1}\big)\equiv F_{1,0}F_{0,0}-(1+q_1q_3^{-1}-q_1^2)F_{0,0}F_{1,0}+q_1F_{0,1}F_{1,-1}\,.$$

Using the quadratic relation

$$[F_{0,1},F_{1,-1}]_{q^2}=[F_{1,1},F_{0,-1}]_{q^2}-(q_1^{-1}+q_3^{-1})[F_{1,0},F_{0,0}]\,,$$

we obtain

$$\begin{aligned}qq_1^2\tilde{\theta}^{-1}\big(\tilde{H}_{0,1}\big)&\equiv q_1\big([F_{1,1},F_{0,-1}]_{q^2}+q_1[F_{0,0},F_{1,0}]_{q^2}\big)\\ &=qq_1^2\big(\theta^{-1}(H_{0,1})+q_1\theta^{-1}(H_{1,1})\big)\,.\end{aligned}$$

6.5 Action of $\mathbf{K}_m^{\pm,=}(z)$ on GZ_{λ^0}: The General Case

To compute projections of $\tilde{\theta}^{-1}(\tilde{H}_{i,\pm r})$ for $r \geqslant 2$, we use the following argument.

Let us summarize our knowledge so far. Inside $\widetilde{\mathcal{E}}_n'$ we have mutually commuting subalgebras $\mathcal{E}_{1,0}'$ and $\mathcal{E}_{n-1}'$

$$\mathcal{E}_{1,0}' \otimes \mathcal{E}_{n-1}^{0|1} \hookrightarrow \widetilde{\mathcal{E}}_n' .$$

Let $\tilde{\tilde{H}}_{0,r} \in \mathcal{E}_{1,0}'$, $\tilde{H}_{i,r} \in \mathcal{E}_{n-1}^{1|0}$ $(0 \leqslant i \leqslant n-2)$ be the respective commuting elements (we concentrate on the case $r \geqslant 1$). Let $\tilde{\tilde{\theta}}$ be the Miki automorphism of $\mathcal{E}_{1,0}'$. We have two subalgebras of $\mathcal{E}_n^0 = U_q^h\widehat{\mathfrak{gl}}_n \subseteq \widetilde{\mathcal{E}}_n'$

$$\mathcal{A} = \mathrm{pr}\left(\mathbb{C}[\tilde{\tilde{\theta}}^{-1}(\tilde{\tilde{H}}_{0,r}) \mid r \geqslant 1] \otimes \mathbb{C}[\tilde{\theta}^{-1}(\tilde{H}_{i,r}) \mid r \geqslant 1, 0 \leqslant i \leqslant n-2]\right),$$

$$\mathcal{B} = \mathbb{C}[\theta^{-1}(H_{i,r}) \mid r \geqslant 1, 0 \leqslant i \leqslant n-1] .$$

Both subalgebras are commutative. Note that both belong to the subalgebra generated by $E_{i,r}$, $0 \leqslant i \leqslant n-1$, $r \in \mathbb{Z}$. From Proposition 6.8 and 6.9, we know that

$$\theta^{-1}(H_{i,1}) \in \mathcal{A} \cap \mathcal{B} \quad (0 \leqslant i \leqslant n-1) .$$

Proposition 6.10 *We have $\mathcal{A} = \mathcal{B}$.* □

Proof We can reduce the problem from $U_q\widehat{\mathfrak{gl}}_n$ to $U_q\widehat{\mathfrak{sl}}_n$. So it suffices to prove the following statement: Let $U_q\mathfrak{n} = \langle e_0, \dots, e_{n-1}\rangle$ be the nilpotent subalgebra of $U_q\widehat{\mathfrak{sl}}_n$, and let $H_{i,r} \in U_q\widehat{\mathfrak{sl}}_n$ $(r \geqslant 1, 1 \leqslant i \leqslant n-1)$ be the Cartan-like commutative elements. Then the commutant of $H_{1,1}, \dots, H_{n-1,1}$ in $U_q\mathfrak{n}$ coincides with $\mathbb{C}[H_{i,r} \mid r \geqslant 1, 1 \leqslant i \leqslant n-1]$. The commutant obviously contains $\mathbb{C}[H_{i,r} \mid r \geqslant 1, 1 \leqslant i \leqslant n-1]$. Also for $q = 1$ the statement is easy to see. Therefore it follows for generic q.

Proposition 6.10 says that the projection $\mathrm{pr}\left(\tilde{\theta}^{-1}(K_{i,\pm}(z))\right)$ belongs to the algebra generated by $K_i^{\pm,=}(z)$. From [5], we know that equalities (6.5) and (6.6) hold on the highest weight vectors of generic tensor products of Fock spaces. Therefore, by analytic continuation, these equalities hold on all highest weight vectors. We note that the ideal of the projection pr is contained in the ideal that annihilates highest weight vectors. Therefore Proposition 6.7 follows.

6.6 The End of the Proof

Applying Proposition 6.7 repeatedly, we obtain (6.4).

To show Theorem 5.3, it remains to show that the joint spectrum of $\mathsf{K}_m^{\pm,=}(z)$ is simple on $\oplus_{\boldsymbol{\lambda}} W(\boldsymbol{\lambda})$. Suppose that the eigenvalues of $\mathsf{K}_m^{\pm,=}(z)$ coincide on some vectors $v \in W(\boldsymbol{\lambda})$ and $w \in W(\boldsymbol{\mu})$. Recall that $\boldsymbol{\lambda}^0$ is generic. Each Fock space entering as a factor in $W(\boldsymbol{\lambda})$ depends on just one $\lambda^0_{i,j}$. It means that for each i, j the eigenvalues should be the same on Fock spaces depending on $\lambda_{i,j}$ and $\mu_{i,j}$. The spectral parameters of these Fock spaces differ by q_2^a with some $a \in \mathbb{Z}$. The eigenvectors in a Fock space are parametrized by partitions, and the corresponding eigenvalues are given by products over concave and convex boxes, see Lemma 3.4 in [5]. With the aid of this formula, it is easy to see that if the eigenvalues are the same on two partitions in two such Fock modules, then $a = 0$ and the partitions are the same.

Acknowledgments The research of BF is supported by the Russian Science Foundation grant project 16-11-10316. MJ is partially supported by JSPS KAKENHI grant number JP16K05183. EM is partially supported by a grant from the Simons Foundation #353831.

We would like to thank the Kyoto University for hospitality during our visit in summer 2017 when part of this work was completed.

References

1. M. Bershtein, B. Feigin, and G. Merzon, *Plane partitions with a "pit": generating functions and representation theory*, Selecta Math. (N.S.) **24** (2018), no. 1, 21–62
2. B. Feigin, E. Feigin, M. Jimbo, T. Miwa, and E. Mukhin,*Quantum continuous* $\mathfrak{gl}_\infty$: *Semi-infinite construction of representations*, Kyoto J. Math **51** (2011), no. 2, 337–364
3. B. Feigin, M. Jimbo, and E. Mukhin, *Towards trigonometric deformation of* $\mathfrak{sl}_2$ *coset VOA*, arXiv:1811.02056, 1–20
4. B. Feigin, M. Jimbo, T. Miwa, and E. Mukhin, *Branching rules for quantum toroidal* $\mathfrak{gl}_N$, Adv. Math. **300** (2016), 229–274
5. B. Feigin, M. Jimbo, T. Miwa, and E. Mukhin, *Representations of quantum toroidal* $\mathfrak{gl}_N$, J. Algebra **380** (2013), 78–108
6. M. Jimbo, *A q-analogue of $U(\mathfrak{gl}_{N+1})$, Hecke algebra, and the Yang-Baxter equation*, Lett. Math. Phys. **11** (1986), no. 3, 247–252
7. M. Jimbo, *Quantum R matrix related to the generalized Toda system: an algebraic approach*, Lecture Notes in Physics **246** (1986) 335–361
8. K. Miki, *Toroidal braid group action and an automorphism of toroidal algebra $U_q(\mathfrak{sl}_{n+1,tor})$* ($n \geq 2$), Lett. Math. Phys. **47** (1999), no.4, 365–378
9. K. Miki, *Toroidal and level 0 $U'_q\widehat{sl_{n+1}}$ actions on $U_q\widehat{gl_{n+1}}$ modules*, J. Math. Phys., **40** (1999), no. 6, 3191–3210
10. E. Mukhin and C. Young, *Affinization of category $\mathcal{O}$ for quantum groups*, Trans. Amer. Math. Soc. **366** (2014), no. 9, 4815–4847
11. A. Negut, *Quantum toroidal and shuffle algebras*, arXiv:1302.6202
12. M. Nazarov and V. Tarasov, *Yangians and Gelfand-Zetlin bases*, Publ. RIMS, Kyoto Univ. **30** (1994) 459–478

Dynamical Quantum Determinants and Pfaffians

Naihuan Jing and Jian Zhang

Dedicated to Vyjayanthi Chari in honor of her 60th birthday

Abstract We introduce the dynamical quantum Pfaffian on the dynamical quantum general linear group and prove its fundamental transformation identity. Hyper quantum dynamical Pfaffian is also introduced and formulas connecting them are given.

1 Introduction

Dynamical quantum groups are important generalization of quantum groups introduced by Etingof and Varchenko [3] in connection with the elliptic quantum groups [5, 6]. See the review [2] for the background and related literature as well as comparison with usual quantum groups (see [1]). The dynamical quantum group is in fact some quantum groupoid, thus also related to the deformation of the Poisson groupoid [16, 19, 21]. In this paper we essentially follow [3] to define the dynamical quantum group with some modification [14].

As discussed in [4] in general, given an R-matrix one can associate certain quantum semigroup $A(R)$ via the RTT formulation. Let V be the complex n-dimensional vector space with basis v_i and dual basis λ_i for V^*. We consider the dynamical R-matrix $R(\underline{\lambda})$ defined on $V \otimes V$ as follows:

N. Jing (✉)
Department of Mathematics, North Carolina State University, Raleigh, NC, USA
e-mail: jing@math.ncsu.edu

J. Zhang
School of Mathematics and Statistics, Central China Normal University, Wuhan, Hubei, China
e-mail: jzhang@mail.ccnu.edu.cn

J. Greenstein et al. (eds.), *Interactions of Quantum Affine Algebras with Cluster Algebras, Current Algebras and Categorification*, Progress in Mathematics 337,
https://doi.org/10.1007/978-3-030-63849-8_13

$$R(\underline{\lambda}) = q\sum_{i=1}^{n} e_{ii}\otimes e_{ii} + \sum_{i<j}^{n} e_{ii}\otimes e_{jj} + \sum_{i>j}^{n} g(\lambda_i-\lambda_j)e_{ii}\otimes e_{jj}$$
$$+\sum_{i\neq j}^{n} h(\lambda_i-\lambda_j)e_{ii}\otimes e_{jj}, \tag{1.1}$$

where $\underline{\lambda} = (\lambda_1, \ldots, \lambda_n)$, e_{ij} are the unit matrix elements inside $\mathrm{End}(V)$ such that $e_{ij}v_k = \delta_{jk}v_i$, and g, h are certain q-analog functions on V (see (2.1)). The corresponding bialgebra $\mathcal{F}_R(M(n))$ is called the dynamical quantum group in the general linear type, which generalizes the usual quantum general linear semigroup $M_q(n)$. The non-dynamical quantum semigroup $M_q(n)$ becomes the quantum group $\mathrm{GL}_q(n)$ with the help of a special central element called the quantum determinant $\det_q$.

On the quantum semigroup $M_q(n)$, one can develop a theory of quantum linear algebra [8] and introduce quantum determinants and minors. One can prove key equations such as the quantum Cramer identity, Cayley–Hamilton identity etc. [17, 22]. For a unified treatment using Manin's quadratic algebras, see [11, 13].

Correspondingly, on the dynamical quantum general linear semigroupoid $\mathcal{F}_R(M(n))$ we can also introduce the dynamical quantum determinant, minors and prove that they also enjoy similar favorable properties [9, 14, 15]. It turns out that the quantum dynamical determinant is also a central group-like element, and the Laplace expansions for quantum dynamical minors are also satisfied in a manner similar to the non-dynamical quantum situation. In particular the quantum dynamical determinant also turns $\mathcal{F}_R(M(n))$ into a dynamical quantum groupoid [18, 20].

The goal of this paper is to introduce the quantum dynamical Pfaffian and show that it enjoys favorable properties similar to the quantum group situation [10]. Our main technique is to use quadratic algebras or quantum de Rham complexes [17] to study quantum determinants and quantum Pfaffians and express them as the scaling constants of quantum differential forms (cf. [11]). In particular, we prove that the dynamical quantum Pfaffian satisfies the transformation property:

$$\mathrm{Pf}(ABA^t) = \det(A)\mathrm{Pf}(B) \tag{1.2}$$

even though the identity $\mathrm{Pf}(A) = \sqrt{\det(A)}$ no longer holds for the quantum anti-symmetric matrix.

The paper is organized as follows. In section two, we introduce the dynamical quantum general linear group via the generalized quantum Yang–Baxter R-matrix and review the basic information on quantum dynamical minors and determinants. In section three, we give a factorization formula for the dynamical quantum determinant in terms of quasi-determinant of Gelfand and Retakh. In section four, we study the dynamical quantum Pfaffians using q-forms. In the last section, quantum dynamical hyper-Pfaffians are given and their fundamental properties and identities are discussed.

2 Dynamical Analogue of the Quantum Algebra M(n)

In this section, we recall some basic facts about dynamical quantum groups [14].

Let $\mathfrak{h}^*$ be the dual space of the n-dimensional commutative Lie algebra $\mathfrak{h}$, and we fix a linear basis $\{e_i\}$ of $\mathfrak{h}^*$, so $\mathfrak{h}^*$ can be identified with $\mathbb{C}^n$. For $[1, n] = \{1, 2, \ldots, n\}$, define $\omega : [1, n] \to \mathfrak{h}^*$ by $\omega(i) = e_i$.

Fix a generic $q \in \mathbb{C}^\times$. For $\lambda \in \mathfrak{h}^*$, the functional $q^\lambda : \mathfrak{h} \longrightarrow \mathbb{C}$ is defined as usual by $v \mapsto q^{\lambda(v)}$, $v \in \mathfrak{h}$. We denote by $h(\lambda)$ and $g(\lambda)$ the following special meromorphic functionals on $\mathfrak{h}$:

$$\begin{aligned} h(\lambda) &= q\frac{q^{-2\lambda} - q^{-2}}{q^{-2\lambda} - 1}, \\ g(\lambda) &= h(\lambda)h(-\lambda) = \frac{(q^{-2\lambda} - q^{-2})(q^{-2\lambda} - q^{2})}{(q^{-2\lambda} - 1)^2}. \end{aligned} \tag{2.1}$$

Let $M_{\mathfrak{h}^*}$ be the space of meromorphic functionals on $\mathfrak{h}^*$. In particular, the above $f(\lambda)$, $g(\lambda)$ are elements inside $M_{\mathfrak{h}^*}$. Let $\mathfrak{h}$-algebra $\mathcal{F}_R(M(n))$ be the associative algebra generated by the elements t_{ij}, $1 \leqslant i, j \leqslant n$, together with two copies of $M_{\mathfrak{h}^*}$. The elements of the two copies $M_{\mathfrak{h}^*}$ are $f(\underline{\lambda}) = f(\lambda_1, \ldots, \lambda_n)$ and $f(\underline{\mu}) = f(\mu_1, \ldots, \mu_n)$, embedded as subalgebras. Here λ_i (resp., μ_i) is a function on $\mathfrak{h}$. The defining relations of $\mathcal{F}_R(M(n))$ consist of two types. The first group of relations are given by

$$\begin{aligned} f_1(\underline{\lambda}) f_2(\underline{\mu}) &= f_2(\underline{\mu}) f_1(\underline{\lambda}), \\ f(\underline{\lambda}) t_{ij} &= t_{ij} f(\underline{\lambda} + \omega(i)), \\ f(\underline{\mu}) t_{ij} &= t_{ij} f(\underline{\mu} + \omega(j)), \end{aligned} \tag{2.2}$$

where $f, f_1, f_2 \in M_{\mathfrak{h}^*}$. The second set of relations are

$$\begin{aligned} h(\mu_k - \mu_l) t_{ik} t_{il} &= t_{il} t_{ik}, \quad k < l \\ h(\lambda_j - \lambda_i) t_{jk} t_{ik} &= t_{ik} t_{jk}, \quad i < j \\ t_{ik} t_{jl} &= t_{jl} t_{ik} + \big(h(\lambda_j - \lambda_i) - h(\mu_k - \mu_l)\big)\, t_{jk} t_{il}, \quad i < j, k < l \\ g(\mu_k - \mu_l) t_{ik} t_{jl} &= g(\lambda_i - \lambda_j) t_{jl} t_{ik} \\ &\quad + \big(h(\mu_l - \mu_k) - h(\lambda_i - \lambda_j)\big)\, t_{il} t_{jk}, i < j, k < l. \end{aligned} \tag{2.3}$$

The algebra $\mathcal{F}_R(M(n))$ has a bigradation defined as follows. Let

$$deg(t_{ij}) = (\omega(i), \omega(j)) = (e_i, e_j) \in \mathbb{N}^n \times \mathbb{N}^n$$

and extend this multiplicatively. Then,

$$\mathcal{F}_R(M(n)) = \bigoplus_{(m,p)\in\mathbb{N}^n\times\mathbb{N}^n\cup(0,0)} \mathcal{F}_{m,p},$$

where the summand $f(\underline{\lambda}), f(\underline{\mu}) \in \mathcal{F}_{00} = M_{\mathfrak{h}}^{\otimes 2}$, and $\mathcal{F}_{m,p} = \{t|deg(t) = (m, p) \in \mathbb{N}^n \times \mathbb{N}^n\}$, and the moment maps are given by $\mu_l(f) = f(\lambda), \mu_r(f) = f(\mu)$.

For $\alpha \in \mathfrak{h}^*$, we denote by $T_\alpha : M_{\mathfrak{h}^*} \to M_{\mathfrak{h}^*}$ the automorphism $(T_\alpha f)(\lambda) = f(\lambda + \alpha)$ for all $\lambda \in \mathfrak{h}^*$. The algebra $\mathcal{F}_R(M(n))$ has a comultiplication $\Delta : \mathcal{F}_R(M(n)) \longrightarrow \mathcal{F}_R(M(n)) \otimes \mathcal{F}_R(M(n))$ given by

$$\begin{aligned}\Delta(t_{ij}) &= \sum_k t_{ik} \otimes t_{kj},\\ \Delta(f(\underline{\lambda})) &= f(\underline{\lambda}) \otimes 1,\\ \Delta(f(\underline{\mu})) &= 1 \otimes f(\underline{\mu}),\end{aligned} \tag{2.4}$$

and the counit ε given by $\varepsilon(t_{ij}) = \delta_{ij} T_{-\omega(i)}$, $\varepsilon(f(\underline{\lambda})) = \varepsilon(f(\underline{\mu})) = f$, and the map is extended as a homomorphism, and we equip $\mathcal{F}_R(M(n))$ with the structure of an $\mathfrak{h}$-bialgebroid.

Definition 2.1 *An $\mathfrak{h}$-space V is a vector space over $M_{\mathfrak{h}^*}$ equipped with a diagonalizable $\mathfrak{h}$-module, i.e., $V = \sum_{\alpha\in\mathfrak{h}^*} V_\alpha$, with $M_{\mathfrak{h}^*} V_\alpha \in V_\alpha$ for all $\alpha \in \mathfrak{h}^*$. A morphism of $\mathfrak{h}$-spaces is an $\mathfrak{h}$-invariant $\mathfrak{h}^*$-linear map.* □

We next define the tensor product of an $\mathfrak{h}$-bialgebroid $\mathcal{A}$ and an $\mathfrak{h}$-space V. Put $V\widetilde{\otimes}\mathcal{A} = \bigoplus_{\alpha,\beta\in\mathfrak{h}^*}(V_\alpha \otimes_{\mathfrak{h}^*} \mathcal{A}_{\alpha\beta})$, where $\otimes_{\mathfrak{h}^*}$ denotes the usual tensor product modulo the relations $v \otimes \mu_l(f)a = fv \otimes a$. The grading $V_\alpha \otimes_{\mathfrak{h}^*} \mathcal{A}_{\alpha\beta} \subseteq (V \otimes \mathcal{A})_\beta$ for all α and $f(v \otimes a) = v \otimes \mu_r(f)a$ makes $V\widetilde{\otimes}\mathcal{A}$ into an $\mathfrak{h}$-space. Analogously $\mathcal{A}\widetilde{\otimes}V = \bigoplus_{\alpha,\beta\in\mathfrak{h}^*}(\mathcal{A}_{\alpha\beta} \otimes_{\mathfrak{h}^*} V_\beta)$, where $\otimes_{\mathfrak{h}^*}$ denotes the usual tensor product modulo the relations $\mu_r(f)a \otimes v = a \otimes fv$. The grading $\mathcal{A}_{\alpha\beta} \otimes_{\mathfrak{h}^*} V_\beta \subseteq (\mathcal{A} \otimes V)_\alpha$ and $f(a \otimes v) = \mu_l(f)a \otimes v$ makes $\mathcal{A}\widetilde{\otimes}V$ into an $\mathfrak{h}$-space.

We now construct two special $\mathcal{F}_R(M(n))$-comodules. Let $W = M_{\mathfrak{h}}\langle w_i\rangle$ be the unital associative algebra generated by the elements w_i, $1 \leqslant i \leqslant n$, and $M_{\mathfrak{h}^*}$, its elements denoted by $f(\underline{\lambda})$, subject to the relations

$$\begin{aligned}w_i^2 &= 0, 1 \leqslant i \leqslant n,\\ w_j w_i &= -h(\lambda_j - \lambda_i) w_i w_j, 1 \leqslant i < j \leqslant n,\end{aligned} \tag{2.5}$$

as well as the relation $f(\underline{\lambda})w_i = w_i f(\underline{\lambda} + \omega(i))$ for all $f \in M_{\mathfrak{h}^*}$.

Let $V = M_{\mathfrak{h}}\langle v_i\rangle$ be the unital associative algebra generated by the elements v_i, $1 \leqslant i \leqslant n$, and a copy of $M_{\mathfrak{h}^*}$ embedded as a subalgebra, with its elements denoted by $f(\underline{\lambda})$ subject to the relations

$$\begin{aligned}v_i^2 &= 0, 1 \leqslant i \leqslant n,\\ v_i v_j &= -h(\lambda_i - \lambda_j) v_j v_i, 1 \leqslant i < j \leqslant n,\end{aligned} \tag{2.6}$$

plus that $f(\underline{\lambda})v_i = v_i f(\underline{\lambda}+\omega(i))$ for all $f \in M_{\mathfrak{h}^*}$.

The following result is easy to see.

Theorem 2.2 ([14]) *Define* $\alpha_R(1) = 1 \otimes 1$, $\alpha_R(w_i) = \sum_{j=1}^{n} w_j \otimes t_{ji}$, $\alpha_L(1) = 1 \otimes 1$, *and* $\alpha_L(v_i) = \sum_{j=1}^{n} t_{ij} \otimes v_j$. *Then* α_L *extends uniquely to* $\alpha_L : V \to \mathcal{F}_R(M(n)) \otimes V$ *such that* V *is a left* $\mathfrak{h}$*-comodule algebra for* $\mathcal{F}_R(M(n))$ *and* α_R *extends uniquely to* $\alpha_R : W \to W \otimes \mathcal{F}_R(M(n))$ *such that* W *is a right* $\mathfrak{h}$*-comodule algebra for* $\mathcal{F}_R(M(n))$. □

Let I be a subset of $[1,n]$ with entries $i_1 < i_2 < \cdots < i_r$ and S_r be the symmetric group in r letters. For an element $\sigma \in S_r$, we use $l(\sigma)$ denote the length of σ. The generalized sign functions $S(\sigma, I)$ and $\tilde{S}(\sigma, I)$ are defined as follows:

$$S(\sigma, I)(\underline{\lambda}) = \prod_{1\leqslant k<l\leqslant r:\sigma(k)>\sigma(l)} (-h(\lambda_{i_{\sigma(k)}} - \lambda_{i_{\sigma(l)}}))$$

$$= (-q)^{l(\sigma)} \prod_{1\leqslant k<l\leqslant r:\sigma(k)>\sigma(l)} \frac{q^{-2\lambda_{i_{\sigma(k)}}} - q^{-2} q^{-2\lambda_{i_{\sigma(l)}}}}{q^{-2\lambda_{i_{\sigma(k)}}} - q^{-2\lambda_{i_{\sigma(l)}}}}, \tag{2.7}$$

$$\tilde{S}(\sigma, I)(\underline{\lambda}) = \prod_{k<l:\sigma(k)>\sigma(l)} (-h(\lambda_{i_{\sigma(l)}} - \lambda_{i_{\sigma(k)}})) = \frac{1}{S(\sigma, I)(\underline{\lambda}+\underline{1})}, \tag{2.8}$$

where $\underline{1} = (1, \ldots, 1)$.

For two subsets $I, J \subseteq [1,n]$ with $|I| = |J| = r$, the dynamical quantum column minor determinants ξ_J^I and row minor determinants η_J^I are defined as follows:

$$\xi_J^I = \mu_r(S(\rho, J)^{-1}) \sum_{\sigma\in S_r} \mu_l(S(\sigma, I)) t_{i_{\sigma(1)} j_{\rho(1)}} \cdots t_{i_{\sigma(r)} j_{\rho(r)}}, \tag{2.9}$$

$$\eta_J^I = \mu_l(\tilde{S}(\rho, I)^{-1}) \sum_{\sigma\in S_r} \mu_r(\tilde{S}(\sigma, J)) t_{i_{\rho(r)} j_{\sigma(r)}} \cdots t_{i_{\rho(1)} j_{\sigma(1)}}, \tag{2.10}$$

where $\rho \in S_r$. Using the comodule structures of α_L and α_R, the following result can be easily obtained.

Theorem 2.3 ([14]) $\xi_J^I = \eta_J^I$ *in* $\mathcal{F}_R(M(n))$.

The element $\det = \xi_{\{1,2,\ldots,n\}}^{\{1,2\ldots,n\}}$ will be called the (quantum dynamic) determinant of $\mathcal{F}_R(M(n))$.

For ordered subsets I, J, one has that

$$\alpha_L(v_I) = \sum_{|K|=|I|} \xi_K^I \otimes v_K, \tag{2.11}$$

$$\alpha_R(w_J) = \sum_{|K|=|J|} w_K \otimes \xi_J^K, \tag{2.12}$$

$$\Delta(\xi_J^I) = \sum_{|K|=|I|} \xi_K^I \otimes \xi_J^K. \tag{2.13}$$

For disjoint ordered subsets I_1, I_2 of $\{1, 2, \ldots, n\}$, we introduce the quantum sign element $\text{sign}(I_1, I_2)$ inside $M_{\mathfrak{h}^*}$ by

$$\text{sign}(I_1, I_2) = \prod_{k>l; k\in I_1, l\in I_2} (-h(\lambda_k - \lambda_l)). \tag{2.14}$$

Proposition 2.4 ([14]) *Let I, J_1, J_2 be subsets of $\{1, 2, \ldots, n\}$. If $J = J_1 \cup J_2$, $|I| = |J|$. Then,*

$$\mu_r(\text{sign}(J_1; J_2))\xi_J^I = \sum_{I_1\cup I_2=I} \mu_l(\text{sign}(I_1; I_2))\xi_{J_1}^{I_1}\xi_{J_2}^{I_2}, \tag{2.15}$$

$$\xi_I^J = \sum_{I_1\cup I_2=I} \xi_{I_1}^{J_1} \frac{\mu_l(\text{sign}(J_2; J_1))}{\mu_r(\text{sign}(I_2; I_1))} \xi_{J_2}^{I_2}. \tag{2.16}$$

It is easy to see from Proposition 2.4 that for any $i, j \in [1, n]$

$$\begin{aligned}
\delta_{ij} \det &= \sum_{k=1}^{n} \frac{\text{sign}(\{k\}; \hat{k})(\underline{\lambda})}{\text{sign}(\{i\}; \hat{i})(\underline{\mu})} t_{kj} \xi_{\hat{i}}^{\hat{k}}, \\
\delta_{ij} \det &= \sum_{k=1}^{n} t_{jk} \frac{\text{sign}(\hat{i}; \{i\})(\underline{\lambda})}{\text{sign}(\hat{k}; \{k\})(\underline{\mu})} \xi_{\hat{k}}^{\hat{i}}, \\
\delta_{ij} \det &= \sum_{k=1}^{n} \frac{\text{sign}(\hat{k}; \{k\})(\underline{\lambda})}{\text{sign}(\hat{i}; \{i\})(\underline{\mu})} \xi_{\hat{i}}^{\hat{k}} t_{kj}, \\
\delta_{ij} \det &= \sum_{k=1}^{n} \xi_{\hat{k}}^{\hat{i}} \frac{\text{sign}(\{i\}; \hat{i})(\underline{\lambda})}{\text{sign}(\{k\}; \hat{k})(\underline{\mu})} t_{jk}.
\end{aligned} \tag{2.17}$$

Lemma 2.5 ([14]) *In $\mathcal{F}_R(M(n))$, the determinant commutes with all quantum minor determinants. In particular,* det *commutes with all generators t_{ij}. Moreover,* $\Delta(\det) = \det \otimes \det$ *and* $\varepsilon(\det) = T_{-\underline{1}}$, *with* $\underline{1} = (1, \ldots, 1) \in \mathfrak{h}^*$. □

Proposition 2.6 ([14]) *The $\mathfrak{h}$-bialgebroid $\mathcal{F}_R(M(n))$ is an $\mathfrak{h}$-Hopf algebroid with the antipode S defined on the generators by* $S(\det^{-1}) = \det$, $S(\mu_r(f)) = \mu_l(f)$, *and* $S(\mu_l(f)) = \mu_r(f)$ *for all* $f \in M_{\mathfrak{h}^*}$ *and*

$$S(t_{ij}) = \det^{-1} \frac{\mu_l(\mathrm{sign}(\hat{j}; \{j\}))}{\mu_r(\mathrm{sign}(\hat{i}; \{i\}))} \xi_{\hat{i}}^{\hat{j}}$$

and extended as an algebra antihomomorphism. □

3 Quasideterminants and Dieudonné Determinants

Throughout this section, we work with rings of fractions of noncommutative rings.

Definition 3.1 ([7]) *Let* $\mathrm{X} = (x_{ij})$ *be an* $n \times n$ *matrix over a ring with identity such that its inverse matrix* X^{-1} *exists, and the* (j, i)*th entry of* X^{-1} *is an invertible element of the ring. Then the* (ij)*th quasideterminant of X is defined by the formula*

$$|\mathrm{X}|_{ij} = \begin{vmatrix} x_{11} & \dots & x_{1j} & \dots & x_{1n} \\ & \dots & & \dots & \\ x_{i1} & \dots & \boxed{x_{ij}} & \dots & x_{in} \\ & \dots & & \dots & \\ x_{n1} & \dots & x_{nj} & \dots & x_{nn} \end{vmatrix} = (\mathrm{X}^{-1})_{ji}^{-1},$$

where the first or the second notation with $\boxed{x_{ij}}$ *denotes the quasideterminant.* □

When $n \geqslant 2$, and let X^{ij} be the $(n-1) \times (n-1)$-matrix obtained from X by deleting the ith row and jth column. In general $\mathrm{X}^{i_1 \cdots i_r, j_1 \cdots j_r}$ denotes the submatrix obtained from X by deleting the $i_1, \cdots, i_r$-th rows, and $i_1, \cdots, i_r$-th columns. Then,

$$|\mathrm{X}|_{ij} = x_{ij} - \sum_{i', j'} x_{ii'} (|\mathrm{X}^{ij}|_{j'i'}) x_{j'j},$$

where the sum runs over $i' \notin I \setminus \{i\}, j' \notin J \setminus \{j\}$.

Theorem 3.2 *Let* T *be the matrix of generators* t_{ij} *of* $\mathfrak{F}_R(M(n))$ *and* $\sigma = i_1 \dots i_n$ *and* $\tau = j_1 \dots j_n$ *be two permutations of* S_n*. In the ring of fractions of* $\mathfrak{F}_R(M(n))$*, one has that*

$$\det(T) = \frac{\prod_{k=1}^{n} \mu_l(\mathrm{sign}(I_k{}^c; i_k))}{\prod_{k=1}^{n} \mu_r(\mathrm{sign}(J_k{}^c; j_k))} t_{i_n j_n} \dots |T^{i_1 j_1}|_{i_2 j_2} |T|_{i_1 j_1}, \tag{3.1}$$

where $I_k = \{i_1, \dots, i_k\}$*,* $J_k = \{j_1, \dots, j_k\}$*.* □

Proof By definition the quasi-determinants of T are inverses of the entries of $S(T)$,

$$|T|_{ij} = S(t_{ji})^{-1} = \xi_{\hat{j}}^{\hat{i}^{-1}} \frac{\mu_r(\text{sign}(\hat{j}; \{j\}))}{\mu_l(\text{sign}(\hat{i}; \{i\}))} \det(T), \tag{3.2}$$

then

$$\det(T) = \frac{\mu_l(\text{sign}(\hat{i}; \{i\}))}{\mu_r(\text{sign}(\hat{j}; \{j\}))} \xi_{\hat{j}}^{\hat{i}} |T|_{ij}. \tag{3.3}$$

Equation (3.1) follows from induction on n.

Remark 3.3 *If $i_k = j_k = n + 1 - k$ for any k, all the factors on the right-hand side of 3.1 commute with each other. In general the factors do not commute.* □

4 Dynamical Quantum Pfaffians

First we review the general theory of the Pfaffian [11, 12], and we assume the minimum condition here. Let $\mathcal{B}$ be the algebra generated by the elements b_{ij} for $1 \leqslant i < j \leqslant 2n$, and a copy of $M_{\mathfrak{h}^*}$ embedded as a subalgebra, with its elements denoted by $f(\underline{\lambda})$. The dynamical quantum Pfaffian is defined by

$$\text{Pf}(B) = \sum_{\sigma \in \Pi} S(\sigma) b_{\sigma(1)\sigma(2)} b_{\sigma(3)\sigma(4)} \cdots b_{\sigma(2n-1)\sigma(2n)},$$

where $S(\sigma) = S(\sigma, [1, 2n])$, Π is the set of permutations σ of $2n$ such that $\sigma(2i - 1) < \sigma(2i), i = 1, \ldots, n$.

For any two disjoint subsets I_1, I_2 of $[1, 2n]$, we define the dynamical quantum sign functions $\text{sign}(I_1, I_2)$ and $\widetilde{\text{sign}}(I_1, I_2)$ by

$$\begin{aligned} \text{sign}(I_1, I_2) &= \prod_{k>l; k \in I_1, l \in I_2} (-h(\lambda_k - \lambda_l)), \\ \widetilde{\text{sign}}(I_1, I_2) &= \prod_{k<l; k \in I_1, l \in I_2} (-h(\lambda_k - \lambda_l)). \end{aligned} \tag{4.1}$$

Let $I = \{i_1, i_2, \ldots, i_k\}$ with $1 \leqslant i_1 < i_2 < \ldots, i_k \leqslant 2n$. Denote by B_I the submatrix of B with the rows and columns indexed by I. The following result gives an iterative algorithm to compute the dynamic Pfaffian.

Proposition 4.1 *For each $0 \leqslant t \leqslant n$, we have that*

$$\text{Pf}(B) = \sum_{I} \text{sign}(I; I^c) \text{Pf}(B_I) \text{Pf}(B_{I^c}), \tag{4.2}$$

where the sum runs over all subsets I of $[1, 2n]$ such that $|I|=2t$ and I^c is the complement of I. □

Proof Define the tensor product of $W \otimes \mathcal{B}$ to be the usual tensor product modulo the relations $fw \otimes b = w \otimes fb$. Let $\Omega = \sum_{i<j} w_i w_j \otimes b_{ij}$, then

$$\bigwedge^n \Omega = w_1 \wedge \cdots \wedge w_{2n} \otimes \mathrm{Pf}(B). \tag{4.3}$$

On the other hand,

$$\begin{aligned} \bigwedge^n \Omega &= \Omega^t \bigwedge \Omega^{n-t} \\ &= \sum_{I,J} (w_I \otimes \mathrm{Pf}(B_I))\,(w_J \otimes \mathrm{Pf}(B_J)) \\ &= \sum_{I,J} w_I w_J \otimes \mathrm{Pf}(B_I)\mathrm{Pf}(B_J). \end{aligned} \tag{4.4}$$

Note that $w_I w_J$ vanishes unless $J = I^c$, and therefore we have that

$$\mathrm{Pf}(B) = \sum_I \mathrm{sign}(I; I^c)\mathrm{Pf}(B_I)\mathrm{Pf}(B_{I^c}).$$

The following transformation formula establishes the relation between the quantum dynamic Pfaffian and determinant.

Theorem 4.2 *Denote by $\mathfrak{F}_R(M(2n))\widetilde{\otimes}\mathcal{B}$ the usual tensor product modulo the relations*

$$\mu_r(f)t \otimes b = t \otimes fb \quad \textit{and} \quad f(t \otimes b) = \mu_l(f)t \otimes b.$$

Let ξ^{ij}_{kl} be the 2×2-dynamical quantum minors in $\mathfrak{F}_R(M(2n))$, b_{ij} the generators of $\mathcal{B}$, and $c_{ij} = \sum_{k<l} \xi^{ij}_{kl} \otimes b_{kl}$. Then in $\mathfrak{F}_R(M(2n))\widetilde{\otimes}\mathcal{B}$, we have

$$\mathrm{Pf}(C) = \det(T) \otimes \mathrm{Pf}(B).$$

Proof Let $w \otimes t \otimes b$ be an element of $W\widetilde{\otimes}\mathfrak{F}_R(M(n))\widetilde{\otimes}\mathcal{B}$, then

$$\begin{aligned} fw \otimes t \otimes b &= w \otimes \mu_l(f)t \otimes b = w \otimes f(t \otimes b), \\ f(w \otimes t) \otimes b &= w \otimes \mu_r(f)t \otimes b = w \otimes t \otimes f(b). \end{aligned} \tag{4.5}$$

Let $\delta_i = \sum_{j=1}^{2n} w_j \otimes t_{ji}$, and consider the element $\Omega = \sum_{i,j} w_i w_j \otimes c_{ij}$. It is clear that

$$\Omega^n = w_1 \cdots w_{2n} \otimes \mathrm{Pf}(C), \tag{4.6}$$

where the product is the wedge product among w_i. On the other hand, $\Omega = \sum \delta_i \delta_j \otimes b_{ij}$. Then,

$$\Omega^n = \delta_1 \cdots \delta_{2n} \otimes \mathrm{Pf}(B) = w_1 \cdots w_{2n} \otimes \det(T) \otimes \mathrm{Pf}(B). \tag{4.7}$$

Comparing (4.13) and (4.7), we conclude that

$$\mathrm{Pf}(C) = \det(T) \otimes \mathrm{Pf}(B).$$

Let $\widetilde{\mathcal{B}}$ be the algebra generated by the elements $\widetilde{b}_{ji}$ for $1 \leqslant i < j \leqslant 2n$, and a copy of $M_{\mathfrak{h}^*}$ embedded as a subalgebra, with its elements denoted by $f(\underline{\lambda})$. the dynamical quantum Pfaffian is defined by

$$\widetilde{\mathrm{Pf}}(\widetilde{B}) = \sum_{\sigma \in \Pi} \tilde{S}(\sigma) \widetilde{b}_{\sigma(2n)\sigma(2n-1)} \widetilde{b}_{\sigma(2n-2)\sigma(2n-3)} \cdots \widetilde{b}_{\sigma(2)\sigma(1)},$$

where Π is the set of shuffle permutations σ of $2n$ such that $\sigma(2i-1) < \sigma(2i), i = 1, \ldots, n$.

For any two disjoint subsets I_1, I_2 of $\{1, 2, \ldots, 2n\}$, we define the dynamical quantum sign by

$$\widetilde{\mathrm{sign}}(I_1, I_2) = \prod_{k<l; k \in I_1, l \in I_2} (-h(\lambda_k - \lambda_l)). \tag{4.8}$$

Similarly, we have the following statements.

Proposition 4.3 *For any* $0 \leqslant t \leqslant n$, *we have that*

$$\widetilde{\mathrm{Pf}}(\widetilde{B}) = \sum_I \widetilde{\mathrm{sign}}(I; I^c) \widetilde{\mathrm{Pf}}(\widetilde{B}_I) \widetilde{\mathrm{Pf}}(\widetilde{B}_{I^c}), \tag{4.9}$$

where the sum runs through all subsets I *of* $[1, 2n]$ *such that* $|I|=2t$. □

Proof Define the tensor product of $\widetilde{\mathcal{B}} \otimes V$ to be the usual tensor product modulo the relations $fb \otimes v = b \otimes fv$. Let $\widetilde{\Omega} = \sum_{i>j} b_{ij} \otimes v_i v_j$, then

$$\bigwedge^n \widetilde{\Omega} = \widetilde{\mathrm{Pf}}(\widetilde{B}) \otimes v_n \wedge \cdots \wedge v_1. \tag{4.10}$$

On the other hand,

$$\begin{aligned}\bigwedge^n \widetilde{\Omega} &= \widetilde{\Omega}^t \bigwedge \widetilde{\Omega}^{n-t}\\ &= \sum_{I,J} \left(\widetilde{\mathrm{Pf}}(\widetilde{B}_I) \otimes v_I\right) \left(\widetilde{\mathrm{Pf}}(\widetilde{B}_J) \otimes v_J\right)\\ &= \sum_{I,J} \widetilde{\mathrm{Pf}}(\widetilde{B}_I)\widetilde{\mathrm{Pf}}(\widetilde{B}_J) \otimes v_I v_J. \end{aligned} \tag{4.11}$$

It is easy to see that $v_I v_J$ vanishes unless $J = I^c$. Thus we conclude that

$$\widetilde{\mathrm{Pf}}(\widetilde{B}) = \sum_I \widetilde{\mathrm{sign}}(I;\, I^c)\mathrm{Pf}(\widetilde{B}_I)\mathrm{Pf}(\widetilde{B}_{I^c}).$$

Theorem 4.4 *Denote by* $\mathcal{B}\widetilde{\otimes}\mathcal{F}_R(M(2n))$ *the usual tensor product modulo the relations* $b \otimes \mu_l(f)t = fb \otimes t$ *and* $f(b \otimes t) = b \otimes \mu_r(f)t$. *Let* ξ_{ij}^{kl} *be the dynamical quantum minor in* $\mathcal{F}_R(M(n))$, $c_{ji} = \sum_{k<l} \widetilde{b}_{lk} \otimes \xi_{ij}^{kl}$. *Then in* $\mathcal{B}\widetilde{\otimes}\mathcal{F}_R(M(2n))$, *we have* $\widetilde{\mathrm{Pf}}(C) = \widetilde{\mathrm{Pf}}(\widetilde{B}) \otimes \det(T)$. □

Proof Let $b \otimes t \otimes v$ be an element of $\mathcal{B}\widetilde{\otimes}\mathcal{F}_R(M(n))\widetilde{\otimes}V$, then

$$\begin{aligned} b \otimes t \otimes fv &= b \otimes \mu_r(f)t \otimes v = f(b \otimes t) \otimes v,\\ b \otimes f(t \otimes v) &= b \otimes \mu_l(f)t \otimes v = f(b) \otimes t \otimes v. \end{aligned} \tag{4.12}$$

Let $\delta_i = \sum_{j=1}^{2n} t_{ij} \otimes v_j$, and consider the element $\widetilde{\Omega} = \sum c_{ji} \otimes v_j v_i$. It is clear that

$$\widetilde{\Omega}^n = \widetilde{\mathrm{Pf}}(C) \otimes v_{2n} \cdots v_1. \tag{4.13}$$

On the other hand, $\widetilde{\Omega} = \sum b_{lk} \otimes \delta_l \delta_k$. Then,

$$\widetilde{\Omega}^n = \widetilde{\mathrm{Pf}}(C) \otimes \delta_{2n} \cdots \delta_1 = \widetilde{\mathrm{Pf}}(\widetilde{B}) \otimes \det(T) \otimes \delta_{2n} \cdots \delta_1. \tag{4.14}$$

Comparing (4.13) and (4.7), we conclude that

$$\widetilde{\mathrm{Pf}}(C) = \widetilde{\mathrm{Pf}}(\widetilde{B}) \otimes \det(T).$$

We now generalize the notion of the dynamical quantum Pfaffian to the dynamical quantum hyper-Pfaffian. Let $\mathcal{B}$ be the algebra generated by the elements $b_{i_1 \cdots i_m}, i_1 < i_2 < \cdots < i_m, 1 \leqslant i_k \leqslant mn, k = 1, \ldots, m$, and a copy of $M_{\mathfrak{h}^*}$ embedded as a subalgebra, with its elements denoted by $f(\underline{\lambda})$. The dynamical quantum hyper-Pfaffian is defined by

$$\mathrm{Pf}_m(B) = \sum_{\sigma\in\Pi} S(\sigma) b_{\sigma(1)\cdots\sigma(m)} \cdots b_{\sigma(m(n-1)+1)\cdots\sigma(mn)},$$

where Π is the set of permutations σ of mn such that $\sigma((k-1)m+1) < \sigma((k-1)m+2) < \cdots < \sigma(km)$, $k = 1, \ldots, n$.

Note that the dynamical hyper-Pfaffian uses only the entries $b_{i_1\cdots i_m}$, where $i_1 < \cdots < i_m$. Clearly $\mathrm{Pf}_2(B) = \mathrm{Pf}(B)$, the quantum dynamical Pfaffian.

Proposition 4.5 *For any* $0 \leqslant t \leqslant n$,

$$\mathrm{Pf}_m(B) = \sum_I \mathrm{sign}(I; I^c)\mathrm{Pf}_m(B_I)\mathrm{Pf}_m(B_{I^c}), \tag{4.15}$$

where the sum is taken over all subset I *of* $[1, mn]$ *such that* $|I| = mt$. □

Let $I = \{i_1, i_2, \cdots, i_m\}$ such that $i_1 < i_2 < \cdots < i_m$. We denote by b_I the element $b_{i_1\cdots i_m}$.

Theorem 4.6 *Denote by* $\mathcal{F}_R(M(mn))\widetilde{\otimes}\mathcal{B}$ *the usual tensor product modulo the relations*

$$\mu_r(f)t \otimes b = t \otimes fb \text{ and } f(t \otimes b) = \mu_l(f)t \otimes b.$$

Let ξ^I_J *be the dynamical quantum minor in* $\mathcal{F}_R(M(n))$, $c_I = \sum_J \xi^I_J \otimes b_J$. *Then* $\mathrm{Pf}_m(C) = \det(T) \otimes \mathrm{Pf}_m(B)$ *in* $\mathcal{F}_R(M(mn))\widetilde{\otimes}\mathcal{B}$. □

Let $\widetilde{\mathcal{B}}$ be the algebra generated by the elements $\widetilde{b}_{i_1\cdots i_m}$, $i_1 > i_2 > \cdots > i_m$, $1 \leqslant i_k \leqslant mn$, $k = 1, \ldots, m$. Define dynamical quantum hyper-Pfaffian by

$$\widetilde{\mathrm{Pf}}_m(\widetilde{B}) = \sum_{\sigma\in\Pi} \widetilde{S}(\sigma)\widetilde{b}_{\sigma(mn)\cdots\sigma(mn-m+1)}\widetilde{b}_{\sigma(mn-m)\sigma(mn-2m+1)} \cdots \widetilde{b}_{\sigma(m)\cdots\sigma(1)},$$

where Π is the set of permutations σ of mn such that $\sigma((k-1)m+1) < \sigma((k-1)m+2) < \cdots < \sigma(km)$, $k = 1, \ldots, n$.

Clearly $\widetilde{\mathrm{Pf}}_2(\widetilde{B}) = \widetilde{\mathrm{Pf}}(\widetilde{B})$ as discussed before.

Proposition 4.7 *For any* $0 \leqslant t \leqslant n$,

$$\widetilde{\mathrm{Pf}}_m(\widetilde{B}) = \sum_I \widetilde{\mathrm{sign}}(I; I^c)\widetilde{\mathrm{Pf}}_m(\widetilde{B}_I)\widetilde{\mathrm{Pf}}_m(\widetilde{B}_{I^c}), \tag{4.16}$$

where the sum is taken over all subset I *of* $[1, mn]$ *such that* $|I| = mt$. □

Let $I = \{i_1, i_2, \cdots, i_m\}$ such that $i_1 < i_2 < \cdots < i_m$. We denote by $\widetilde{b}_I$ the element $b_{i_m\cdots i_1}$. The following result is proved by the similar method as in the case of $m = 2$.

Theorem 4.8 *Denote by* $\widetilde{\mathcal{B}}\widetilde{\otimes}\mathcal{F}_R(M(mn))$ *the usual tensor product modulo the relations* $b \otimes \mu_l(f)t = fb \otimes t$ *and* $f(b \otimes t) = b \otimes \mu_r(f)t$*. Let* ξ_I^J *be the dynamical quantum minor in* $\mathcal{F}_R(M(n))$, $c_I = \sum_J \widetilde{b}_J \otimes \xi_I^J$*. Then in* $\widetilde{\mathcal{B}}\widetilde{\otimes}\mathcal{F}_R(M(mn))$*, we have* $\widetilde{\mathrm{Pf}}_m(C) = \widetilde{\mathrm{Pf}}_m(\widetilde{B}) \otimes \det(T)$. □

Acknowledgments The work is supported by National Natural Science Foundation of China (grant no. 11531004), FAPESP (grant no. 2015/05927-0), and Humboldt foundation. Jing acknowledges the support of Max-Planck Institute for Mathematics in the Sciences, Leipzig, during the research.

References

1. V. Chari and A. Pressley, A guide to quantum groups, Cambridge University Press, Cambridge, 1994.
2. P. Etingof and O. Schiffmann, *Lectures on the dynamical Yang-Baxter equations*, Quantum Groups and Lie Theory (Durham, 1999), London Math. Soc. Lecture Note Ser., vol. 290, Cambridge University Press, Cambridge, 2001, pp. 89–129.
3. P. Etingof and A. Varchenko, *Exchange dynamical quantum groups*, Comm. Math. Phys. 205 (1999), 19–52.
4. L. D. Faddeev, N. Yu. Reshetikhin, and L. A. Takhtajan, *Quantization of Lie groups and Lie algebras*, Algebraic analysis, Vol. I, pp. 129–139, Academic Press, Boston, MA, 1988.
5. G. Felder, *Elliptic quantum groups*, XIth Int. Congress of Mathematical Physics (Paris, 1994), pp. 211–218, International Press, Cambridge, MA, 1995.
6. G. Felder and A. Varchenko, *On representations of the elliptic quantum group* $E_{\tau,\eta}(sl_2)$, Commun. Math. Phys. 181 (1996), 741–761.
7. I. Gelfand, V. Retakh, GRI. Gelfand and V. Retakh, *Quasideterminants*, I, Selecta Math. (N.S.) 3 (1997), 517–546.
8. M. Hashimoto and T. Hayashi, *Quantum multilinear algebra*. Tohoku Math. J. (2) 44 (1992), 471–521.
9. L. K. Hadjiivanov, A. P. Isaev, O. V. Ogievetsky, P. N. Pyatov, and I. T. Todorov, *Hecke algebraic properties of dynamical R-matrices*. Application to related quantum matrix algebras, J. Math. Phys. 40 (1999), 427–448.
10. N. Jing and J. Zhang, *Quantum Pfaffians and hyper-Pfaffians*, Adv. Math. 265 (2014), 336–361.
11. N. Jing and J. Zhang, *Multiparameter Quantum Pfaffians*, arXiv:1701.07458.
12. N. Jing and J. Zhang, *Quantum permanents and Hafnians via Pfaffians*, Lett. Math. Phys. 106 (2016), 1451–1464.
13. N. Jing and J. Zhang, *Capelli identity on multiparameter quantum linear groups*, Sci. China Math. 61 (2018), 253–268.
14. E. Koelink and Y. V. Norden, *The dynamical* $U(n)$ *quantum group*, Int. J. Math. Math. Sci. 2006, Art. ID 65279, 30 pp.
15. E. Koelink and Y. V. Norden, *Pairings and actions for dynamical quantum groups*, Adv. Math. 208 (2007), 1–39.
16. J.-H. Lu, *Hopf algebroids and quantum groupoids*, Int. J. Math. 7 (1996), 47–70.
17. Yu. I. Manin, *Notes on quantum groups and quantum de Rham complexes*. Teoret. Mat. Fiz. 92 (1992), 425–450; English transl. in: Theoret. Math. Phys. 92 (1992), 997–1023.
18. J. V. Stokman, *Vertex-IRF transformations, dynamical quantum groups, and harmonic analysis*, Indag. Math. (N. S.) 14 (2003), 545–570.
19. M. Takeuchi, *Groups of algebras over* $A\overline{\otimes}A$, J. Math. Soc. Japan 29 (1977), 459–492.

20. Y. van Norden, *Dynamical quantum groups: duality and special functions*, thesis, TU Delft, Delft, 2005.
21. P. Xu, *Quantum groupoids*, Commun. Math. Phys. 216 (2001), 539–581.
22. E. Taft and J. Towber, *Quantum deformation of flag schemes and Grassmann schemes. I. A q-deformation of the shape-algebra for* $\mathrm{GL}(n)$, J. Algebra 142 (1991), 1–36.

GPSR Compliance
The European Union's (EU) General Product Safety Regulation (GPSR) is a set of rules that requires consumer products to be safe and our obligations to ensure this.

If you have any concerns about our products, you can contact us on

ProductSafety@springernature.com

In case Publisher is established outside the EU, the EU authorized representative is:

Springer Nature Customer Service Center GmbH
Europaplatz 3
69115 Heidelberg, Germany

www.ingramcontent.com/pod-product-compliance
Ingram Content Group UK Ltd.
Pitfield, Milton Keynes, MK11 3LW, UK
UKHW021832270726
14058UKWH00001B/112

* 9 7 8 3 0 3 0 6 3 8 5 1 1 *